VOLUME FOUR HUNDRED AND FORTY-TWO

Methods in
ENZYMOLOGY

Programmed Cell Death, General Principles for Studying Cell Death, Part A

METHODS IN ENZYMOLOGY

Editors-in-Chief

JOHN N. ABELSON AND MELVIN I. SIMON

Division of Biology
California Institute of Technology
Pasadena, California

Founding Editors

SIDNEY P. COLOWICK AND NATHAN O. KAPLAN

VOLUME FOUR HUNDRED AND FORTY-TWO

METHODS IN ENZYMOLOGY

Programmed Cell Death, General Principles for Studying Cell Death, Part A

EDITED BY

ROYA KHOSRAVI-FAR
Harvard Medical School, Department of Pathology, BIDMC Research North, RN 270D
Boston, MA, USA

ZAHRA ZAKERI
Professor of Biology
Queens College of City
University of New York
Flushing, NY, USA

RICHARD A. LOCKSHIN
Department of Biological Sciences
St. John's University
Jamaica, NY, USA

MAURO PIACENTINI
Department of Biology
University of Rome
Rome, Italy

ELSEVIER

AMSTERDAM • BOSTON • HEIDELBERG • LONDON
NEW YORK • OXFORD • PARIS • SAN DIEGO
SAN FRANCISCO • SINGAPORE • SYDNEY • TOKYO
Academic Press is an imprint of Elsevier

Academic Press is an imprint of Elsevier
525 B Street, Suite 1900, San Diego, California 92101-4495, USA
84 Theobald's Road, London WC1X 8RR, UK

This book is printed on acid-free paper. ∞

For information on all Elsevier Academic Press publications
visit our Web site at www.books.elsevier.com

ISBN-13: 978-0-12-374312-1

PRINTED IN THE UNITED STATES OF AMERICA
08 09 10 11   9 8 7 6 5 4 3 2 1

CONTENTS

11. Granzymes and Cell Death   213

Denis Martinvalet, Jerome Thiery, and Dipanjan Chowdhury

12. Investigation of the Proapoptotic Bcl-2 Family Member Bid on the Crossroad of the DNA Damage Response and Apoptosis   231

Sandra S. Zinkel

Contributors

Emad S. Alnemri
Department of Biochemistry and Molecular Biology, Center for Apoptosis Research, Kimmel Cancer Institute, Thomas Jefferson University, Philadelphia, Pennsylvania

Hülya Bayir
Department of Critical Care Medicine, University of Pittsburgh, Pittsburgh, Pennsylvania, and Department of Environmental and Occupational Health, University of Pittsburgh, Pittsburgh, Pennsylvania, and Center for Free Radical and Antioxidant Health, University of Pittsburgh, Pittsburgh, Pennsylvania

Tom Vanden Berghe
Department of Molecular Biology, Ghent University, Ghent, Belgium, and Molecular Signaling and Cell Death Unit, Department for Molecular Biomedical Research, VIB, Ghent, Belgium

Lea Bojič
Department of Biochemistry, Molecular and Structural Biology, J. Stefan Institute, Ljubljana, Slovenia

Dipanjan Chowdhury
Dana Farber Cancer Institute and Department of Radiation Oncology, Harvard Medical School, Boston, Massachusetts

Laura Ciarlo
Therapeutic Research and Medicines Evaluation, Istituto Superiore di Sanità, Rome, Italy

Sean P. Cullen
Molecular Cell Biology Laboratory, Department of Genetics, The Smurfit Institute, Trinity College, Dublin, Ireland

Steven T. DeKosky
Department of Neurology, University of Pittsburgh, Pittsburgh, Pennsylvania

Katharina D'Herde
Department of Anatomy, Embryology, Histology and Medical Physics, Ghent University, Ghent, Belgium

Mauro Degli Esposti
Faculty of Life Sciences, The University of Manchester, Manchester, United Kingdom

Laura Falasca
INMI-IRCCS "L. Spallanzani," Rome, Italy

Christine Feig
The Ben May Department for Cancer Research, University of Chicago, Chicago, Illinois

Teresa Fernandes-Alnemri
Department of Biochemistry and Molecular Biology, Center for Apoptosis Research, Kimmel Cancer Institute, Thomas Jefferson University, Philadelphia, Pennsylvania

Lorenzo Galluzzi
INSERM, U848, Institut Gustave Roussy, and Université Paris–Sud 11, Villejuif, France

Tina Garofalo
Department of Experimental medicine, "Sapienza" University of Rome, Rome, Italy

Giuseppina Di Giacomo
Department of Biology, University of Rome Tor Vergata, Rome, Italy

Anna Maria Giammarioli
Therapeutic Research and Medicines Evaluation, Istituto Superiore di Sanità, Rome, Italy

Marie-Lise Gougeon
INSERM U668, Paris, France, and Antiviral Immunity, Biotherapy and Vaccine Unit, Institut Pasteur, Paris, France

Joel S. Greenberger
Department of Radiation Oncology, University of Pittsburgh, Pittsburgh, Pennsylvania

Ronald Hamilton
Department of Pathology, University of Pittsburgh, Pittsburgh, Pennsylvania

Emilie Hangen
INSERM, U848, Institut Gustave Roussy, and Université Paris–Sud 11, Villejuif, France

Yi-Te Hsu
Department of Biochemistry and Molecular Biology, Medical University of South Carolina, USC, Charleston, South Carolina

Saška Ivanova
Department of Biochemistry, Molecular and Structural Biology, J. Stefan Institute, Ljubljana, Slovenia

Sarah R. Jacobs
Department of Pharmacology and Cancer Biology, Department of Immunology, and Sarah W. Stedman Nutrition and Metabolism Center, Duke University, Durham, North Carolina

Valerian E. Kagan
Department of Environmental and Occupational Health, University of Pittsburgh, Pittsburgh, Pennsylvania, and Center for Free Radical and Antioxidant Health, University of Pittsburgh, Pittsburgh, Pennsylvania

Randal J. Kaufman
Howard Hughes Medical Institute, The University of Michigan Medical Center, Ann Arbor, Michigan, and Department of Internal Medicine, The University of Michigan Medical Center, Ann Arbor, Michigan, and Department of Biological Chemistry, The University of Michigan Medical Center, Ann Arbor, Michigan

Kohki Kawane
Department of Medical Chemistry, Graduate School of Medicine, Kyoto University, Yoshida, Sakyo-ku, Kyoto, Japan

Oliver Kepp
INSERM, U848, Institut Gustave Roussy, and Université Paris-Sud 11, Villejuif, France

Patrick M. Kochanek
Department of Critical Care Medicine, University of Pittsburgh, Pittsburgh, Pennsylvania

Guido Kroemer
INSERM, U848, Institut Gustave Roussy, and Université Paris-Sud 11, Villejuif, France

Dmitri V. Krysko
Department of Molecular Biology, Ghent University, Ghent, Belgium, and Molecular Signaling and Cell Death Unit, Department for Molecular Biomedical Research, VIB, Ghent, Belgium

Hervé Lecoeur
Unité d'Immunophysiologie et Parasitisme Intracellulaire, Paris, France

Richard A. Lockshin
Department of Biological Sciences, St. John's University, Queens, New York

Alexander U. Lüthi
Molecular Cell Biology Laboratory, Department of Genetics, The Smurfit Institute, Trinity College, Dublin, Ireland

Walter Malorni
Therapeutic Research and Medicines Evaluation, and Department of Drug Research and Evaluation, Istituto Superiore di Sanità, Rome, Italy

Valeria Manganelli

Department of Experimental Medicine, "Sapienza" University of Rome, Rome, Italy

Seamus J. Martin

Molecular Cell Biology Laboratory, Department of Genetics, The Smurfit Institute, Trinity College, Dublin, Ireland

Denis Martinvalet

Immune Disease Institute and Department of Pediatrics, Harvard Medical School, Boston, Massachusetts

Pier Giorgio Mastroberardino

Department of Biology, University of Rome Tor Vergata, Rome, Italy

Marie-Thérèse Melki

INSERM U668, Paris, France, and Antiviral Immunity, Biotherapy and Vaccine Unit, Institut Pasteur, Paris, France

Alicia Melendez

Department of Biology, Queens College and Graduate Center of the City University of New York, Flushing, New York

Roberta Misasi

Department of Experimental Medicine, "Sapienza" University of Rome, Rome, Italy

Nazanine Modjtahedi

INSERM, U848, Institut Gustave Roussy, and Université Paris-Sud 11, Villejuif, France

Shigekazu Nagata

Department of Medical Chemistry, Graduate School of Medicine, Kyoto University, Yoshida, Sakyo-ku, Kyoto, Japan

Eef Parthoens

Department of Molecular Biology, Ghent University, Ghent, Belgium, and Microscopy Core, Department for Molecular Biomedical Research, VIB, Ghent, Belgium

Jean-Luc Perfettini

INSERM, U848, Institut Gustave Roussy, and Université Paris-Sud 11, Villejuif, France

Ana Petelin

Department of Biochemistry, Molecular and Structural Biology, J. Stefan Institute, Ljubljana, Slovenia

Marcus E. Peter

The Ben May Department for Cancer Research, University of Chicago, Chicago, Illinois

Mauro Piacentini
INMI-IRCCS "L. Spallanzani," Rome, Italy, and Department of Biology, University of Rome Tor Vergata, Rome, Italy

Urška Repnik
Department of Biochemistry, Molecular and Structural Biology, J. Stefan Institute, Ljubljana, Slovenia

Jeffrey C. Rathmell
Department of Pharmacology and Cancer Biology, Department of Immunology, and Sarah W. Stedman Nutrition and Metabolism Center, Duke University, Durham, North Carolina

Carlo Rodolfo
Department of Biology, University of Rome Tor Vergata, Rome, Italy

Tatiana R. Rosenstock
Department of Pharmacology, Federal University of São Paulo, School of Medicine, São Paulo, Brazil

Héla Saïdi
INSERM U668, Paris, France, and Antiviral Immunity, Biotherapy and Vaccine Unit, Institut Pasteur, Paris, France

Stefan Schütze
Institute of Immunology, University of Kiel, Kiel, German

Claire Séror
INSERM, U848, Institut Gustave Roussy, and Université Paris-Sud 11, Villejuif, France

Yigong Shi
Department of Molecular Biology, Lewis Thomas Laboratory, Princeton University, Princeton, New Jersey

Soraya S. Smaili
Department of Pharmacology, Federal University of São Paulo, School of Medicine, São Paulo, Brazil

Maurizio Sorice
Department of Experimental Medicine, "Sapienza" University of Rome, Rome, Italy

Vladimir Tchikov
Institute of Immunology, University of Kiel, Kiel, German

Jerome Thiery
Immune Disease Institute and Department of Pediatrics, Harvard Medical School, Boston, Massachusetts

Antonella Tinari
Department of Technology and Health, Istituto Superiore di Sanità, Rome, Italy

Boris Turk
Department of Biochemistry, Molecular and Structural Biology, J. Stefan Institute, Ljubljana, Slovenia

Vito Turk
Department of Biochemistry, Molecular and Structural Biology, J. Stefan Institute, Ljubljana, Slovenia

Vladimir A. Tyurin
Department of Environmental and Occupational Health, University of Pittsburgh, Pittsburgh, Pennsylvania, and Center for Free Radical and Antioxidant Health, University of Pittsburgh, Pittsburgh, Pennsylvania

Yulia Y. Tyurina
Department of Environmental and Occupational Health, University of Pittsburgh, Pittsburgh, Pennsylvania, and Center for Free Radical and Antioxidant Health, University of Pittsburgh, Pittsburgh, Pennsylvania

Alena Vaculova
Institute of Environmental Medicine, Division of Toxicology, Karolinska Institute, Stockholm, Sweden

Peter Vandenabeele
Department of Molecular Biology, Ghent University, Ghent, Belgium, and Molecular Signaling and Cell Death Unit, Department for Molecular Biomedical Research, VIB, Ghent

Ilio Vitale
INSERM, U848, Institut Gustave Roussy, and Université Paris-Sud 11, Villejuif, France

Heather L. Wieman
Department of Pharmacology and Cancer Biology, Department of Immunology, and Sarah W. Stedman Nutrition and Metabolism Center, Duke University, Durham, North Carolina

Zahra Zakeri
Department of Biology, Queens College and Graduate Center of the City University of New York, Flushing, New York

Kezhong Zhang
Department of Biological Chemistry, The University of Michigan Medical Center, Ann Arbor, Michigan

Yuxing Zhao
Department of Pharmacology and Cancer Biology, Department of Immunology, and Sarah W. Stedman Nutrition and Metabolism Center, Duke University, Durham, North Carolina

Boris Zhivotovsky
Institute of Environmental Medicine, Division of Toxicology, Karolinska Institute, Stockholm, Sweden

Sandra S. Zinkel
Medicine, Vanderbilt University School of Medicine, Nashville, Tennessee

PREFACE

The idea of assembling a *Methods in Enzymology* book on cell death at first struck us as not very different from assembling a *Methods* book on Biology. There are so many aspects to cover and such a diversity of techniques that it seemed impossible to put together a manual of direct laboratory use such as one might do by offering a volume on, for instance, the isolation and study of mitochondria. After some reflection, we realized that the complexity was the point: it would be useful to compile a work that would, first, clearly define the types of cell death; second, indicate that not all forms of death were apoptosis; third, indicate the major means of documenting apoptosis; and fourth, explain how to document forms of cell death that are not apoptosis. In this manner we would provide a guide for the junior investigator and define criteria in a manner that would help readers recognize and avoid the confusion that exists in the literature because of the lack of clear definitions.

There are many means of organizing these overlapping topics. We have chosen to group the more biologic and general topics into the first volume, with the more clinical and specific topics in the second volume. A particular reader will find it useful to pick and choose information, perhaps in several chapters. Nevertheless, readers can expect to find in the first volume: comparisons of the characteristics of apoptosis, necrosis, autophagy, and other forms of cell death (Chapters 1, 12, 13, 19, and 20); means of studying the often unusual forms of cell death in nonmammalian organisms (Chapters 21 and 22); more general and commonly used methods for studying apoptosis in mammalian cells (Chapters 2, 3, 10); and means for studying specific events or components of apoptosis: The CD95 DISC (Chapter 4); the TNF receptosome (Chapter 5); the apoptosome (Chapter 6); and four types of enzymes active in apoptosis and other forms of cell death: caspases (Chapter 7); lysosomal cathepsins (Chapter 8); granzymes (Chapter 9); and nucleases (Chapter 11). In addition, we include discussions of the metabolic components of apoptosis such as mitochondrial membrane permeabilization (Chapter 14); oxidative lipidomics (Chapter 15); endoplasmic reticulum stress-induced apoptosis (Chapter 16); membrane scrambling (Chapter 17); and glucose metabolism (Chapter 18). In the second volume, readers can find chapters on studying apoptosis in different model systems (Chapters 1, 2, 3, 4, 5), investigation of several of the key promoters and signaling pathways that regulate programmed cell death (Chapters 6, 7, 8, 9, 10,11, and 21), analysis of several mechanisms by which apoptotic pathways are regulated (Chapters 12, 13, 14), analysis of programmed cell death in different tissue and organ systems (Chapters 15, 16, and 17), methods for generation and analysis of death ligands (Chapters 18 and 19), methods for identifying caspase activity and substrates (Chapter 21), generation and analysis of BCL-2 stabilized peptides (Chapters 22 and 23), and, finally, methods for a genetic screening strategy that could be potentially be used in identifying novel factors regulating programmed cell death (Chapter 24).

Although each reader will undoubtedly identify particular items that he or she would have preferred to see, or a different order of presentation, we were very pleased that almost all of the contributors that we originally invited accepted our invitation and delivered to us manuscripts conforming to our goals. Thus, this work represents what we intended—a broad spectrum of approaches to apoptosis and programmed cell death by recognized leaders in their respective fields. Thus, we hope that it will contribute by serving as a guide and a means of helping researchers to ask the most meaningful questions and to design their experiments for the highest clarity.

METHODS IN ENZYMOLOGY

CHAPTER ONE

Analyzing Morphological and Ultrastructural Features in Cell Death

Antonella Tinari,* Anna Maria Giammarioli,[†] Valeria Manganelli,[‡] Laura Ciarlo,[†] *and* Walter Malorni[†]

Contents

* Department of Technology and Health, Istituto Superiore di Sanità, Rome, Italy
[†] Therapeutic Research and Medicines Evaluation, Istituto Superiore di Sanità, Rome, Italy
[‡] Department of Experimental Medicine, "Sapienza" University of Rome, Italy

Methods in Enzymology, Volume 442
ISSN 0076-6879, DOI: 10.1016/S0076-6879(08)01401-8

Abstract

Diverse forms of cell death have initially been described thanks to their observation at the electron microscope. Morphological and ultrastructural features of necrosis, apoptosis, and autophagy, considered here as prototypic cell death processes, allow one to characterize and quantify early and late cytopathological changes occurring in cells undergoing degeneration. Both light microscopy and scanning electron microscopy can provide useful insights, for example, to quantitatively evaluate cell death or to characterize cell surface changes of the cells, respectively. However, transmission electron microscopy preparation allows distinguishing among different forms of cell death. This chapter describes in brief the methods used to characterize cell death forms, including membrane, nucleus, and organelle changes, and shows paradigmatic micrographs. In particular, morphogenetic changes occurring in mitochondria during apoptosis, that is, fission process or, conversely, vacuole formation during autophagy, are shown. Possible artifacts are also described. Ultrastructural analysis seems still to provide essential information for studies on cell death.

1. INTRODUCTION

A balance between cell proliferation and cell death regulates and controls the cell number in an organism. Characterization of cell death mainly derives from morphological and ultrastructural observations. In fact, since the pivotal work of Kerr *et al.* (1972), studies on structural features of injured cells have been recognized as crucial factors in the identification of cells undergoing death and in the comprehension of the mechanisms of cell death (Huerta *et al.*, 2007). Biochemical, molecular, or ultrastructural approaches can be used to analyze the complex sequence of modifications ultimately leading to cell death. However, the cascade of biochemical and physiological events leading to changes of macromolecule synthesis, in cellular homeostasis, in cell volume regulation, and finally in loss of cell viability, are intimately related to definite morphological changes. Changes in nuclear morphology, in organelle ultrastructure as well as specific phenomena at the cell surface level, are often considered as markers associated with particular cell death programs.

At least three types of cell death have been distinguished in mammalian cells by morphological and ultrastructural criteria: (1) necrosis, (2) apoptosis, or programmed cell death, and (3) autophagy. These are described very briefly later. An updated and comprehensive description of these and other particular forms of cell death process can be found elsewhere (Galluzzi *et al.*, 2007; Otsuki *et al.*, 2003; Ziegler and Groscurth, 2004).

Necrosis, caused by infarction, infections, poisoning, and so on, usually affects contiguous groups of cells and causes an inflammatory response, presumably through the liberation of factors from dead cells. Cellular swelling, plasma membrane ruptures, breakdown of cell organelles, and denaturation of cytoplasmic proteins represent the morphological features of this kind of cell death.

Apoptosis is a form of cell death in which a programmed sequence of events leads to the elimination of cells without releasing harmful substances into the surrounding area. Apoptosis plays a crucial role in developing and maintaining health by eliminating old, unnecessary, and unhealthy cells. Chromatin condensation (pyknosis) and fragmentation (karyorrhexis), membrane preservation, overall cell shrinkage, blebbing of the plasma membrane, and formation of apoptotic bodies that contain nuclear or cytoplasmic material and their subsequent phagocytosis are the cytological characteristics of apoptosis. The process of apoptosis is highly regulated. Too much or too little apoptosis can have pathological consequences, such as the onset of autoimmune diseases, neurodegenerative diseases, and developmental defects or, conversely, when apoptosis is impaired, cancer (Galluzzi *et al.*, 2007; Takemura and Fujiwara, 2006).

Finally, the so-called mitotic catastrophe (MC) has also been described. MC indicates the type of mammalian cell death caused by aberrant mitosis. In a physiological condition, the M phase, characterized by sister chromatid alignment, is followed by cytokinesis, in which the cytoplasm and its contents, including chromosomes, are partitioned into two daughter cells. MC has thus been associated with the formation of multinucleated, giant cells containing uncondensed chromosomes. A defect in cyclin complex inactivation has been indicated as a possible key event in the occurrence of MC. It has been observed that agents that perturb the cell cytoskeleton, specifically those that damage microtubules and disrupt the mitotic spindle, can cause MC. Interestingly, it has also been hypothesized that MC could be unconnected to apoptosis (Castedo *et al.*, 2004).

Autophagy is a catabolic process involving degradation of a cell's own components through the lysosomal machinery. It is a tightly regulated process that participates in organelle turnover and in bioenergetics management of starvation. Autophagy resulting in the total destruction of the cell is also known as cytoplasmic cell death or type II programmed cell death. It may help to halt the progression of some diseases and plays a protective role against infection. However, in some situations, it may actually contribute to the development of a disease. Methods to evaluate autophagy have been described elsewhere (Klionsky *et al.*, 2007).

Different methods for studying cell death are reported in Table 1.1, whereas morphologic characteristics of cell death are reported in Table 1.2.

Table 1.1   Morphologic characteristics of cell death

	Nucleus	Plasma Membrane	Cytoplasm	Characteristics
Apoptosis	Chromatin condensation Nuclear fragmentation DNA laddering	Blebbing Smoothing	Fragmentation (apoptotic body formation)	Caspase dependent
Autophagy	Partial chromatin condensation No DNA laddering	Blebbing	Increased number of autophagic vacuoles (double membraned)	Caspase independent Increased lysosomal activity
Necrosis	Clumping Casual degradation of nuclear DNA	Swelling Ruptures	Vacuoles increasing Organelle degeneration Mitochondrial swelling	

Table 1.2   Methods for studying cell death

Cell death types	Methods
Apoptosis	Electron microscopy TUNEL Annexin V Activity of caspases DNA laddering Mitochondrial membrane potential
Autophagy	Electron microscopy Determination of protein degradation Determination of protein markers on autophagic membranes Monodansylcadaverine staining
Necrosis	Electron microscopy Nuclear staining Evaluation of inflammation and damage in the surrounding tissues

2. THE CONTRIBUTION OF MORPHOLOGICAL ANALYSES

In the last years, numerous different morphological techniques have been considered for studying the complex sequence of structural modifications ultimately leading to cell death. These are mainly qualitative and quantitative analyses that can be carried out by means of light (LM) and electron microscopy (EM) techniques. Cytometric and morphometric analyses can be performed using light microscopy, fluorescence microscopy, and confocal microscopy, as well as scanning and transmission electron microscopy (TEM). Some of these techniques are of widespread use in laboratory practice, whereas others are employed specifically in certain experimental conditions. For example, conventional TEM analysis is the only reliable method for monitoring autophagy *in situ*.

3. LIGHT MICROSCOPY

Light microscopy is a simple and efficient way to specifically identify and quantify cell death. For instance, *in situ* labeling of cell death versus proliferating cells allows high sensitivity and precise identification of the cell population involved in these different phenomena. In view of the importance of LM, which should always precede EM analyses, a very brief description of this approach is provided later.

Apoptotic cells are not easily distinguishable by LM from other elements with condensed chromatin, such as mitotic cells in telophase. Hence, several different techniques have been developed. For example, cell death detection kits, based on the labeling of DNA strand breaks, have been considered as markers of apoptosis. Terminal deoxynucleotidyl transferase-mediated dUTP nick end labeling (TUNEL) is a method for the rapid identification and quantification of apoptotic cell fractions (Labat-Moleur *et al.*, 1998).

3.1. TUNEL

The TUNEL technique relies on the use of endogenous enzymes that allow the incorporation of labeled nucleotides into the 3′-hydroxyl (3′OH) recessed termini of DNA breaks. The added value in the use of *in situ* immunocychemical techniques resides in the possibility of evaluating both morphological and staining features in the same sample. This chapter describes the TUNEL method implemented with alkaline phosphatase-antialkaline phosphatase (APAAP) and peroxidase-anti-peroxidase (PAP) immunocytochemistry methods (see later).

3.2. Fluorescence microscopy

This technique is capable of imaging the distribution of a single molecular species based solely on the properties of fluorescence emission. Chromatin condensation, nuclear shrinkage, and formation of apoptotic bodies can be observed easily under fluorescence microscopy, after appropriate staining of nuclei with DNA-specific fluorochromes [4′,6-diamidino-2-phenylindole (DAPI), Hoechst 33258, see later] (Fig. 1.1). Using fluorescence micros-copy, the precise location of intracellular components labeled with specific fluorophores can also be monitored. This has been used to assess the presence of autophagic vacuoles. In particular, some fluorescent dyes (LC3, lysotracker) have been used for the morphological detection of cells undergoing autophagy, also called self-cannibalism (Mormone *et al.*, 2006). Specimens prepared for conventional fluorescence microscopy can also be observed by confocal microscopy.

4. Sample Processing for Light Microscopy

4.1. Cell viability

The intact membrane of living cells excludes cationic dyes such as Trypan blue or propidium iodide (PI). Because of their extensive membrane dam-age, necrotic cells are quickly stained by short incubation with these dyes. Staining techniques involve either dye exclusion, in which the membrane

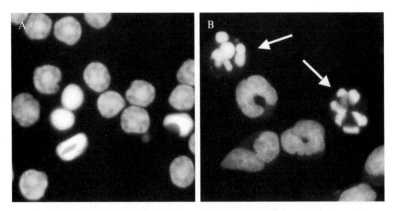

Figure 1.1 Morphological analysis by fluorescence microscopy of Hoechst-stained cells. Note different aspects of apoptotic chromatin aggregation (arrows) in thymocytes (A) and CEM lymphoblastoid cells (B). Nuclei of apoptotic thymocytes show typical morphological features of chromatin aggregation, whereas other aggregation figures, as well as clumping and typical fragmentation, are shown by apoptotic nuclei of lymphoblastoid CEM cells.

of a viable cell excludes the dye, or specific uptake of a vital dye (viable cells appear brightly refringent while dead cells are dark under a phase-contrast microscope).

4.1.2. Methods
All procedures should be performed rapidly (not longer than 1 h before viewing) and should use a low concentration (0.01%) solution of dye in physiological buffer to minimize cell damage.

4.1.2.1. Methylene blue: Vital stain

- 1.5 g methylene blue
- 30 ml 95% alcohol
- 2 ml 0.1 potassium hydroxide

1. Dissolve dye in alcohol.
2. Add potassium hydroxide.
3. Store in dark at 4°.

4.1.2.2. Trypan blue: Dye exclusion   Stock 2% (w/v) solution in distilled water, filter, add 1 volume solution to 9 volumes of cell suspension, and incubate for 30 min at 37°.

4.2. In situ methods
Chemical fixation
Generally, fixation should not take less than 15 min or exceed 24 h duration, except in the case of formaldehyde. Fixation procedures should be mild to leave the antigen unaltered. The fixative can be coagulant, cross-linking, additive, linking to the substance to be fixed with different penetration rates. Penetration rates of some common fixatives (in order of fastest to slowest) are ethanol, methanol, acetone > acetic acid > formaldehyde > osmium tetroxide.

Living cells can be stained by using vital stains (e.g., lysotracker, mitotracker) and then fixed, but, generally, cells are fixed and then stained or otherwise processed. When detecting intracellular markers of cell death, which cannot be stained by using vital stains, permeabilization of a fixed cell membrane is necessary in order to allow entry of the antibody molecules to detect intracellular antigens.

The choice of fixative and/or permeabilization depends on the sample and staining technique to be used.

Staining

As described previously, staining involves the selective uptake and interaction of reagents from a soluble phase with a solid specimen (e.g., attachment of a dye or binding of an antibody). In general, dye molecules contain a chromogen, the colored component, and an auxochrome, which attaches the dye to the specimen. The chromogen contains a chromophore that is involved in the absorption of certain wavelengths of light to give color (immunocytochemistry) or emit light of longer wavelength (immunofluorescence) with respect to absorbed light.

4.3. Protocol of cultured cells

Both adherent cell lines and cells growing in suspension can be examined. A short paragraph of protocols dealing with fixation requirements, staining procedures, and time of observation is included. In general, adherent cells can be fixed simply by adding fixative solution to the monolayer, whereas cell suspensions can be prepared by adding the fixative solution to the formed pellet after centrifugation.

The following protocols consider different staining procedures: (i) observing an apoptotic nucleus under fluorescence microscopy (Hoechst 33258 or DAPI) or by light microscopy (TUNEL) and (ii) analyzing mitochondria.

4.3.1. Hoechst 33258

Nuclei can be stained with different fluorochromes, depending on the experimental requirements. For example, if a simultaneous analysis of membrane antigens with fluorescein–isothiocyanate (FITC) and/or tetramethylrhodamine is performed, nuclei should be stained with an ultraviolet-excited, blue-emitting fluorochrome such as Hoechst 33258 (excitation 365 nm, emission 465 nm) or DAPI (excitation 358 nm, emission 461 nm). Propidium iodide (excitation 536 nm, emission 617 nm) can be used for routine analysis of nuclear morphology, especially for necrotic cells. The following protocols describe Hoechst 33258 staining.

Preparation of paraformaldehyde (PFA) solution (100 ml)

- Dissolve 3.7 g of paraformaldehyde in 50 ml H_2O, add 2 drops of 10 N NaOH, and heat 30 min at 60°. Add 50 ml of phosphate-buffered saline (PBS) 2×. Adjust pH to 7.2.
- The PFA solution must be freshly prepared.

Preparation of Hoechst Dissolve 10 mg of Hoechst 33258 in 10 ml of ethanol. Clarify by centrifugation (5 min at 5000 rpm); store this solution at 4° in the dark. Hoechst staining solutions can be stored at 4° for several weeks protected from the light.

4.3.1.1. Cells growing on coverslips (or remaining adhering to the substrate).

Sterilization of coverslips (12 mm)

a. Sonicate the coverslips in acetone for 15 min at room temperature.
b. Wash the coverslips with dH$_2$O to remove the acetone completely.
c. Place the coverslips on a clean paper sheet separating each other. Air dry the coverslips.
d. Put the coverslips into a heat-resistant glass bottle.
e. Sterilize the coverslips in an autoclave for 15 min at 121° (heat-resistant glass bottle must remain partially opened during sterilization).
f. Close the heat-resistant glass bottle and leave it in a sterile hood.

Methodology

1. Remove the medium and wash the cells twice with PBS.
2. Directly add to the sample the 3.7% paraformaldehyde in PBS. Leave for 20 min at room temperature.
3. Remove the fixative and rinse three times in PBS.
4. Directly add to the sample 0.5% Triton X-100 and leave for 5 min at room temperature. Rinse three times with PBS.
5. Incubate the coverslip with 1 μg/ml Hoechst for 30 min at 37°.
6. Rinse the sample with PBS and mount on the slide with glycerol: PBS (1:1).
7. Observe the sample with a fluorescence microscope (ultraviolet light).

   Slides can be kept in the dark at 4° for several days.

4.3.1.2. Cells growing in suspension

Polylysine-coated coverslips

• Sonicate the coverslips (12 mm) in acetone for 15 min at room temperature.
• Wash the coverslips with distilled water (dH$_2$O) to remove the acetone completely.
• Place the coverslips on a clean paper sheet separating each other. Air dry the coverslips.
• Gently spread 20 μl polylysine solution (0.1 mg/ml dH$_2$O) on the surface of the coverslips.
• Allow the polylysine drops on the coverslips to dry at room temperature.

Methodology

1. Centrifuge cells (2–5 $\times$ 10^5) at 1000 rpm for 5 min and wash once with PBS.
2. Remove the supernatant and resuspend the pellet in 5 ml 3.7% paraformaldehyde in PBS. Leave the sample at room temperature for 15 min.
3. Centrifuge cells at 1000 rpm for 5 min.
4. Remove the supernatant and resuspend the pellet in 5 ml PBS. Centrifuge sample at 1000 rpm for 5 min.

5. Resuspend the pellet in 20 to 30 μl PBS (depending on the number of cells) and deposit it on a polylysine-coated coverslip. Allow the cells to adhere on the coverslip for 30 min at room temperature.
6. Wash the coverslip twice with PBS. Continue from step 4 of previous paragraph (adhering cells).

Slides can be kept in the dark at 4° for several days.

4.3.1.3. Cells growing in monolayer: detached cells (i.e., following apoptotic treatments)

Methodology
1. Collect the culture supernatant, centrifuge at 1500 rpm for 10 min, and wash once with PBS. Resuspend the pellet in 4 ml 3.7% paraformaldehyde in PBS. Leave the sample at room temperature for 20 min.
2. Centrifuge the detached cells at 1500 rpm for 10 min. Wash once with PBS. Centrifuge sample at 1500 rpm for 10 min.
3. Continue from step 3 of previous paragraph *(Cells Growing In Suspension)*

Slides can be kept in the dark at 4° for several days.

4.3.2. TUNEL (*in situ* immunocytochemistry)

Extensive DNA degradation is a characteristic event that occurs in the late stages of apoptosis. Cleavage of the DNA may yield double-stranded, low molecular weight DNA fragments (mono- and oligonucleosomes), as well as single strand breaks ("nicks") in high molecular weight DNA. This technique relies on the use of endogenous enzymes that allow incorporation of labeled nucleotides into the 3'-OH termini of DNA breaks. The pretreatment of biological samples to obtain suitable access to DNA breaks for enzymatic reactions is highly impaired by fixation procedures, such as cold absolute acetone treatment. Actually, the incubation of samples with detergents, such as Triton X-100, did not highly improve TUNEL sensitivity. In this protocol, TUNEL-positive cells are stained using either APAAP or PAP immunocytochemistry methods (Labat-Moleur *et al.*, 1998).

Methodology

Cytospin preparation
1. Centrifuge 2 to 8 × 10^4 cells/slide at 400 to 600 rpm for 5 min.
2. Carefully wipe slide around the cell spot and dry at room temperature.
3. Fix the slide sample in absolute acetone for 10 min at 4°.
4. Wash with Tris buffer solution (TBS) (pH 7.6) twice.
5. Dry and saturate the sample with 10% rabbit serum or fetal calf serum or AB human serum as appropriate.
6. Cover with TUNEL solution and incubate for 60 min in a 37° humid atmosphere.

7. Wash with TBS twice.
8. Dry and cover with the mouse anti–FITC Ab and incubate for 60 min at room temperature in a dark condition.
9. Wash twice with TBS.
10. Dry, cover with linking antibody (usually a rabbit antimouse for both the PAP and the APAAP method), and incubate for 30 min at room temperature.
11. Wash twice with TBS.
12. Dry, cover with APAAP or PAP as appropriate (for TUNEL staining we prefer APAAP, allowing the use of a wider panel of chromogens, particularly dark chromogen as NBT or Fast blue), and incubate for 45 min at room temperature.
13. Wash twice with TBS.
14. Dry, cover specimens with developing solution with the substrate chromogen standard composition suitable for either PAP (0.05% 3.3-diaminobenzidine tetrahydrochloride color reaction, DAB) or APAAP (Fast blue or BCIP/NBT for dark blue staining, Fast red for red staining + N,N-dimethyl-formamide), and incubate 10 to 30 min.
15. Block reaction with TBS for 10 min.
16. Cover coverslip with aqueous mounting media (Aquamount).

4.4. Mitochondria

After mitochondria have been visualized successfully by vital staining with rhodamine 123, numerous other fluorescent stains that target mitochondria have become available. Although conventional fluorescent stains for mito- chondria, such as tetramethylrosamine and rhodamine 123, are readily sequestered by functioning mitochondria, these stains are easily washed out of cells once mitochondria experience a loss in membrane potential. This characteristic limits the use of such conventional stains in experiments that require cells to be treated with aldehyde fixatives or with other agents that affect the energetic state of the mitochondria. To overcome this limitation, MitoTracker probes can be used—a series of mitochondrion- selective stains that are concentrated by active mitochondria and well retained during cell fixation (Keij *et al.*, 2000). To label mitochondria, cells are simply incubated in submicromolar concentrations of a Mito- Tracker probe, which passively diffuses across the plasma membrane and accumulates in active mitochondria. Once their mitochondria are labeled, cells can be treated with an aldehyde-based fixative to allow further proces- sing of the sample.

4.4.1. Dye stock solution preparation

This preparation can be utilized for rhodamine 123, MitoTracker red, MitoTracker green, and nonyl acridine orange.

Depending on the purpose, a 0.5 or 0.25 mM stock solution of the dye(s) should be prepared in high-quality, anhydrous dimethyl sulfoxide or absolute ethanol. As fluorescent dyes generally decompose if illuminated, it is necessary to keep stock solutions in well-sealed dark reagent bottle. All dye stock solutions can be kept at $-20°$.

4.4.2. Cell preparation and staining

Dilute the MitoTracker stock solution to the final working concentration in growth medium, with or without serum to match the medium that the cells were grown in. For live cell staining, the recommended working concentrations are 25 to 500 nM. For staining cells that are to be fixed and permeabilized (see Fixation and Permeabilization after Staining), a working concentration of 100 to 500 nM can be used. To reduce potential artifacts from overloading, the concentration of dye should be kept as low as possible. For MitoTracker green probes, a slightly lower concentration (20–200 nM) is suggested.

4.4.2.1. Staining adhering cells

1. Grow cells on coverslips inside a petri dish filled with the appropriate culture medium.
2. Remove the medium from the dish and add the prewarmed (37°) growth medium containing the MitoTracker probe.
3. Incubate the cells for 15 to 45 min under growth conditions appropriate for the particular cell type.
4. Replace the loading solution with fresh prewarmed medium
5. After staining, wash the cells in fresh, prewarmed growth medium.
6. Remove the growth medium covering the cells
7. Replace medium with freshly prepared, prewarmed growth medium containing 3.7% formaldehyde
8. Incubate at 37° for 15 min.
9. Rinse the cells several times in PBS.
10. Put cells onto the slide and observe them under a fluorescence microscope.

When cells are going to be subsequently labeled with an antibody, a permeabilization step is usually required to enhance the accessibility of the antigen.

Cell permeation

1. Incubate the fixed cells in PBS containing 0.5% Triton X-100 at room temperature for 5 min.

2. Rinse the cells in PBS.
3. Proceed with antibody labeling protocols.

Alternatively, the cells may be permeabilized by incubating them in ice-cold acetone for 5 min and then washing in PBS.

Staining of cells growing in suspension

1. Centrifuge to obtain a cell pellet and aspirate the supernatant.
2. Resuspend the cells in prewarmed (37°) medium containing the Mito-Tracker. Incubate the cells for 15 to 45 min under growth conditions appropriate for the particular cell type.
3. Repellet the cells by centrifugation.
4. Replace medium with freshly prepared, prewarmed growth medium containing 3.7% formaldehyde.
5. Incubate at 37° for 15 min.
6. Rinse the cells several times in PBS.
7. Resuspend the pellet in 20 to 30 μl PBS (depending on the number of cells) and deposit it on a polylysine-coated coverslip.
8. Leave the cells to adhere on coverslips for 30 min at room temperature.
9. Wash the coverslips twice with PBS.
10. Put the cells on a slide and observe them under a fluorescence microscope.

When the cells are going to be labeled with an antibody, continue from step 11 of the previous paragraph.

5. ELECTRON MICROSCOPY

As it is not possible to study living cells by transmission electron microscopy, it is necessary to fix and process them in order to obtain sections of 80 nm thick, or even thinner, from original samples. Thus, biological material is chemically fixed, dehydrated through an acetone or ethanol series, and then embedded in epoxy or acrylic resins. The sample is then thin sectioned by an ultramicrotome. Sectioning the sample allows one to look at cross sections through samples to view internal structural features. Sections of 80 to 100 nm thickness are collected on copper grids and counterstained with electron-dense stains before observation.

The EM study of cell death is essentially focused on the following goals: (A) concerning *necrosis*, ruptures of the plasma membrane; swelling of mitochondria accompanied by matrix rarefaction and disruption of cristae; nuclear chromatin dispersion (Fig. 1.2); (B) concerning *apoptosis*, cell surface blebbing (Fig. 1.3), chromatin aggregation (Fig. 1.4), detection of MC

(Fig. 1.5) and mitochondrial changes, including fission (Fig. 1.6) and mitoptosis (Fig. 1.7) (Castedo *et al.*, 2004; Matarrese *et al.*, 2005; Tinari *et al.*, 2007) (Figs. 1.3–1.5); and (C) concerning *autophagy*, double membraned vacuoles (Kroemer *et al.*, 2005; Malorni *et al.*, 2007) (Fig. 1.8).

5.1. Specimens

Different protocols can be used depending on the type of the sample considered: cell pellets or monolayers need shorter exposure time to the fixative, dehydrating agent, and embedding resin with respect to tissue samples, which need longer times.

Cell monolayers can be fixed in glutaraldehyde and postfixed in OsO_4 simply by adding fixative solution to the monolayer, whereas cell suspensions can be prepared by adding the fixative solution to the formed pellet after centrifugation. Cell monolayers can also be embedded directly in the culture flasks. Tissue samples should be excised, immediately dipped into the primary fixative, and gently cut into small pieces. The tissue should generally be cut into pieces smaller than 1 mm to assure rapid fixation of the entire piece of tissue. Tissues can also be fixed by perfusion procedure in living anesthetized animals to minimize the effects of cutting on unfixed tissues and of anoxia resulting from cutting off circulation.

5.2. Fixation

The aim of fixation is to preserve every detail of the cellular ultrastructure right down to the molecular level exactly as it was in life the instant before cell death so that it can resist the effects of subsequent steps in the preparative procedure. A two-stage fixation procedure, using glutaraldehyde followed by OsO_4, is now almost used universally. Fixation steps are generally performed in a fume hood, especially when OsO_4 is used, as it is extremely volatile and toxic.

Glutaraldehyde is the preferred aldehyde for primary fixation. Rapid penetration of tissues resulting in less distortion, good preservation of structural relationships, and low tendency to extract cytoplasmic components are the major characteristics of glutaraldehyde, resulting in improved fidelity of ultrastructure. However, glutaraldehyde is capable of penetrating very poorly inside tissues, about 2 mm, so the specimen has to be no more than 1 to 2 mm^3.

Although glutaraldehyde preserves many cellular structures, it does not preserve lipids that would be extracted completely by the alcohol used in the dehydrating steps. Moreover, it has the property of stabilizing glycogen, thus preventing its subsequent loss due to OsO_4. Furthermore, it preserves structures such as microtubules that are poorly maintained by OsO_4 and stabilizes nucleoproteins better than OsO_4. For these reasons,

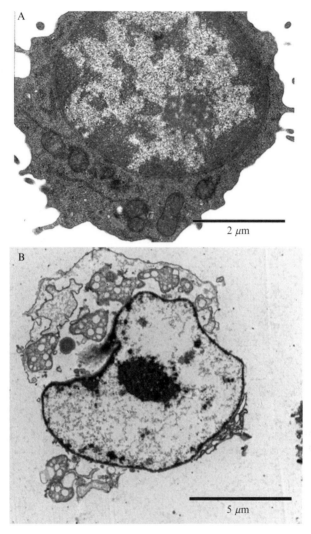

Figure 1.2   A necrotic cell as it appears at TEM observation. A normal lymphocyte (A) and a lymphocyte displaying characteristic features of necrosis (B), i.e., plasma membrane ruptures, cytoplasm swelling, and organelle degradation. At the nuclear level, the degradation of nuclear chromatin can be appreciated.

glutaraldehyde is used in conjunction with OsO_4 to give better specimen stability during subsequent dehydration and embedding.

Osmium tetroxide (OsO_4) reacts with unsaturated lipids and with tryptophan and histidine in proteins, thus cross-linking polypeptide chains together. OsO_4 stabilizes cellular proteins that form the matrix of the cytoplasm by forming cross links. Osmium is good for rendering lipids

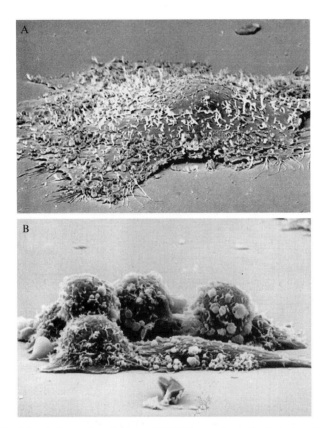

Figure 1.3 Morphological analysis by SEM. The surface of substrate adhering cells (astrocytoma cells) is characterized mainly by numerous microvillous structures and ruffles (A). Exposure to an apoptotic stimulus (B) induces marked morphological changes represented by cell retraction and loss of the cell–substrate interaction, cell rounding and shrinking, and the formation of surface protrusion such as blebs and blisters.

insoluble but has no effect on carbohydrates such as glycogen. It preserves lipids by forming addition compounds with unsaturated fatty acid chains. It is used for preserving cellular membranes and nucleoproteins. Because it combines chemically with most of the cellular constituents, osmium metal delineates them almost perfectly, also enhancing the image contrast at the EM. It penetrates tissue very slowly due to the low diffusion rate of this large molecule, and the time of fixation is usually 60 min. Prolonged fixation for periods over 90 min can occasionally lead to the solubilization of some tissue components. OsO_4 is usually prepared in a buffered solution at near neutral pH and is maintained at $4°$ in a dark glass bottle.

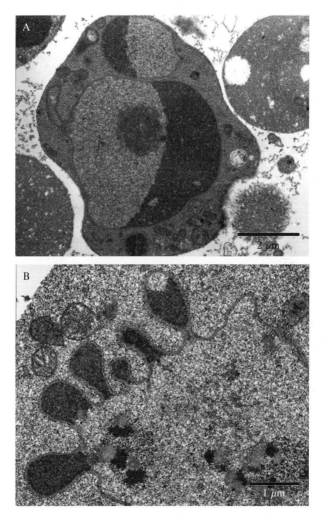

Figure 1.4   Electron micrographs of two different apoptotic cells. (A) A cell treated with an apoptotic stimulus showing typical morphological changes of apoptosis: plasma membrane integrity, cell shrinking, mitochondrial changes (such as the increase of electron density corresponding to changes of mitochondrial membrane potential), chromatin marginalization, and micronuclei formation. (B) A mouse bone marrow cell displaying quite different features of early apoptosis: note the chromatin clumping marginalized at choked nuclear regions and nucleolar degradation.

5.3. Buffers

Because of osmotic pressure, the fixative solution may cause disruptive effects such as shrinkage and swelling of the organelles. Thus, an isotonic buffer is used to maintain the pH of the fixative solution at the physiological value. The most commonly used buffer is cacodylate, which contains two

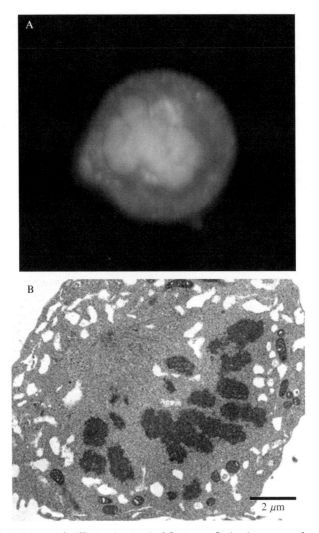

Figure 1.5 Micrographs illustrating typical features of mitotic catastrophe as detected by fluorescence microscopy (A) and electron microscopy (B). (A) Hoechst staining (in blue) indicates the chromatin derangement, whereas microtubules are labeled in green. (B) Note signs of defective arrangement of chromosomes and cell damage. (See color insert.)

other components: sodium acetate to provide buffering capacity in the acid range and sodium barbiturate for the alkaline range.

5.4. Dehydration

The aim of dehydration is to replace all the free water contained in the specimen with a fluid that is miscible with the embedding monomer. Agents such as ethyl alcohol, methyl alcohol, isopropyl alcohol, and acetone

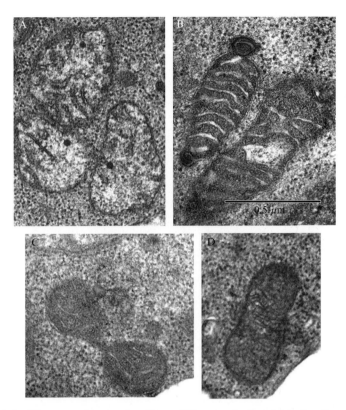

Figure 1.6 Electron micrographs of two different mitochondrial alterations detectable in lymphocytes and corresponding to (A) the rarefaction of mitochondrial matrix and ruptures of internal cristae usually detectable in necrotic cells; (B) early stages of mitochondrial fission (budding) usually associated with the apoptotic process: note the condensation of internal cristae (corresponding to the increase of mitochondrial membrane potential detectable soon after apoptotic stimulation) (Matarrese *et al.*, 2005); (C) mitochondria undergoing fission; and (D) normal features of a lymphocyte mitochondrion.

are used to dehydrate the specimen. Ethanol is the most widely used dehydration agent because it does not harden the specimen and make it too brittle for subsequent ultrathin sectioning. Other agents can be used depending on the nature of the embedding material being used. The time of dehydration steps is normally kept short to prevent the extraction of cell components and subsequent shrinkage. The specimen sample is transferred into mixtures of ethanol and decreasing water percentage. It may then be transferred to a fluid, propylene oxide that is completely miscible with both alcohol and epoxy resin monomer, which is almost invariably used for general embedding purposes.

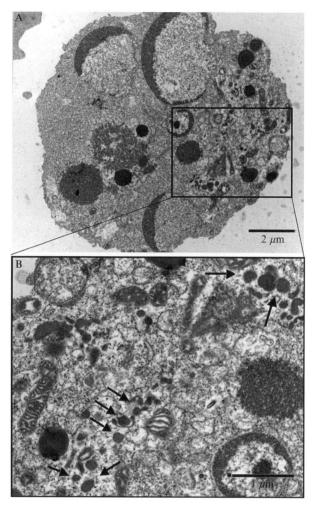

Figure 1.7   A cell in late apoptosis. (A) Nuclear and mitochondria fragmentations are observable at this magnification. (B) An enlargement of the EM image above (boxed area) displays outer membrane mitoptosis (OMM), a degenerative process of mitochondria in which condensation and collapse of internal cristae precede disruption of the outer membrane with the consequent spreading of mitochondria fragments (arrows) throughout the cell cytoplasm.

5.5. "En block" staining

Further differential electron contrast can be added after osmium fixation and before the tissue is embedded. The washing and dehydrating fluids can be used to add further stains, such as uranyl acetate, phosphotungstic acid, or potassium permanganate. The most commonly used substance is uranyl acetate, which is usually added to cell samples during the ethanol 70% step.

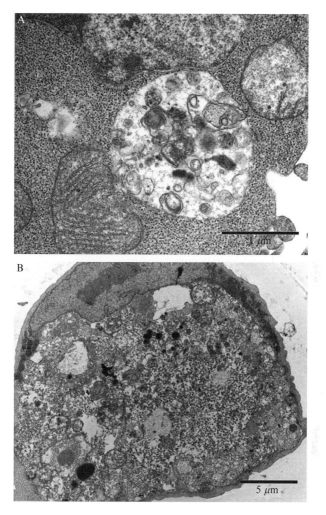

Figure 1.8   Autophagy. (A) A cell in which a large autophagic vacuole is evident: note cellular debris inside the vacuole, indicating an advanced digestion status. (B) A TEM picture of a lymphocyte dying by autophagy. Note the large amount of cytoplasm remnants, ribosomes, and organelles inside the vacuole.

5.6. Embedding

Properties of an ideal embedding medium are that it should be soluble in ethanol (or acetone) before polymerization, it should not modify the specimen by itself, it should not physically disrupt or distort the specimen, it should have uniform hardness, it should produce a block hard and plastic enough to cut ultrathin sections, and it should be stable under electron irradiation. Unfortunately, none of the embedding media possess all these properties together. A satisfactory compromise is represented by epoxy resins, which are used commonly for ultrstructural analyses.

5.7. Epoxy resins

These are a family of synthetic resins that polymerize at 40 to 60° producing yellow-brown solid blocks. The resin has two different chemically reactive groups: epoxide end groups and hydroxyl groups spaced along the length of the chain. If a mixture of resins, amines that form long chain polymers, and anhydride, which forms cross bridges between resin molecules via the hydroxyl groups, is heated, polymerization takes place in three dimensions, forming a stable, inert substance very resistant to heat and solvents.

The cutting properties of the final polymer depend on the components of the resin monomer mixture. The mixture consists of an epoxy resin, a hardener, an accelerator that controls the rate of hardening, and a plasticizer that controls hardness of the block. The mixed components are polymerized by heating to 60° for 48 h. Epon 812 is perhaps the most widely used embedding media. With Epon 812, a mixture of the resin and the two hardeners is made up in the correct proportions, leaving out the catalyst that should be added after the first three elements are mixed. The proportion of the two hardeners used in the monomer mixtures controls the hardness of the final block, thus influencing the subsequent ultrathin section cutting step.

The processes of dehydration and infiltration should be carried out as rapidly as possible because all the reagents used are powerful lipid solvents and remove a significant amount of lipid even after it has been fixed with OsO_4.

5.8. Thin sectioning

Sections are cut using a diamond knife, with the ultramicrotome set to cut at around 80 to 100 nm. The sections are picked up onto 200 mesh copper grids unless they are used for immunocytochemistry, in which case gold or nickel grids are used.

The thickness of sections can be estimated from their color in reflected light. Sections of about 60 nm appear dark gray to gray in color, sections of 60 to 90 nm in thickness appear silver, and sections of 90 to 150 nm will be gold. The best range of section thickness in obtaining good resolution to study cell death, as well as a good contrast for TEM observation, is from a silver to light gold color.

5.9. Thin section staining

Thin sections are usually stained with solutions of heavy metal salts with the aim of enhancing the contrast of specimens. The metal ions of the staining solutions react chemically with some cellular components, thus increasing their density. The most commonly used heavy metals for section staining are lead, as lead citrate, and uranium, as uranyl acetate. Because of the different staining properties of uranium and lead, it is common practice to double stain sections, first using uranyl acetate for 10 min followed by a 10-min incubation with lead citrate. Thin sections should be washed with distilled water after each staining to prevent heavy metal crystals contaminating the surface of the section when it dries.

Uranyl acetate is prepared for the moment as a saturated solution in water or 70% ethanol, filtered through 0.22-μm filters, and maintained in the dark to prevent crystal formation. Grids are placed with sections floating on the surface of a 50- to 100-μl droplet of the staining solution. The reaction should be carried out in the dark.

5.10. Artifacts

These procedures allow the analysis of chromatin clumps and chromatin aggregation during apoptosis, mitochondria disruption during necrosis, or mitochondria remodeling during apoptosis (Fig. 1.6). The analysis of autophagy, characterized by the presence of double-membraned large or small vacuoles containing cell organelles or debris (self-cannibalism), as well as entire cells (xeno-cannibalism), can be performed easily. However, several artifacts can derive from inappropriate sample preparation, for example, by inappropriate buffering or fixation procedures. Among these are (i) rupture of plasma membrane, (ii) swelling of organelles, including mitochondria or endoplasmic reticulum, (iii) rarefaction of mitochondrial matrix or disruption of mitochondrial cristae, (iv) surface blebbing, (v) cell shrinking or cytoplasm rarefaction, (vi) artifactual formation of vacuoles and cytoplasmic inclusions, and (vii) loss of cell–cell and cell–substrate contacts, that is, cell retraction and shrinking. Thus, apart from chromatin aggregation, all other alterations occurring in diverse cell death forms can also be the consequence of failures in preparation procedures. Hence, a careful EM analysis takes into consideration all these possible artifacts.

5.11. Embedding protocol

1. Primary fixation may be performed for 20 min or more, depending on the nature, the size, and the permeability of the sample, at room temperature. Pelleted cells and cell monolayers need 20 min of exposure to the fixative, whereas tissue blocks, cut into approximately 1-mm cubes, need longer times. Fix tissue in 2.5% glutaraldehyde in 0.1 M sodium cacodylate buffer, pH 7.2 to 7.4.
2. Wash in 0.1 M cacodylate buffer twice. Samples may be stored at 4° at this stage.
3. Postfix in 1% osmium tetroxide in 0.2 M cacodylate buffer for 1 h at room temperature.

4. Thoroughly rinse in 0.2 *M* cacodylate buffer three times. This is done to avoid precipitate formation due to the osmium–alcohol reaction.
5. Dehydrate in 50% ethanol for 10 min.
6. Dehydrate in 70% ethanol for 10 min. Samples may be stored at this stage.
7. Dehydrate in 90% ethanol for 10 min twice.
8. Dehydrate in 100% ethanol for 15 min twice.
9. Propylene oxide, for 10 min twice, should be used for tissue samples. In such a case, proceed utilizing propylene oxide as the resin vehicle.
10. Infiltrate with 100% ethanol/epoxy resin mixture (2:1) for 1.5 h.
11. Infiltrate with 100% ethanol/epoxy resin mixture (1:1) for 2 h.
12. Infiltrate with 100% ethanol/epoxy resin mixture (1:2) overnight.
13. Infiltrate with absolute resin for 6 to 8 hours. Remove caps from vials to allow any remaining ethanol to evaporate.
14. Embed in labeled capsules with freshly prepared resin. In particular, embed cell monolayers directly in the culture flasks.
15. Polymerize at 60° for 48 h.

Suspended cells may produce a pellet that is cohesive enough to process after spinning but if not they should be embedded in 2% agar gel as follows.

- Centrifuge cells at 1000 rpm for 5 min. Remove the supernatant and resuspend the pellet in 1 ml 0.1 *M* cacodylate buffer.
- Centrifuge sample at 1000 rpm for 5 min. Remove the supernatant and resuspend the pellet in 1 ml 2.5% glutaraldehyde in 0.1 *M* cacodylate buffer. Let the cells fix for 20 min at room temperature.
- Centrifuge cells at 1000 rpm for 5 min. Remove the supernatant and wash the cells twice with 1 ml 0.1 *M* cacodylate buffer. Centrifuge cells at 1000 rpm for 5 min. Remove the supernatant and resuspend the pellet in 1 ml of 1% osmium tetroxide (OsO_4) in 0.2 *M* cacodylate buffer. Let the cells fix for 1 h at room temperature.
- Centrifuge cells at 1000 rpm for 5 min. Remove the supernatant and wash the cells three times with 1 ml 0.2 *M* cacodylate buffer.
- Centrifuge cells at 1000 rpm for 5 min.
- Prepare a 2% solution of agar in distilled water at a temperature of 37°. Fill each tube with agar solution, resuspend the samples, and spin them at full speed for 30 s.
- Cool the tubes in a beaker of cold water to set the agar.
- Cut the embedded pellet in pieces of about 1 mm^3. Put the pieces in a small glass bottle and continue with embedding protocol step 5.

6. Scanning Electron Microscopy (SEM)

Scanning electron microscopy allows one to analyze the cell surface. The use of SEM in the analysis of cell death is mainly referred to the study of cell surface alterations such as smoothing, loss of microvillous structures, blebbing, and shrinking, which are important signs of cell injury (Fig. 1.3). They, however, cannot be considered as specific markers of cell death.

Cell lines that grow in monolayer or dispersed cells, seeded on polylysine-coated coverslips, are fixed as for the TEM method. This fixation procedure could induce some changes on the cell surface. To avoid artifacts, the fixative has to be dropped gently on the samples. After cells have been dehydrated through the graded ethanol solution (see TEM protocol), they are critically point dried in CO_2. The critical point-drying (CPD) technique provides a much better means of avoiding the damaging effects of surface tension forces than air drying. The specimen is thus transferred into a closed pressure vessel with the aim of substituting the 100° ethanol present in the cells with liquid CO_2. To allow SEM observation, the specimens are then usually coated with a conducting layer to inhibit charging. Gold sputtering has the advantage of requiring only a low vacuum and is the most commonly used coat.

6.1. Protocol

1. Let the cells grow on coverslips and proceed to fixation and dehydration as for the TEM protocol (from step 1 to step 8). Never let the samples dry.
2. Dehydrate samples by CPD.
3. Glue the coverslips onto stubs. Let them dry about 20 min at room temperature and overnight at 37°.
4. Cover the samples with gold (thickness about 20 nm) using the sputter coater.
5. Conserve the samples at 37° in a specimen desiccator.

If cells are grown in suspension:

- Proceed to the primary fixation as for TEM protocol point 1.
- Resuspend the pellet in 20 to 30 ml 0.1 M cacodylate buffer and deposit it on polylysine-coated (0.1 mg/ml dH_2O) coverslips (12 mm). Let cells adhere on the coverslip for 20 min at room temperature.
- Gently add 0.1 M cacodylate buffer. Remove 0.1 M cacodylate buffer and proceed to postfixation as for TEM protocol step 3.
- Proceed as for SEM protocol.

REFERENCES

Castedo, M., Perfettini, J. L., Roumier, T., Andreau, K., Medema, R., and Kroemer, G. (2004). Cell death by mitotic catastrophe: A molecular definition. *Oncogene* **23,** 2825–2837.

Galluzzi, L., Maiuri, M. C., Vitale, I., Zischka, H., Castedo, M., Zitvogel, L., and Kroemer, G. (2007). Cell death modalities: Classification and pathophysiological implications. *Cell Death Differ.* **14,** 1237–1243.

Huerta, S., Goulet, E. J., Huerta-Yepez, S., and Livingston, E. H. (2007). Screening and detection of apoptosis. *J. Surg. Res.* **139,** 143–156.

Keij, J. F., Bell-Prince, C., and Steinkamp, J. A. (2000). Staining of mitochondrial membranes with 10-nonyl acridine orange, MitoFluor Green, and MitoTracker Green is affected by mitochondrial membrane potential altering drugs. *Cytometry* **39,** 203–210.

Kerr, J. F., Wyllie, A. H., and Currie, A. R. (1972). Apoptosis: A basic biological henomenon with wide-ranging implications in tissue kinetics. *Br. J. Cancer* **26,** 239–257.

Klionsky, D. J., Cuervo, A. M., and Seglen, P. O. (2007). Methods for monitoring autophagy from yeast to human. *Autophagy* **3,** 181–206.

Kroemer, G., El-Deiry, W. S., Golstein, P., Peter, M. E., Vaux, D., Vandenabeele, P., Zhivotovsky, B., Blagosklonny, M. V., Malorni, W., Knight, R. A., Piacentini, M., Nagata, S., *et al.* (2005). Classification of cell death: Recommendations of the Nomenclature Committee on Cell Death. *Cell Death Differ* **12,** 1463–1467.

Labat-Moleur, F., Guillermet, C., Lorimier, P., Robert, C., Lantuejoul, S., Brambilla, E., and Negoescu, A. (1998). TUNEL apoptotic cell detection in tissue sections: Critical evaluation and improvement. *J. Histochem. Cytochem.* **46,** 327–334.

Malorni, W., Matarrese, P., Tinari, A., Farrace, M. G., and Piacentini, M. (2007). Xeno-cannibalism: A survival "escamotage." *Autophagy* **3,** 75–77.

Matarrese, P., Tinari, A., Mormone, E., Bianco, G. A., Toscano, M. A., Ascione, B., Rabinovich, G. A., and Malorni, W. (2005). Galectin-1 sensitizes resting human T lymphocytes to Fas (CD95)-mediated cell death via mitochondrial hyperpolarization, budding, and fission. *J. Biol. Chem.* **280,** 6969–6985.

Mormone, E., Matarrese, P., Tinari, A., Cannella, M., Maglione, V., Farrace, M. G., Piacentini, M., Frati, L., Malorni, W., and Squitieri, F. (2006). Genotype-dependent priming to self- and xeno-cannibalism in heterozygous and homozygous lymphoblasts from patients with Huntington's disease. *J. Neurochem.* **98,** 1090–1099.

Otsuki, Y., Li, Z., and Shibata, M. A. (2003). Apoptotic detection methods: From morphology to gene. *Prog. Histochem. Cytochem.* **38,** 275–339.

Takemura, G., and Fujiwara, H (2006). Morphological aspects of apoptosis in heart diseases. *J. Cell Mol. Med.* **10,** 56–75.

Tinari, A., Garofalo, T., Sorice, M., Esposti, M. D., and Malorni, W. (2007). Mitoptosis: Different pathways for mitochondrial execution. *Autophagy* **3,** 282–284.

Ziegler, U., and Groscurth, P. (2004). Morphological features of cell death. *News Physiol. Sci.* **19,** 124–128.

CHAPTER TWO

EVALUATION OF SOME CELL DEATH FEATURES BY REAL TIME REAL SPACE MICROSCOPY

Soraya S. Smaili,* Tatiana R. Rosenstock,* *and* Yi-Te Hsu[†]

Contents

Abstract

In the past few years, the investigation of many cell death features, especially ones associated with changes in $\Delta\Psi_m$, have gained an important insights with the development of high-resolution fluorescence microscopy. With the use of real time real space measurements, it was possible to perform dynamic studies not only to investigate the location of the organelles, but also to follow changes in transport mechanism, such as Ca^{2+} concentration in different subcellular compartments. In addition, this technique has been used for the simultaneous tracking of organelle location, ion measurements, and $\Delta\Psi_m$, which clearly contributed to further understanding mechanisms related to the control of cell death. This chapter describes the methodology employed to study changes in $\Delta\Psi_m$, Bax translocation, and Ca^{2+} measurements upon apoptotic induction. It also details the new technique developed and employed in our laboratory to

* Department of Pharmacology, Federal University of São Paulo, School of Medicine, São Paulo, Brazil
† Department of Biochemistry and Molecular Biology, Medical University of South Carolina, USC, Charleston, South Carolina

Methods in Enzymology, Volume 442
ISSN 0076-6879, DOI: 10.1016/S0076-6879(08)01402-X

measure Ca^{2+} signaling in brain slices by confocal microscopy. This method has been applied to investigate real time real space studies in different models of neurodegenerative processes, such as Huntington's disease and aging.

1. MITOCHONDRIA AND CELL DEATH

Transient increases in cytosolic (Ca_c^{2+}) and mitochondrial Ca^{2+} (Ca_m^{2+}) are essential elements in the control of most physiological processes. In the presence of a physiological stimulus, increases in Ca_c^{2+}, as a consequence of the influx from extracellular medium or fluxes from organelles, may activate Ca_m^{2+} uptake system, mitochondrial dehydrogenases, and ATP production (Robb-Gaspers *et al.*, 1998). Mitochondria, along with the endoplasmic reticulum (ER), play pivotal roles in regulating intracellular Ca^{2+} contents by activating the Ca^{2+} uptake mechanism and modulating Ca_c^{2+} during physiological stimuli (Rizzuto and Pozzan, 2006; Rizzuto *et al.*, 1998; Robb-Gapers *et al.*, 1998; Smaili *et al.*, 2001b). Mitochondria also release Ca^{2+} by the Na–Ca exchanger or by the permeability transition pore (PTP) operating in two conductance states: high conductance with full opening or low conductance with pore flickering (Jouaville *et al.*, 1998; Petronilli *et al.*, 1998; Smaili and Russell, 1999; Zoratti and Zsabo, 1995). The opening of the PTP in a low conductance state is related to transient openings of the pore (Jouaville *et al.*, 1998), which may shape Ca_c^{2+} signals (Smaili and Russell, 1999; Smaili *et al.*, 2001b) and reduce oxygen utilization and the formation of superoxide due to leakage of electrons from the mitochondrial electron transport chain (Skulachev, 1997). However, a sustained increase in Ca_c^{2+} and Ca_m^{2+} may contribute to oxidative stress, regulation, and activation of a number of Ca^{2+}-dependent enzymes, such as phospholipases, proteases, and nucleases (Smaili *et al.*, 2003).

In the presence of some apoptotic stimuli, the activation of mitochondrial processes may lead to the release of cytochrome *c*. This is followed by the activation of caspases, nuclear fragmentation, and apoptotic cell death (Kroemer *et al.*, 2007). Mitochondrial membrane potential $(\Delta\Psi_m)$ has been qualified as an important hallmark associated with apoptotic cell death. During apoptosis, nuclear disintegration can be preceded by loss or entire dissipation of the $\Delta\Psi_m$ (Marchetti *et al.*, 1996). In many cases, this loss of $\Delta\Psi_m$ during apoptosis could be prevented by cyclosporine A or other PTP inhibitors, indicating that permeability transition and the full opening of the PTP are involved in the release of cytochrome *c* and apoptosis induction (Marzo *et al.*, 1998). However, the complete loss of $\Delta\Psi_m$ may not directly involve the opening of the PTP (Kuwana *et al.*, 2002).

Either with or without the involvement of the PTP, the loss or the collapse of $\Delta\Psi_m$ might be a reflex of changes in permeabilization of the mitochondrial membranes. As a consequence of this permeabilization, mitochondria release apoptotic factors such as cytochrome c, AIF, caspases, and SmacDiablo, among others (Kroemer et $al.$, 2007). In addition, because mitochondria are important reservoirs for Ca^{2+} and serve to maintain calcium homeostasis (Smaili and Russell, 1999; Smaili et $al.$, 2001b), alterations in the bioenergetics will contribute to elevate reactive oxygen species (ROS) production, decrease the ability of mitochondria to take up calcium, and increase cell death (Carvalho et $al.$, 2004; Hoek et $al.$, 1995; Kowaltowski et $al.$, 2001; Pacher and Hajnoczky, 2001; Smaili et $al.$, 2003).

Thus, investigation of $\Delta\Psi_m$ changes became a necessary step in apoptosis studies. These may be carried out successfully by using high–resolution fluorescence microscopy, especially with real time real space measurements, allowing dynamic studies to follow mitochondrial location and changes in transport mechanisms that cause $\Delta\Psi_m$ dissipation. These measurements may become more interesting if tracking of location or other measurements together with $\Delta\Psi_m$ are performed simultaneously. This technique can certainly contribute to the further understanding of the mechanisms related to the permeabilization of the mitochondrial membranes, activation of the PTP, release of apoptotic factors, and Ca^{2+} measurements among other steps in the apoptotic signaling.

1.1. Single cell measurements of $\Delta\Psi_m$

The $\Delta\Psi_m$ measurements can be performed using cells in culture and observed with a high–resolution confocal microscope or a microscope coupled to a digital charge-coupled device (CCD) camera. This type of microscopy allows precise measurements on single isolated cells where it is possible to visualize subcellular compartments, movements, and dynamic changes of intracellular proteins or elements. Studies could be performed in primary or immortalized cells were plated on coverslips coated (e.g., poly-L-lysine; poly-ornithine) 1 or 2 days before the actual experiment. Right before the experiment, a coverslip is removed from the culture dish, placed in a Leiden chamber, and washed three times with microscopy buffer containing (mM): 130 NaCl, 5.36 KCl, 0.8 $MgSO_4$, 1 Na_2HPO_4, 25 glucose, 20 HEPES, 1 Na–pyruvate, 1.50 $CaCl_2$, 1 ascorbic acid, pH 7.3. The Leiden chamber is transferred to a thermostatically controlled temperature chamber at $37°$ and incubated with potentiometric dyes from rhodamine family (e.g., rhodamine123, TMRM, and TMRE). TMRE, tetramethylrhodamine ethyl ester (Molecular Probes/Invitrogen, Eugene, OR), is a cationic potentiometric dye that accumulates preferentially in energized mitochondria. This is driven by $\Delta\Psi_m$ following the Nerst equation. The dye is reversibly taken up by live cells and provides not only a qualitative, but also a semiquantitative measurement. By using TMRE it is possible to follow $\Delta\Psi_m$ dissipation, as indicated by a loss of dye accumulation in the mitochondria to the cytosol; it is also possible to verify an increase in $\Delta\Psi_m$ as indicated by the an increase in fluorescence at the organelle level. For more quantitative measurements of the $\Delta\Psi_m$, TMRE signal is calibrated by using nigericin and FCCP plus oligomycin at the end of the experiment. According to the type of cell used, the TMRE concentration and incubation period may vary from 20 to 50 nM and from 10 to 30 min, respectively. To avoid TMRE consumption as a consequence of photobleaching, the dye is not washed out and is kept in contact with the cells throughout the image acquisition. In addition, before starting the experiment, TMRE fluorescence stability should be tested by acquiring few images until a stable baseline is achieved. When a stable baseline is not obtained, due to photobleaching or photodynamic effects, the cells are discarded. TMRE fluorescence (548 nm excitation and 585 nm emission) can be acquired using a confocal microscope or a microscope coupled to a CCD camera. Fluorescence is extracted, normalized for comparison, and expressed in arbitrary units.

Figure 2.1 shows primarily cultured murine astrocytes loaded with TMRE (50 nM for 15 min) and visualized using a TE300 Nikon inverted microscope (Nikon Osaka, Japan) coupled to a cooled CCD camera Micro-Max 512BFT (Roper Sci, Princeton Instruments, USA) controlled by imaging software BioIP. Because of the high-resolution of the microscope used, individual mitochondria are localized and it is possible to identify some morphological characteristics of the mitochondria. The mitochondria appeared elongated and dispersed near the cell border and more concentrated around the nucleus. Cells are then stimulated with 3-nitropropionic acid (3NP, 1 mM), a classic inhibitor of the mitochondrial complex II. 3NP causes an increase in ROS production, opens the PTP, and is an apoptotic inducer used to establish pharmacological models for Huntington's disease (HD) (Rosenstock *et al.*, 2004). Real time microscopy shows that 3NP causes some fluctuations in $\Delta\Psi_m$ in most of the mitochondria analyzed, although a small percentage showed an increase in $\Delta\Psi_m$ or no change at all. After the addiction of 3NP, cells were treated with FCCP (5-8 μM) plus oligomycin (1 μg/ml), which are used to induce a rapid collapse of $\Delta\Psi_m$ without excessive ATP consumption. To analyze fluorescence changes, single mitochondrion is analyzed using the region of interest (ROI) tool drawn around each organelle and fluorescence is extracted using a specific software (BioIP Anderson Engineering, Wilmington, USA). In the image shown in Figure 2.1, circles indicate mitochondria that show a loss of $\Delta\Psi_m$, and arrows indicate regions with high density of mitochondria that went out of focus due to an increase in TMRE concentration in the cytosol. Fluorescence intensities, were in arbitrary units, were normalized ($\Delta F/F_o$) and plotted against time.

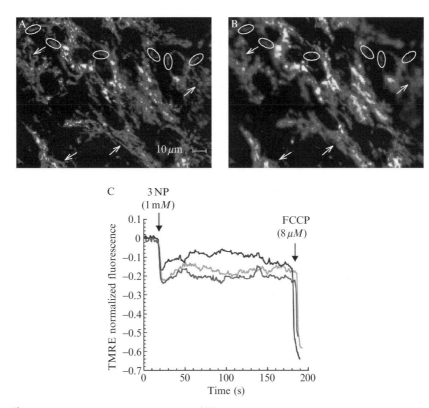

Figure 2.1 3NP induces a decrease in $\Delta\Psi_m$. Astrocyte cultures from B6xCBA/F1 mice loaded with TMRE (50 nM) for 15 min, before (A) and after (B) the addition of 1 mM 3NP. Images were acquired using a CCD camera and $40\times$ oil immersion objective with a delay of 6 s. Circles indicate mitochondria that lost $\Delta\Psi_m$, and arrows are regions that previously presented high density of mitochondria that went out of focus due to the increase in TMRE in the cytosol. (C) $\Delta\Psi_m$ fluorescence traces extracted from digital images and during stimulation of cells with 3NP followed by FCCP (8 μM). Traces presented were obtained from three different mitochondria. Similar measurements were made in five different experiments, and results were compiled with the percentage of mitochondria showing decrease, increase, or no change in $\Delta\Psi_m$ (adapted from Rosenstock *et al.*, 2004). (See color insert.)

2. BAX, MITOCHONDRIA, AND CALCIUM SIGNALING

Members of the Bcl-2 family of proteins are a group of proteins that play important roles in apoptosis regulation. Members of this family can promote either cell survival or cell death. Bax is a proapoptotic member of the family first identified as a Bcl-2-binding partner (Oltvai *et al.*, 1993), and its overexpression accelerates cell death in response to various stimuli (Yang and Korsmeyer, 1996). Upregulation of Bax has been reported in promoting neuronal and cardiomyocytic cell death present in certain pathological conditions (Krajewski *et al.*, 1995; Rampino *et al.*, 1997). The deficiency has been implicated in neuronal development and spermatogenesis (Knudson *et al.*, 1995; Rodriguez *et al.*, 1997).

Bax, like other members of the Bcl-2 family, has three conserved regions known as BH (Bcl-2 homology) domains 1–3 (Yin *et al.*, 1994; Zha *et al.*, 1996), which are important for its apoptotic functions. In healthy cells, Bax is predominantly a soluble monomeric protein (Hsu *et al.*, 1997), despite the fact that it possesses a C-terminal hydrophobic segment, which, unlike those of Bcl-2 and Bcl-X_L, is sequestered inside a hydrophobic cleft (Suzuki *et al.*, 2000). Upon apoptosis induction by a variety of stimuli, especially in Bax-overexpressing systems, a significant amount of Bax translocates from the cytosol to the membrane fractions and inserts into mitochondria (Hsu *et al.*, 1997; Hsu and Smaili, 2002). The exact mechanism of Bax translocation is still a matter of intense debate and study. Up to now, several hypotheses have been proposed to explain Bax insertion, which leads to the release of cytochrome *c* and other apoptotic factors stored within the mitochondrial intermembranes space (for review, see Kroemer *et al.*, 2007). Some evidence points to the formation of pores by the assembly of Bax and other proapoptotic proteins, which could permeabilize the outer mitochondrial membrane (Kuwana *et al.*, 2002). In certain cases, pore formation was associated with the formation of Bax oligomers, when Bax is in more elevated levels which could lead to the permeabilization of the inner membrane (Pastorino *et al.*, 1999). In other studies, Bax was shown to destabilize mitochondrial membranes (Basanez *et al.*, 2002), to change the lipid–protein interaction or to modulate fission process (Kroemer *et al.*, 2007). It remains uncertain if Bax acts on the opening of the PTP. Some reports showed no involvement of the PTP (Eskes *et al.*, 1998; Kuwana *et al.*, 2002), while others have shown that Bax may interact with few components of the PTP (for review, see Kroemer *et al.*, 2007).

An important feature associated with the translocation of Bax to mitochondria and the release of cytochrome *c* is the loss of $\Delta\Psi_m$ (Pastorino *et al.*, 1999; Smaili *et al.*, 2001b). Cytochrome *c* activates caspases, leading to the proteolysis of the cell, whereas the loss of $\Delta\Psi_m$ corresponds to phenomena that may include an increase in ROS production, bioenergetic defects, and a decrease in cellular energy production (Kowaltowski *et al.*, 2002; Smaili *et al.*, 2003). The proapoptotic activity of Bax can be counteracted by coexpression with prosurvival factors such as Bcl-X_L, which can block Bax translocation to mitochondria during apoptosis (Finucane *et al.*, 1999).

2.1. Generation of GFP–Bax constructs and evaluation Bax movement

As mentioned earlier, in healthy cells, Bax is mainly located in the cytosol and in the presence of apoptotic stimuli, it translocates to membranes, in particular mitochondrial membranes. Therefore, it is very important to investigate the intracellular distribution of Bax. Intracellular distribution of Bax can be determined by subcellular fractionation or by immunofluorescence microscopy using anti-Bax antibodies, for detailed information on this topic, readers are referred to Hsu and Youle (1998) and Hsu and Smaili (2002).

Another way to study Bax distribution is by using real time measurements, which became employed extensively after the development of digital microscopy equipment and software. This technique was yet strengthened by its association with molecular biology approaches that included the use of the green fluorescent protein (GFP) constructs. The tagging of GFP to Bax allows one to monitor the intracellular redistribution of Bax during apoptotic stimulation. For this purpose, the gene encoding human Bax was subcloned into the $3'$ end of a GFP expression vector. The GFP–Bax construct can be transfected into cells and visualization using high-resolution digital fluorescence microscopy and Bax movement is followed in real time GFP–Bax movement can also be evaluated with other important parameters such as the locations of organelles like mitochondria or ER using specific dyes, for instance, Mitotracker red or ER-Tracker. Cells transfected with GFP–Bax can also be monitored simultaneously for $\Delta\Psi_m$ or calcium changes using potentiometric dyes such as TMRE or calcium dyes such as Fura-2 (all dyes can be obtained from Molecular Probes/Invitrogen, USA).

GFP–Bax constructs can be generated by polymerase chain reaction (PCR) of human Bax using flanking primers containing the appropriate restriction sites for subcloning into the C3-EGFP vector (Clontech-Invitrogen, USA). The PCR fragment was inserted between the *Hind*III and the *Eco*RI sites of the polycloning regions of the vector. Then, *Escherichia coli* is transformed with the ligated products and plated in LB agar plates. Colonies were picked and grown overnight for the plasmid purification with the Wizard Plus Minipreps DNA purification system (Promega). The plasmid preparations were digested with restriction enzymes, and the digested samples were analyzed by agarose gel (1.5%). After determining the clone with the correct insert using the restriction digest and DNA sequencing, a large-scale plasmid DNA preparation was performed with the Qiafilter Maxiprep kit (Qiagen). The final DNA pellet was ressuspended in TRIS ethylenediamine tetraacetic acid (TE), buffer and the DNA concentration and purity determined by measuring the A_{260} of the sample.

For GFP–Bax transfection, cells (primary or immortalized) in culture must grow on coverslips or chamber slides coated with poly-L-lysine (5 μg/ml). GFP–Bax constructs were transfected successfully in different cell lines, such as

L929, Cos-7, MCF-7, or primarily cultured rat astrocytes. Cells are usually maintained at 37° and 5% CO_2 and plated on coverslips 2 days prior to the experiment. The day before the actual experiment, cells must be transfected with 0.5 μg C3–GFP–Bax plasmid per well with lipid reagents using the protocols described by the manufacturers. LipofectAmine (Invitrogen) or FuGENE (Roche) are both suitable for GFP–Bax transfections in many different cells. For cotransfection studies used to determine how the expression of other proteins (e.g., Bcl-X_L) affects Bax translocation, the transfection procedure can be done with 0.5 μg of GFP–Bax plasmid and 1.5 μg of the plasmid encoding the protein of interest. It may be necessary to determine the optimal amount of lipid reagents to use in different cell types. After transfection, cells are incubated overnight and used the next day (in general, cells transfected with Bax are used within 15–20 h post transfection).

For the tracking of GFP–Bax movement and colocalization with mitochondria, cells are first washed with the microscopy buffer (see above for composition) three times and incubated with 20 nM Mitotracker red CMXRos (MTR, Molecular Probes/Invitrogen, EUA) for 10 min. After that, coverslips are placed in a Leiden chamber, and cells are visualized using a confocal microscope or a microscope coupled to a digital CCD camera. Figure 2.2 shows rat astrocytes transfected with GFP–Bax (pseudocolor green), loaded with MTR (pseudocolor red), treated with 500 nM of staurosporine (STS), and tracked over time. Images are collected using a confocal microscope (Zeiss LSM510 Meta, Heidelberg, Germany) with a Plan-Neofluor 40× oil-immersion 1.3 NA lens, with excitation at the 488-nm laser line argon/krypton, 543-nm laser line He/Ne and a two-photon titanium sapphire laser module (Coherent, USA). Initially, GFP–Bax appeared with a diffused pattern, indicating its cytosolic location. Upon apoptosis induction with STS, GFP-Bax shifted to a punctated pattern largely colocalized with MTR, as shown in the overlaid image.

2.2. Simultaneous measurements of Bax movement and $\Delta\Psi_m$

Slightly before or simultaneously with massive Bax translocation into mitochondria, a decrease or a complete dissipation of the $\Delta\Psi_m$ is observed (Smaili *et al.*, 2001a), and is associated with cytochrome c release (Carvalho *et al.*, 2004). Evaluation of the $\Delta\Psi_m$ in circumstances where Bax overexpression occurs is considered an important feature associated with its translocation and, consequently, with its toxicity. The simultaneous measurements of Bax distribution and the dynamic changes of $\Delta\Psi_m$ allow studies on the mechanisms of Bax translocation, insertion, and permeabilization of the mitochondrial membranes.

For these measurements, cells plated on coverslips and transfected with GFP-Bax, as described earlier, are washed with the microscopy buffer three times and incubated with the potentiometric dye TMRE for 10 to 20 min

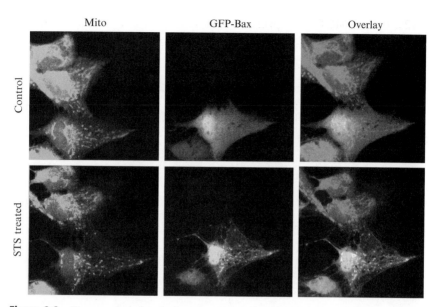

Figure 2.2   Bax translocation during apoptosis induction. Primary cultured astrocytes transfected previously with GFP–Bax were treated with 500 nM staurosporine (STS), loaded with Mitotracker red CMXRos (MTR, 20 nM), and tracked over time by confocal microscopy for 4 h. The GFP–Bax overexpressing cell displayed a diffused pattern, indicative of its cytosolic location. Upon apoptosis induction with STS, GFP–Bax shifted from a diffuse to a punctate pattern that mainly colocalized with mitochondria (MTR staining) as shown in the overlay panels. (See color insert.)

(this depends on the type of cells used as described earlier). TMRE is a cationic potentiometric dye that preferentially accumulates in the mitochondria and, in the concentration range from 20 to 50 nM, does not exhibit toxic effects at the electron transport chain level, as observed with other dyes from the same family (Bernardi *et al.*, 2002). Energized mitochondria take up the dye rapidly and fluctuations of the $\Delta\Psi_m$ can be observed.

Figure 2.3 shows Cos-7 cells transfected with GFP–Bax and were visualized with a CCD camera. Images were acquired every 120 s, for an average of 4 h, using a filter set fitted with a dual-band dichroic mirror (520 and 575 nm) and bandpass filters (520 and 600 nm) from Chroma Technologies. The cells were excited using 485- and 530-nm excitation filters and treated with 1 μM of STS. GFP–Bax and TMRE were monitored simultaneously in real time real space experiments. Over time, the GFP–Bax fluorescence pattern shifted from a diffused to a punctated one. The cell transfected with GFP–Bax, differently from the nontransfected ones, showed a dissipation of the $\Delta\Psi_m$, as indicated by the loss in TMRE fluorescence, which may occur slightly before or simultaneously with Bax location into mitochondria (Carvalho *et al.*, 2004; Smaili *et al.*, 2001b, 2003).

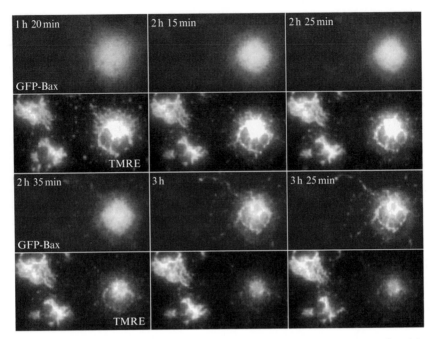

Figure 2.3 Dissipation of $\Delta\Psi_m$ is associated with Bax translocation to mitochondria. Cos-7 cells transfected previously with the GFP–Bax construct were stained with TMRE (50 nM) for $\Delta\Psi_m$ measurements and treated with 1 μM STS. In healthy cells, Bax was soluble in the cytosol, mitochondria were energized, and $\Delta\Psi_m$ was well maintained. Upon the induction of apoptosis with STS, Bax translocated to membranes (panels in row 1 and 3) and dissipation of $\Delta\Psi_m$ occurred (panels in row 2 and 4) immediately before or simultaneously with translocation of Bax. Most of the GFP–Bax is colocalized at the previous mitochondrial location showed by TMRE staining. Rows 1 and 3 show one GFP–Bax overexpressing cell, and rows 2 and 4 show the same cell (overexpressing GFP–Bax) and two other cells (on the left) stained with TMRE that were not transfected with GFP–Bax. Note that only the cell that was expressing GFP–Bax showed a loss in the $\Delta\Psi_m$ (Smaili *et al.*, 2003).

In another set of experiments, the cells were cotransfected with GFP–Bax and Bcl-X_L, and STS was tested using the same approach described in Fig. 2.3 to evaluate the effect of antiapoptotic Bcl-X_L on Bax translocation and Bax-induced loss of $\Delta\Psi_m$. Figure 2.4 shows that STS was not able to induce either $\Delta\Psi_m$ loss or Bax translocation in the Bcl-X_L overexpressing system.

2.3. Bax effect on calcium signaling

More recently, it has been shown that Bcl-2 proteins may regulate Ca_c^{2+} levels or are regulated by the intracellular Ca^{2+} stores. Studies have shown that Bax and other proapoptotic members of the Bcl-2 family modulate

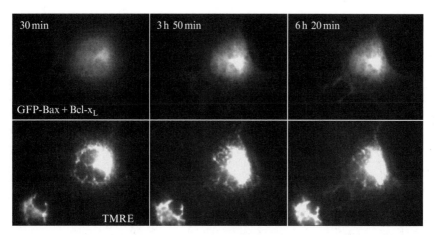

Figure 2.4 Bcl-X$_L$ protects cells from STS-induced GFP–Bax translocation and $\Delta\Psi_m$ dissipation. Cos-7 cells were cotransfected with GFP–Bax and Bcl-X$_L$ plasmids and analyzed using the same approaches described in Fig. 2.3 in order to evaluate the effect of the antiapoptotic protein Bcl-X$_L$ on Bax translocation and Bax-induced loss of $\Delta\Psi_m$. Images show that STS was not able to induce either $\Delta\Psi_m$ loss or Bax translocation in a Bcl-X$_L$ co-overexpressing system.

intracellular Ca^{2+} stores (Nutt *et al.*, 2001; Oakes *et al.*, 2005; Scorrano *et al.*, 2003). In fact, the relationship between ER and mitochondria seems to determine the amount of Ca^{2+} in the system and may regulate cell death machinery. At the ER level, Bax was shown to regulate Ca^{2+} fluxes and activated caspase-12, leading to apoptosis (Zong *et al.*, 2003). In Bax and Bak knockout cells (Bax$^{-/-}$/Bak$^{-/-}$), a decrease in ER Ca^{2+} content and actually a decrease in apoptosis induction were observed (Scorrano *et al.*, 2003). At the mitochondrial level, Bax may activate the PTP opening (Marzo *et al.*, 1998; Pastorino *et al.*, 1998; Carvalho *et al.*, 2004) and permeabilize the inner mitochondrial membrane (Pastorino *et al.*, 1999), to cause the loss of the $\Delta\Psi_m$, the inhibition of respiratory rates, and the release of cytochrome *c* and Ca^{2+}_m (Carvalho *et al.*, 2004). As mentioned earlier, Bax effects on the permeabilization of mitochondrial membranes are still not clearly understood (for review, see Kroemer *el al.*, 2007), as well as its effects on Ca^{2+} contents and fluxes. Thus, further studies are required to yield a deeper understanding of the dynamic role of Ca^{2+} and its modulation by Bcl-2 proteins in the control of apoptosis.

In this context, the use of real time real space measurements of Ca^{2+} signaling coupled to a microinjection system is an additional feature that contributes to directly test the effect of recombinant Bax. For this purpose, cells are plated on coverslips and loaded with Fura-2 AM (10 μM), a ratiometric Ca^{2+} dye, for 20 to 30 min in a regular microscopy buffer (described earlier). Before the experiment, cells are washed with the same buffer to remove external dye not loaded into cells. Ca_c^{2+} levels in isolated intact cells are measured using high-resolution digital microscopy coupled to a cooled CCD camera and controlled by computer software. To examined the direct effect of recombinant Bax, a semiautomatic programmable micromanipulator InjectMan NI 2 and a FemtoJet injector (Eppendorf, Hamburg, Germany), especially suitable for microinjection of adherent cells, can be placed in same scope stage used for digital imaging. With the FemtoJet injector the axial movement is precisely controlled and fast (6000 μm/s), contributing to efficiently set the parameters and decrease the mortality rate caused by microinjection manipulation (less than 5% of the cells). The parameters for microinjection are different from cell to cell and should be fully tested using fluorescent dyes that load the cytosolic compartment. Full-length and mutant Bax were obtained as described previously (Suzuki *et al.*, 2000), produced in *E. coli* as a chitin fusion-biding protein using the pTYB1 plasmid, and are purified by affinity chromatography followed by ion-exchange chromatography. The peak protein fraction is concentrated, aliquoted, and stored at $-80°$. Bax oligomerization is induced by a 1-h incubation of the protein with 1% octylglucoside.

Figure 2.5 shows rat astrocytes from primary cultures loaded with Fura-2 AM for 20 min at room temperature and then washed three times with microscopy buffer. After placing the coverslips in the Leiden chamber and the temperature controller unit, images were acquired with a TE300 Nikon-inverted microscope (Nikon Osaka, Japan) coupled with a cooled CCD camera MicroMax 512BFT (Roper Sci, Princeton Instruments, USA) controlled by an imaging software (BioIP, Anderson Engineering, Wilmington, USA), with 340- and 380-nm excitation filters and a 510-nm emission cube (Chroma Technologies). Then, one cell in the field was chosen for microinjection, the injector was placed in the right position and the cell attached. The precise angles for microinjection were set at $45°$ for the cytoplasm of these cells. The parameters used were for injections were 50 hPa for compensatory pressure, 100 hPa for injection pressure in 0.2 s, with injection volume about 40 to 70 fL, through a glass microcapilary (Femtotips II) with a 0.5-μm internal diameter. For each experiment, one of the cells in the field was microinjected with recombinant Bax (10 ng/ml). In control experiments, cells were microinjected with the buffer used for Bax dilution or with Bax mutants lacking toxicity (Carvalho *et al.*, 2004). Images were collected at 3-s intervals for 5 to 6 min. After the experiments, cells were analyzed using the ROI tool. The fluorescence intensity extracted, and ratio (340/380) calculated and plotted using BioIP software. Recombinant Bax induced an increase in Ca_c^{2+} in the microinjected cells, and the Ca^{2+} wave propagated immediately after injection into the adjacent cells with a peak reached after 2 to 4 s. Different wave patterns were observed, and in certain cells the transient peak was followed by a more sustained response that persisted for several seconds. Cells were numbered

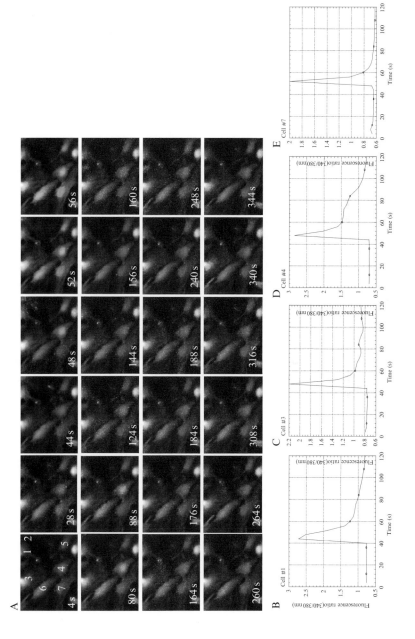

according to the location of the microinjected cell (1) to show that the increase in Ca_c^{2+} started more rapidly in cells that were closer to the microjected one, as evidenced by a smaller time lag from the beginning of the response. In some cells Bax induced not only one wave but also Ca^{2+} oscillations with a smaller amplitude and lower frequency when compared to the first wave.

3. CALCIUM SIGNALING AND NEURODEGENERATIVE PROCESSES

The mechanisms of pathogenesis and cell death in neurodegenerative diseases are still poorly understood, despite the intense investigation over the last decades. At the cellular and signaling level, several common processes were found to be involved, such as the deficit of energy supplies, alteration of Ca^{2+} homeostasis, alterations of the $\Delta\Psi_m$, and increase in oxidative stress (Halliwell and Gutteridge, 1985; Kukreja *et al.*, 1990; Rosenstock *et al.*, 2004; Smaili *et al.*, 2003). In many neurodegenerative diseases, mitochondria play a pivotal role, where mitochondrial defects are observed, especially at the level of the electron transport chain associated with deficits in energy production and an increase in ROS production. Many toxic agents that inhibit the mitochondrial respiratory chain and the transport of electrons are actually inducers of a selective neurodegeneration process. Among them, antimycin A and myxothiazol act as inhibitors of complex III (Brouillet *et al.*, 1993; Green and Reed, 1998; Smith *et al.*, 1997). The 1-Methyl-4-phenylpyridine, the product of 1-methyl-4-phenyl-1,2,3,6-tetrahydropyridine (MPTP), the MPTP itself and rotenone are inhibitors of complex I (Storey *et al.*, 1992), which induce alterations that mimic Parkinson's disease in animals. Other toxic agents, such as methylmalonate and 3-nitropropionic acid (3NP), are inhibitors of complex II (Ludolph, 1992; Palfi *et al.*, 1996) and were shown to cause anatomic,

Figure 2.5 Bax microinjected induced Ca^{2+} waves, which propagate throughout the cells. Astrocytes plated on coverslips were loaded with Fura-2 AM ($10 \mu M$) in the microscopy buffer. Cytosolic Ca^{2+} levels in isolated cells were analyzed using high-resolution digital microscopy with an inverted microscope coupled to a cooled CCD camera and controlled by a computer software. For each experiment, one cell in the field was used for microinjection of recombinant Bax (rBax, 10 ng/ml) that was injected in bolus. (A) Images were collected at 3-s intervals and rBax was injected in cell #1 at 40 s when changes in fluorescence intensity were observed (red pseudocolor represents an increase in the 340/380 ratio, indicating an elevation in cytosolic Ca^{2+}). (B) Graphs show that the increase in cytosolic Ca^{2+} after the rBax injection occurred not only in injected cell #1, but also in adjacent cells #2 (C), #3 (D), and #4 (E) at different intervals. Cells showed a transient peak, which reached a maximum response after 2 to 4 s from the beginning of the effect (adapted from Carvalho *et al.*, 2004). (See color insert.)

neurochemical changes, and neuronal death similar to those occurring in HD (Beal *et al.*, 1993; Brouillet *et al.*, 1995).

Huntington's disease is a hereditary neurodegenerative disorder caused by an expansion of CAG repeats in chromosome 4 (Gusella *et al.*, 1983). Patients with this disorder have severe motor symptoms, psychic disorders, and cognitive deficits as the most common signs. Several cellular and molecular mechanisms are associated with HD, with one being a bioenergetic alteration that includes inhibition of the succinate dehydrogenase (SDH), an enzyme associated with complex II. Animals treated with 3NP, the metabolic product of 3-nitropropanol, became a relevant tool to study this disease at the mitochondrial level. (Brouillet *et al.*, 1999; Rosenstock *et al.*, 2004). Actually, data have shown that correlations exist among 3NP toxicity, metabolic dysfunctions, and motor deficits present in HD (Guyot *et al.*, 1997). 3NP induces alterations in mitochondrial homeostasis (Rosenstock *et al.*, 2004) that lead to an impairment of energy production (Beal, 1995) and orofacial dyskinesia, one of the first motor signs of HD (Brouillet *et al.*, 1999). In rodents this appears as a vacuous chewing movement (Andreassen and Jorgensen, 1995; Brouillet *et al.*, 1999; Rosenstock *et al.*, 2004). The effect of 3NP on $\Delta\Psi_m$ and the decrease in this parameter observed in the majority of the mitochondria studied were discussed earlier. This effect was correlated with an increase in ROS production and opening of the PTP (Rosenstock *et al.*, 2004).

Many of the multifactorial features associated with neurodegenerative diseases, such as increased ROS production, mitochondrial DNA damage, and mitochondrial dysfunctions (Lopes *et al.*, 2004; Meccocci *et al.*, 1994; Sato and Tauchi, 1982), were also associated with the neurodegeneration observed in the aging process. As in HD, these affect the energy production and cause a decline of the physiological conditions (Beckman and Ames, 1998; Mattson, 2000; Ozawa, 1997; Szibor and Holtz, 2003). Thus, taken together these elements strongly contribute to accelerate cell death and degeneration (Cadenas and Davies, 2000; Harman, 1999).

3.1. Huntington's disease and animal models

Although the neurotoxic agents that inhibit mitochondrial functions are important tools, their use is not enough to answer several questions concerning the neurodegenerative mechanisms involved in these disorders. Thus, transgenic mice were developed as animal models, especially for the study of HD. Among them, a successful and frequently employed model is mice with the transgene of the promoter region of human HD, with approximately 1 kb, that carries an expansion of the trinucleotide CAG between 115 and 156 units (Mangiarini *et al.*, 1996). There are three different strains of these mice, R6/1, R6/2, and R6/5, with R6/1 being the most commonly used, with 115 CAGs, and R6/2, with 150 CAGs, since they mimic the adult and the juvenile forms of HD, respectively (Carter *et al.*, 1999; Mangiarini *et al.*, 1996). Other models include mice that express huntingtin (htt), a protein related to HD development, encoded by 18 CAG repeats; mice with mutant htt, encoded by 46 CAGs (Hodgson *et al.*, 1998); mice that codify the first third part of the htt cDNA (Laforet *et al.*, 1998); and mice with 71 and 94 CAG repeats (Levine *et al.*, 1998). Each mutant present specific features related to HD, such as phenotypic altera-tions, neuronal loss, intranuclear inclusions of htt, and high sensitivity of *N*-methyl-D-aspartic acid receptor activation.

3.2. Use of real time measurements in brain slices from transgenic mice

Although many models allow investigation of the processes involved in neurodegenerative disorders, cell cultures from these animals are not always possible, which may limit live measurements. Neuronal and glial cell cultures are usually obtained from embryonic or newborn animals, which are not ideal to study the adult neurodegenerative disorder develop-ment. Furthermore, neurons or glia isolated in cultures do not permit the investigation of the neuron–glia interaction, which is an important parame-ter of neurotransmission in the central nervous system. The cross-talk between these two cells is usually underevaluated and many mechanisms related to them are still unknown. One possible tool for these studies which is described in the literature refers to the use of brain slices in culture (Newell *et al.*, 1995; Tasker *et al.*, 1992). However, by keeping the slices in culture, a change in the morphological characteristic and synaptic trans-mition usually occurs. Therefore, we developed a technique where fresh brain slices from mice or rats can be used successfully for real time real space microscopy. These slices are employed to verify calcium, $\Delta\Psi_m$ dynamic changes, or other measurements such as ROS production and cell viability. Because such intracellular mechanisms are common features associated with a large variety of neurological disorders, this method might be used in several different studies related to cell degeneration and/or protection.

The use of brain slices in real time studies was first performed using adult R6/1 mice, which develop the disease around 5 to 6 months of age. For these studies, animals were sacrificed quickly, in accordance with the guidelines of the Committee on Care and Use of Laboratory Animal Resources, National Research Council, USA, and their brains removed. In this procedure, which may not exceed 2 min, the cranial hubcap has to be cut bilaterally with pliers or small scissors from the caudal to frontal brain with minimum damage to the brain tissues. The cranial hubcap is then removed with tweezers and the brain is pushed with a spatula to a container kept at $-10°$ to decrease metabolism and avoid further damages to cells. Because we are interested in specific brain regions, such as cortex, corpus callosum, and striatum, the cerebellum is removed at this time. The brain is then gently attached (glued) to the base of a Vibratome slicer (Vibroslices, Campdem Instruments, USA), which is used to cut fresh samples by horizontal vibration of a blade brought near the tissue with a specified speed. Hence, after gluing the brain with plastic glue (cyanoacrylate ester), it is transferred to the inner compartment of the Vibratome filled with a transfer solution that allows correct conservation and maintenance of brain cells. The transfer solution is composed of 1.25 mM NaH$_2$PO$_4$H$_2$O, 2.0 mM MgSO$_4$, 2.0 mM MgCl$_2$, 2.0 mM CaCl$_2$, 3.5 mM KCl, 10 mM glucose, 220 mM sucrose, and 21 mM NaHCO$_3$ (pH 7.38). The transfer solution is kept with constant oxygenation and low temperature (around 4°) while the brains are cut in coronal (frontal) slices of 200 μm using 9.5 vibration and 0.5 units of speed. Depending on the animal and tissue used, these parameters can be changed. It is very important to stress that the whole procedure, from mouse euthanasia to the transfer of the glued brain to the inner compartment of the Vibratome, should take no longer than 5 min. To remove slices from the Vibratome, a plastic pipette is used to decrease impacts and transfer them to a recipient containing transfer solution at 4° and oxygen where they are equilibrated for at least 30 min prior the start of the experiments.

After the equilibration period, slices can be loaded, *ex vivo*, with the fluorescent probe either at room temperature or at 37°. For calcium measurements, dyes such as Fura-2 AM (10 μM) or Fluo-3 AM (10 μM) are used with 20% Pluronic F127 (Molecular Probes/Invitrogen, Frederick, MD), which increases the membrane permeation and allows entrance of the dyes into cells. The temperature and incubation time to reach the visualization of the dye will depend on tissue characteristics and the properties of the fluorescence indicator, which must be tested before the experiment. Figure 2.6 shows a brain slice from a R6/1 mouse loaded with Fluo-3 AM (10 μM) (Molecular Probes/Invitrogen) for 2 h at room temperature in the presence of Pluronic F127. This dye is used to evaluate the Ca^{2+} homeostasis upon stimulation in real time measurements and is well suited for measurements by confocal microscopy chosen for better visualization of thicker samples. After loading, the slice is transferred to 25-mm–diameter glass coverslips with a plastic pipette where it is stretched gently with a little tweezer. Subsequently, solution remaining on the coverslip is removed with filter paper, and the borders of the slice are glued (glue is applied with a small needle) at the coverslip. At this point, the slice is flatted and fixed, which allows the experiment to proceed without out-of-focus artifacts. Thus, the coverslip was placed in a Leiden chamber filled with oxygenized transfer solution at 4° and in the absence of a fluorescent probe. Slices are visualized in a confocal microscope (Zeiss LSM510 Meta, Heidelberg, Germany) with a 10× objective and excited with a argon/krypton laser (488 nm). For real time real space experiments, images are acquired in one focal plane (Z axis) with 18 to 20 μm of optical thickness collected from the

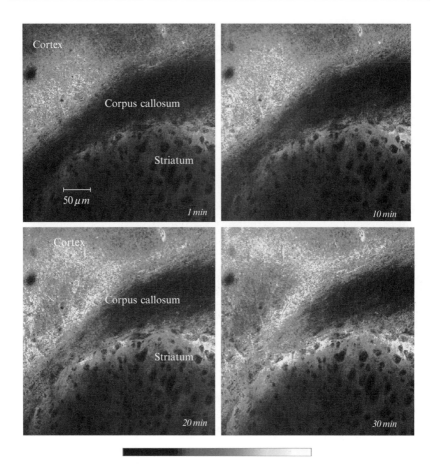

Figure 2.6   Glutamate induced an increase in Ca^{2+} in brain slices. A brain slice (200 μm thickness) from a 2-month-old B6xCBA/F1 male mouse was loaded with calcium dye, Fluo-3 AM. Images acquired by confocal microscope showed few details of the brain structures of interest (cortex and striatum). An increase in cytosolic Ca^{2+} occurred a few minutes after glutamate addition. In addition, there was a Ca^{2+} wave from the cortex and striatum to the corpus callosum direction. The bar represents the scale of fluorescence intensities (0–255). The colors were attributed. (See color insert.)

internal surface to avoid areas possibly damaged as a consequence of the physical slicing process. Before starting the experiment, and consequently before the addition of any drug, slices are exposed to light (2–3 min) to ensure a stable baseline. The collected images were analyzed and included the global area of the brain section or specific brain areas of interest, such as striatum, corpus callosum, and cortex. Fluorescence were extracted from digital images through the ROI tool using BioIP (Anderson Engineering, Wilmington, USA). The data extracted were normalized and plotted over time. Calibrations were performed at the end of each experiment using digitonin (100 μg/ml) to obtain the maximum fluorescence intensity (after the addition of the drug of interest) in relation to the baseline.

Figure 2.7 shows the effect of glutamate (1 mM) over time in brain slices of 3-month-old R6/1 male mice loaded with Fluo-3 AM. The different regions of the brain investigated, include cortex, corpus callosum, and striatum. After glutamate there was an increase in the fluorescence intensity, that represent an increase in Ca^{2+} concentration. It was also possible to observe that there was a Ca^{2+} wave that initiated at the striatum and traveled

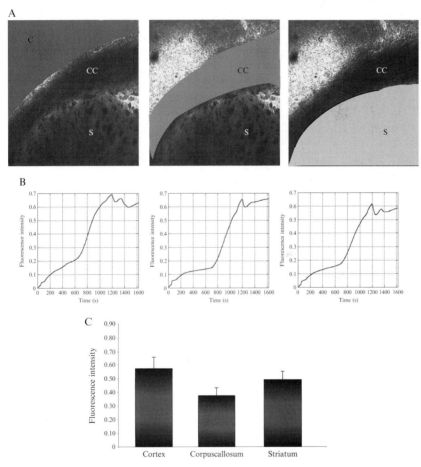

Figure 2.7 Analysis of fluorescence intensities extracted from brain slices loaded with Fluo-3 AM. (A) Images were delineated from three different regions (cortex, C; corpus callosum, CC; striatum, S) using the ROI tool. (B) Fluorescence intensities from each brain area were extracted, normalized, and plotted. Graphics represent the fluorescence of C, CC, and S, from right to left, respectively. (C) Histogram shows the mean fluorescence (± standard deviation) of five experiments.

in the direction of the corpus callosum. After the extraction of fluorescence from the different regions, data were normalized and plotted in a histogram for comparison and statistical analyses.

ACKNOWLEDGMENTS

The authors thank Paul Anderson for his excellent technical support with the imaging system. This work was supported by Fundação de Amparo à Pesquisa do Estado de São Paulo (FAPESP). T.R.R. is a fellow of Coordenação Aperfeiçoamento Pessoal de Ensino Superior (CAPES), and S.S.S. is supported by Conselho Nacional de Desenvolvimento Científico e Tecnológico (CNPq) fellowship.

REFERENCES

Andreassen, O. A., and Jorgensen, H. A. (1995). The mitochondrial toxin 3-nitropropionic acid induces vacuous chewing movements in rats: Implications for the tardive dyskinesia? *Psychopharmacology (Berl.)* **119,** 474–476.

Basañez, G., Sharpe, J. C., Galanis, J., Brandt, T. B., Hardwick, J. M., and Zimmerberg, J. (2002). Bax-type apoptotic proteins porate pure lipid bilayers through a mechanism sensitive to intrinsic monolayer curvature. *J. Biol. Chem.* **277,** 49360–49365.

Beal, M. F. (1995). Aging, energy and oxidative stress in neurodegenerative diseases. *Ann. Neurol.* **38,** 357–366.

Beal, M. F., Brouillet, E., Jenkins, B. G., Ferrante, R. J., Kowall, N. W., Miller, J. M., Storey, E., Srivastava, R., Rosen, B., and Hyman, B. T. (1993). Neurochemical and histologic characterization of striatal excitotoxic lesions produced by the mitochondrial toxin 3-nitropropionic acid. *J. Neurosci.* **13,** 4181–4192.

Beckman, K. B., and Ames, B. N. (1998). The free radical theory of aging matures. *Physiol. Rev.* **78,** 547–581.

Bernardi, P. (2002). Mitochondrial transport of cations: Channels, exchangers, and permeability transition. *Physiol. Rev.* **79,** 1127–1255.

Brouillet, E., Condé, F., Beal, M. F., and Hantraye, P. (1999). Replicating Huntington's disease phenotype in experimental animals. *Progress Neurobiol.* **59,** 427–468.

Brouillet, E., Hantraye, P., Ferrante, R. J., Dolan, R., Leroy-Willig, A., Kowall, N. W., and Beal, M. F. (1995). Chronic mitochondrial energy impairment produces selective striatal degeneration and abnormal choreiform movements in primates. *Proc. Natl. Acad. Sci. USA* **92,** 7105–7109.

Brouillet, E., Jenkins, B. G., Hyman, B. T., Ferrante, R. J., Kowall, N. W., Srivastava, R., Roy, D. S., Rosen, B. R., and Beal, M. F. (1993). Age-dependent vulnerability of the striatum to the mitochondrial toxin 3-nitropropioni acid. *J. Neurochem.* **60,** 356–359.

Cadenas, E., and Davies, K. J. A. (2000). Mitochondrial free radical generation, oxidative stress, and aging. *Free Radic. Biol. Med.* **29,** 222–230.

Carter, R. J., Lione, L. A., Humby, T., Mangiarini, L., Mahal, A., Bates, G. P., Dunnett, S. B., and Morton, A. J. (1999). Characterization of progressive motor deficits in mice transgenic for the human Huntington's disease mutation. *J. Neurosci.* **19,** 3248–3257.

Carvalho, A. C. P., Sharpe, J., Rosenstock, T. R., Teles, A. V. F., Youle, R. J., and Smaili, S. S. (2004). Bax affects intracellular Ca^{2+} stores and induces Ca^{2+} wave propagation. *Cell Death Differ.* **11,** 1265–1276.

Eskes, R., Antonsson, B., Osen-Sand, A., Montessuit, S., Richter, C., Sadoul, R., Mazzei, G., Nichols, A., and Martinou, J. C. (1998). Bax-induced cytochrome C release from mitochondria is independent of the permeability transition pore but highly dependent on Mg^{2+} ions. *J. Cell Biol.* **143**, 217–224.

Finucane, D. M., Bossy-Wetzel, E., Waterhouse, N. J., Cotter, T. G., and Green, D. R. (1999). Bax-induced caspase activation and apoptosis via cytochrome *c* release from mitochondria is inhibitable by Bcl-xL. *J. Biol. Chem.* **274**, 2225–2233.

Green, D. R., and Reed, J. C. (1998). Mitochondria and apoptosis. *Science* **281**, 1309–1311.

Gusella, J. F., Wexler, N. S., Conneally, P. M., Naylor, S. L., Anderson, M. A., Tanzi, R. E., Watkins, P. C., Ottina, K., Wallace, M. R., Sakaguchi, A. Y., *et al.* (1983). A polymorphic DNA marker genetically linked to Huntington's disease. *Nature* **306**, 234–238.

Guyot, M. C., Hantraye, P., Dolan, R., Palfi, S., Maziere, M., and Brouillet, E. (1997). Quantifiable bradykinesia, gait abnormalities and Huntington's disease-like striatal lesions in rats chronically treated with 3-nitropropionic acid. *Neuroscience* **79**, 45–56.

Halliwell, B., and Gutteridge, J. M. C. (1998). "Free Radicals in Biology and Medicine." Oxford Univ. Press, Oxford.

Harman, D. (1999). Free radical theory of aging: Increasing the average life expectancy at birth and the maximum life span. *J. Anti. Aging Med.* **2**, 199–208.

Hodgson, G., Agopyan, N., Smith, D., Lepiane, F., Mccutcheon, K., O'kuskey, J. R., Bissada, N., Jamot, L., Roder, J., Rubin, E. M., and Hayden, M. R. (1998). YAC transgenic mice expressing mutant human huntington show deficits in hippocampal long-term potentiation. *Soc. Neurosci. Abstr.* **127**, 1.

Hoek, J. B., Farber, J. L., Thomas, A. P., and Wang, X. (1995). Calcium ion-dependent signalling and mitochondrial dysfunction: Mitochondrial calcium uptake during hormonal stimulation in intact liver cells and its implication for the mitochondrial permeability transition. *Biochem. Biophys. Acta* **1271**, 93–102.

Hsu, Y. T., and Smaili, S. S. (2002). Molecular characterization of the proapoptotic protein Bax. *Neuromethods* **37**, 1–20.

Hsu, Y. T., Wolter, K. G., and Youle, R. J. (1997). Cytosol-to-membrane redistribution of Bax and Bcl-X_L during apoptosis. *Proc. Natl. Acad. Sci.* **94**, 3668–3672.

Hsu, Y. T., and Youle, R. J. (1998). Nonionic detergents induce dimerization among members of the Bcl-2 family. *J. Biol. Chem.* **273**, 10777–10783.

Jouaville, L. S., Ichas, F., and Mazat, J. P. (1998). Modulation of cell calcium signals by mitochondria. *Mol. Cell Biochem.* **184**, 371–376.

Knudson, C. M., Tung, K. S. K., Tourtellote, W. G., Brown, G. A. J., and Korsmeyer, S. J. (1995). Bax-deficient mice with lymphoid hyperplasia and male germ cell death. *Science* **270**, 96–99.

Kowaltowski, A. J., Castilho, R. F., and Vercesi, A. E. (2001). Mitochondrial permeability transition and oxidative stress. *FEBS Lett.* **495**, 12–15.

Krajewski, S., Mai, J. K., Krajewska, M., Sikorska, M., Mossakowski, M. J., and Reed, J. C. (1995). Upregulation of Bax protein levels in neurons following cerebral ischemia. *J. Neurosci.* **15**, 6364–6376.

Kroemer, K., Galluzzi, L., and Brenner, C. (2007). Mitochondrial membrane permeabilization in cell death. *Physiol. Rev.* **87**, 99–163.

Kukreja, R. C., Weaver, A. B., and Hess, M. L. (1990). Sarcolemmal Na(+)-K(+)-ATPase: Inactivation by neutrophil-derived free radicals and oxidants. *Am. J. Physiol.* **259**, H1330–H1336.

Kuwana, T., Mackey, M. R., Perkins, G., Ellisman, M. H., Latterich, M., Schneiter, R., Green, D. R., and Newmeyer, D. D. (2002). Bid, Bax, lipids cooperate to form supramolecular openings in the outer mitochondrial membrane. *Cell* **111**, 331–342.

Laforet, G. A., Lee, H. S., Cadigan, B., Chang, P., Chase, K. O., Sapp, E., Martin, E. M., Mcintyre, C., Williams, M., Reddy, P. H., Tagle, D., Stein, J. S., *et al.* (1998).

Development and characterization of a novel transgenic model of Huntington's disease which recapitulates features of the human illness. *Soc. Neurosci. Abstr.* **380,** 8.

Levine, M. S., Chesselet, M.-F., Koppel, A., Gruen, E., Cepada, C., Carpenter, E. M., Zanjani, H., Hurst, R. S., Altenus, K. L., Murai, J. S., Efstratiadis, A., and Zeitlin, S. (1998). Enhanced sensitivity to glutamate receptor activation in mouse models of Huntington's disease. *Soc. Neurosci. Abstr.* **380,** 7.

Lopes, G. S., Mora, O. A., Cerri, P., Faria, F. P., Jurkiewicz, N. H., Jurkiewicz, A., and Smaili, S. S. (2004). Mitochondrial alterations and apoptosis in smooth muscle from age rats.. *Biochim. Biophys. Acta* **1658,** 187–194.

Ludolph, A. C., Seeling, M., Ludolph, A., Novitt, P., Allen, C. N., Spencer, P. S., and Sabri, M. I. (1992). 3-Nitropropionic acid decreases cellular energy levels and causes neurodegeneration in cortical explants. *Neurodegeneration* **1,** 155–161.

Mangiarini, L., Sathasivam, K., Seller, M., Cozens, B., Harper, A., Hetherington, C., Lawton, M., Trottier, Y., Lehrach, H., Davies, S. W., and Bates, G. P. (1996). Exon 1 of the HD gene with an expanded CAG repeat is sufficient to cause a progressive neurological phenotype in transgenic mice. *Cell* **87,** 493–506.

Marchetti, P., Hirsch, T., Zamzami, N., Castedo, M., Decaudin, D., Susin, S. A., Masse, B., and Kroemer, G. (1996). Mitochondrial permeability transition triggers lymphocyte apoptosis. *J. Immunol.* **157,** 4830–4836.

Marzo, I., Brenner, C., Zamzami, N., Jurgensmeier, J. M., Susin, S. A., Vieira, H. L., Prevost, M. C., Xie, Z., Matsuyama, S., Reed, J. C., and Kroemer, G. (1998). Bax and adenine nucleotide translocator cooperate in the mitochondrial control of apoptosis. *Science* **281,** 2027–2031.

Mattson, M. P. (2000). Apoptosis in neurodegenerative disorders. *Nat. Rev. Mol. Cell. Biol.* **1,** 120–129.

Meccocci, P., MacGArvey, U., and Beal, M. F. (1994). Oxidative damage to mitochondrial DNA is increased in Alzheimer's disease. *Ann. Neurol.* **36,** 747–751.

Newell, D. W., Barth, A., Papermaster, V., and Malouf, A. T. (1995). Glutamate and non-glutamate receptor mediated toxicity caused by oxygen and glucose deprivation in organotypic hippocampal cultures. *J. Neurosci.* **15,** 7702–7711.

Oakes, S. A., Scorrano, L., Opferman, J. T., Bassik, M. C., Nishino, M., Pozzan, T., and Korsmeyer, S. J. (2005). Proapoptotic BAX and BAK regulate the type 1 inositol trisphosphate receptor and calcium leak from the endoplasmic reticulum. *Proc. Natl. Acad. Sci. USA* **102,** 105–110.

Oltvai, Z. N., Milliman, C. L., and Korsmeyer, S. J. (1993). Bcl-2 heterodimerizes in vivo with a conserved homolog, Bax, that accelerates programmed cell death. *Cell* **74,** 609–619.

Ozawa, T. (1997). Genetic and functional changes in mitochondria associated with aging. *Physiol. Rev.* **77,** 425–464.

Pacher, P., and Hajnoczky, G. (2001). Propagation of the apoptotic signal by mitochondrial waves. *EMBO J.* **20,** 4107–4121.

Palfi, S., Ferrante, R. J., Brouillet, E., Beal, M. F., Dolan, R., Guyot, M.-C., Peschanski, M., and Hantraye, P. (1996). Chronic 3-nitropropionic acid treatment in baboons replicates the cognitive and motor deficits of Huntington's disease. *J. Neurosci.* **16,** 3019–3025.

Pastorino, J. G., Tafani, M., Rothman, R. J., Marcineviciute, A., Hoek, J. B., and Farber, J. L. (1999). Functional consequences of the sustained or transient activation by Bax of the mitochondrial permeability transition pore. *J. Biol. Chem.* **274,** 31734–31739.

Petronilli, V., Miotto, G., Canton, M., Brini, M., Colonna, M., Bernardi, P., and Di Lisa, F. (1999). Transient and long-lasting openings of the mitochondrial permeability transition pore can be monitored directly in intact cells by changes in mitochondrial calcein fluorescence.. *Biophys. J.* **76,** 725–734.

Rampino, N., Yamamoto, H., Ionov, Y., Li, Y., Sawai, H., Reed, J. C., and Perucho, M. (1997). Somatic frameshift mutations in the BAX gene in colon cancers of the microsatellite mutator phenotype. *Science* **275**, 967–969.

Rizzuto, R., and Pozzan, T. (2006). Microdomains of intracellular Ca^{2+}: Molecular determinants and functional consequences. *Physiol. Rev.* **86**, 369–408.

Robb-Gaspers, L. D., Burnett, P., Rutter, G. A., Denton, R. A., Rizzuto, R., and Thomas, A. P. (1998). Integrating cytosolic calcium signals into mitochondrial metabolic responses. *EMBO J.* **17**, 4987–5000.

Rodriguez, S. J., Tung, K. S. K., Tourtellote, W. G., Brown, G. A. J., and Korsmeyer, S. J. (1995). Bax-deficient mice with lymphoic hyperplasia and male germ cell death. *Science* **270**, 96–99.

Rosenstock, T. R., Carvalho, A. C. P., Jurkiewicz, A., Frussa-Filho, R., and Smaili, S. S. (2004). Mitochondrial calcium, oxidative stress and apoptosis in a neurodegenerative disease model induced by 3-nitropropionic acid. *J. Neurochem.* **88**, 1220–1228.

Sato, T., and Tauchi, H. (1982). Age changes of mitochondria of rat kidney. *Mech. Aging. Dev.* **20**, 111–126.

Scorrano, L., Oakes, S. A., Opferman, J. T., Cheng, E. H., Sorcinelli, M. D., Pozzan, T., and Korsmeyer, S. J. (2003). BAX and BAK regulation of endoplasmic reticulum Ca^{2+}: A control point for apoptosis. *Science* **300**, 135–139.

Skulachev, V. P. (1997). Membrane-linked systems preventing superoxide formation. *Biosci. Rep.* **17**, 347–366.

Smaili, S. S., Hsu, Y.-T., Carvalho, A. C. P., Rosenstock, T. R., Sharpe, J. C., and Youle, R. J. (2003). Mitochondria, calcium and pro-apoptotic proteins as mediators in cell death signaling. *Braz. J. Med. Biol. Res.* **36**, 183–190.

Smaili, S. S., Hsu, Y. T., Sanders, K., Russell, J. T., and Youle, R. J. (2001a). Bax translocation to mitochondria subsequent to a rapid loss of mitochondrial membrane potential. *Cell Death Differ.* **8**, 909–920.

Smaili, S. S., and Russell, J. T. (1999). Permeability transition pore regulates both mitochondrial membrane potential and agonist-evoked Ca^{2+} signals in oligodendrocytes progenitors. *Cell Calcium* **26**, 121–130.

Smaili, S. S., Stelatto, K. A., Burnett, P., Thomas, A. P., and Gaspers, L. D. (2001b). Cyclosporin A inhibits inositol 1,4,5-trisphosphate-dependent Ca^{2+} signals by enhancing Ca^{2+} uptake into the endoplasmic reticulum and mitochondria. *J. Biol. Chem.* **29**, 23329–23340.

Smith, T. S., and Bennett, J. P., Jr. (1997). Mitochondrial toxins in neurodegenerative diseases: *In vivo* brain hydroxyl radical production during systemic MPTP treatment or following microdialysis infusion of methylpyridinium or azide ions. *Brain Res.* **765**, 183–186.

Storey, E., Hyman, B. T., Jenkins, B. T., Brouillet, E., Miller, J. M., Rosen, B. R., and Beal, M. F. (1992). MPP^{+} produces excitotoxic lesions in rat striatum due to impairment of oxidative metabolism. *J. Neurochem.* **58**, 1975–1978.

Suzuki, M., Youle, R. J., and Tjandra, N. (2000). Structure of Bax: Coregulation of dimer formation and intracellular localization. *Cell* **103**, 645–654.

Szibor, M., and Holtz, J. (2003). Mitochondrial ageing. *Basic Res. Cardiol.* **98**(4), 210–218.

Tasker, R. C., Coyle, J. T., and Vornov, J. J. (1992). The regional vulnerability to hypoglycemia induced neurotoxicity in organotypic hippocampal culture: Protection by early tetrodotoxin or delayed MK 801. *J. Neurosci.* **12**, 4298–4308.

Toescu, E. C., Myronova, N., and Verkhratsky, A. (2000). Age-related structural and functional changes of brain mitochondria. *Cell Calcium* **28**, 329–338.

Yang, E., and Korsmeyer, S. J. (1996). Molecular apoptosis: A discourse on the Bcl-2 family and cell death. *Blood* **88**, 386–401.

Yin, X., Oltvai, Z. N., and Korsmeyer, S. J. (1994). BH1 and BH2 domains of Bcl-2 are required for inhibition of apoptosis and heterodimerization with Bax. *Nature* **369,** 321–323.

Zha, H., Kisk, H. A., Yaffe, M. P., Mahajan, N., Herman, B., and Reed, J. C. (1996). Structure-function comparisons of the proapoptotic protein Bax in yeast and mammalian cells. *Mol. Cell. Biol.* **16,** 6494–6508.

Zong, W. X., Li, C., Hatzivassiliou, G., Lindsten, T., Yu, Q. C., Yuan, J., and Thompson, C. B. (2003). Bax and Bak can localize to the endoplasmic reticulum to initiate apoptosis. *J. Cell Biol.* **162,** 59–69.

Zoratti, M., and Szabo, I. (1995). The mitochondrial permeability transition. *Biochim. Biophys. Acta* **1241,** 139–176.

CHAPTER THREE

Analysis of Apoptotic Pathways by Multiparametric Flow Cytometry: Application to HIV Infection

Hervé Lecoeur,* Marie-Thérèse Melki,[†,‡] Héla Saïdi,[†,‡] and Marie-Lise Gougeon[†,‡]

Contents

* Unité d'Immunophysiologie et Parasitisme Intracellulaire, Paris, France
† Antiviral Immunity, Biotherapy and Vaccine Unit, Institut Pasteur, Paris, France
‡ INSERM U668, Paris, France

Methods in Enzymology, Volume 442
ISSN 0076-6879, DOI: 10.1016/S0076-6879(08)01403-1

Abstract

Flow cytometry analysis of apoptosis allows the detection, at the single cell level, of essential features of apoptotic cells. They include alterations in plasma membrane integrity, detected with the 7-aminoactinomycin D assay, translocation of phosphatidylserine from the inner to the outer layer of the plasma membrane analyzed with the annexin-V/PI assay, DNA strand breaks in apoptotic nuclei measured with the *in situ* nick translation and terminal deoxynucleotidyl transferase dUTP-mediated nick end labeling assays, and morphological modifications evidenced with FSC/SSC criteria. In addition, mitochondrial events such as the drop in transmembrane potential $\Delta\Psi_m$ can be detected with the cationic lipophilic dye 3,3'-dihexyloxacarbocyanine iodide and down-regulation of the Bcl-2 molecule by specific intracellular staining. Multiparametric flow cytometry combines all these approaches for a thorough sequential analysis of apoptosis, especially for heterogenous populations such as human peripheral mononuclear cells. Several examples of combined staining of apoptotic cells are shown on peripheral blood lymphocytes from chronically HIV-infected patients, prone to undergo premature apoptosis.

1. INTRODUCTION

Apoptosis (programmed cell death) is an essential and evolutionary conserved process for normal embryogenesis, organ development, tissue homeostasis, and the elimination of deleterious cells from multicellular organisms. Any deregulation or aberrant activation of apoptosis can be involved in the pathogenesis of human diseases, such as cancer, AIDS, autoimmune diseases, or neurodegenerative disorders. Apoptosis is a multi-component programmed cell death process characterized by the redistribution of phosphatidylserine (PS) from the inner to the outer leaflet of the plasma membrane, cellular morphological patterns such as chromatin condensation, nuclear fragmentation, cytoplasmic shrinkage, membrane blebbing, the formation of apoptotic vesicles, and consequent phagocytosis by immune cells (Strasser et al., 2000).

Two main apoptotic pathways have been described in mammalian cells. The intrinsic (mitochondrial) apoptosis pathway is activated upon cellular and genotoxic stresses such as lipid peroxidation and oxidative stress, growth factor withdrawal, or ultraviolet radiation. It involves members of Bcl-2 family proteins. Multidomain proapoptotic Bcl-2 proteins Bak and Bax form channels on mitochondria and facilitate the release of apoptosis regulator proteins [cytochrome c, Smac/DIABLO, apoptosis inducing factor (AIF), Omi/HtrA2, or endonuclease Gs], whereas multidomain antiapoptotic Bcl-2 proteins (Bcl-2, Bcl-xL) inhibit the release of these apoptosis regulator proteins. BH3-only proapoptotic Bcl-2 proteins (Bad, Bim, Bid, Noxa, and Puma) selectively interact with either multidomain proapoptotic (direct activators) or antiapoptotic (sensitizers) Bcl-2 proteins and promote apoptosis. The release of cytochrome c into the cytosol triggers caspase-3 activation through formation of the apoptosome (cytochrome c/Apaf-1/caspase-9-containing complex), whereas Smac/DIABLO and Omi/HtrA2 promote caspase activation through neutralizing the inhibitors of apoptosis proteins (Saelens et al., 2004).

The extrinsic pathway is activated upon stimulation of death receptors of the tumor necrosis factor (TNF) receptor superfamily such as CD95 (Fas) or TNF-related apoptosis-inducing ligand (TRAIL) receptors by the CD95 ligand (CD95-L) or TRAIL. It results in receptor aggregation and recruitment of the adaptor molecule Fas-associated death domain (FADD) and caspase-8. Upon recruitment, caspase-8 becomes activated and initiates apoptosis by direct cleavage of downstream effector caspases. The receptor and the mitochondrial pathway can be interconnected at different levels. Upon death receptor triggering, activation of caspase-8 may result in cleavage of Bid, a Bcl-2 family protein with a BH3 domain only, which in turn translocates to mitochondria to release cytochrome c, thereby initiating a mitochondrial amplification loop. In addition, cleavage of caspase-6 downstream of mitochondria may feed back to the receptor pathway by cleaving caspase-8 (Krammer, 2000).

Most of these events can be analyzed by flow cytometry. For example, early events such as PS externalization are shown with the annexin-V assay, the drop in mitochondrial transmembrane potential measured with 3,3'-dihexyloxacarbocyanine iodide [$DiOC_6(3)$], morphological changes identified by FSC/SSC analysis, plasma membrane disruption detected by the incorporation of 7-aminoactinomycin D (7-AAD), intracellular expression of Bcl-2 family members quantified with specific monoclonal antibodies (mAbs), and DNA fragmentation detected with the terminal deoxynucleotidyl transferase dUTP-mediated nick end labeling (TUNEL) or *in situ* nick translation (ISNT) assays. This chapter reports in detail how to detect and quantify these events by flow cytometry and demonstrates the possible interconnection of some of these events at the single cell level through the multiparametric analysis of primary lymphocytes primed prematurely for apoptosis, that is, blood lymphocytes from HIV-infected persons.

2. INTRACELLULAR EXPRESSION OF BCL-2 FAMILY MEMBERS AND RELATIONSHIP WITH APOPTOSIS

In mammalian cells, Bcl-2 family proteins are one of the main "apoptotic sensors" and they act primarily on mitochondria, where they regulate the survival or death signals in a preventive or provocative fashion. The Bcl-2 family proteins can be classified into three groups based on their structural and functional properties (Borner, 2003). The first group involves the multidomain antiapoptotic members Bcl-2, Bcl-xL, Bcl-w, Mcl-1, A1/ Bfl-1, Boo/Diva, and NR-13. They exhibit all four Bcl-2 homology domains (BH1–4), which are essential for their survival function through mediating the protein–protein interactions and a transmembrane domain, which is formed by a stretch of hydrophobic amino acids near their C-terminal. Bcl-2 associates directly with the mitochondrial permeability transition pore [via the voltage-dependent anion channel (VDAC), for example] and permits its stabilization. The second group of Bcl-2 protein family mainly involves Bax, Bak, and Bok/Mtd. Bax is localized mainly in the cytosol or loosely attached to the outer membrane of mitochondria or endoplasmic reticulum as a monomer. Following an apoptotic stimulus, Bax undergoes a unique conformational change exposing its C-terminal hydrophobic domain, which is involved in its anchorage to the mitochondrial membrane. In the mitochondrial membrane, Bax forms dimers, oligomers, or high-order multimers (Antignani and Youle, 2006), induces outer mitochondrial membrane permeabilization, and permits the release of different molecules involved in the apoptosis cascade (see earlier discussion).

The third group of family involves BH3-only proteins such as Bid, Bad, Bim, Bik, Blk, Hrk, BNIP3, Nix, BMF, Noxa, and Puma. These proteins share only the amphipathic α-helical BH2 homology domain and mainly act through inhibition of Bcl-2/Bcl-xL and activation of Bak and Bax. They act as the sentinels of cell death sensing machinery and coordinate the fine-tuning of apoptotic response through their interactions with pro- and antiapoptotic Bcl-2 members (Green, 2007).

Flow cytometry detection of intracellular expression of Bcl-2 family members not only determines the intensity of expression of these proteins at the single cell level, it also permits characterizing the phenotype of cells expressing these proteins in complex biological samples and assessing the relationship between Bcl-2 expression intensity and priming for apoptosis.

2.1. Protocol for intracellular Bcl-2 staining, combined with cell surface staining and apoptosis detection

1. This technique requires from 2 to 5 × 10^5 cells per sample. Immunostaining can be performed in 96-well plates.
2. Wash the cells in phosphate-buffered saline (PBS)-bovine serum albumin (BSA)(1%)-sodium azide (0.01%) (5-min centrifugation at 300g).
3. Stain the cell surface antigens with the corresponding mAbs for 20 min in PBS-BSA-azide at 4° in the dark.
4. Wash the cells in 200 μl PBS-BSA-azide and fix them for 15 min at 4° in 100 μl PBS-BSA-azide containing 1% paraformaldehyde (PFA). A stock solution of PFA at 4% in PBS can be stored at −20°.
5. Wash cells in 200 μl PBS-BSA-azide and discard the supernatant.
6. Add 100 μl of mAbs specific for Bcl-2-related proteins diluted appropriately in PBS-BSA-azide containing 0.01% saponin to permeabilize the cells.
7. Incubate stained cells for 30 min at 4° in the dark. Any additional intracellular staining (to detect cytokines, caspases, cyclins, and so on) can be performed simultaneously with the corresponding mAbs.
8. Centrifuge and wash cells in 200 μl PBS-BSA-azide containing 0.01% saponin.
9. Centrifuge and fix cell samples in 100 μl PBS-BSA-azide containing 1% PFA.
10. Analyze 10,000 stained cells with the appropriate software.

2.2. Comments, cautions, and pitfalls

2.2.1. Control samples required for analysis of Bcl-2 expression

When Bcl-2 expression is analyzed in peripheral blood mononuclear cells (PBMC) from patients (such as HIV-infected subjects), PBMC from a control donor should be analyzed in the same experiment. This control sample is used to normalize the level of Bcl-2 expression and to adequately identify cells with "normal" expression of the protein (Bcl2^{No}). It allows the identification of cell subsets expressing low (Bcl-2^{low}) and high (Bcl-2^{high}) Bcl-2 levels, as exemplified in Fig. 3.1A (Boudet et al., 1996).

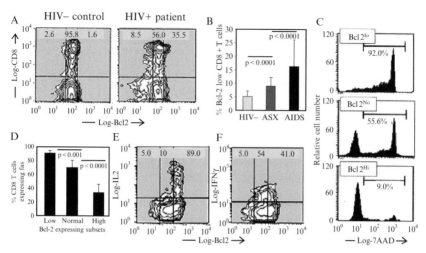

Figure 3.1 Downregulation of Bcl-2 expression primes CD8 T cells from HIV-infected subjects for apoptosis. (A) *Ex vivo* expression of Bcl-2 in CD8[+] T cells from a control subject and an HIV[+] subject. Biparametric dot plots of intracellular Bcl-2/extracellular CD8 stainings. The identification of low, normal, and high-Bcl-2-expressing subsets was performed in comparison with Bcl-2 staining in the control subject. (B) Proportion of CD8[+] T cells expressing a low level of Bcl-2 in control subjects ($n = 13$) and HIV[+] subjects who did not receive any antiretroviral therapy, either asymptomatic (ASX, $n = 8$) or at the AIDS stage ($n = 15$). The mean percentage and standard deviation are indicated. (C) Triple staining CD8/Bcl-2/7-AAD of PBMC from an AIDS patient following 18 h of culture in medium (spontaneous apoptosis). CD8[+] cells were specifically gated, and apoptosis (7-AAD staining) was analyzed in the three Bcl-2 expressing subsets. The percentage of apoptosis in each subset is indicated. (D) Triple staining CD8/Bcl-2/Fas (CD95) of freshly isolated PBMC from HIV[+] subjects ($n = 10$). The percentage of CD8[+] T cells expressing Fas in the three Bcl-2 expressing subsets is shown (mean and standard deviation). (E and F) Triple staining CD3/Bcl-2/IL-2 (E) or CD3/Bcl-2/IFN-γ (F) of PBMC from an HIV[+] subject following 16 h of activation with 50 ng/ml PMA, 300 ng/ml ionomycin, and 100 ng/ml PHA-A in the presence of 10 μg/ml brefeldin A (during the last 12 h of stimulation) to maintain produced cytokines in the cytoplasm. CD3 T cells were gated and the percentage of low-, normal-, and high-expressing Bcl-2 subsets producing IL-2 or IFN-γ is indicated.

2.2.2. Discrimination of cell subsets expressing variable levels of Bcl-2

Cell permeabilization is a mild and labile process, which requires the presence of saponin in all the staining and washing procedures. This protocol allows the detection of subtle variations of Bcl-2 expression. In case the three Bcl-2 subsets are difficult to identify, other permeabilizing approaches could be used. For example, rapid and one-step treatment with Triton X-100 or with 70% (v/v) ethanol could be tested. However, as shown previously (Lecoeur et al., 1998), these approaches should be used with caution as they can significantly alter the cell surface staining, when multiparametric analyses are performed.

2.2.3. Combined detection of Bcl-2 and apoptosis at the single cell level

The detection of Bcl-2–related proteins could be combined with apoptosis detection, using, for example, the 7-AAD assay (Lecoeur *et al.*, 1998). The corresponding protocol is described later.

2.2.4. Limitations of flow cytometry for the analysis of Bcl-2-related proteins intracellular location

Flow cytometry is not adapted to determine the intracellular location of Bcl-2–related proteins and it cannot monitor their translocation into the mitochondrial membrane. Complementary analyses by confocal micros-copy are required to perform these studies.

2.3. *Ex vivo* analysis of Bcl-2 expression in primary lymphocytes: Correlation with their priming for apoptosis

Peripheral T cells from HIV–infected subjects are primed prematurely for apoptosis. Inappropriate cell death is detected not only in infected CD4 T cells, but also in noninfected lymphocytes (B and CD8 T lymphocytes). Several nonexclusive mechanisms contribute to this priming, including the proapoptotic effect of some HIV proteins and the aberrant chronic immune activation because of persistent viral replication, responsible for the exacerbation of physiological apoptosis required for the elimination of autoreactive lymphocytes (Gougeon, 2003).

Ex vivo intracytoplasmic analysis of Bcl-2 expression in CD8$^+$ T cells from HIV–infected subjects as compared to healthy donors revealed striking differences (Fig. 3.1A). Whereas CD8 T lymphocytes from healthy donors showed homogeneous expression of Bcl-2 (Bcl-2^{No} cells), CD8 T cells from patients showed two additional cell subsets: Bcl-2^{Lo} cells and Bcl-2^{hi} cells. A dramatic increase in the percentage of Bcl-2^{Lo} cells was observed in patients, correlat-ing with disease evolution (Fig. 3.1B). Priming for apoptosis of Bcl-2^{Lo} cells is demonstrated by culturing PBMC overnight without stimulation (spontane-ous apoptosis). Indeed, as shown in Fig. 3.1C, a gradient of susceptibility to spontaneous apoptosis is detected by the costaining Bcl-2/7-AAD, with Bcl-2^{Lo} cells being more prone to apoptosis than Bcl-2^{No} or Bcl-2^{hi} cells. The susceptibility of Bcl-2^{Lo} cells to spontaneous apoptosis was correlated with their *in vivo* activation state, characterized by the expression of activation markers such as CD38, HLA-DR, or CD95 (Fas) (Fig. 3.1D). Bcl-2^{Lo} cells were further characterized as cytotoxic T lymphocytes (Boudet *et al.*, 1996). Figures 3.1E and 3.1F show a dual intracellular staining of gated CD8 T cells with the goal of identifying the pattern of cytokine production in relation with Bcl-2 expression. It appears that, following polyclonal stimulation, the most potent producers of interleukin (IL-2) or interferon (IFN)-γ express high levels of Bcl-2 (Ledru *et al.*, 1998).

3. ANALYSIS OF MITOCHONDRIAL TRANSMEMBRANE POTENTIAL BY DiOC$_6$(3) STAINING

The permeability transition pore (PTP) complex is a large polyprotein channel mainly formed by the mitochondrial outer membrane VDAC and the mitochondrial inner membrane protein adenine nucleotide translocase (ANT). The PTP structure also includes regulatory components, such as mitochondrial matrix protein cyclophilin D, peripheral benzodiazepine receptor, hexokinase II, and creatine kinase. PTP is suggested to span both outer and inner mitochondrial membranes and its sustained opening is characterized by mitochondrial depolarization, depletion of ATP, and the release of Ca^{2+} from the mitochondrial matrix, followed by swelling of the mitochondrial matrix (Reed and Kroemer, 2000). Finally, mitochondrial alterations lead to the release of apoptogenic proteins, normally confined in the intermembrane space. These factors include cytochrome *c*, AIF, certain procaspases, Endo G, Diablo/Smac, and Omi/HtrA2. Their release induces an irreversible commitment toward cell death.

Thus, an early drop in mitochondrial membrane potential ($\Delta\Psi_m$) occurs during intrinsic apoptosis (Petit *et al.*, 1995). Variations in $\Delta\Psi_m$ can be analyzed following uptake of the cationic lipophilic dye DiOC$_6$(3) by apoptotic cells. DiOC$_6$(3) diffuses through the plasma membrane, enters the cytoplasm, and finally accumulates inside the mitochondrion matrix. DiOC$_6$(3) accumulation is driven by the $\Delta\Psi_m$ according to the Nerst equation, and any significant loss of this potential is evidenced by a decrease in DiOC$_6$(3) staining. After an excitation at 488 nm, the emission spectrum of DiOC$_6$(3), close to that of FITC, permits performing combined surface immunostaining.

3.1. Protocol for cytofluorimetric analysis of $\Delta\Psi_m$ and combined detection of apoptosis

1. This technique requires 10^5 to 2×10^5 cells per sample. Immunostaining can be performed in 96–well plates.
2. A DiOC$_6$(3) stock solution at 5 μM in ethanol should be prepared and maintained at $-20°$. DiOC$_6$(3) staining is performed in PBS containing 1% BSA.
3. Centrifuge the cell sample (300g, 5 min) and incubate pellets for 20 min with 1 to 40 nM of DiOC$_6$(3) at 37° in the dark. The appropriate concentration of DiOC$_6$(3) should be determined for each cell subset in preliminary tests. A 20-min incubation allows reaching the near-equilibrium distribution of the dye in the cytoplasm and the mitochondrial matrix.

4. Immediately acquire the stained cells and analyze them by flow cytometry.
5. When surface staining is performed on the same sample (with antibodies conjugated to PE or PE-Cy5, for example), it should be performed in PBS-BSA without NaN_3 for 10 min at $4°$ in the dark before $DiOC_6(3)$ staining. Then wash the cells in PBS-BSA at room temperature and perform $DiOC_6(3)$ staining as described earlier.

3.2. Comments, cautions, and pitfalls

3.2.1. Positive controls for mitochondrial depolarization

In every experiment, including $DiOC_6(3)$ staining, a positive control is performed for mitochondrial depolarization. This control may be cells treated by the uncoupling agent carbonyl cyanide m-chlorophenylhydra-zone. A concentration of 20 to 50 μM of mCICCP generally induces a 100% decrease in $\Delta\Psi_m$. This control is required to validate the reliability of $DiOC_6(3)$ staining.

3.2.2. DiOC₆(3) and endoplasmic reticulum staining

It has been shown that high doses of $DiOC_6(3)$ may stain the membranes of the endoplasmic reticulum (Terasaki, 1989). In order to check that the $DiOC_6(3)$ concentration used specifically stains mitochondria, stained cells may be observed by epifluorescence microscopy.

3.2.3. DiOC₆(3) staining and plasma membrane potential ($\Delta\Psi_p$)

As observed for other lipophilic cations, uptake of $DiOC_6(3)$ from the extracellular medium to the mitochondrion depends on both the plasma membrane potential ($\Delta\Psi_p$) and the mitochondrial transmembrane potential ($\Delta\Psi_m$). At doses higher than 40 nM, the mitochondrial-bound dye is quenched almost completely and the uptake of cyanin dye becomes highly dependent on the plasma membrane potential ($\Delta\Psi_p$) (Wilson et al., 1985). A low concentration of cyanin dye ($<$40 nM) and a low dye-to-cell ratio are therefore recommended to favor the contribution of $\Delta\Psi_m$ to cell fluorescence (Rottenberg and Wu, 1998; Wilson et al., 1985).

3.2.4. DiOC₆(3) staining and mitochondrial respiration

High concentrations of $DiOC_6(3)$ can inhibit mitochondrial respiration by inhibiting NADH dehydrogenase. Therefore, they should not be used for $\Delta\Psi_m$ analyses (Rottenberg and Wu, 1998).

3.2.5. DiOC₆(3) uptake and multidrug-resistance pump (MDR)

$DiOC_6(3)$ is a fluorescent probe whose accumulation inside the cell can be reduced by efficient extrusion by the MDR. Consequently, the choice of $DiOC_6(3)$ concentration should take into account the characteristics of the biological sample studied.

3.2.6. DiOC$_6$(3) staining and primary necrotic cells

Primary necrosis and apoptosis are now considered as the extremes of a continuum. Moreover, it is known that these two types of cell death can occur within a cell culture or a biological sample. During primary necrosis (accidental cell death), the whole cell structure is destroyed rapidly. Mitochondria undergo a drastic matrix swelling that can occur before any detectable rupture of the plasma membrane. In some instances, primary necrosis can be also triggered by the opening of the permeability transition pore (for review, see Kroemer *et al.*, 1998). This process is associated with a $\Delta\Psi_m$ disruption and reduced DiOC$_6$(3) staining. Thus, caution must be taken when DiOC$_6$(3) is only used for cytometric analyses. Microscopic observations of morphological features of dying cells should confirm the type of cell death involved in the samples.

3.2.7. Combination of DiOC$_6$(3) staining with other stainings

Staining of surface antigens can be performed in combination with DiOC$_6$(3) staining. However, surface staining should be made in the absence of sodium azide, an inhibitor of cytochrome oxidase that may introduce artifacts in $\Delta\Psi_m$ analyses.

The loss of plasma membrane integrity, detected by 7–AAD staining, can be analyzed in combination with the loss of $\Delta\Psi_m$. In that case, both dyes [7–AAD at 20 μg/ml and DiOC$_6$(3), diluted in PBS-BSA], should be added together on the cell sample and incubation should last 20 min at 37° in the dark. When the staining is performed on freshly isolated primary cells, it should be performed immediately after cell isolation. Data acquisition on the flow cytometer should be performed right after the staining.

3.3. A drop in $\Delta\Psi_m$ is part of the apoptotic process induced in primary lymphocytes from HIV-infected subjects

Figure 3.2A shows a DiOC$_6$(3) staining on PBMC from a chronically infected HIV-infected subject following overnight culture either in medium (spontaneous apoptosis) or in the presence of calcium ionophore (ionomycin) or coated anti-CD3 antibodies (activation- induced cell death, AICD). DiOC$_6$(3) fluorescence/FSC dot plots show three populations: living cells, that is, DiOC$_6$(3)^{high}/normal FSC (FSC^{No}), apoptotic cells with $\Delta\Psi_m$ dissipation and decreased size, that is, DiOC$_6$(3)^{low}/FSC^{No}, apoptotic cells with $\Delta\Psi_m$ dissipation, and normal size, that is, DiOC$_6$(3)^{low}/FSC^{No} (red boxes in Fig. 3.2). DiOC$_6$(3)^{low}/FSC^{No} cells are not detected *ex vivo* (data not shown) and correspond to a transitory apoptotic subset. The combination of DiOC$_6$(3) staining with 7–AAD can discriminate early from late apoptotic cells, according to the fluorescence intensity of

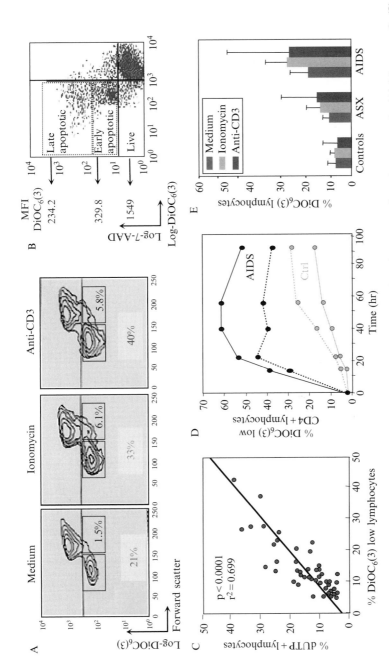

Figure 3.2 Analysis of mitochondrial membrane potential ($\triangle \Psi_m$) in primary lymphocytes undergoing apoptosis. (A) Measure of $\triangle \Psi_m$ collapse in PBMC from an AIDS patient, cultured overnight in medium or stimulated with ionomycin or anti-CD3 mAbs. Biparametric FSC/DiOC$_6$(3) density plot are shown. The percentage of DiOC$_6$(3) low cells is indicated. Red boxes correspond to the fraction of lymphocytes that dropped $\triangle \Psi_m$ but did not show any cell shrinkage (first apoptosis step). Black boxes correspond to cells more engaged toward the

7-AAD+ cells (Fig. 3.2B). These apoptotic subsets show progressive $\Delta\Psi_m$ dissipation, as shown by the fluorescence intensity of $DiOC_6(3)$ staining (Fig. 3.2B).

$DiOC_6(3)$ staining is a reliable approach to detect and quantify apoptosis in lymphocytes. As shown in Fig. 3.2C, a strong correlation was observed between the percentage of $DiOC_6(3)^{low}$ cells and the percentage of dUTP$^+$ cells with fragmented DNA (detected with the TUNEL assay, detailed later). Figure 3.2D shows the kinetics of CD4 T cell apoptosis, detected with $DiOC_6(3)$ following an overnight culture of patient's PBMC in medium (spontaneous apoptosis) or in the presence of anti-CD3 antibodies. The level of apoptosis was higher in patient's CD4 T cells as compared to the control subject, and the percentage of cells with a drop in $\Delta\Psi_m$ reached a plateau after 24 h of culture. These data indicate that a fraction of lymphocytes from HIV$^+$ subjects are prone to die following *ex vivo* stimulation (extrinsic pathway) or following short-term culture in the absence of growth factors (intrinsic pathway). Figure 3.2E confirms in a cohort of untreated patients, either asymptomatic (ASX) or at the AIDS stage, that HIV infection is associated with an increased priming of peripheral lymphocytes for spontaneous or activation-induced cell death ($p > 0.05$ for all culture conditions when patients are compared to controls), as shown previously (reviewed in Gougeon, 2003).

4. ANALYSIS OF DNA FRAGMENTATION BY ISNT AND TUNEL ASSAYS

Apoptosis is associated with characteristic nuclear modifications, including perinuclear chromatin condensation, segmentation of the nucleus into dense nuclear bodies, and progressive DNA fragmentation (Andrew

apoptotic process, with both a collapsed $\Delta\Psi_m$ and cell shrinkage. (B) Combined analysis of $\Delta\Psi_m$ collapse and loss of PM integrity in PBMC from an AIDS patient stimulated overnight with ionomycin. A biparametric $DiOC_6(3)/7$-AAD dot plot is presented. Early and late apoptotic cells are defined according to the extent of 7-AAD staining (7-AAD^{Lo} and 7-AAD^{Hi}, respectively). The mean fluorescence intensity (MFI) of the $DiOC_6(3)$ staining is indicated in blue. (C) Correlation between the percentage of $DiOC_6(3)^{Lo}$ and dUTP$^+$ (i.e., with fragmented DNA, TUNEL assay) in CD4$^+$ T cells from an AIDS patient cultured overnight in medium. A combined CD4 staining was performed for each assay, and apoptosis was determined on gated CD4$^+$ T cells. (D) Kinetics of apoptosis induction in PBMC from a control donor (Ctrl, green lines) and an AIDS patient (brown lines). PBMC were cultured in medium (plain lines) or in the presence of anti-CD3 mAbs (red line). A dual CD4/$DiOC_6(3)$ staining was performed. CD4$^+$ cells were gated, and $\Delta\Psi_m$ collapse was analyzed in this subset. (E) Quantification of $DiOC_6(3)^{Lo}$ cells in CD4$^+$ lymphocytes from controls ($n = 10$), ASX ($n = 37$), and AIDS ($n = 13$) subjects following overnight culture in medium, ionomycin, or anti-CD3 mAbs. Mean and standard deviations are shown. (See color insert.)

and Willie, 1980; Brown *et al.*, 1992). An early DNA cleavage into high molecular weight fragments results from the cleavage at the base of 50-kb DNA loops associated with the nuclear scaffold (Filipski *et al.*, 1990). This process can be followed by a more pronounced DNA degradation occurring at internucleosomal levels, which generate low molecular weight fragments (Brown *et al.*, 1992). Several flow cytometric approaches were proposed to evidence DNA fragmentation, through the detection of 3'OH termini, DNA condensation, DNA strand breaks, or DNA sensitivity to denaturation (reviewed in Lecoeur, 2002). DNA cleavage during apoptosis can be mediated by different and specific endonucleases (Samejima and Earnshaw, 2005). Except for acid DNases, enzymes involved in this complex process generate DNA fragments with 3'OH and 5'P ends. 3'OH DNA fragments can be evidenced by different approaches, such as ISNT or TUNEL assays.

4.1. The *in situ* nick translation assay

The ISNT was first reported by Jonker and colleagues (1993). In this assay, DNA polymerase I catalyzes the incorporation of deoxyribonucleotides along a DNA strand and evidences 3'OH recessed ends and single DNA breaks. The ISNT assay can be combined with the phenotyping of surface or intracellular molecules. However, it is important to check that the epitopes recognized by the corresponding antibodies are not denaturized after ethanol permeabilization, which is required in this assay (Lecoeur *et al.*, 1997). In that case, immunophenotyping should precede the ISNT assay.

1. This assay requires 5×10^5 to 1×10^6 cells per sample. Surface phenotyping, if required, should be performed first: centrifuge the cells (300g, 5 min), stain them with the appropriate mAb diluted in 100 μl PBS-BSA-azide, and incubate for 20 min at 4° in the dark. Wash stained cells with 200 μl PBS-BSA-azide.

2. Fix cell samples for 15 min at 4° in 200 μl PBS containing 1% PFA (a stock solution of PFA at 4% can be stored at −20°).

3. Wash the cells with 200 μl PBS and permeabilize them in 200 μl of 70% (v/v) cold ethanol for 4 min at 4°. Ethanol is required for cell permeabilization and DNA Pol I access to the nucleus. Centrifuge the cells immediately at 300g for 5 min at 4°.

4. Wash the cells with 200 μl PBS and incubate the pellet for 90 min at 37° in 50 μl of "nick translation buffer" (composed of 50 mM Tris-HCl, 5 mM MgCl$_2$, 10 mM β-mercaptoethanol, and 10 μg/ml serum albumin bovine, pH 7.8) containing the following mix: 19 μM of unlabeled nucleotides [D-adenosine 5'-triphosphate (dATP), D-cytosine 5'-triphosphate (dCTP), D-guanosine 5'-triphosphate (dGTP) (Boehringer

Mannhein)]–55 μM biotin-16-dUTP– 100 U/ml DNA–polymerase I (Promega Corp).

5. Wash the cells with 200 μl PBS. Incubate the pellet for 30 min at room temperature in 40 μl of staining buffer [standard saline citrate (SSC) containing 0.1% Triton X-100 and 5% nonfat milk powder] containing 20 μg/ml RNase DNase free and 5 μg/ml avidin-FITC.
6. Wash the cells with 200 μl PBS and fix them with 200 ml of 1% PFA in PBS.
7. Immediately acquire stained cells in a flow cytometer and analyze fluorescence with the appropriate software.

4.2. Comments, cautions, and pitfalls

4.2.1. Control samples for DNA fragmentation

A positive control for the ISNT assay should be prepared by treating cells with 20 μl of DNase I at 1 μg/ml for 1 h. A negative control may be prepared by treating cells with 20 μl of DNase I at 1 μg/ml for 1 h and performing the ISNT assay, omitting DNA Pol I.

4.2.2. Artifacts because of cell permeabilization

Intranuclear nick translation requires permeabilization of plasma and nuclear membranes. Ethanol treatment may induce drastic modifications of cell morphology. Consequently, this key step may induce artifacts that skew the phenotypic analysis of cells with fragmented DNA (Lecoeur *et al.*, 1998). In case the impact of ethanol treatment is too drastic, altering concomitant phenotyping, permeabilization may be performed for 3 min in PBS containing 0.1% sodium citrate and 0.1% Triton X-100. However, one has to check that this procedure is still adapted to appropriate ISNT staining.

Ethanol could also alter the antigenicity of certain surface epitopes. Consequently, such modified epitopes cannot be recognized any more by specific antibodies (Lecoeur *et al.*, 1998). It is therefore important to compare the expression of the surface antigens of interest before and after cell permeabilization.

4.3. Analysis of DNA fragmentation with the ISNT assay in peripheral lymphocytes prone to premature cell death

The ISNT assay was applied to quantify spontaneous apoptosis in PBMC from HIV[+] subjects cultured for 24 h in the absence of stimulation. A phenotypic analysis of lymphocytes undergoing apoptosis was performed. Specific monoclonal antibodies directed against CD4 (T cell marker) and CD19 (B cell marker) were used. Figures 3.3A and 3.3B show the impact of the ISNT assay on surface detection of both markers. A drop in mean fluorescence intensity was observed for both CD4 and CD19 molecules

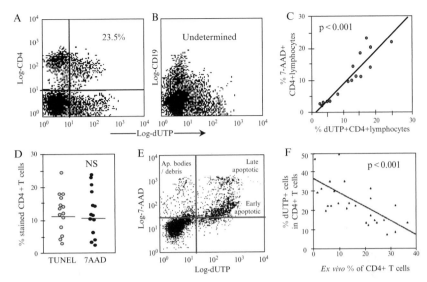

Figure 3.3 Analysis of DNA fragmenation in primary lymphocytes undergoing apoptosis. (A and B) Analysis of DNA fragmentation by the ISNT assay in lymphocytes from an ASX HIV-infected patient prone to undergo spontaneous apoptosis. Biparametric dUTP-FITC/CD4-PE (A) and dUTP-FITC/CD19-PE (B) dot plots obtained after a 24-h culture. The percentage of dUTP-positive cells among CD4$^+$ T cells is indicated (A). Quantification of apoptosis in the CD19$^+$ B cell subset was not reliable with this assay. (C– F) Analysis of DNA fragmentation by the TUNEL assay. (C) Correlation between the percentage of dUTP$^+$ (i.e., with fragmented DNA, TUNEL assay) and 7-AAD$^+$ cells in CD4$^+$ T cells undergoing spontaneous apoptosis ($n = 18$ HIV$^+$ patients). (D) Comparison of the percentage of apoptotic cells determined by TUNEL (white dots) and 7-AAD (black dots) assays ($n = 18$ patients). The median values are indicated. (E) Combination of the TUNEL and 7-AAD assays on PBMC from an ASX subject undergoing spontaneous apoptosis. (F) The percentage of CD4$^+$ T cells lymphocytes with fragmented DNA (dUTP$^+$) in response to AICD (coated anti-CD3 mAbs) correlates with disease evolution, evaluated as the *ex vivo* percentage of CD4$^+$ T cells ($n = 29$ patients).

on dUTP$^+$ cells. Indeed, during apoptosis, cells undergo drastic structural and biochemical modifications, including alteration of the plasma membrane, cell shrinkage, and disintegration into apoptotic bodies. These cell processes induce a "natural" loss in the antibody-binding capacity (ABC) of surface antigens (Lecoeur *et al.*, 1997, 1998). Despite the decreased ABC, CD4 detection was still possible in both live (dUTP⁻) and apoptotic (dUTP$^+$) cells (Fig. 3.3A). In contrast, CD19 staining, initially low in live cells, was undetectable in dUTP$^+$ cells (Fig. 3.3B). This artifact significantly potentiates the natural decrease in the ABC of apoptotic cells and results in an inability to detect the CD19 antigen on these cells (Fig. 3.3B).

Figure 3.3C shows that a very high correlation was found between two features of apoptotic cells (alteration of plasma membrane, detected with the 7-AAD assay, and DNA fragmentation, detected with the ISNT).

4.4. The terminal deoxynucleotidyl transferase dUTP-mediated nick end labeling assay

The TUNEL assay is based on the use of terminal deoxynucleotidyl transferase (TdT) (Gorczyca *et al.*, 1993). This enzyme, in the presence of cobalt ions (Co^{2+}), allows tailing every $3'OH$ terminus available in cleaved DNA (blunt, overhang, recessive ends, and single strand breaks). Moreover, it evidences $3'OH$ ends not only in oligonucleosomal fragments but also in high molecular weight fragments, that is, in early apoptotic cells (for review, see Lecoeur, 2002). Cells are first treated with PFA to fix DNA fragments in the nucleus and to avoid DNA leakage into the supernatant during ethanol permeabilization. TdT then catalyzes the incorporation of oligonucleotides conjugated to fluorochrome (dUTP-FITC) (Gorczyca *et al.*, 1993). Detection of DNA fragmentation with the TUNEL assay can be combined to cell phenotyping. However, as already discussed for the ISNT assay, one should check that ethanol permeabilization still allows the detection of molecules of interest.

1. This assay requires 5×10^5 to 1×10^6 cells per sample. Surface phenotyping, if required, should be performed first: centrifuge the cells ($300g$, 5 min), stain them with the appropriate mAb diluted in 100 μl PBS-BSA-azide, and incubate for 20 min at 4° in the dark. Wash stained cells with 200 μl PBS-BSA-azide.
2. Fix cell samples for 15 min at 4° in 200 μl PBS containing 1% PFA (a stock solution of PFA at 4% can be stored at −20°).
3. Wash cells with 200 μl PBS and permeabilize them in 200 μl of 70% (v/v) cold ethanol for 1 min at 4°.
4. Centrifuge immediately at $300g$ for 5 min at 4° and wash the cells with 200 μl PBS.
5. Resuspend the pellet in 10 μl of *staining solution* containing the following mix for 100 μl: 20 μl reaction buffer [composed of 1 M potassium cacodylate, 125 mM Tris-HCl (pH 6.6), and 1.25 mg/ml BSA], 10 μl of 25 mM cobalt chloride, 1 μl dUTP-FITC diluted at the appropriate dilution (manufacturer's conditions), 1 μl of 25 U/μl TdT (stock solution at 25 U/μl), and 68 μl distilled water. Incubate the cells at 37° for 60 min.
6. Add 150 μl washing buffer [0.1% Triton X-100 (v/v), 4 mg/ml BSA] and centrifuge the cells and fix them for 15 min at room temperature in the dark in 200 μl of 1% PFA in PBS. Analyze cell samples by flow cytometry.

4.5. Comments, cautions, and pitfalls

4.5.1. Compared sensitivity of TUNEL and ISNT assays

The TUNEL assay is more sensitive than the ISNT assay (Gorczyka *et al.*, 1993). Indeed, TdT permits a faster incorporation of dUTP than DNA Pol I and also allows the tailing of every 3′OH recessed end (Lecoeur, 2002; Roychoudhury *et al.*, 1976). These features explain why so many commercial kits recommend the TUNEL procedure.

4.5.2. Artifacts because of cell permeabilization: See Section 4.2

4.5.3. Lack of specificity of the TUNEL assay

The TUNEL assay also detects primary necrosis. This cell death is characterized by a nuclear dissolution (karyolysis) that is associated to random DNA cleavage. When DNA from necrotic cells is analyzed by gel electrophoresis, DNA fragments appear as a smear, which contrasts with the classical DNA ladder observed with apoptotic cells. However, DNA fragments from necrotic cells can harbor 3′OH ends that are evidenced by the TdT (Grasl-Kraup *et al.*, 1995). This should be kept in mind, particularly for biological samples where both types of cell death coexist. Microscopic examination of morphological features should be done on every tested sample.

4.6. Analysis with the TUNEL assay of DNA fragmentation in peripheral lymphocytes prone to premature cell death

The TUNEL assay was applied on PBMC from ASX HIV-infected patients and compared to the 7-AAD assay. Following overnight incubation in medium, PBMC were double stained for the CD4 molecule and either for DNA fragmentation with the TUNEL assay or for the loss of plasma membrane (PM) integrity with 7-AAD. Gated CD4 T cells were analyzed for the percentage of either dUTP$^+$ or 7-AAD$^+$ cells (Fig. 3.3D). No differences were observed between the two assays, and a strong correlation was found between both (Fig. 3.3C). The dot plot in Fig. 3.3E shows that cells with a cleaved DNA showed either a moderate loss of PM integrity (early apoptotic cells, 7-AAD^{Lo} cells) or a pronounced loss of PM integrity (late apoptotic cells, 7-AAD^{Hi} cells). Therefore, the sensitivity of the TUNEL assay is comparable to that of the 7-AAD assay on primary lymphocytes.

Apoptosis susceptibility of patients' CD4 T cells was found correlated with disease evolution, as measured by the *in vivo* percentage of CD4 T cells (Fig. 3.3F). This correlation was observed for spontaneous cell but also for AICD.

5. ANALYSIS OF THE LOSS OF PLASMA MEMBRANE ASYMMETRY WITH THE ANNEXIN-V ASSAY

Phosphatidylserine is an anionic phospholipid located in the inner leaflet to the Plasma Membrane (PM) of living cells. During apoptosis, PS residues are translocated to the external leaflet of the PM through complex mechanisms, involving at least in part the increase in scramblase activity (Fadeel *et al.*, 1999; Frasch *et al.*, 2000). *In vivo*, those residues are recognized by the PS receptor (OSR) on macrophages that phagocyte apoptotic cells (Fadok *et al.*, 1992). The translocation of PS residues can be evidenced through the fixation of annexin-V conjugated to several fluorochromes. The use of annexin-V in combination with propidium iodide (PI) allows one to discriminate early (annexin-V$^+$, IP$^-$) from late apoptotic/necrotic lymphocytes (annexin-V$^+$, IP$^+$). This assay is still considered as the reference assay for apoptosis detection and quantification.

5.1. The annexin-V assay

1. This assay requires 2 to 5 × 10^5 cells per sample. Surface phenotyping, if required, should be performed first: centrifuge the cells (300g, 5 min), stain them with the appropriate mAb diluted in 100 μl PBS-BSA-azide, and incubate for 20 min at 4° in the dark. Wash stained cells with 200 μl PBS-BSA-azide.
2. Resuspend the pellet in 100 μl staining buffer (140 mM NaCl, 5 mM Ca^{2+}, and 10 mM HEPES in distilled water) containing the appropriate concentration of annexin-V (according to the manufacturer's recommendations) and 1 μl PI (stock solution of PI at 50 μg/ml kept at 4°).
3. Incubate for 20 min at room temperature in the dark.
4. Fix the cells in staining buffer containing 1% PFA.
5. For intracellular staining, permeabilize the cells for 15 min with the staining buffer containing 0.01% saponin and incubate them with the appropriate concentration of antibodies diluted in this buffer.
6. Fix the cells in staining buffer containing 1% PFA for 20 min in the dark at 4°.
7. Analyze 10,000 events by flow cytometry.

5.2. Comments, cautions, and pitfalls

5.2.1. Annexin-V staining and Ca^{2+} dependence

The absence of any covalent bond between annexin-V and PS residues, and the Ca^{2+} ion dependence of staining, implies that Ca^{2+} must be present during each step of the procedure. It is even required during the fixation and permeabilization steps.

5.2.2. Translocation of PS residues during oncosis

Oncosis is defined by an early stage of primary necrosis, during which the cell swells (Majno and Joris, 1995). Several studies have pointed out that cells undergoing oncosis, in absence of any drastic disruption of the plasma membrane, expose PS residues to the outer leaflet of the PM. Consequently oncotic cells harbor an "apoptotic-like" annexin V^+ IP^- phenotype, initially considered to be specific for early apoptotic cells (Krysko *et al.*, 2004; Lecoeur *et al.*, 2001a; Waring *et al.*, 1999). These studies underline the absence of specificity of the PS assay. Observations by optical or electron microscopy are required to determine the features of dying cells stained by annexin-V.

5.2.3. Translocation of PS residues in absence of cell death

Phosphatidylserine residue translocation to the outer leaflet of the PM is not systematically associated to apoptosis or oncosis. This process can be induced by a treatment with calcium ionophore, as shown on thymocytes (Marguet *et al.*, 1999) or human sperm cells (Martin *et al.*, 2005). Consequently, annexin-V staining should be validated by other classical apoptosis assays.

5.3. Application of the annexin-V assay to the analysis of AICD

As shown previously, a correlation exists between the percentage of annexin V^+ PI^- cells and the percentage of 7-AAD^{lo} cells in Jurkat cells when 7-AAD was used at 20 $\mu g/ml$, suggesting that both subsets included early apoptotic cells (Lecoeur *et al.*, 2002). This is confirmed in Fig. 3.4A on another cell type, that is, murine thymocytes incubated overnight at $37°$, showing that early apoptotic cells (7-AAD^{lo}) are annexin-V^+. Commercial annexin-V/7-AAD kits use a much lower concentration of 7-AAD (around 1 $\mu g/ml$). It is noteworthy that, under these conditions, early apoptotic cells become 7-AAD negative (Fig. 3.4A). Therefore, the characteristics of early apoptotic cells depend on 7-AAD concentrations used: a low 7-AAD concentration does not stain early apoptotic cells (annexin-V^+), whereas a high 7-AAD concentration does. This observation was confirmed by multi-spectral imaging cytometry (Figure 3.4B, C, and D). At high dose of 7-AAD, live cells (Gate R1) are 7-AAD negative, whereas annexin-V+ early apoptotic cells (Gate R2) weakly incorporate the DNA dye (see representative cells in channel 6, 7-AAD, Fig. 3.4C). Late apoptotic cells (Gate R3) harbor a more pronounced 7-AAD staining (Fig. 3.4D). These observations clearly indicate that these two annexin-V+ subsets (R2 and R3) represent a continuum in the cell death process, and both should be considered for apoptosis quantitation (Lecoeur *et al.*, 2002). Figure 3.4E shows that annexin-V staining allows the phenotyping of apoptotic cells: PBMC from an ASX patient were incubated overnight in medium and the proportion of apoptosis in CD4 T cells was quantified. Under these conditions, annexin-V stains both early and late apoptotic cells. Figure 3.4F shows that data obtained with annexin-V and 7-AAD assays are highly correlated.

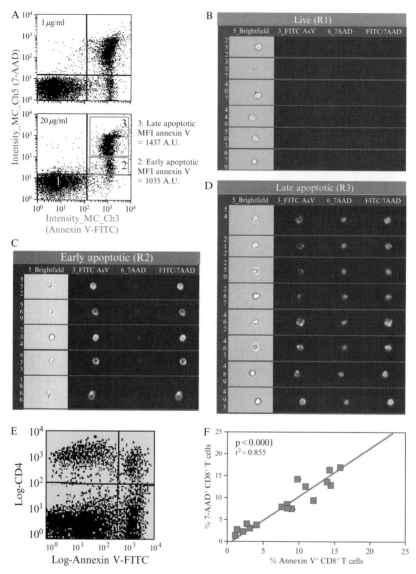

Figure 3.4 Annexin-V staining of apoptotic cells and correlation with 7-AAD staining. (A) Biparametric annexin-V/7-AAD dot plot of murine thymocytes incubated overnight in culture medium at 37 °C. The co-staining annexin-V-FITC / 7-AAD has been performed under two experimental conditions: 1 μg/ml and 20 μg /ml of 7-AAD. Early apoptotic cells are annexin-V$^+$ and 7-AAD^{low}. (B, C, D) Analysis of such thymocytes by multi-spectral imaging in flow. Cells were stained by annexin-V-FITC and 20 μg / ml 7-AAD, unfixed and were analysed with an Image Stream 100 (Amnis). Live, early and late apoptotic cells were gated (gates R1, R2 and R3 respectively) and representative thymocytes from these gates are presented in parts B, C and D. Bright field, annexin-V staining (channel 3), 7-AAD staining (channel 6) and the overlapping of these two parameters are presented for each cell. (E) PBMC from an ASX HIV-infected patient were incubated overnight in culture medium à 37 °C. Apoptosis of CD4 T cells was quantified following a co-staining with annexin-V

6. Analysis of the Loss of Plasma Membrane Integrity with the 7-Amino Actinomycin D Assay

During the apoptotic process, cells show a progressive increase in PM permeability (Schmid *et al.*, 1992). Consequently, apoptotic cells become permissive to dyes such as PI or 7-AAD. 7-AAD is an analog of actinomycin D (AD), which has been initially synthesized in an attempt to obtain compounds that might be useful for chromosome analysis and to obtain biological derivatives of actinomycin D (Modest and Sengupta, 1974). Initially described for the detection of apoptosis in lymphoid cells (Schmid *et al.*, 1992), 7-AAD staining is compatible with a multiparametric flow cytometric analysis of apoptotic cells from diverse cell lineages: PBMC (Lecoeur *et al.*, 1997, 1998, 2001), adherent hepatic MMHD3 cells (Lecoeur *et al.*, 2002), or adherent primary neurons (Lecoeur *et al.*, 2004). When excited at 488 nm, its emission spectrum (>600 nm) permits the combined detection of fluorochromes emitting green and orange fluorescence, with no major problem of compensation. This molecule exhibits other numerous advantages when compared to other dyes (Lecoeur *et al.*, 2002). For example, a high concentration of 7-AAD (20 μM) allows detection of two major subsets of apoptotic cells, that is, early apoptotic cells, with a moderate loss of PM integrity, and late apoptotic cells, with a disrupted PM. As shown earlier, early apoptotic cells detected with 7-AAD are stained concomitantly with annexin-V.

6.1. The 7-AAD assay

1. This assay requires 1 to 2 × 10⁵ cells per sample.
2. Wash cells (300*g*, 5 min). Resuspend cell pellets in 100 μl PBS–1% BSA–0.01% sodium azide (PBS-BSA-azide) containing 20 μg/ml 7-AAD and incubate them for 20 min at room temperature in the dark.
3. Centrifuge cells at 300*g* for 5 min and resuspend the pellet in 100 μl PBS-BSA-azide containing 20 μg/ml AD.
4. Immediate acquisition of data on a flow cytometer is preferable. If not possible, fix the cells in PBS-BSA-azide containing 1% PFA and AD at 20 μg/ml. Cell samples can be kept at 4° in the dark for 1 night.
5. The combined detection of surface antigens can be done before 7-AAD staining. For combined intracellular stainings, 7-AAD staining should be performed first, before fixing the cells in PBS-BSA-azide containing 1%

and anti-CD4 mAb. The percentage of apoptosis in CD4 T cells in indicated on the figure. (F) PBMC from ASX HIV-infected patients (*n* = 17) were incubated overnight in culture medium à 37 °C. Apoptosis of CD8 T+ cells was quantified following a co-staining with annexin-V, or 7-AAD, and anti-CD8 mAb. The linear regression curve between both variables is shown. The coefficient of correlation and the *p*-value are indicated. (See color insert.)

PFA and AD at 20 μg/ml. Centrifuge and permeabilize the cells for 15 min with PBS–BSA–azide containing 0.01% saponin and AD at 20 μg/ml. Intracellular staining should be done in this same buffer for 20 min.
6. After centrifugation, fix the cells in PBS–BSA–azide containing 1% PFA and AD at 20 μg/ml.
7. Analyze 10,000 events by flow cytometry.

6.2. Comments, cautions, and pitfalls

6.2.1. Advantages of the 7-AAD assay
This assay is rapid and inexpensive. It is noninvasive and allows the combined detection of both intracellular (mitochondrial proteins) and extracellular (phenotyping) molecules. It is as sensitive as the other assays, such as the annexin-V assay or the TUNEL assay (Lecoeur *et al.*, 1996, 1997). In addition, with only one molecule, it is possible to detect two steps of the apoptotic process, in contrast to the annexin V/PI assay, which requires a combination of two stainings.

6.2.2. Lack of specificity for apoptosis detection
Like most of the apoptosis assays, the 7-AAD assay also stains necrotic cells (Lecoeur *et al.*, 2002).

6.3. Detection with the 7-AAD assay of early and late apoptotic cells on primary cells prone to premature cell death analysis

7-Aminoactinomycin D is a marker of choice for the analysis of apoptosis in numerous cell types and has been a useful tool for the thorough analysis of apoptotic pathways triggered by HIV (Gougeon, 2003; Lecoeur *et al.*, 2002). 7-AAD staining allows the distinction of sequential steps of the apoptotic process in peripheral lymphocytes prone to undergo cell death. The biparametric 7-AAD/FSC dot plot discriminates live lymphocytes from lymphocytes with a moderate (early apoptotic) or a pronounced (late apoptotic) loss of PM integrity (Fig. 3.5A). In addition, apoptotic bodies/debris can be evidenced as low FSC and variable but positive 7-AAD staining (because of the low amount of DNA in these bodies). Thus, 7-AAD staining allows exclusion from the analysis of apoptotic bodies, which is impossible to do with annexin-V staining, which requires costaining with PI or 7-AAD at low doses to discriminate early from late apoptotic cells. Figures 3.5B and 3.5C show the combined detection of surface (CD4) or intracellular (IL-2) molecules on 7-AAD[+] cells. Figure 3.5C demonstrates that priming for apoptosis of T cells producing a given cytokine can be analyzed with the 7-AAD assay. This experiment was performed on PBMC from an HIV-infected patient following overnight

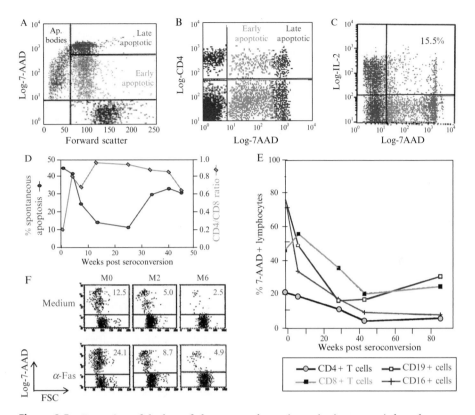

Figure 3.5 Detection of the loss of plasma membrane integrity in apoptotic lymphocytes by the 7-AAD assay. (A) Biparametric 7-AAD/FSC dot plot of PBMC from an AIDS patient undergoing spontaneous apoptosis (48-h culture). Early apoptotic (green dots), late apoptotic (red dots) lymphocytes, and apoptotic bodies / debris (orange dots) can be discriminated easily from live lymphocytes (blue dots) according to combined cell size and 7-AAD staining. (B) Biparametric 7-AAD/CD4 dot plot PBMC from an AIDS patient undergoing spontaneous apoptosis (24-h culture). Apoptotic bodies/cell debris were discarded from the analysis. Early and late apoptotic lymphocytes were discriminated easily through the intensity of 7-AAD incorporation. (C) Triple staining CD3/IL-2/7-AAD of PBMC from an HIV^{+} subject stimulated overnight in the presence of PMA (50 ng/ml), ionomycin (300 ng/ml), and PHA-A (100 ng/ml). Brefeldin A (10 μg/ml) was added during the last 12 h of culture. Analysis was performed on gated CD3 T cells. The percentage of apoptosis in IL-2-secreting CD3^{+} T cells is indicated. (D) Kinetic of spontaneous apoptosis and CD4/CD8 ratio in a subject following HIV-specific seroconversion. Spontaneous apoptosis (24-h culture) was quantified in total PBMC with the 7-AAD assay (red line), the CD4/CD8 ratio was determined *ex vivo* by flow cytometry (green line). (E) Kinetic of spontaneous apoptosis in different lymphocyte subsets from the same donor. (F) Spontaneous and Fas-induced apoptosis in CD4^{+} T cells from an HIV-infected patient at baseline and 2 and 6 months following initiation of antiretroviral therapy. FSC/7-AAD dot plots on gated CD4^{+} T cells are shown, and the percentage of apoptotic CD4 T cells is indicated. (See color insert.)

polyclonal stimulation with PMA and ionomycin. Quantification of CD3 T cells producing IL-2 within this subset enumeration of apoptotic cells was performed as described earlier after costaining with specific antibodies, and as reported previously. Note that the majority of apoptotic IL-2 producing cells are localized in the late apoptotic subset (7-AAD^{Hi} cells).

The 7-AAD assay is perfectly adapted for the monitoring of apoptosis during HIV infection. A few examples are given. Figure 3.5D shows that spontaneous T-cell apoptosis, measured after overnight incubation in culture medium of PBMC from a patient who recently seroconverted, is very high and inversely correlated with the CD4/CD8 ratio. At the first time point of analysis (i.e., at diagnosis of seroconversion), the CD4/CD8 ratio was low (0.2) while spontaneous apoptosis was observed in 45% of T cells. Five weeks later, apoptosis was reduced dramatically to 25% while the CD4/CD8 ratio increased to 1.0. At week 25, the level of apoptosis increased again and was associated with a moderate decrease of the CD4/CD8 ratio. Identification of cell subsets dying of apoptosis postseroconversion is shown in Fig. 3.5E. All lymphocytes subsets, including CD4 T cells, CD8 T, CD19 B cells, and CD16 NK cells, were highly apoptotic shortly after primary HIV infection. The chronic phase of the infection was then associated with a decrease in apoptosis levels to values usually observed in untreated patients (Gougeon *et al.*, 1996). Highly active antiretroviral therapy (HAART), combining reverse transcriptase and protease inhibitors, was found to be highly efficient in the suppression of plasmatic viral load of patients. Figure 3.5F shows the impact of HAART on spontaneous and Fas-induced apoptosis in patient's T cells 2 and 4 months following therapy initiation. Both intrinsic and extrinsic apoptosis were suppressed by HAART, which was concomitant with an increase in the absolute number of CD4$^+$ T cells (data not shown) (Ledru *et al.*, 1998). However, as reported, some antiretroviral combination did not suppress apoptosis while the restoration of CD4 T cell numbers occurred (De Oliveira Pinto *et al.*, 2002).

7. A FLOW CYTOMETRY ASSAY TO QUANTIFY CELL-MEDIATED CYTOTOXICITY

Cell-mediated cytotoxicity is a major effector pathway of the immune response against pathogens or tumors and a fundamental mechanism to maintain homeostasis of the immune system. Different cell effectors, such as cytolytic CD4 or CD8 T lymphocytes and NK cells, can mediate this process. Two main pathways are involved in target cell killing. The first one involves recognition by the T-cell receptor on effector cells of peptide/MHC complexes on the surface of target cells, and target killing is mediated by the perforin/granzyme pathway. The second one involves ligands specific for death receptors of the TNF–receptor family (FasL, TNF-α, TRAIL), which deliver death signals to target cells expressing the corresponding receptors (Depraetere and Golstein, 1997). Under certain circumstances, both pathways can be mediated by the same cytotoxic T cell (Garcia *et al.*, 1997; Kagi *et al.*, 1994).

The reference method for the quantification of cell-mediated lysis has long been the ^{51}Cr release assay, which requires a radioactive isotope, is not highly sensitive, and does not permit characterization of the two partners involved in the reaction. The CFSE/7-AAD cytofluorimetric assay is based on the combination of two dyes, that is, 5- (and 6-) carboxyfluorescein diacetate succinimydyl ester (CFSE) to label the effectors and 7-AAD to stain apoptotic targets. The advantages of the CFSE/7-AAD assay as compared to the classical ^{51}Cr release assay are the following: (1) the use of nonradioactive markers. (2) CFSE is a "universal marker" able to stain any population of effector cells, thus avoiding the requirement for specific mAbs used in other flow cytometry assays (Godoy-Ramirez *et al.*, 2000; Golberg *et al.*, 1999). (3) CFSE binding is stable, making this test usable for long-term cytotoxicity assays. (4) CFSE is nontoxic, has no effect on the lytic properties of the effectors, and has no influence on other effector functions, such as cytokine production, as reported previously (Lecoeur *et al.*, 2001). (5) Quantitation of cell lysis following 7-AAD staining is highly reproducible and reliable. (6) In contrast to staining with other probes for apoptosis detection, 7-AAD staining is resistant to fixation and permeabilization procedures required for intracellular staining of target cells. Thus, it allows the concomitant detection of intracellular molecules involved in the apoptotic process.

7.1. The CFSE/7-AAD cytotoxicity assay

1. It is important to determine the appropriate concentration of CFSE for effector cell staining. To do so, stain 10^6 effector cells with CFSE in PBS for 15 min at 37° in a volume of 1 ml. A stock solution of CFSE at 5 mM in dimethyl sulfoxide can be prepared and kept at −20°. CFSE green fluorescence should be located between 10^2 and 10^3 AU (see Figure 3.6B).

2. Wash the cells in PBS containing 10% fetal calf serum. Centrifuge at 300g for 5 min and discard the supernatant.

3. Repeat this washing step to get rid of the remaining soluble CFSE.

4. Immediately coculture CFSE-stained effectors with the target.

5. Seed a constant number of target cells in 48-flat well microplates and add CFSE-labeled effector cells at different E:T ratios (2:1, 7:1, 20:1 and 60:1). In parallel, incubate target cells alone to measure basal apoptosis.

6. Incubate the cocultures in a total volume of 150 μl of complete medium for 4 h in a 5% CO_2 atmosphere at 37° .

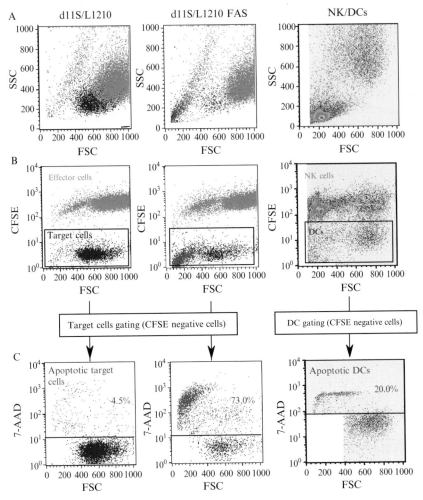

Figure 3.6 Analysis of cell-mediated cytotoxicity by the cytofluorimetric CFSE/
7-AAD assay. (A– C) Cytotoxicity assay between d11s effector and L1210 or L1210Fas target cells. (A) Biparametric FSC/SSC dot plots of the cell coculture (4-h contact)
corresponding to an E:T ratio of 7:1. Effectors (CFSE-stained d11s cells) appear in green,
live target cells in blue, and apoptotic target cells in red (7-AAD positive cells). (B)
Biparametric FSC/CFSE dot plots of the coculture. Target cells (CFSE negative) can be
specifically gated. (C) Biparametric FSC/7-AAD dot plots of selected target cells. Target
cell lysis is quantified by the percentage of 7-AAD positive cells (red dots). (D– F) Cytotoxicity assay between human NK cells and DC. After CFSE staining activated NK
cells have been cocultured with DC. Gated DC (CFSE negative) showed 20% lysis,
quantified by 7-AAD staining. (See color insert.)

7. Wash cell mixtures in PBS-1% BSA containing 0.1% sodium azide and resuspend pellets in the same buffer containing 20 μg/ml 7-AAD (Sigma, France) for 20 min at 4° in the dark as described previously (Lecoeur et al., 2001).

8. Wash cell samples in PBS-BSA-NaN$_3$ containing 20 μg/ml of actinomycin D and fix samples in the same buffer containing 1% PFA for 20 min.

9. If required, the combined detection of surface antigens should be done before 7-AAD staining. For combined intracellular staining, 7-AAD staining should be performed first and then fix cells in PBS-BSA-azide containing 1% PFA and AD at 20 μg/ml (see step 5 in Section 6.1).

10. Acquire stained cells on a flow cytometry. Analyze 10,000 cells by flow cytometry.

7.2. Comments, cautions, and pitfalls

7.2.1. High sensitivity of the CFSE/7-AAD assay

This assay is more sensitive than the ^{51}Cr release assay, particularly at low E:T ratios (Lecoeur et al., 2001). This is probably because of the possible detection of early apoptotic cells with 7-AAD staining that exhibit weak membrane alteration and do not release ^{51}Cr. However, at high E:T ratios, a possible limitation of the CFSE/7-AAD assay is the lack of consideration of apoptotic bodies/debris during flow cytometry analysis, in contrast to the cumulative ^{51}Cr release assay. Therefore, when an effector/target cell system has to be set up, it is important to define the experimental conditions for optimal lysis detection.

7.2.2. Possible analysis of processes associated with cell-mediated cytotoxicity

This assay allows characterizing some events associated with cell-mediated cytotoxicity at the single cell level in both effector and target cells. For example, perforin release by the effector can be studied on gated CFSE$^+$ cells, whereas caspase activation in the target can be evidenced on CFSE$^-$ cells (Lecoeur et al., 2001). This assay should contribute to a better understanding of the mechanisms involved in cell-mediated cytotoxicity in normal and pathological situations.

7.2.3. Definition of appropriate conditions for CFSE staining

CFSE concentrations should be adjusted to the type of effectors. It is important that the fluorescence intensity of CFSE$^+$ cells is high enough to easily distinguish them from target cells (CFSE$^-$). Also, CFSE staining should be calibrated in term of minutes of incubation at 37°, as little variations can induce strong differences in staining intensity. Finally, note that no fetal calf serum should be present during the CFSE staining procedure.

7.2.4. Limitation of the CFSE/7-AAD assay

One major limitation of this assay is the high number of effector cells at the high E/T cell ratio.

7.3. Analysis of cell-mediated cytotoxicity by the CFSE/7-AAD assay

The cell-mediated cytotoxicity assay is illustrated in Fig 3.6 through two examples. In the first one, d11S cells (a cytotoxic murine hybridoma cell line) are used as effectors, and L1210 cells (a murine leukemic cell line) or L1210Fas (its CD95-transfected counterpart) are used as targets. The induction of CD95L on the effectors is performed by a 3-h PMA (50 ng/ml) + ionomycin (0.3 μg/ml) activation, as reported (Lecoeur *et al.*, 2001). d11S cells are stained for 10 min at 37° with 0.2 μM of CFSE. Following effector–target coculture, the percentage of cytotoxicity in T cells is determined in a three-step flow cytometry analysis. *First step*: Gate the target cells as CFSE negative on the FSC-CFSE dot plot (Fig. 3.6B). *Second step*: Gate and discard apoptotic bodies/debris according to their low cell size (FSC^{lo}) and weak 7-AAD incorporation (7-AAD^{lo}). *Third step*: Determine the percentage of lysis in target cells on the FSC/7-AAD dot plot. It is given directly by the percentage of apoptotic cells (7-AAD^{lo} + 7-AAD^{hi}) within gated CFSE–negative cells (Fig. 3.6C). Compare this value to the basal cell apoptosis (obtained on target cells not incubated with effectors). Note that a specific target lysis was observed only for L1210 cells that expressed the Fas receptor (L1210Fas cells) (Fig. 3.6C).

In the second example, the cytotoxic activity of natural killer (NK) cells on dendritic cells (DC) has been studied. NK cells are obtained from the monocyte-depleted fraction of PBMC by negative selection of CD56$^+$ cells, and their purity is superior to 90% as assessed by flow cytometry. Forty-eight hours before coculture, NK cells are activated with PHA (5 μg/ml) and rhIL-2 (10 μg/ml). DC are prepared from monocytes enriched in CD14$^+$ cells with a specific kit and differentiated by a 6-day culture in the presence of rhGM-CSF and rhIL- 4 (10 ng/ml each). At initiation of the coculture, activated NK cells, at a concentration of 10^6 cells/ml, are stained for 10 min at 37° with 1 μM of CFSE. Stained NK cells are then cocultured with DC at a ratio of 5:1 for 24 h. The lysis of DC is quantified with the 7-AAD test (as detailed earlier). The analysis of DC requires (1) gating of CFSE-negative cells (Fig. 3.6E), (2) discarding of apoptotic debris according to their low cell size (FSC^{lo}) and weak 7-AAD incorporation (7-AAD^{lo}), and (3) determining the percentage of lysis in DC on the FSC/7-AAD dot plot (Fig. 3.6F). Note that in these conditions the majority of dead target cells are late apoptotic ones (7-AAD^{Hi} DCs).

8. CONCLUSIONS

Flow cytometry is a very potent technology adapted to the identification of apoptotic subsets in complex populations such as human PBMC. The multiparametric analyses described in this study allow the detection of several features associated with apoptotic death, including cell shrinking, alteration of PM integrity, translocation of PS residues to the external leaflet of the PM, drop in mitochondrial membrane potential, downregulation of the intracellular expression of Bcl-2 family members, and DNA fragmentation. With this approach, the molecular mechanisms involved in apoptotic death induced by some pathogens into their hosts could be studied thoroughly in specific subsets and correlated with disease evolution. For example, the *in vivo*-reduced expression of Bcl-2 in T cells from HIV-infected subjects has been detected specifically in CD8$^+$ T cells with a cytotoxic phenotype and has been correlated with disease evolution (Boudet *et al.*, 1996). These observations suggested that inappropriate apoptosis induced by chronic HIV infection was responsible for the destruction of CTL through the intrinsic cell death pathway (reviewed in Gougeon, 2003). Studies have shown the possible combined detection of viral proteins and intracellular antigens (Bax, Bcl-2) (Goosby *et al.*, 2005). Regarding the analysis of mitochondrial membrane potential, a correlation was found between the decreased *ex vivo* DiOC$_6$(3) staining PBMC from HIV-infected subjects and apoptosis generated after a short-term culture and assessed by the TUNEL assay (Macho *et al.*, 1995). The decrease in $\Delta\Psi_m$ was also shown to be a good marker of apoptosis in T-cell acute lymphoblastic leukemia (Ozgen *et al.*, 2000). This chapter also reported for the first time a new cytotoxic assay that easily quantifies the lysis of DC induced by NK cells. DC and NK cells play a critical role in early defenses against cancer and infections. They specialize in complementary functions, including IL-12 or IFN-α/β secretion and antigen presentation for DC, and IFN-γ secretion and killing of infected or tumor cells for NK cells. Evidence of NK-DC cross talk has accumulated, and this interaction may lead to NK cell activation, DC maturation, or apoptosis depending on the activation status of both cell types (Walzer *et al.*, 2005). The assay reported in this chapter should help in identifying the pathway(s) activated in the destruction of DC by NK cells and the receptors and effector cytokines involved.

ACKNOWLEDGMENTS

This work was supported by grants from ANRS (Agence National de Recherche sur le SIDA), Sidaction, EU Grant AUTOROME (STREP, LSHM-CT-2004-005264), Institut Pasteur and INSERM. MTM is supported by a research fellowship from the Research Ministry, France, HS is supported by a research fellowship from Sidaction.

REFERENCES

Antignani, A., and Youle, R. J. (2006). How do Bax and Bak lead to permeabilization of the outer mitochondrial membrane? *Curr. Opin. Cell Biol.* **18**, 685–689.

Borner, C. (2003). The Bcl-2 protein family: Sensors and checkpoints for life-or-death decisions. *Mol. Immunol.* **39**, 615–647.

Boudet, F., Lecoeur, H., and Gougeon, M. L. (1996). Apoptosis associated with *ex vivo* down-regulation of Bcl-2 and Fas in potential cytotoxic CD8+ lymphocytes during HIV infection. *J. Immunol.* **156**, 2282–2293.

Brown, D. G., Sun, X. M., and Cohen, G. M. (1992). Dexamethasone-induced apoptosis involves cleavage of DNA to large fragments prior to internucleosomal fragmentation. *J. Biol. Chem.* **268**, 3037–3039.

de Oliveira Pinto, L. M., Lecoeur, H., Ledru, E., Rapp, C., Patey, O., and Gougeon, M. L. (2001). Lack of control of T cell apoptosis under HAART: Influence of therapy regimen *in vivo* and *in vitro*. *AIDS* **16**, 329–339.

Depraetere, V., and Golstein, P. (1997). Fas and other cell death signaling pathways. *Semin. Immunol.* **9**, 93.

Fadeel, B., Gleiss, B., Högstrand, K., Chandra, J., Wiedmer, T., Sims, P. J., Henter, J. I., Orrenius, S., and Samali, A. (1999). Phosphatidylserine exposure during apoptosis is a cell-type-specific event and does not correlate with plasma membrane phospholipid scramblase expression. *Biochem. Biophys. Res. Commun.* **266**, 504–511.

Fadok, V. A., Voelker, D. R., Campbell, P. A., Cohen, J. J., Bratton, D. L., and Henson, P. M. (1992). Exposure of phosphatidylserine on the surface of apoptotic lymphocytes triggers specific recognition and removal by macrophages. *J. Immunol.* **7**, 2207–2216.

Frasch, S. C., Henson, P. M., Kailey, J. M., Richter, D. A., Janes, M. S., Fadok, V. A., and Bratton, D. L. (2000). Regulation of phospholipid scramblase activity during apoptosis and cell activation by protein kinase Cdelta. *J. Biol. Chem.* **275**, 23065–23073.

Filipski, J., Leblanc, J., Youdale, T., Sikorska, M., and Walker, P. R. (1990). Periodicity of DNA folding in higher order chromatin structures. *EMBO J.* **9**, 1319–1327.

Garcia, S., Février, M., Dadaglio, G., Lecoeur, H., Rivière, Y., and Gougeon, M. L. (1997). Potential deleterious effect of anti-viral cytotoxic lymphocyte through the CD95 (FAS/APO-1)-mediated pathway during chronic HIV infection. *Immunol. Lett.* **57**, 53–58.

Godoy-Ramirez, K., Franck, K., and Gaines, H. (2000). A novel method for the simultaneous assessment of natural killer cell conjugate formation and cytotoxicity at the single cell level by multiparameter flow cytometry. *J. Immunol. Methods.* **239**, 35.

Golberg, J. E., Sherwood, S. W., and Clayberger, C. (1999). A novel method for measuring CTL and NK cytotoxicity using annexin-V and two color flow cytometry. *J. Immunol. Methods.* **224**, 1.

Goosby, C., Paniagua, M., Tallman, M., and Gartenhaus, R. B. (2005). Bcl-2 regulatory pathway is functional in chronic lymphocytic leukemia. *Cytometry B Clin. Cytom.* **63**, 36–46.

Gorczyca, W., Gong, J., and Darzynkiewicz, Z. (1993). Detection of DNA strand breaks in individual apoptotic cells by the *in situ* terminal deoxynucleotidyl transferase and nick translation assays. *Cancer Res.* **53**, 1945–1951.

Gougeon, M. L. (2003). Apoptosis as an HIV strategy to escape immune attack. *Nat. Rev. Immunol.* **3**, 392–404.

Grasl-kraup, B., Ruttkay-Nedecky, B., Koudelka, H., Bukowska, K., Bursch, W., and Schulte-Hermann, R. (1995). In situ denaturation of fragmented DNA (TUNEL assay) fails to discriminate among apoptosis, necrosis and autolytic cell death: A cautionary note. *Hepatology.* **21**, 1465–1468.

Green, D. R. (2007). Life, death, BH3 profiles, and the salmon mousse. *Cancer Cell*. **12**, 97–99.

Jonker, R. (1993). *In* "Detection of Apoptosis Using Fluorescent in Situ Nick Translation," p. 355. Springer Verlag, Berlin.

Kagi, D., Vignaux, F., Ledermann, B., Burki, K., Depraetere, V., Nagata, S., Hengartner, H., and Golstein, P. (1994). Fas and perforin pathways as major mechanisms of T cell mediated cytotoxicity. *Science* **265**, 528.

Krammer, P. H. (2000). CD95's deadly mission in the immune system. *Nature* **407**, 789–795.

Kroemer, G., Dallaporta, B., and Resche-Rigon, M. (1998). The mitochondrial death/life regulator in apoptosis and necrosis. *Annu. Rev. Physiol*. **60**, 619–642.

Lecoeur, H. (2002). Nuclear apoptosis detection by flow cytometry: Influence of endogenous endonucleases. *Exp. Cell Res*. **277**, 1–14.

Lecoeur, H., Chauvier, D., Langonné, A., Rebouillat, D., Brugg, B., Mariani, J., Edelman, L., and Jacotot, E. (2004). Fixed- and real-time cytofluorometric technologies for dynamic analysis of apoptosis in primary cortical neurons. *Apoptosis* **9**, 157–169.

Lecoeur, H., de Oliveira Pinto, L. M., and Gougeon, M. L. (2002). Multiparametric flow cytometric analysis of biochemical and functional events associated with apoptosis and oncosis using the 7-aminoactinomycin D assay. *J. Immunol. Methods* **265**, 81–86.

Lecoeur, H., Février, M., Garcia, S., Rivière, Y., and Gougeon, M. L. (2001). A novel flow cytometric assay for quantitation and multiparametric characterization of cell-mediated cytotoxicity. *J. Immunol. Methods* **253**, 177–187.

Lecoeur, H., and Gougeon, M. L. (1996). Comparative analysis of flow cytometric methods for apoptosis quantitation in murine thymocytes and human peripheral lymphocytes from controls and HIV-infected persons: Evidence for interference by granulocytes and erythrocytes. *J. Immunol. Methods* **198**, 87–99.

Lecoeur, H., Ledru, E., and Gougeon, M. L. (1998). A flowcytometric method for the simultaneous detection of both intracellular and extracellular antigens of peripheral lymphocytes. *J. Immuol. Methods* **217**, 11–26.

Lecoeur, H., Ledru, E., Prevost, M. C., and Gougeon, M. L. (1997). Strategies to phenotype apoptotic peripheral lymphocytes comparing 7-AAD, annexinV and ISNT assays. *J. Immunol. Methods* **209**, 111–123.

Lecoeur, H., Prévost, M. C., and Gougeon, M. L. (2001). Oncosis is associated to exposure of phosphatidylserine residues on the outside layer of the plasma membrane: A reconsideration of the specificity of the annexin-V/propidium iodide assay. *Cytometry*. **44**, 65–72.

Ledru, E., Lecoeur, H., Garcia, S., Roué, R., and Gougeon, M. L. (1998). Differential susceptibility to activation-induced apoptosis among peripheral Th1 subsets as revealed by single cell analysis: Correlation with Bcl-2 expression and consequences for AIDS pathogenesis. *J. Immunol*. **160**, 3194–3206.

Macho, A., Castedo, M., Marchetti, P., Aguilar, J. J., Decaudin, D., Zamzami, N., Girard, P. M., Uriel, J., and Kroemer, G. (1995). Mitochondrial dysfunctions in circulating T lymphocytes from human immunodeficiency virus-1 carriers. *Blood*. **86**, 2481–2487.

Majno, G., and Joris, I. (1995). Apoptosis, oncosis, and necrosis: An overview of cell death. *Am. J. Pathol*. **146**, 3–15.

Marguet, D., Luciani, M. F., Moynault, A., Williamson, P., and Chimini, G. (1999). Engulfment of apoptotic cells involves the redistribution of membrane phosphatidylserine on phagocyte and prey. *Nat. Cell Biol*. **1**(7), 454–456.

Martin, G., Sabido, O., Durand, P., and Levy, R. (2005). Phosphatidylserine externalization in human sperm induced by calcium ionophore A23187: Relationship with apoptosis, membrane scrambling and the acrosome reaction. *Hum. Reprod*. **20**, 3459–3568.

Modest, E. J., and Sengupta, S. K. (1974). 7-Substituted actinomycin D (NSC-3053) analogous as fluorescent DNA-binding and antitumor agents. *Cancer Chemother. Rep.* **58,** 35.

Ozgen, U., Sava.an, S., Buck, S., and Ravindranath, Y. (2000). Comparison of $DiOC_6(3)$ uptake and annexin V labeling for quantification of apoptosis in leukemia cells and non-malignant T lymphocytes from children. *Cytometry* **42,** 74–78.

Petit, P. X., Lecoeur, H., Zorn, E., Dauguet, C., Mignotte, B., and Gougeon, M. L. (1995). Alterations in mitochondrial structure and function are early events in dexamethasone-induced thymocytes apoptosis. *J. Cell Biol.* **130,** 157–167.

Reed, J. C., and Kroemer, G. (2000). Mechanisms of mitochondrial membrane permeabilization. *Cell Death Differ.* **7**(12), 1145.

Rottenberg, H., and Wu, S. (1998). Quantitative assay by flow cytometry of the mitochondrial membrane potential in intact cells. *Biochim. Biophys. Acta.* **1404,** 394–404.

Roychoudhury, R., Jay, E., and Wu, R. (1976). Terminal labeling and addition of homopolymer tracts to duplex DNA fragments by terminal deoxynucleotidyl transferase. *Nucleic Acid Res..* **3,** 101–116.

Saelens, X., Festjens, N., Walle, L. V., van Gurp, M., van Loo, G., and Vandenabeele, P. (2004). Toxic proteins released from mitochondria in cell death. *Oncogene* **23,** 2861–2874.

Samejima, K., and Earnshaw, W. C. (2005). Trashing the genome: The role of nucleases during apoptosis. *Nat. Rev. Mol. Cell Biol.* **6,** 677–688.

Schmid, I., Krall, W. J., Uittenbogaart, C. H., Braun, J., and Giorgi, J. V. (1992). Dead cell discrimination with 7-aminoactinomycin D in combination with dual color immunofluorescence in single laser flow cytometry. *Cytometry* **13,** 204.

Strasser, A., O'Connor, L., and Dixit, V. M. (2000). Apoptosis signaling. *Annu. Rev. Biochem.* **69,** 217–245.

Terasaki, M. (1989). Fluorescent labeling of endoplasmic reticulum. *Methods Cell Biol.* **29,** 125–135.

Walzer, T., Dalod, M., Robbins, S. H., Zitvogel, L., and Vivier, E. (2005). Natural-killer cells and dendritic cells: "l'union fait la force." *Blood* **106,** 2252–2258.

Waring, P., Lambert, D., Sjaarda, A., Hurne, A., and Beaver, J. (1999). Increased cell surface exposure of phosphatidylserine on propidium iodide negative thymocytes undergoing death by necrosis. *Cell Death Differ.* **6,** 624–637.

Wilson, H. A., Seligmann, B. E., and Chused, T. M. (1985). Voltage-sensitive cyanine dye fluorescence signals in lymphocytes: Plasma membrane and mitochondrial components. *J. Cell. Physiol.* **125,** 61–71.

CHAPTER FOUR

METHODS TO ANALYZE THE PALMITOYLATED CD95 HIGH MOLECULAR WEIGHT DEATH-INDUCING SIGNALING COMPLEX

Christine Feig *and* Marcus E. Peter

Contents

Abstract

Major work elucidating the structure and function of the CD95 death inducing signaling complex (DISC) was carried in the late 1990s. Since then the DISC has become a paradigm for multiprotein signaling complexes in the apoptosis literature. We analyzed the earliest events of DISC formation with a set of

The Ben May Department for Cancer Research, University of Chicago, Chicago, Illinois

Methods in Enzymology, Volume 442
ISSN 0076-6879, DOI: 10.1016/S0076-6879(08)01404-3

different techniques and found a surprising new characteristic of the CD95 DISC. Data revealed that CD95 is palmitoylated on cysteine 199. This lipid modification enhances receptor aggregation to the high molecular weight DISC.

1. INTRODUCTION

CD95 (APO-1, Fas) is, along with tumor necrosis factor (TNF)-R1 and TNF-related apoptosis-inducing ligand receptors, a member of the death receptor family, which itself belongs to the TNF-receptor/nerve growth factor receptor superfamily (Schulze-Osthoff et al., 1998). Death receptors are characterized by the presence of a conserved region in the intracellular domain, termed death domain (DD), which is crucial for the initiation of apoptotic signaling. Lacking enzymatic activity this domain serves as a docking site recruiting the adapter molecule FADD via homotypic DD–DD interactions. Once bound to CD95, FADD recruits procaspases-8/10 and the caspase-8 regulator c-FLIP into the growing receptor complex, generating the death inducing signaling complex (DISC) (Kischkel et al., 1995; Peter and Krammer, 2003). The DISC thus serves as a platform, enabling procaspase activation by enforcing their dimerization (Boatright et al., 2003; Donepudi et al., 2003). Subsequent cleavage in *trans* stabilizes and releases the heterotetrameric initiator caspases-8 and -10 into the cytosol, where they cleave downstream targets, most importantly caspase-3 (Chang et al., 2003).

In the traditional view, the DISC forms after stimulation of CD95 by its cognate ligand or agonistic antibodies at the cell surface. However, we have shown that the DISC forms during the internalization of CD95. We found that FADD, as well as caspases-8 and -10, were maximally recruited/activated when the receptor was localized to an endosomal compartment. Preventing internalization of CD95 efficiently blocked the apoptotic program, thus demonstrating a requirement for receptor internalization in executing the apoptotic signal (Lee et al., 2006).

In the past, activation of CD95 was commonly demonstrated by immunoprecipitation of the DISC and subsequent Western blotting for its major constituents: CD95, FADD, caspase-8, caspase-10, and c-FLIP (Scaffidi et al., 1999). While traditional DISC immunoprecipitation presents a snapshot of a highly dynamic complex, nondetergent-based isolation of CD95 receptosomes is a superior technology that has allowed us to study the maturation of the DISC with high spatial and temporal resolution. Second, receptor aggregation, observed as SDS- and β-mercaptoethanol-stable high molecular weight CD95 complexes (CD95[hi]) in SDS-PAGE, was demonstrated to be one of the first measurable features of CD95 activation (Kischkel *et al.*, 1995). However, their relevance for apoptotic signaling and even their mere existence had been questioned (Lee and Shacter, 2001; Legembre *et al.*, 2003). By using sucrose density centrifugation, indirect DISC immunoprecipitation through caspase-8 and investigations into the biochemical nature of CD95^{hi} we discovered that after stimulation a small percentage of total CD95 takes on an SDS-stable nature (CD95^{hi}). CD95^{hi} further aggregates to form mega dalton complexes, which also contain FADD and caspase-8. Caspase-8 is exclusively activated in the high molecular weight DISC (hiDISC). Furthermore, we demonstrated that CD95^{hi} and hiDISC formation are facilitated by palmitoylation of CD95 (Feig *et al.*, 2007). The goal of this chapter is to give detailed technical guidance to those wishing to study the hiDISC.

2. USE OF SUCROSE DENSITY GRADIENTS FOR ANALYSIS OF HiDISC

We applied continuous sucrose density gradient centrifugation to determine the size of the native CD95 DISC. During the ultracentrifugation process, proteins and protein complexes are kept in a native conformation and are separated according to their Stokes radii (Fig. 4.1), which is a direct consequence of their molecular weight and size. Afterward, the individual fractions are prepared for SDS-PAGE, which results in the disruption of noncovalent bonds. Therefore, the resulting Western blot image has to be interpreted in two dimensions (Fig. 4.1C). First, as the sedimentation occurs under nonreducing conditions, one can discern the molecular weight of a native protein/protein complex in the first dimension, where the molecular weight decreases with ascending fraction number. Second, because of the reducing environment during the SDS-PAGE disulfide bonds as well as noncovalent interactions are disrupted and protein complexes dissociate into their individual subunits. Therefore, the apparent molecular weight of complex subunits and individual proteins is determined in the second dimension.

2.1. Equipment

Hoefer SG15 gradient maker (Amersham)
Magnetic stir plate with miniflea
Peristaltic low flow pump (Fisher) with tubing size I.D.: 3/32"

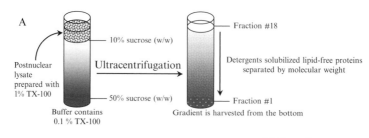

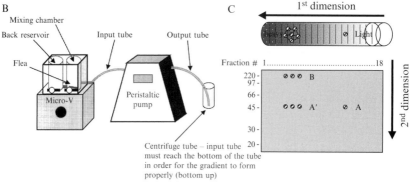

Figure 4.1   Illustration of continuous sucrose gradient centrifugation. (A) Schematic of gradient setup. The continuous gradient from 10 to 50% sucrose (w/w) was prepared using a gradient mixer. The protein mixture is carefully overlaid on top of the gradient, and 18 fractions are taken from the bottom up after ultracentrifugation. (B) Scheme illustrating the equipment setup for the sucrose density gradient preparation. (C) Hypothetical Western blot result. The sucrose gradient analysis contains information in two dimensions. The first dimension results from the separation of small and large native proteins/protein complexes during ultracentrifugation. A larger heavier protein will sediment into a denser gradient area and thus appear in a lower fraction number than a small protein. The second dimension results from resolution of the individual fractions on SDS-PAGE, providing information about the apparent molecular mass of a protein or protein complex subunit. Several scenarios are imaginable. For example, protein A has an apparent molecular mass of 45 kDa, but it sediments into two major gradient regions. One in the lighter fractions corresponds to the molecular mass of the monomeric protein (A) and one in the much denser fractions, indicating that it is aggregated either with itself or with other proteins (A′/B). If this larger complex is aggregated by noncovalent bonds, it will dissociate during SDS-PAGE and thus appear with an apparent molecular mass of the individual subunits of 45 kDa (scenario A′). If protein A is part of a complex with covalent bonds, or in a way aggregated so that SDS cannot access this complex, it will shift up in the second dimension and have an apparent molecular mass corresponding to the native size of the complex (scenario B).

Beckman centrifuge with rotor SW55Ti plus polyallomer tubes
BD spinal needle 18G × 3-1/2 (ref# 405184)
Transparent silicon tubing, I.D. 3/32″
Eppendorf centrifuge

2.2. Reagents

HMW calibration kit for native electrophoresis (Amersham)
Nitrocellulose membrane "Hybond C" (Amersham)
ECL (Amersham)
SuperSignal Dura West extended ECL (Pierce)
Coomassie GelCode Blue reagent (Pierce)

2.3. Buffers

Cell lysis buffer (1% Triton X-100): 30 mM Tris-HCl (pH 7.5), 150 mM
   NaCl, 2.5 mM EDTA, 10% (v/v) glycerol, 1% (v/v) Triton X-100,
   1 mM phenylmethylsulfonyl fluoride (PMSF), protease inhibitor
   cocktail tablet (Roche)

Buffer for sucrose solutions (0.1% Triton X-100): 30 mM Tris-HCl (pH
   7.5), 150 mM NaCl, 2.5 mM EDTA, 10% (v/v) glycerol, 0.1% (v/v)
   Triton X-100, 1 mM PMSF, protease inhibitor cocktail tablet (Roche)

50% (w/w) sucrose: 50 g sucrose dissolved in 50 ml 0.1% Triton X-100
   buffer

10% (w/w) sucrose: 10 g sucrose dissolved in 90 ml 0.1% Triton X-100
   buffer

5× reducing sample buffer (RSB): 50 mM Tris-HCl (pH 7.5), 50% (v/v)
   glycerol, 10% (w/v) SDS, 12.5% (v/v) β-mercaptoethanol, 0.025%
   (w/v) bromophenol blue

Buffers for SDS-PAGE

Separating gel: 375 mM Tris-HCl (pH 8.8), 12% (v/v) acrylamide/bis
   37.5:1(2.6%C) (Bio-Rad), 0.1% (w/v) SDS, 0.05% (w/v) ammonium
   persulfate (APS), 0.2% (v/v) N,N,N,N'-tetramethylethylenediamine
   (TEMED)

SDS-PAGE stacking gel: 127 mM Tris-HCl (pH 6.8), 4.6% (v/v) acryl-
   amide/bis 37.5:1(2.6%C), (0.1% (w/v) SDS, 0.01% (w/v) APS, 0.13%
   (v/v) TEMED

SDS-PAGE running buffer: 25 mM Tris base, 192 mM glycine, 0.1%
   (w/v) SDS

Gel fixing buffer: 10% (v/v) glacial acetic acid, 20% (v/v) methanol

Protein transfer buffer (for Bio-Rad wet transfer systems): 25 mM Tris
   base, 190 mM glycine, 20% (v/v) methanol

Ponceau solution: 0.1% (w/v) Ponceau S, 5% (v/v) glacial acetic acid

1× phosphate-buffered saline (PBS; pH 7.5): 4.3 mM Na$_2$HPO$_4$ 7 H$_2$O,
   1.5 mM KH$_2$PO$_4$, 2.7 mM KCl, 137 mM NaCl

1× PBS/T: 0.05% (v/v) Tween 20 in 1× PBS

Western blot blocking buffer: 5% (w/v) dry nonfat milk in 1× PBS/T

2.4. Antibodies and ligands (for all methods)

Antigen (clone)	Species/ isotype	Source/reference	Dilution in WB
Actin	Mouse IgG1	Sigma	1:50,000
Caspase-10	Mouse IgG1	MBL	1:1000
Caspase-8 (C15)	Mouse IgG2b	Scaffidi et al. (1997)/Alexis	1:70
Caspase-8 (N2)	Mouse IgG1	Scaffidi et al. (1997)/Alexis	Undiluted
CD95 (anti-APO-1)	Mouse IgG3	Trauth et al. (1989)	
CD95 (C20)	Rabbit	Santa Cruz	1:1000
CD95 (3D5)	Mouse IgG1	Alexis Biochemicals	1:500
FADD	Mouse IgG1	BD Pharmingen	1:500
Flag M2	Mouse IgG1	Sigma	
c-FLIP (NF6)	Mouse IgG1	Scaffidi et al. (1999)/Alexis	1:5
Antimouse IgG1	Goat	Southern Biotech	
Antimouse IgG2b	Goat	Southern Biotech	
Antimouse IgG	Goat	Santa Cruz	
Antirabbit	Goat	Southern Biotech	
Leucine zipper- tagged CD95 ligand		Walczak et al. (1997)	

2.5. Method

2.5.1. Gradient setup

The pump is cleaned with deionized water for about 5 min. Meanwhile, 2.3 ml 50% sucrose is added to the back reservoir of the gradient mixer while all stopcocks remain closed. The stopcock, which connects the back reservoir to the mixing chamber, is opened briefly to fill the tube between both chambers with 50% sucrose, not allowing any solution to enter the mixing chamber. Now the mixing chamber, already containing a stir bar, is filled with an equal volume (2.3 ml) of 10% sucrose. The input tube is connected to the peristaltic pump, and a spinal needle is placed on the end of the output tube so that it touches the center at the bottom of a centrifuge tube. The magnetic stirrer is turned on, and now the anterior stopcock connecting the mixing chamber to the input tube is opened. Finally, the pump is started at a flow rate of 0.6 ml/min (setting "slow 3" on Fisher model). Once the liquid front (at this point only 10% sucrose) has reached an arbitrarily designated spot on the input tube, the stopcock connecting the back reservoir and the mixing chamber is opened to begin mixing both solutions. The arbitrarily designated spot should be marked on the silicone tubing to ensure consistency in the gradient setup. The pump speed can now be increased to 0.85 ml/min ("slow 7" on Fisher model). Toward the end when the sucrose solutions are almost drained, the pump speed is lowered to 0.6 ml/min, and the gradient mixer is tilted slightly to allow all liquid to flow into the input tube. The pump is stopped once the first air bubble reaches an arbitrarily designated spot on the output tube. Slowly and steadily the needle is pulled out, and the centrifuge tube is transferred to ice to cool for about 20 min. The pump is washed with deionized water between each gradient setup for 5 min. It is of paramount importance that no air bubbles form during the entire procedure, as this will disrupt the gradient.

2.5.2. Postnuclear lysate preparation

Depending on the cell size, 2×10^7 SKW6.4 or 1×10^7 MCF7 (FB) cells are used for each stimulation point. The cells are either left untreated or stimulated with 1 μg/ml anti-APO-1 [unconjugated, biotinylated as described (Lee *et al.*, 2006)] or 200 ng/ml LzCD95 ligand (LzCD95L), respectively. To stop the stimulation, ice-cold PBS is added, and the cell pellet is washed once more with ice-cold PBS. The cells are lysed in a small volume (300 μl) of 1% Triton X-100 lysis buffer for 10 min on ice. Nuclei and other cellular debris are removed by a 10-min centrifugation at 14,000 rpm in a cold tabletop centrifuge. Of the postnuclear lysate, 270 μl is carefully loaded onto a prechilled sucrose gradient prepared as described earlier. Therefore, it is recommended that the tip of a 1-ml pipette touches the side of the centrifuge tube just above the gradient surface level. The lysate should be ejected in a very slow but continuous flow. The cell lysate and the gradient do *not* mix. A clear interphase line should be visible. Loading the gradient can be practiced easily by overlaying with lysis buffer containing a dye. All the following steps should be carried out with ultimate care to not disturb the gradient. The residual 30 μl of lysate is kept as a loading control, mixed with 7.5 μl 5× RSB and boiled for 5 min.

2.5.3. Ultracentrifugation

The centrifuge tubes are placed in the Beckman SW55Ti rotor adaptors and spun for 16 h at 37,500 rpm at 4°.

2.5.4. Harvesting fractions

Eighteen fractions of 250 μl are harvested starting at the bottom of the gradient. As for the gradient setup, air bubbles are disastrous in this step! A new spinal needle is connected to the input tube of the pump and slowly inserted into the gradient without disturbing it. The pump is turned on at a flow rate of 0.6 ml/min, and for each fraction 11 drops are collected at the output tube. For SDS-PAGE 100 μl of each fraction is mixed with 25 μl of 5× RSB and boiled for 5 min. The total volume of each sample is loaded onto a 12% SDS gel. Because of the high percentage of sucrose in the protein samples, it is not recommended to perform the electrophoresis at high voltage. It is recommended to run the gel overnight at 60 to 80 V. The pump is rinsed with deionized water for 5 min between each gradient harvest.

2.5.5. Protein transfer and Western blot

Proteins are transferred onto a nitrocellulose membrane using the Bio-Rad wet transfer system. The transfer success can be verified quickly by immersing the membrane in Ponceau S solution. Membranes are blocked for 1 h at room temperature in blocking buffer, washed twice in PBS/T, and probed with the indicated primary antibodies in PBS/T for 1 h at room temperature (exception: anticaspase-8 N2: overnight at 4°). After three washes in PBS/T, membranes are incubated with horseradish peroxide-labeled secondary antibodies diluted in blocking buffer for 1 h at room temperature. The membranes are washed again three times with PBS/T, and proteins are visualized by enhanced chemiluminescence (Amersham or, for antibodies that give a weaker signal, e.g., anticaspase-8 N2: SuperSignal Dura West extended ECL, Pierce).

2.5.6. Gradient calibration

The gradient will first need to be calibrated. One aliquot of lyophilized proteins (Amershams HMW calibration kit for native electrophoresis) is rehydrated in 300 μl 1% Triton X-100 lysis buffer, and 270 μl is loaded onto a continuous gradient. One hundred microliters of each fraction is added to 25 μl 5× RSB, boiled for 5 min, and analyzed on 12% SDS-PAGE. After electrophoresis the gel is directly fixed, washed three times with deionized water, and soaked in Coomassie staining solution for 20 min. Staining is enhanced by destaining in deionized water overnight, after which the gel is dried (see Feig et al., 2007).

To evaluate calibration, the molecular weight of each protein is then plotted against the sucrose concentration corresponding to the peak elution fraction. Note that because the sucrose solutions are prepared as w/w the molecular weight increases exponentially. The linear regression curve shown in Feig et al. (2007) indicates that the gradient can separate protein/protein complexes in the range of 10 kDa (fraction #18) to about 7000 kDa (fraction #1).

2.6. Merits of the technique and conclusions

The major conclusions drawn from the sucrose gradient analysis of the CD95 DISC are as follow.

1. The CD95 DISC aggregates to high molecular weight complexes (hiDISC).
2. The hiDISC contains a single species of SDS-stable receptor aggregates, the 180-kDa CD95hi.
3. Caspase-8 is exclusively activated in the hiDISC. Therefore, only CD95hi is the active receptor species, even though it represents a small percentage of total CD95.
4. The hiDISC is a highly dynamic complex. It actively evolves after receptor stimulation, gaining in size and caspase-8 processing capability over time.
5. The hiDISC forms autonomously after receptor stimulation. The stimulating antibody does not contribute to the complex.

Thus, the analysis of receptor complexes via sucrose density gradient centrifugation revealed an unexpected property of the DISC, aggregation to the megadalton hiDISC (Fig. 4.2). Furthermore, it may offer a resolution to the conflicting data on CD95hi found in the literature. Although some studies indicated a possible involvement of CD95hi in the apoptotic process (Kamitani et al., 1997; Kischkel et al., 1995), others questioned this notion (Lee and Shacter, 2001; Legembre et al., 2003). Several SDS-stable CD95 species (97, 180, and 200 kDa) exist and cannot be sufficiently separated on traditional SDS-PAGE. Using sucrose gradients we could demonstrate that only one, the 180-kDa CD95hi species, constitutes the hiDISC. Therefore, we encourage the careful interpretation of data generated by one-dimensional SDS-PAGE. This technique is simply not sufficient to reveal the true nature of the DISC and the contribution of CD95hi to apoptotic signaling. Sucrose density gradient centrifugation allowed us to clearly separate CD95 receptor complexes triggering the apoptotic signaling cascade from uninvolved bystanders.

3. Association of CD95HI with Active Caspase-8 (Reverse DISC-IP)

While a classical DISC-IP is a powerful technique to query the activation of CD95 in general obtaining a yes/no answer, it does not provide detailed information about the activation status of the receptor. As illustrated in Fig. 4.3A, the immunoprecipitation of CD95 will result in the isolation of fully activated receptor complexes, as well as partially activated ones. However, we asked whether isolation of caspase-8 would

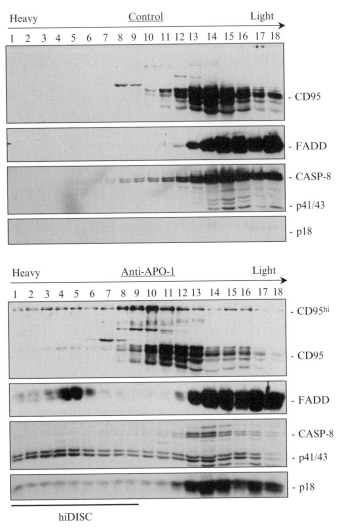

Figure 4.2 Analysis of the hiDISC. Sucrose density gradient centrifugation. SKW6.4 cells (2×10^7) were left untreated (top) or stimulated with 1 μg/ml anti-APO-1 for 120 min (bottom). Postnuclear lysates were separated on a continuous sucrose gradient ranging from 10 to 50% (w/w) sucrose. After ultracentrifugation, 18 fractions were harvested and analyzed on a 12% SDS-PAGE. hiDISC formation is observed in the high-density fractions and contains CD95hi and FADD, as well as the caspase-8 prodomain.

tell us more about the active receptor complexes. This experimental approach is based on the fact that caspase-8 is only recruited to the CD95 after its stimulation. For instance, what is the precise participation of CD95 versus CD95hi in the complex, and what percentage of total cellular CD95 does in fact become activated?

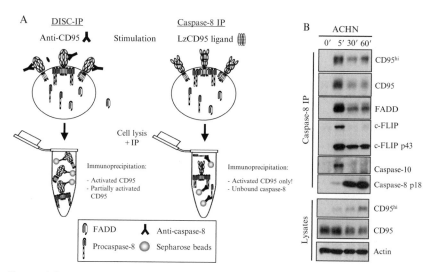

Figure 4.3 Caspase-8 immunoprecipitation offers a new look at the CD95-DISC. (A) Comparison of classical DISC-IP and inverse (caspase-8-dependent) DISC-IP. While a classical DISC-IP results in the precipitation of stimulated, but also partially and possibly nonstimulated receptor complexes, caspase-8 immunoprecipitation offers the advantage of solely isolating activated receptor complexes. CD95 is immunoprecipitated indirectly isolating an unmanipulated receptor. Moreover, in this experimental setting, CD95 is stimulated with a recombinant ligand, more closely mimicking the stimulation in a physiological context. (B) Example of caspase-8 immunoprecipitation. ACHN cells were stimulated with 200 ng/ml LzCD95L for the indicated periods of time. Postnuclear lysates were subjected to immunoprecipitation with 10 μg anticaspase-8 C15 antibody/sample and analyzed for the association with the DISC components CD95, FADD, c-FLIP$_L$, and caspases-8/-10, as well as β-actin.

3.1. Reagents

Protein A-Sepharose as 50% slurry in PBS (Sigma-Aldrich)
Buffers, antibodies: see Sections 2.3 and 2.4

3.2. Method

3.2.1. Cell stimulation, lysis, and immunoprecipitation

SKW6.4/H9/ACHN cells (2×10^7) per data point are stimulated for up to 60 min with 200 ng/ml LzCD95L. Postnuclear lysates are generated as described in Section 2.5.2. Briefly, cells are washed once in ice-cold PBS at the end of the stimulation kinetics and resuspended in 1 ml 1% Triton X-100 lysis buffer. Cell lysis is carried out for 10 min on ice, and the lysates are cleared of debris by centrifugation at 14,000 rpm for 10 min in a cold Eppendorf centrifuge. Thirty microliters of the supernatant is kept as a control and prepared for SDS-PAGE by adding 7.5 μl 5 × RSB and boiled for 5 min. The residual lysate is subjected to caspase-8 immunoprecipitation with

10 μg anticaspase-8 C15 antibody and 30 μl protein A–Sepharose. After rotating incubation for 2 h at 4°, the Sepharose beads are pelleted by centrifugation for 1 min at 6,000 rpm and washed four times with 1 ml 1% Triton X-100 lysis buffer each. After the last wash, the wash buffer is completely aspirated with a 27.5-gauge needle fitted to a vacuum pump, and the Sepharose pellet is resuspended quickly in 50 μl 1× RSB. Samples are boiled for 5 min and either frozen at −20° or analyzed directly on a 12% SDS-PAGE.

3.2.2. SDS-PAGE and Western blot
Please refer to Section 2.5.

3.3. Merits of the technique and conclusions

Caspase-8 immunoprecipitation offers a few advantages over a DISC-IP (Fig. 4.3A). Instead of relying on stimulation with an agonistic antibody, which has been suggested to cause artifactual aggregation of CD95 (Legembre et al., 2003), cells are stimulated with a recombinant ligand preparation. This mimics more closely the stimulation occurring in a physiological context. Second, CD95 is precipitated only indirectly and therefore not aggregated secondarily by antibody cross-linking. It is the DISC in an unmanipulated state. Finally, as caspase-8 is recruited to CD95 only after receptor activation its immunoprecipitation will only isolate receptors that have fully assembled the DISC and are committed to apoptosis. This is in contrast to the DISC-IP, in which all antibody-bound receptors are isolated regardless of whether the apoptotic signaling cascade has been initiated.

Figure 4.3B shows a typical result of caspase-8 immunoprecipitation. None of the DISC components are isolated in association with caspase-8 from unstimulated cells, which verifies (a) the specificity of this method for DISC isolation and (b) proves the assumption on which this technique is based on.

With this method we have gained new insight into the CD95 DISC. Namely, we showed that CD95^{hi} is the activated receptor species to which caspase-8 binds preferentially and where caspase activation occurs. This was further supported by the sucrose density gradient studies of the hiDISC.

4. Role of Palmitoylation for CD95 Signaling

Palmitoylation (chemical terminology: S-acylation) is the covalent attachment of a 16-carbon fatty acid (palmitate) to a cysteine residue via a thioester bond, changing the hydrophobicity of the target protein. In most cases, this results in localization of the modified protein to cellular membranes (e.g., Golgi, endoplasmic reticulum, plasma membrane). However, palmitoylation can also occur on transmembrane proteins, which often results in the recruitment into lipid rafts. Although no general consensus motif exists, there are a few sequence guidelines predicting palmitoylation of transmembrane proteins. Transmembrane proteins are primarily palmitoylated when the cysteine residue resides within the 10 intracellular juxtamembrane amino acids, preferably when surrounded by basic amino acids. Palmitoylation is unique among the lipid modifications as it is reversible, suggesting that it could regulate protein function in a stimulation-dependent manner similar to phosphorylation (Resh, 2006).

To this day, the only proof for palmitoylation of a protein is either through *in vivo* metabolic labeling with ^{3}[H]palmitic acid or acyl-biotinyl-exchange chemistry followed by mass spectrometry, which are very time-consuming or require expensive technology (Wan *et al.*, 2007). The major obstacle to developing alternative techniques is that the precise biochemical mechanism of palmitoylation *in vivo* is debated. Attempts to identify bona fide palmitoyl-acyl transferases (PAT) have proven difficult because of (a) the instability of the isolated activity and (b) the lack of a consensus palmitoylation sequence in target proteins. Furthermore, the existence of PATs has been questioned in general because spontaneous palmitoylation occurs in cell-free *in vitro* assays, albeit with a slower kinetics (Duncan and Gilman, 1996; Veit, 2000).

To investigate the palmitoylation of CD95, a protocol was developed based on the instructions on labeling cells with [^{3}H]palmitic acid given in Jones (2004).

4.1. Reagents

2-Bromohexadecanoic (palmitic) acid (2-BrPA) (Sigma-Aldrich): 300 mM
    stock in dimethyl sulfoxide (DMSO), aliquoted and stored at $-20°$
[9,10(n)^{3}H]palmitic acid (specific activity 50 Ci/mmol), Amersham
Amplify (Amersham)
Serum-free cell culture medium

ProteinA/G-Sepharose (Sigma-Aldrich, protein G-Sepharose)

Buffers, antibodies: see previous sections

4.2. Method

4.2.1. *In vivo* metabolic labeling with [9,10(n)^{3}H]palmitic acid

4.2.1.1. Preparation of enriched [^{3}H]palmitic acid [^{3}H]Palmitic acid is provided as an ethanol solution at a concentration of 1 mCi/ml but is used at 0.25 to 0.5 mCi/ml for *in vivo* metabolic labeling. Because ethanol is toxic to cells it has to be removed by concentrating the solution in a vacuum centrifuge to a volume equivalent to <0.3% (v/v) of the final labeling medium. To enhance uptake into the cells, DMSO is added to the concentrated [³H]palmitic acid to obtain a final concentration of 1% (v//v) in the labeling medium.

For example, 1 mCi is sufficient to label cells in a volume of 4 ml, which will result in a concentration of 0.25 mCi/ml. Therefore, 1 ml of [³H] palmitic is concentrated to 12 μl, 40 μl DMSO is added, and the residual volume, up to 4 ml, is supplied by serum-free cell culture medium.

4.2.1.2. In vivo *metabolic labeling, immunoprecipitation, and fluorography*

For labeling of exogenous CD95, [9,10(n)³H]palmitic acid is used at 0.25 mCi/ml. 293T cells (1 × 10⁶) are seeded on 10-cm dishes 2 days prior to transfection. They are transfected with 3 μg pcDNA3, pcDNA3-Flag-CD95 wt, and pcDNA3-Flag-CD95 C199S using the calcium phosphate method. Twenty hours after transfection, cells are harvested, washed in serum-free Dulbecco's modified Eagle medium (DMEM), starved for 1 h in serum-free DMEM, and labeled for 1 h in 570 μl (each sample) 0.25 mCi/ml [9,10(n)³H]palmitic acid provided in serum-free DMEM. If inhibition of palmitoylation is desired simultaneously, 100 μM 2-BrPA is added for the entire duration of the labeling including starvation (for details, see later). Cells are washed once in ice-cold PBS and lysed in 1 ml 1% Triton X-100 lysis buffer. Flag-tagged CD95 is immunoprecipitated from the postnuclear lysates with 10 μg anti-Flag M2 and 30 μl protein G-Sepharose (IP for 2 h at 4° with constant rotation). As protein G-Sepharose is provided as an ethanol solution, it is washed three times with PBS prior to usage. Following immunoprecipitation, the Sepharose pellet is washed four times with 1 ml 1% Triton X-100 lysis buffer, dried quickly after the last wash, and resuspended in 50 μl 1× RSB. It is crucial at this step to be precise in drying the individual pellets equally and adding exactly 50 μl reducing sample buffer, as 10% of the precipitate is used for confirmation of successful immunoprecipitation of Flag-tagged CD95 in Western blot (detection with anti-CD95 C20 antibody) while the residual 90% is loaded onto a separate 10% SDS-PAGE for fluorography. Differences in volume would undermine the significance of the experiment. The gel intended for fluorography is fixed for 30 min, immersed directly in "Amplify" for 30 min, dried, and exposed to a Kodak film at −80°.

Considering the lower expression levels of endogenous CD95 in cells (when compared to exogenous CD95), [9,10(n)³H]palmitic acid is used at 0.5 mCi/ml to increase the efficiency of labeling endogenous protein. SKW6.4 (4 × 10⁷) are washed once with serum-free RPMI and starved for 1 h in 7 ml serum-free RPMI. Metabolic labeling is carried out in 1.2 ml 0.5 mCi/ml [9,10(n)³H]palmitic acid (in serum-free RPMI) for 3 h. Cells are washed once in 1 ml ice-cold PBS and lysed in 1 ml 1% Triton X-100 lysis buffer. After removal of cellular debris, the postnuclear lysates are precleared with 8 μg normal rabbit IgG and 30 μl protein A–Sepharose (IP for 2 h at 4° with constant rotation). Sepharose beads are sedimented by centrifugation at 6000 rpm, and the precleared lysate (supernatant) is removed with a Hamilton pipette. This supernatant is subsequently subjected to CD95 immunoprecipitation with 4 μg anti-CD95 (C20), while the Sepharose pellet from the preclear is washed four times and prepared for SDS-PAGE as described earlier. After immunoprecipitation of CD95, the Sepharose pellet is treated in the same manner as described earlier. Both preclear and CD95-IP are analyzed on a 12% SDS-PAGE. Again, 10% of each IP is used for Western blot (the anti-CD95 3D5 clone is used for detection of CD95) while the residual 90% is subjected to fluorography as described previously.

Both serum starvation and labeling in a small culture volume cause stress to the cells. Therefore, incubation times should be kept to a minimum. This needs to be determined individually. Using more cells will give a stronger radioactive signal, but that may affect the viability of the cells during the metabolic labeling. Depending on the number of cells used, the abundance of the protein (exogenous versus endogenous), and the degree of palmitoylation, the exposure times for fluorography will vary substantially (between weeks and months).

4.2.2. Inhibition of palmitoylation with 2-bromopalmitic acid (2-BrPA)

Because cells are differentially sensitive to 2-BrPA, the optimal concentration, as well as treatment times, has to be determined by titration for every cell line. While (Chakrabandhu *et al.*, 2007) successfully inhibited palmitoylation with 300 μM, this amount was toxic to SKW6.4 and H9 cells, respectively. A concentration of 100 μM was sufficient to inhibit palmitoylation in our experiments.

2-Bromopalmitic acid is a white powder and can be dissolved in either ethanol or DMSO. A 300 mM stock is prepared in DMSO and aliquoted, and each aliquot is thawed only once. Solubilization of 2-BrPA in culture medium can be a bit tricky. The DMSO stock is first diluted 1:100 in 37° culture medium in an Eppendorf tube. To be completely dissolved it will need to be vortexed vigorously and kept at 37° consistently. This culture medium stock is then further diluted with warm culture medium. As opposed to *in vivo* metabolic labeling, there is no need to use serum-free medium.

5. CONCLUSIONS

Figure 4.4 contains an example illustrating the palmitoylation of CD95 on cysteine 199. This *in vivo* metabolic labeling experiment conducted in 293T cells expressing Flag-tagged CD95 wt or Cys199→Ser (CS)

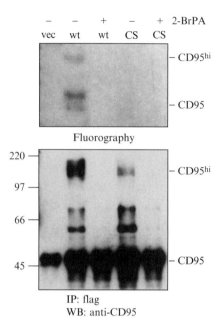

Figure 4.4 Palmitoylation of CD95. CD95 is palmitoylated on cysteine 199. 293T cells were transfected with empty vector, Flag-tagged CD95 wt and C199S mutant, respectively. Cells were labeled metabolically with [9,10(n)³H]palmitic acid, and CD95 was immunoprecipitated with anti-Flag M2. 2-BrPA was added during the labeling procedure at 100 μM.

mutant demonstrates that only wild-type CD95 can be labeled with radioactive palmitate. This incorporation can be prevented by simultaneous exposure to the palmitoylation inhibitor 2-BrPA. However, 2-BrPA did not have an effect on the CS CD95 mutant, which did not incorporate [9,10(n)³H] palmitic acid in the first place. Furthermore, it shows that monomeric as well as SDS-stable CD95 (CD95^{hi}) are palmitoylated.

We and others have shown that palmitoylation affects CD95 signaling such that a deficiency in palmitoylation impairs apoptotic signaling (Chakrabandhu et al., 2007; Feig et al., 2007). We demonstrated that this is because of a block in CD95^{hi} and hiDISC formation. The corresponding Western blot in Fig. 4.4 supports this statement. While CD95^{hi} is abundant in wild-type CD95 expressing cells, it is strongly reduced in the palmitoylation-deficient CS mutant. Furthermore, treatment with the palmitoylation inhibitor completely abrogated the formation of CD95^{hi} in both wild-type and CS CD95.

ACKNOWLEDGMENT

The work was supported by the National Institutes of Health (R01 CA93519).

REFERENCES

Boatright, K. M., Renatus, M., Scott, F. L., Sperandio, S., Shin, H., Pedersen, I. M., Ricci, J. E., Edris, W. A., Sutherlin, D. P., Green, D. R., and Salvesen, G. S. (2003). A unified model for apical caspase activation. *Mol. Cell* **11**, 529–541.

Chakrabandhu, K., Herincs, Z., Huault, S., Dost, B., Peng, L., Conchonaud, F., Marguet, D., He, H. T., and Hueber, A. O. (2007). Palmitoylation is required for efficient Fas cell death signaling. *EMBO J.* **26**, 209–220.

Chang, D. W., Xing, Z., Capacio, V. L., Peter, M. E., and Yang, X. (2003). Interdimer processing mechanism of procaspase-8 activation. *EMBO J.* **22**, 4132–4142.

Donepudi, M., Mac Sweeney, A., Briand, C., and Grutter, M. G. (2003). Insights into the regulatory mechanism for caspase-8 activation. *Mol. Cell* **11**, 543–549.

Duncan, J. A., and Gilman, A. G. (1996). Autoacylation of G protein alpha subunits. *J. Biol. Chem.* **271**, 23594–23600.

Feig, C., Tchikov, V., Schutze, S., and Peter, M. E. (2007). Palmitoylation of CD95 facilitates formation of SDS-stable receptor aggregates that initiate apoptosis signaling. *EMBO J.* **26**, 221–231.

Jones, T. L. (2004). Role of palmitoylation in RGS protein function. *Methods Enzymol.* **389**, 33–55.

Kamitani, T., Nguyen, H. P., and Yeh, E. T. (1997). Activation-induced aggregation and processing of the human Fas antigen: Detection with cytoplasmic domain-specific antibodies. *J. Biol. Chem.* **272**, 22307–22314.

Kischkel, F. C., Hellbardt, S., Behrmann, I., Germer, M., Pawlita, M., Krammer, P. H., and Peter, M. E. (1995). Cytotoxicity-dependent APO-1 (Fas/CD95)-associated proteins form a death-inducing signaling complex (DISC) with the receptor. *EMBO J.* **14**, 5579–5588.

Lee, K. H., Feig, C., Tchikov, V., Schickel, R., Hallas, C., Schutze, S., Peter, M. E., and Chan, A. C. (2006). The role of receptor internalization in CD95 signaling. *EMBO J.* **25**, 1009–1023.

Lee, Y., and Shacter, E. (2001). Fas aggregation does not correlate with Fas-mediated apoptosis. *J. Immunol.* **167**, 82–89.

Legembre, P., Beneteau, M., Daburon, S., Moreau, J. F., and Taupin, J. L. (2003). Cutting edge: SDS-stable Fas microaggregates: An early event of Fas activation occurring with agonistic anti-Fas antibody but not with Fas ligand. *J. Immunol.* **171**, 5659–5662.

Peter, M. E., and Krammer, P. H. (2003). The CD95(APO-1/Fas) DISC and beyond. *Cell Death Differ.* **10**, 26–35.

Resh, M. D. (2006). Palmitoylation of ligands, receptors, and intracellular signaling molecules. *Sci. STKE* **2006**, re14.

Scaffidi, C., Medema, J. P., Krammer, P. H., and Peter, M. E. (1997). FLICE is predominantly expressed as two functionally active isoforms, caspase-8/a and caspase-8/b. *J. Biol. Chem.* **272**, 26953–26958.

Scaffidi, C., Schmitz, I., Krammer, P. H., and Peter, M. E. (1999). The role of c-FLIP in modulation of CD95-induced apoptosis. *J. Biol. Chem.* **274**, 1541–1548.

Schulze-Osthoff, K., Ferrari, D., Los, M., Wesselborg, S., and Peter, M. E. (1998). Apoptosis signaling by death receptors. *Eur. J. Biochem.* **254**, 439–459.

Trauth, B. C., Klas, C., Peters, A. M., Matzku, S., Moller, P., Falk, W., Debatin, K. M., and Krammer, P. H. (1989). Monoclonal antibody-mediated tumor regression by induction of apoptosis. *Science* **245**, 301–305.

Veit, M. (2000). Palmitoylation of the 25-kDa synaptosomal protein (SNAP-25) *in vitro* occurs in the absence of an enzyme, but is stimulated by binding to syntaxin. *Biochem. J.* **345**(Pt 1), 145–151.

Walczak, H., Degli-Esposti, M. A., Johnson, R. S., Smolak, P. J., Waugh, J. Y., Boiani, N., Timour, M. S., Gerhart, M. J., Schooley, K. A., Smith, C. A., Goodwin, R. G., and Rauch, C. T. (1997). TRAIL-R2: A novel apoptosis-mediating receptor for TRAIL. *EMBO J.* **16,** 5386–5397.

Wan, J., Roth, A. F., Bailey, A. O., and Davis, N. G. (2007). Palmitoylated proteins: Purification and identification. *Nat. Protoc.* **2,** 1573–1584.

CHAPTER FIVE

Immunomagnetic Isolation of Tumor Necrosis Factor Receptosomes

Vladimir Tchikov *and* Stefan Schütze

Contents

Abstract

Internalized tumor necrosis factor (TNF) receptor-1 (TNF-R1) recruits the adaptor proteins TRADD and FADD, as well as caspase-8, to establish the "death-inducing signaling complex" (DISC). DISC formation and apoptosis depend strictly on TNF-R1 internalization, whereas recruitment of TRAF-2 and RIP-1 to signal for NF-κB activation occurs from TNF-R1 at the cell surface. Findings revealed that

Institute of Immunology, University of Kiel, Kiel, German

Methods in Enzymology, Volume 442
ISSN 0076-6879, DOI: 10.1016/S0076–6879(08)01405-5

TNF-R1 establishes divergent TNF signaling pathways depending on compartmentalization of TNF-R1 to the plasma membrane or to plasma membrane-derived endocytic vesicles harboring the TNF-R1-associated DISC. These data were obtained by a novel technique for the isolation of morphologically intact endocytic vesicles containing magnetically labeled TNF-R1 complexes (termed TNF receptosomes) using a custom-made high gradient magnetic chamber. This chapter describes the protocol of immunomagnetic labeling using biologically active biotin TNF as a ligand coupled to magnetic streptavidin nanobeads, followed by a gentle mechanical homogenization procedure to preserve the morphological structure of membrane vesicles containing activated TNF-R1 complexes. Isolation of the magnetized receptosomes in a high magnetic gradient is described, and the kinetics of TNF-R1 internalization and endosomal trafficking/maturation of the receptosomes is characterized. Using a biotinylated anti-CD95 antibody as ligand and streptavidin-coated magnetic nanobeads for separation in the high gradient magnetic chamber, the immunomagnetic separation approach was additionally applied to characterize the internalization and maturation of CD95 receptosomes.

1. INTRODUCTION

Tumor necrosis factor (TNF) is a highly pleiotropic cytokine that elicits diverse cellular responses ranging from proliferation and differentiation to activation of apoptosis (Locksley et al., 2001; Wajant et al., 2003). The different biological activities are mediated by two distinct cell surface receptors: TNF receptor-1 (TNF-R1) and TNF receptor-2 (TNF-R2). TNF-R1 appears to be the key mediator of TNF signaling.

Elements of the signaling pathways of TNF-R1-mediated apoptosis (the TNF-R1 adapter proteins TRADD and FADD, as well as caspase-8) and of NF-κB activation (TRADD, RIP-1, TRAFs) are well defined. However, the molecular mechanisms that regulate formation of the initial signaling complexes at the activated TNF-R1 to selectively transmit specific signal transduction events were poorly understood. Using a pharmacological approach by inhibiting TNF-R1 endocytosis, it was shown that selected TNF-R1 death domain signaling pathways, including those leading to apoptosis, were dependent on TNF receptor internalization, while others were not (Schütze et al., 1999). These findings suggested a regulatory role of TNF-R1 compartmentalization for selective internalization-dependent (proapoptotic) and internalization-independent (mitogenic, proinflammatory) signaling. Based on these observations, the endosomal compartment was recognized as a novel signaling organelle involved in selectively transmitting death signals from TNF-R1.

2. COMPARTMENTALIZATION OF TNF-R1 APOPTOSIS SIGNALING

Endocytosis of cell surface receptors has long been regarded as a mechanism to switch off membrane-derived receptor signaling. However, more recent studies demonstrate that signaling continues in the endocytic pathway and that certain signaling events appear to require endocytosis for full activation to occur (reviewed in Di Fiore and De Camilli, 2001; McPherson *et al.*, 2001).

Earlier reports suggested that formation of the TNF-mediated death-inducing signaling complex (DISC) follows a different mechanism than the one utilized by CD95L or TRAIL (Harper *et al.*, 2003; Kischkel *et al.*, 1995; Medema *et al.*, 1997; Sprick *et al.*, 2000). Two reports by Schneider-Brachert *et al.* (2004, 2006) demonstrated the important role of internalized TNF-R1 as the essential platform for recruiting the DISC to so-called TNF-R1 receptosomes. Recruitment of TNF-R1 death domain adaptor proteins TRADD, FADD, and caspase-8 to form the DISC occurred within 3 min after TNF stimulation and was still associated with TNF-R1 after 60 min. Inhibition of TNF-R1 internalization blocked DISC recruitment and apoptosis but still allowed the recruitment of RIP-1 and TRAF-2 to signal for NF-κB activation.

Studies by Schneider-Brachert *et al.* (2004, 2006) were based on a novel experimental approach by labeling of TNF receptors with biotin TNF coupled to streptavidin–coated magnetic nanobeads and isolation of intact TNF/receptor complexes within their native membrane environment using a specialized magnetic device. Immunomagnetic isolation of morphological intact vesicles from different stages of TNF receptosome trafficking and maturation revealed association with marker proteins of early endosomes, such as clathrin and Rab4, which dissociate after 30 min of internalization, as well as Rab5, which remained at TNF receptosomes up to 60 min. After 30 min of internalization, fusion of TNF receptosomes and trans–Golgi vesicles was observed (indicated by markers of trans-Golgi membranes p47A and Vti-1b, as well as GRP 78, syntaxin-6, and Rab8), resulting in the formation of multivesicular endosomes. At later times, TNF receptosomes accumulated lysosomal proteins such as CTSD and LAMP-1. Analysis of TNF-R1 death domain adaptor proteins in the magnetic isolates revealed recruitment of TRADD after 3 min to isolated TNF receptosomes, as well as FADD and caspase-8.

Deletion of a domain within the cytoplasmic TNF-R1 sequence termed TRID (for TNF receptor internalization domain) (amino acids 234–243 according to the translated amino acid sequence of human TNF-R1, Accession No. GI:435959) or point mutations within the YQRW internalization motif (amino acids 236–239) resulted in complete elimination of TNF-R1 internalization, preventing the recruitment of TRADD, FADD, and caspase-8, and led to an almost complete inhibition of TNF-induced apoptosis. Of note, the DD of TNF-R1 ΔTRID can still recruit RIP-1 and TRAF-2 for the activation of NF-κB even in the absence of TRADD as an assembly platform (Hsu *et al.*, 1996; Zheng *et al.*, 2006).

A previous study by Micheau and Tschopp (2003) proposed a model in which TNF-R1 signaling involves the assembly of two distinct signaling complexes that sequentially activate NF-κB and caspases. However, the role of TNF-R1 endocytosis as the essential mechanism for the immediate recruitment of the TNF-R1-associated DISC and the impact of the endosomal trafficking on apoptosis signaling were not addressed.

The current view of the compartmentalization of TNF-R1 signaling is summarized in Figure 5.1: TNF-R1 at the cell surface, probably located in lipid rafts, signals for NF-κB activation via RIP-1 and TRAF-2. This first step can occur in the absence of TRADD. In the second step, and depending on TNF-R1 internalization, the DISC proteins TRADD, FADD, and caspase-8 are recruited together with the activated TNF receptor as the "carrier" within this complex. In a third step, along the endocytic pathway, TNF receptosomes fuse with trans-Golgi vesicles containing acid sphingomyelinase (A-SMase) and the proapoptotic aspartate protease cathepsin D (CTSD) to form multivesicular endosomes (Raiborg *et al.*, 2003). Within these multivesicular endosomes, activated caspase-8 mediates stimulation of the A-SMase/ceramide/CTSD cascade, which is capable of mediating apoptosis via Bid cleavage and caspase-9 activation (Heinrich *et al.*, 1999, 2004).

The experimental approach used to characterize receptor-associated protein complexes was also used successfully to analyze CD95 internalization and DISC formation (Feig *et al.*, 2007; Lee *et al.*, 2006), demonstrating that the method is not restricted to the TNF receptor system, but can be applied more generally to isolate different ligand/receptor complexes in their native and functional state from whole cell lysates.

The following sections describe the experimental approach that has led to these novel findings in the field of death receptor signaling.

3. METHODS FOR MAGNETIC LABELING AND ISOLATION OF SUBCELLULAR SIGNALING STRUCTURES

3.1. Magnetic labeling

Basically, cell surface proteins such as receptors can be labeled with streptavidin-coated magnetic nanobeads (SA-MNB) either by using the biotinylated ligand as a connecting agent (Fig. 5.2A) or by using a receptor-specific antibody linked to a secondary isotype-specific antibody, tagged with

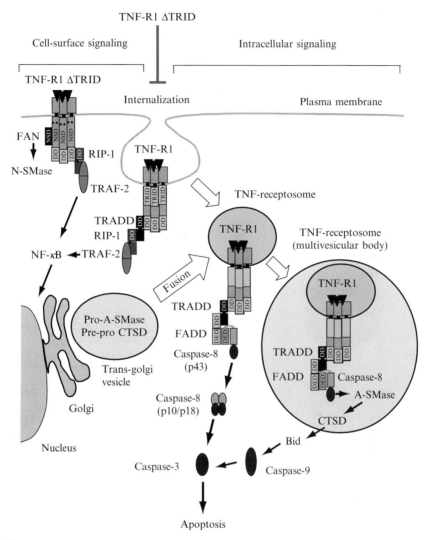

Figure 5.1 Compartmentalization of TNF-R1 signaling. TNF-R1 at the cell surface, probably located in lipid rafts, signals for NF-κB activation. When internalization of the receptor is blocked, the recruitment of RIP-1 and TRAF-2 is sufficient for NF-κB signaling. This can occur in the absence of TRADD. In a second step, and apparently from non-rafts, TNF-R1 is internalized by the formation of clathin-coated pits within minutes. In a third step, the DISC proteins TRADD, FADD, and caspase-8 are recruited to TNF-R1 at the receptosomes containing the activated TNF receptor as carrier within this complex. In a fourth step, along the endocytic pathway, TNF receptosomes fuse with positive trans-Golgi vesicles containing pro-acid sphingomyelinase (pro-A-SMase) and pre-pro cathepsin D (pre-pro CTSD) to form multivesicular bodies. In the multivesicular bodies, activated caspase-8 stimulates the A-SMase/ceramide/CTSD cascade, which is capable of mediating apoptosis via Bid cleavage and caspase-9 activation.

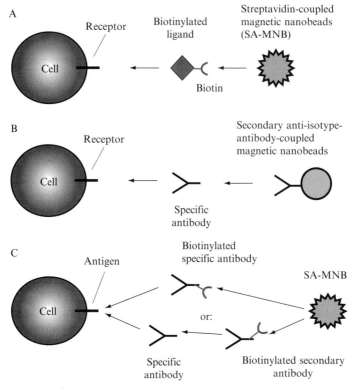

Figure 5.2   Magnetic labeling of cell surface receptors. Cell surface protein-like receptors can be labeled with streptavidin-coated magnetic nanobeads (SA-MNB) either by
using the biotinylated ligand as a connecting agent (A) or by using a specific antibody
directed against the receptor to which a secondary anti-isotype-specific antibody, coupled to magnetic nanobeads, is linked (B) or by a biotinylated primary or secondary
antibody used as a target for SA-MNB (C).

magnetic nanobeads (Fig. 5.2B). If such nanobeads, carrying the respective
secondary antibody, are not available, the specific antireceptor antibody can
be biotinylated and used as a target for SA-MNB or a soluble biotinylated
secondary antibody can be used instead (Fig. 5.2C).

3.2. Receptor internalization and trafficking

For investigating receptor-mediated signaling pathways, the initial signaling
complex between the cell surface receptor and the respective ligand is
labeled by magnetic nanobeads (MNB), termed the "magnetic signaling
complex" (MSC) in Fig. 5.3A. Binding of the labeled ligand to the receptor
at physiological temperature leads to the formation of secondary signaling
structures at the intracellular part of the receptor.

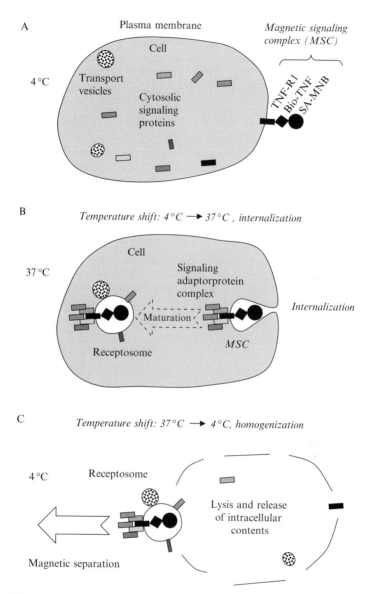

Figure 5.3 Isolation of magnetized signaling complexes. (A) The magnetic signaling complex (MSC) composed of the receptor (i.e., TNF-R1), the specific ligand or antibody (i.e., biotinylated TNF), and the magnetic label [streptavidin–coupled magnetic nano-beads (MNB)] is formed at the plasma membrane. (B) Internalization of the MSC is accompanied by the recruitment of adaptor protein complexes, capable of transmitting ligand–induced signaling into the cytoplasm. After pinching off from the cell surface, the endosomal MSC (receptosome) trafficks into the cell, subjected to endosomal maturation and fusion with other intracellular membrane compartments. (C) The maturation and signaling process is terminated by cooling down the cells and, after gentle mechanical homogenization of the cells, the magnetized receptosomes complexed with adaptor and regulator proteins can be isolated in a magnetic chamber.

Internalization of the MSC results in the formation of endosomes containing the activated ligand/receptor complex, so-called "receptosomes." Internalization is initiated by the association of cytoplasmic proteins, which mediate the invagination of the surrounding plasma membrane (such as AP-2 and clathrin), followed by pinching off of the receptosome from the cell surface, again mediated by special proteins such as dynamin-1 and -2. Receptosomes then travel through the cytoplasm, guided by regulatory proteins of the rab and SNARE family (Fig. 5.3B). Receptosomes may also fuse with other intracellular membrane compartments, such as late endosomes and trans-Golgi vesicles, as well as lysosomes. When the cell surface receptor is labeled with a fluorescent dye instead of the magnetic nanobeads (i.e., streptavidin–FITC), internalization, trafficking, and colocalization of receptosomes with intracellular compartments or regulatory proteins can be visualized (Schneider-Brachert et al., 2004).

3.3. Isolation of magnetically labeled signaling complexes

Based on co-internalization of the magnetic label bound to the ligand/receptor complex, internalized receptosomes can be isolated by application of a magnetic field and analyzed together with associated protein complexes at various time points during endocytosis and intracellular trafficking. These analyses include identification of the adapter proteins bound to the cytoplasmic tail of the receptor being involved in signal transduction of the activated ligand/receptor complex, as well as the recruited regulator proteins involved in receptosome trafficking and fusion events. To follow the kinetics of the assembly of adaptor– and regulator–protein complexes to receptosomes, the cells are cooled down to 4 °C, fixing the maturation state of the receptosomes by "freezing" the receptosomes at a distinct signaling stage. After gentle mechanical disruption of the plasma membranes, the receptosomes can be isolated by a combination of magnetic and centrifugal separation methods, depicted in Figs. 5.3C and 5.4.

The following sections describe the isolation of receptosomes containing TNF-R1 activated by biotinylated TNF and labeled with magnetic nanobeads. This protocol has allowed (1) the identification of TNF-R1 receptosomes as death-signaling vesicles (Schneider-Brachert et al., 2004) and (2) the discovery of a novel adenoviral immune escape mechanism (Schneider-Brachert et al., 2006).

 4. Experimental Procedures

4.1. Cells

Cell cultures in suspension can be used directly for the labeling procedures described here. Using trypsin for detaching adherent cell cultures from the surface of cell culture flasks will result in a loss of TNF receptors. We therefore detach adherent cells by scrubbing with a rubber policemen or washing by a flow stream. Some cell lines can be even detatched simply by cooling the cell culture flasks at 4 °C; for 1 h and gentle shaking. Subsequently, cells are washed twice with phosphate-buffered saline (PBS), and cell numbers are counted. Viability is checked by staining with trypan blue. Finally, cells are centrifuged and resuspended in a minimum possible volume, for example, 250 μl.

4.2. Example protocol for U937 cells

In the example described here, experiments are performed using the human leukemic monocyte lymphoma cell line U937. Cells are maintained in CLIK's RPMI culture medium (Biochrom) supplemented with 5% fetal calf serum. Cells are collected in 50 ml Falcon tubes and washed three times with PBS by centrifugation at 4 °C at $100 \times g$ for 10 min. The procedure of TNF receptor labeling, internalization of the labeled TNF/TNF receptor complex, homogenization, magnetic separation, and characterization of isolated subcellular fractions (as depicted in Fig. 5.4) are described in the following sections.

4.3. Labeling of TNF receptors with biotin-TNF and streptavidin magnetic nanobeads

Biotinylation of the ligand is the first requirement for the formation of the MSC, composed of the magnetized ligand (TNF) and the TNF receptor (TNF-R). Biotinylation of TNF should not interfere with the binding of the ligand to its receptor and should possess the same biological activity as the nonbiotinylated native TNF. Unfortunately, in our hands, with a low amount of TNF applied, biotinylation of TNF resulted in a significant loss of its binding capacity to the TNF receptor as measured by FACS analysis or biological assays. Therefore, we screened various commercially available TNF detection kits and found the biotinylated TNF (bio-TNF) included in the Fluorokine-Kit (R&D Systems, Wiesbaden, Germany) suitable for this approach, as the biological activity (cytotoxicity on L929 cells) of the R&D bio-TNF was identical to native human TNF. The binding capacity of bio-TNF to U937 cells was monitored by FACS analysis using streptavidin-FITC for detection, revealing that TNF receptors were saturated with Bio-TNF (Tchikov et al., 2001).

For the subsequent isolation of internalized receptosomes, the magnetic bead coupled to the ligand should be as small as possible to minimize nonspecific effects on the internalization and intracellular trafficking process. We chose streptavidin-coated 50 nm MACS-MicroBeads from Miltenyi Biotec (Bergisch Gladbach, Germany) for magnetic labeling of bio-TNF. The incubation of cells with bio-TNF is performed according to recommendations of the manufacturer, including control for the specificity by

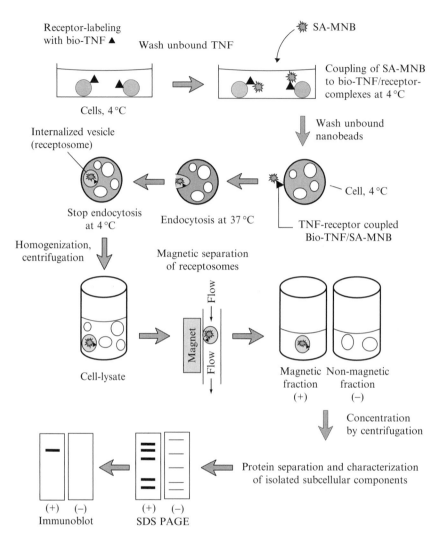

Figure 5.4 Immunomagnetic isolation of TNF receptosomes. TNF receptors on the cell surface are labeled with bio–TNF followed by incubation with SA-MNP at 4 °C. Synchronized internalization of the magnetically labeled TNF receptor complex was initiated by shifting the temperature to 37 °C. Intracellular trafficking of TNF receptosomes was stopped by chilling the cells on ice, and the labeled receptosomes were released from the cells by gentle mechanical homogenization. Separation of the magnetic TNF receptosomes from nonlabeled material was achieved by magnetic separation in the high gradient magnetic separation chamber, and the resulting fractions were analyzed by SDS-PAGE and Western blotting. For details of the separation procedure, see text.

competition with a 100-fold molar excess of unconjugated TNF, which leads to a dramatic reduction in the labeling of cells with magnetic beads as checked by magnetophoretic measurements (see later). For binding of TNF to the TNF receptor, the washed cells are pelleted by centrifugation at $100 \times g$ for 10 min at 4 °C and resuspended in 350 μl of cold PBS containing 200 ng of bio-TNF. Incubation is performed for 1 h on ice. In the next step, magnetic nanobeads are attached to the receptor-bound bio-TNF by incubating the cells with 200 μl of streptavidin-coated 50 nm MACS-MicroBeads again for 1 h on ice. Unbound bio-TNF and MACS-MicroBeads are separated from labeled cells by washing in 30 ml PBS at $100 \times g$.

4.4. Checking the efficiency of labeling with magnetic nanobeads

The labeling efficiency of cells with magnetic nanobeads depends on experimental conditions such as incubation time, concentration of cells, reactants, and the reaction volume. Ideally, the best performance of cell labeling with ligands or magnetic nanobeads is achieved in the situation when each receptor on the cell surface is coupled with the corresponding ligand and one nanoparticle. This can be achieved by providing optimal conditions to reach the maximum rate of binding between cells and reactants. The reaction between cells and reactants is carried out under so-called "crowded conditions" in a minimal volume.

We calculated that 10^8 U937 cells express on the order of $\sim 10^{11}$ TNF receptors (trimers) on their surfaces, as it was reported for $\sim 10^3$ to 10^4 TNF receptors per cell from FACS data (Patton et al., 1989). As 100 μl of bio-TNF added to each cell sample contains 200 ng of bio-TNF with a molecular mass of 17 kDa, we calculated that the given amount of bio-TNF corresponds to 7×10^{12} individual TNF molecules or $\sim 2 \times 10^{12}$ trimers (as the active form of TNF) and $2 \times 10^{12}/10^{11} = 20$ trimers per trimeric receptor. Thus, for the saturation of TNF receptors on cell surfaces with bio-TNF, the incubation sample contains an excess of ~ 20 molecules of bio-TNF per receptor. MACS-MicroBeads contain roughly 2×10^{12} beads/ 2 ml (personal communication). We calculated that 200 μl contain 10^{11} individual beads, corresponding to 10^3 particles per cell in our sample or one bead per one receptor. Unfortunately, the kinetics and equilibrium for reactions between cells and magnetic beads are poorly investigated and the calculation just made cannot be proof for the real labeling of cells with beads (Tchikov et al., 2001). Therefore, experimental proofing and daily checking of the efficiency of cell labeling are required. For this we use a cell magnetophoresis approach, which was reported previously in detail (Tchikov et al., 1999). Briefly, cell magnetophoresis describes a cell motion due to a driving force from an inhomogenous magnetic field. By microscopic analysis of cell magnetophoresis, the number of cells labeled with magnetic beads can be estimated. We determined a 90% labeling efficiency when 200 μl of MACS streptavidin MicroBeads were added to a number of 10^8 U937 cells pretargeted with bio-TNF.

4.5. Internalization

As the next step after washing of labeled cells at $100\times g$ (see earlier discussion), the cell pellet (approximately 350 μl) is stored on ice. To induce receptor internalization, the pellet is resuspended in 20 ml of prewarmed medium and incubated at 37 °C for various time points. Shifting the cells from 4 °C to the physiological temperature induces synchronized internalization of the labeled TNF/receptor complex.

Electron microscopic analysis revealed that the temperature shift induces internalization of magnetically labeled TNF receptors, forming receptosomes within 3 min, fusion of receptosomes with intracellular membrane compartments after 30 to 60 min, and formation of multivesicular structures, including the labeled TNF receptosomes (Fig. 5.5A).

The trafficking and maturation of labeled TNF receptosomes (and the recruitment of signaling adaptor proteins) are stopped by the addition of 20 ml ice-cold RPMI medium followed by washing the samples with ice-cold PBS.

Figure 5.5B shows a typical preparation of TNF-R1 magnetic membrane fractions derived prior to the temperatur shift (0 min) or after incubation of the cells for 10, 30, and 60 min at 37 °C. The magnetic isolation of these fractions is performed as follows.

4.6. Gentle mechanical homogenization of cells by a sequential "destruction–centrifugation" protocol

For isolation of intact subcellular compartments including labeled receptosomes, the plasma membranes of the cells are disrupted by a gentle mechanical procedure (Fig. 5.6A). Cells are resuspended in 30 ml of homogenization buffer composed of 0.25 M sucrose, 15 mM HEPES, and 0.5 mM MgCl$_2$ (pH 7.4) and centrifuged at 150g for 10 min, and the pellet is supplemented with protease inhibitors and pepstatin (i.e., protease inhibitor set from Roche Diagnostics, Manheim, Germany) according to instructions of the manufacturer. Cell disruption is performed by shaking cells and glass beads at 4 °C. We developed a microhomogenization device with a loading chamber of 350 μl containing seven glass beads 3 mm in diameter loaded with the cell suspension in homogenization buffer without bubbles (Fig. 5.6A). The device is subjected to an excentric motor shaker (vortex). Shaking the cells together with the beads results in disruption of the cells by hydrodynamic shear forces. Usually, approximately 30% of U937 cells are disrupted after 5 min, but the efficiency may vary among different cell types.

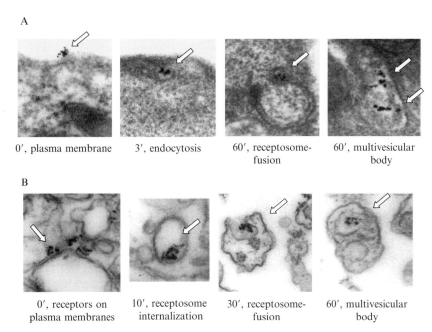

Figure 5.5 Trafficking of TNF receptosomes in U937 cells and magnetic isolates from various stages of TNF receptosome maturation. (A) Internalization and intracellular trafficking/fusion of magnetically labeled TNF receptors. Electron micrographs were taken from U937 cells labeled with bio-TNF/SA-MNP and biotin-gold after various time points after shifting the incubation temperature from 4 to 37 °C. (B) Magnetically isolated fractions were prepared from U937 cells at the indicated time points and analyzed by electron microscopy for morphological integrity.

To preserve the morphological integrity of membrane vesicles in the homogenate, we developed a "destruction–centrifugation protocol," which meets the contrasting requirements as the cells should be destroyed efficiently, while the integrity of subcellular compartments, such as organelles or vesicles, should be kept intact. This is achieved by repetitive homogenization–centrifugation steps. In this discontinuous protocol, homogenization is essentially incomplete at each step and the following centrifugation at a relatively low speed is utilized for separating the released organells from the intact cells. After each step, the supernatant, containing the organelles, is collected into a separate tube. The pellet, containing unbroken cells, is resuspended and homogenized again. This procedure assures that the already released organelles are subjected only once to the physical shear forces for a limited time and are kept safely on ice, subjecting only the remaining unbroken cells to further homogenization stress. Theoretically, shorter homogenization times with a higher number of repetitions of this protocol will result in less damage to released organelles.

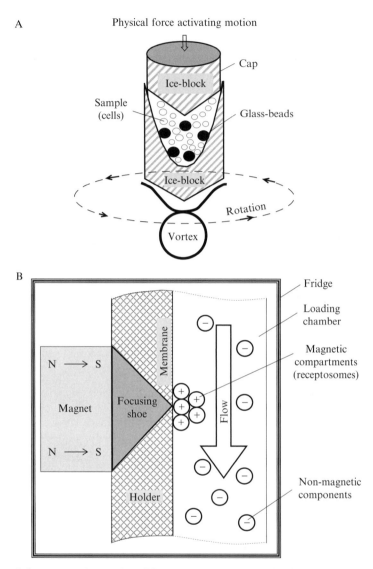

Figure 5.6 Devices for gentle cell homogenization (A) and isolation of magnetically labeled receptosomes in a high gradient magnetic chamber (B). See text for details.

Practically, a three-step method is acceptable, depending on the cell type. As an example, the pellet of magnetically labeled U937 cells is subjected to lysis in the microhomogenization device for 5 min. Thereafter the homogenate is collected into a separate tube and centrifuged at 4 °C and $100 \times g$ for 2 min, and the supernatant is placed in a new tube. Subsequently, the pellet is resuspended in 350 µl of homogenization buffer and subjected again to homogenization. After three repetitions, supernatants are collected and centrifuged at $150 \times g$ and 4 °C for 5 min to eliminate residual debris or intact cells. With this protocol, approximately 700 μl of lysate ($\approx$5 mg protein) is recovered from 10^8 U937 cells, ready for further purification of labeled receptosomes.

4.7. Magnetic isolation of labeled membrane compartments

Magnetic isolation of labeled membrane compartments is performed in a custom-built magnetic separation flow chamber at 4 °C (Fig. 5.6B). The procedure includes the following three steps.

4.7.1. Loading of samples and vesicle magnetophoresis

The flow chamber is loaded with the homogenate (0.7 ml), and the spatial separation of magnetic (signed with plus in Fig. 5.6B) and nonmagnetic components (signed with minus) is achieved in a high gradient magnetic field (see later).

The MACS-MicroBeads used for labeling are of 50 nm in diameter, containing approximately 57% (w/w) of iron oxides embedded in a dextran shell, including magnetite, which exhibits strong magnetism (Kantor et al., 1998). Because the total magnetic moment of a single bead decreases with decreasing bead diameter, the magnetic moment of individual 50 nm beads is relatively poor. As described by the manufacturer, purification of these beads requires a high-gradient magnetic field (Miltenyi et al., 1990).

Based on the low magnetic moment of the small magnetic beads enclosed in the receptosomes, a special design is required for the magnetic system to achieve a high gradient of the magnetic field strong enough to move the labeled vesicles (which are between 100 and 500 nm in diameter) by the magnetic force. Our magnetic chamber construction is based on a high-intensity magnetic field (1 Tesla) from strong permanent magnets (neodymium–iron–boron magnets), which is enhanced further by focusing on the iron tips (a few millimeters or less in altitude, Fig. 5.6B), thus yielding a high concentration of the magnetic field on top of the tips, creating the gradient. Application of cell lysates to this high gradient magnetic field results in movement of the magnetically labeled vesicles (vesicle magneto-phoresis) toward the top of the tips.

After incubation of the homogenates for 3 h, the magnetized components are sedimented at the wall of the magnetic chamber while the nonmagnetic components remain free in suspension.

4.7.2. Washing

As the second step, nonmagnetic material has to be removed from the magnetic chamber. This is performed by applying a gentle washing flow (e.g., 100 μl/min) by pumping the homogenization buffer through the flow chamber. Nonmagnetic components will be removed, leaving the magnetic components detained by magnetic forces at the chamber wall.

4.7.3. Elution

The washed magnetic material is recovered by removing the chamber from the magnetic field. The isolated material is eluted and sedimented by centrifugation at 20,000×g for 1 h. The pellet is resuspended in 200 μl homogenization buffer, frozen, and stored at −20 °C for further analysis. The isolated material represents the magnetic fraction containing the magnetic nanobeads within the receptosomes, as well as receptor-associated signaling proteins and subcellular structures recruited to the initial receptosome during intracellular trafficking and maturation.

Our custom-built free-flow magnetic chamber (Fig. 5.6B; Schütze *et al.*, 2003) is currently under development as an automatic system and will be available on the market soon. For further information, please contact the authors.

4.8. Characterization of the magnetic separation procedure

4.8.1. Washing of the magnetic chamber, purity, and specificity of the magnetic fractions

The washing and magnetic fractions are collected and analyzed for protein content before and after concentration by high-speed centrifugation (Fig. 5.7). Both the nonmagnetic washing fraction and the magnetic fraction are subjected to centrifugation at 20,000×g for 1 h at 4 °C. The remaining 10 μl on the bottom of each centrifuge tube is resuspended in 200 μl of homogenization buffer and is analyzed for protein content (Figs. 5.7A and 5.7B).

Washing with the minimum of 1 volume of loading chamber is already sufficient to remove nonmagnetic protein to a nondetectable level (Fig. 5.7A). However, some residual protein can still be detected after concentration of the samples. Greater washing volumes (5–10 volumes of loading chamber) are then sufficient to remove nonmagnetic protein to nondetectable levels even after concentration (C in Fig. 5.7B). Before concentration, the magnetic fraction contains approximately 0.1 mg/ml protein (NC in Fig. 5.7B), and 0.45 mg/ml protein ($\cong$50 μg of total protein) is detected in the magnetic fraction after concentration. Thus, in this example, a 100-fold enrichment in magnetically labeled subcellular compartments is achieved from a starting lysate derived from 10^8 U937 cells containing 5 mg protein.

To analyze the enrichment of a specific protein in the magnetic fraction compared to the initial lysate, the samples are subjected to Western blot analysis for detection of caspase-8 (H134, sc-7890, Santa Cruz Biotech, Inc., CA). In whole cells, this protein is expressed as full-length (55 kDa procaspase) in untreated cells or as a cleaved (43 kDa receptor-bound

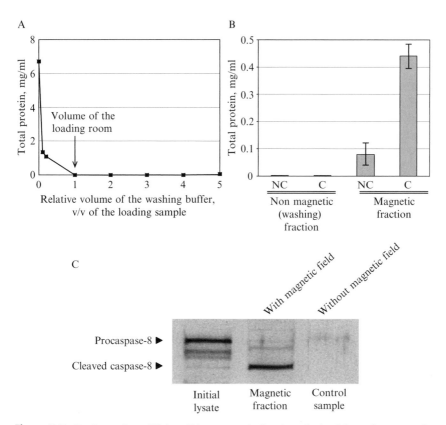

Figure 5.7 Purity and specificity of the magnetic fractions derived from the magnetic chamber. The washing effect was evaluated by the amount of protein present in the eluted washing fractions after binding of the magnetic components to the magnetic chamber wall (A). The amount of protein in the nonmagnetic washing fraction and the magnetic fraction eluted from the chamber before (nonconcentrated, NC) and after concentration (C) of the samples (B). Enrichment of cleaved caspase-8 in the TNF-R1 magnetic membrane fraction compared with initial lysate and in samples derived from the chamber in the absence of a magnetic field (C).

activated) caspase-8 form after TNF treatment. As depicted in Fig. 5.7C, initial lysates exhibit predominantly the full-length caspase-8 proform, whereas the cleaved (activated) form is almost invisible. In contrast, the magnetic fraction contains mainly cleaved caspase-8, bound to TNF-R1. These results demonstrate that TNF-R1 receptosomes contain activated caspase-8, which can be highly enriched in TNF receptosomes by magnetic separation.

Without the magnetic field, no caspase-8 protein is detectable after washing of the chamber, demonstrating that this treatment efficiently removes all protein and that the isolation of receptosomes critically depends on the application of the high gradient magnetic field.

4.8.2. Characterization of TNF receptosome preparations from U937 cells

Tumor necrosis factor receptors on U937 cells are labeled with bio-TNF and magnetic SA-MNB as described earlier. Synchronized TNF receptor internalization is induced by incubation at 37 °C and is then stopped immediately (0 min) and after 3, 10, 30, and 60 min. Lysates are prepared as described; a small part (e.g., 20 μl) is placed in a separate tube as lysate control. Subsequently, the magnetic fractions are isolated as described earlier. Samples from the initial lysate and from the magnetic fractions obtained from various time points are subjected to SDS-PAGE and analyzed by Western blotting.

The amount of protein loaded on the gels can vary from a few micrograms up to a hundred micrograms per lane depending on the expression level of the protein of interest, sensitivity of antibodies, and methods of detection. Based on the limited linearity of the detection systems for Western blotting, it is important not to "overload" the gel, because for analyzing the kinetics of the recruitment of a specific protein, the differences in the content of this protein could be be lost. However, "underloading" may lead to a loss of the signal, although the protein is present at very low amounts. We aimed at a best resolution of time-dependent protein distributions between the single lanes of a given gel. Usually 10 μg from each sample are loaded.

As an example to analyze the trafficking and endosomal maturation of TNF receptosomes, blots derived from the gels are analyzed for (1) anti-clathrin as a marker for the formation of clathrin-coated pits and vesicles, (2) anti-Rab5 as an early endosomal marker, and (3) anti-LAMP-1 as a marker for maturation of late endosomes to the lysosomal compartment in which proteins are degraded. As a control for the purity of the receptosome preparation, the presence of nucleoporin is analyzed (Fig. 5.8A).

4.8.3. Western blotting of isolated TNF receptosome proteins

1. Clathrin is a major protein component in the coat formed during endocytosis. As shown in Fig. 5.8A, clathrin is present in the initial lysate and is recruited to the TNF receptor-associated membrane compartment directly after formation of the MSC even at 4 °C (0-min time point). The amount of clathrin recruited to TNF receptosomes is enhanced strongly during internalization after shifting the temperature to 37 °C for 3 and 10 min. Clathrin is reduced after 60 min of internalization, indicating the uncoating of the maturating vesicles.

2. Rab5 is a protein that regulates the fusion of plasma membrane-derived clathrin-coated vesicles with early endosomes and homotypic fusion among early endosomes. Figure 5.8A shows that this protein is absent at 0 min but is recruited to the magnetic TNF receptor fraction within

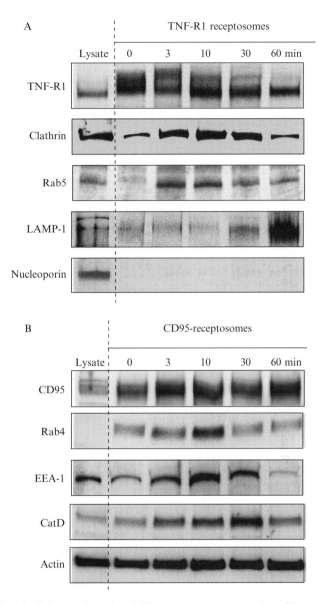

Figure 5.8 Trafficking of TNF and CD95 receptosomes monitored by Western blotting for signature proteins of vesicular maturation. (A) TNF receptosomes were isolated immunomagnetically from U937 cells based on bio-TNF/SA-MNB labeling as described in the text and analyzed by Western blotting for the presence of TNF-R1 and the signature proteins of vesicular maturation, clathrin, rab5, LAMP-1, and nucleoporin. (B) CD95 receptosomes were isolated from SKW6.4 cells after labeling with biotinylated APO-1-3/SA-MNB and analyzed for CD95, rab4, EEA-1, cathepsin D, and actin content.

3 min after incubation at 37 °C. After 30 min the amount of receptosome-bound Rab5 is decreasing as maturation progresses.

3. In accordance with the function as a lysosome-associated membrane protein, LAMP-1 appears after 30 to 60 min, indicating that the receptosome reaches the lysosomal compartment.

4. Nucleoporin is detected in the lysate but is absent in all magnetic membrane fractions, indicating that the receptosome preparations are not contaminated with nuclei. The overall differences in the amounts of specific proteins recruited at various time points of the preparations indicate the high level of purity and the efficiency of the washing procedure.

The recruitment of TNF receptor adaptor proteins of the DISC, such as TRAF-2, RIP-1, TRADD, FADD, and caspase-8 and other signaling proteins to TNF receptosomes isolated applying this method, has been demonstrated in Schneider-Brachert et al. (2006, 2007). The specificity of binding of these adaptor proteins to TNF-R1 was shown by analyzing TNF receptosomes isolated from TNF receptor-deficient cells expressing recombinant TNF-R1 constructs lacking the death domain or the TRID.

4.9. Labeling and isolation of CD95 receptosomes

As a second example for the application of the immunomagnetic receptor isolation approach described earlier for TNF-R1, the isolation and characterization of CD95 receptosomes are described in the following section.

CD95 (Fas/APO-1) is another member of the "death receptor family," sharing its cytoplasmic death domain with TNF-R1 and TRAIL receptor-1 and -2. CD95 internalization in type I tumor cells also plays a previously unrecognized role in CD95 ligand-induced activation of apoptotic pathways (Algeciras-Schimnich, 2002, 2003; Lee et al., 2006). In contrast, engagement of CD95 without receptor internalization results in the activation of non-apoptotic signaling pathways. Hence, as shown for the TNF-R1 system, the subcellular compartment of CD95 signaling also activates divergent pathways to promote distinct survival or apoptotic cellular fates (see also Chapter 4).

As an example, the time course of intracellular CD95 receptosome trafficking is analyzed using SKW6.4 cells (type I cell line). Magnetic fractions are derived after stimulation with biotinylated agonistic anti-CD95 antibody (APO-1-3, IgG3) and streptavidin-coated magnetic nanobeads (see protocol in Fig. 5.2B). The APO-1-3 antibody was a gift from M. E. Peter (Chicago), also available from Alexis Biochemicals [ALX-805-020 Monoclonal Antibody to Fas (human) (APO-1-3)]. Biotinylation of APO-1-3 is performed as described (Algeciras-Schimnich et al., 2003; Lee et al., 2006).

For labeling of CD95 at the plasma membrane, 10^8 SKW6.4 cells from a suspension culture are washed with PBS and incubated in a total volume of 0.25 ml ice-cold PBS with 0.1 ml (0.2 μg) of biotinylated APO-1-3 monoclonal antibody for 1 h on ice. Then cells are washed with 30 ml of ice-cold PBS, and 0.2 ml of 50 nm streptavidin-coated magnetic nanobeads (Miltenyi MACS MicroBeads) solution is added to the cell suspension and incubated for 1 h on ice. Cells are washed with ice-cold PBS, and internalization of magnetically labeled CD95 receptosomes is induced by incubation of the labeled cells at 37 °C for 0, 3, 10, 30, and 60 min. Cells are then homogenized mechanically at 4 °C, and supernatants are subjected to magnetic separation of CD95 membrane fractions as described previously for TNF receptosomes.

The Western blot in Fig. 5.8B reveals that CD95 receptosomes transiently recruit the early endosomal marker protein Rab4 with a similar kinetic as demonstrated for TNF-R1 recruitment of Rab5 (Fig. 5.8A). Rab4 recruitment to CD95 receptosomes is paralleled by the kinetics of receptosome-bound early endosomal antigen-1 (EEA-1). This protein is involved in the fusion between early and late endosomes. The acidic aspartate protease cathepsin D (CTSD) is also transiently recruited with a maximum after 30 min, indicating lysosomal maturation of the CD95-containing vesicles. Actin is constantly associated with CD95 magnetic membrane fractions.

5. CONCLUSIONS

The method of immunomagnetic isolation of cell surface receptors allows for monitoring the temporary maturation and spatial distribution of specific proteins involved in the regulation of vesicular transport of the receptosomes and in intracellular signaling from intracellular compartments. The remarkable feature of this approach is that receptors are isolated in their native membrane environment in the absence of any detergent that might interfere with the association of adaptor proteins. Based on the gentle physical homogenization procedure, the morphology and functional integrity of the compartments are preserved, allowing for functional analysis, such as estimating the enzymatic activity of proteins contained in the vesicles. Most importantly, the method is highly specific for the receptor system of choice, based on the specificity of the ligand or antibody used for labeling the receptosomes. Finally, the isolated samples are of high purity based on the dramatic difference in magnetism of the labeled structures (receptosomes) compared to unlabeled structures within the cell lysates.

The high gradient magnetic chamber described in this chapter can also be used to immunomagnetically isolate phagosomes containing magnetically labeled bacteria from infected cells as well as intact organells or soluble proteins from cell homogenates after posthomogenization labeling with high efficiency. Protocols for these applications will be published elsewhere.

ACKNOWLEDGMENTS

This work was supported by grants from the Bundesministerium für Bildung und Forschung (BMBF), BioChancePlus program, and from the Deutsche Forschungs-gemeinschaft (DFG), SFB 415, project A11; DFG SCHU 733/8-1, and DFG SCHU 733/9-1 given to S. Schütze. We appreciate the continuous support by Dieter Kabelitz.

REFERENCES

Algeciras-Schimnich, A., Shen, L., Barnhart, B. C., Murmann, A. E., Burkhardt, J. K., and Peter, M. E. (2002). Molecular ordering of the initial signaling events of CD95. *Mol. Cell. Biol.* **22,** 207–220.

Algeciras-Schimnich, A., and Peter, M. E. (2003). Actin dependent CD95 internalization is specific for type I cells. *FEBS Lett.* **546,** 185–188.

Di Fiore, P. P., and De Camilli, P. (2001). Endocytosis and signaling: An inseparable partnership. *Cell.* **106,** 1–4.

Feig, C., Tchikov, V., Schütze, S., and Peter, M. E. (2007). Palmitoylation of CD95 facilitates formation of SDS-stable receptor aggregates that initiate apoptosis signaling. *EMBO J.* **26,** 221–231.

Harper, N., Hughes, M., MacFarlane, M., and Cohen, G. M. (2003). Fas-associated death domain protein and caspase-8 are not recruited to the tumor necrosis factor receptor 1 signaling complex during tumor necrosis factor-induced apoptosis. *J. Biol. Chem.* **278,** 25534–25541.

Heinrich, M., Wickel, M., Schneider-Brachert, W., Sandberg, C., Gahr, J., Schwandner, R., Weber, T., Saftig, P., Peters, C., Brunner, J., Krönke, M., and Schütze, S. (1999). Cathepsin D targeted by acid sphingomyelinase-derived ceramide. *EMBO J.* **18,** 5252–5263.

Heinrich, M., Neumeyer, J., Jakob, M., Hallas, C., Tchikov, V., Winoto-Morbach, S., Trauzold, A., Hethke, A., and Schütze, S. (2004). Cathepsin D links TNF-induced acid sphingomyelinase to Bid-mediated caspase-9 and caspase-3 activation. *Cell Death. Differ.* **11,** 550–563.

Hsu, H., Huang, J., Shu, H. B., Baichwal, V., and Goeddel, D. V. (1996). TNF-dependent recruitment of the protein kinase RIP-1 to the TNF receptor-1 signaling complex. *Immunity.* **4,** 387–396.

Kantor, A. B., Gibbons, I., Miltenyi, S., and Schmitz, J. (1998). Magnetic cell sorting with colloidal superparamagnetic particles. *In* "Cell Separation Methods and Applications" (D. Recktenwald and A. Radbruch, eds.), p. 153. Dekker, New York.

Kischkel, F. C., Hellbardt, S., Behrmann, I., Germer, M., Pawlita, M., Krammer, P. H., and Peter, M. E. (1995). Cytotoxicity-dependent APO-1 (Fas/CD95)-associated proteins form a death-inducing signaling complex (DISC) with the receptor. *EMBO J.* **14,** 5579–5588.

Lee, K. H., Feig, C., Tchikov, V., Schickel, R., Hallas, C., Schütze, S., Peter, M. E., and Chan, A. C. (2006). The role of receptor internalization in CD95 signaling. *EMBO J.* **25,** 1009–1023.

Locksley, R. M., Killeen, N., and Lenardo, M. J. (2001). The TNF and TNF receptor superfamilies: Integrating mammalian biology. *Cell.* **104,** 487–501.

McPherson, P. S., Kay, B. K., and Hussain, N. K. (2001). Signaling on the endocytic pathway. *Traffic.* **2,** 375–384.

Medema, J. P., Scaffidi, C., Kischkel, F. C., Shevchenko, A., Mann, M., Krammer, P. H., and Peter, M. E. (1997). FLICE is activated by association with the CD95 death-inducing signaling complex (DISC). *EMBO J.* **16,** 2794–2804.

Micheau, O., and Tschopp, J. (2003). Induction of TNF receptor 1-mediated apoptosis via two sequential signaling complexes. *Cell.* **114,** 181–190.

Miltenyi, S., Müller, W., Weichel, W., and Radbruch, A. (1990). High gradient magnetic cell separation with MACS. *Cytometry.* **11,** 231–238.

Patton, J. S., Rice, G. C., Ranges, G. E., and Palladino, M. A., Jr. (1989). Biology of the tumor necrosis factors. *In* "Macrophage-Derived Cell Regulatory Factors" (C. Sorg, ed.), p. 89. Karger AG, Basel, Switzerland.

Raiborg, C., Rusten, T. E., and Stenmark, H. (2003). Protein sorting into multivesicular endosomes. *Curr. Opin. Cell Biol.* **15,** 446–455.

Schneider-Brachert, W., Tchikov, V., Kruse, M. L., Lehn, A., Jakob, M., Hildt, E., Held-Feindt, J., Kabelitz, D., Krönke, M., and Schütze, S. (2006). Adenovirus E3-14.7K protein inhibits TNF-induced apoptosis by targeting TNF-R1 endocytosis and DISC assembly. *J. Clin. Invest.* **116,** 2901–2913.

Schneider-Brachert, W., Tchikov, V., Neumeyer, J., Jakob, M., Winoto-Morbach, S., Held-Feindt, J., Heinrich, M., Merkel, O., Ehrenschwender, M., Adam, D., Mentlein, R., Kabelitz, D., and Schütze, S. (2004). Compartmentalization of TNF receptor 1 signaling: Internalized TNF receptosomes as death signaling vesicles. *Immunity.* **21,** 415–428.

Schütze, S., Machleidt, T., Adam, D., Schwandner, R., Wiegmann, K., Kruse, M. L., Heinrich, M., Wickel, M., and Krönke, M. (1999). Inhibition of receptor internalization by monodansylcadaverine selectively blocks p55 tumor necrosis factor receptor death domain signaling. *J. Biol. Chem.* **274,** 10203–10212.

Schütze, S., Tchikov, V.M, Kabelitz, D., and Krönke, M. (2003). German patent DE 101 44 291C2.

Sprick, M. R., Weigand, M. A., Rieser, E., Rauch, C. T., Juo, P., Blenis, J., Krammer, P. H., and Walczak, H. (2000). FADD/MORT1 and caspase-8 are recruited to TRAIL receptors 1 and 2 and are essential for apoptosis mediated by TRAIL receptor 2. *Immunity.* **12,** 599–609.

Tchikov, V., Winoto-Morbach, S., Krönke, M., Kabelitz, D., and Schütze, S. (2001). Adhesion of immunomagnetic particles targeted to antigens and cytokine receptors on tumor cells determined by magnetophoresis. *J. Magnet. Magn. Materials.* **225,** 285–293.

Tchikov, V., Schütze, S., and Krönke, M. (1999). Comparison between imunofluorescence and immunomagnetic techniques of cytometry. *J. Magnet. Magn. Materials.* **194,** 242.

Wajant, H., Pfizenmaier, K., and Scheurich, P. (2003). Tumor necrosis factor signaling. *Cell Death Differ.* **10,** 45–65.

Zheng, L., Bidere, N., Staudt, D., Cubre, A., Orenstein, J., Chan, F. K., and Lenardo, M. (2006). Competitive control of independent programs of tumor necrosis factor receptor-induced cell death by TRADD and RIP1. *Mol. Cell. Biol.* **26,** 3505–3513.

CHAPTER SIX

ANALYZING LIPID RAFT DYNAMICS DURING CELL APOPTOSIS

Walter Malorni,* Tina Garofalo,[†] Antonella Tinari,[‡]
Valeria Manganelli,[†] Roberta Misasi,[†] and Maurizio Sorice[†]

Contents

* Department of Drug Research and Evaluation, Istituto Superiore di Sanità, Rome, Italy
† Department of Experimental Medicine, "Sapienza" University of Rome, Rome, Italy
‡ Department of Technology and Health, Istituto Superiore di Sanità, Rome, Italy

Methods in Enzymology, Volume 442
ISSN 0076-6879, DOI: 10.1016/S0076-6879(08)01406-7

Abstract

Increasing lines of evidence suggest a role for lipid rafts, glycosphingolipid-enriched microdomains, in cell life and death. This chapter describes in brief the methods used to analyze raft interactions with proteins involved in apoptosis. This chapter focuses mainly on coimmunoprecipitation methods, which represent a useful tool in analyzing raft dynamics during apoptosis. Glycosphingolipid analysis in the immunoprecipitates is performed by thin-layer chromatography. Moreover, methods for the analysis of mitochondrial raft-like microdomains are also described. Detergent (Triton X-100)-insoluble material from isolated mitochondria can be analyzed by Western blot. Further insights can also come from both light and electron microscopy analyses. These can provide useful information as concerning lipid raft distribution at the cell surface or in the cell cytoplasm. Paradigmatic micrographs are shown. The combined use of all these different approaches appears to be mandatory for analyzing the role of lipid raft dynamics during apoptosis.

1. INTRODUCTION

Several investigations have been carried out since the late 1990s in order to address the function of lipid rafts in cell life and death. It is commonly accepted that the plasma membrane is a complex system in which constituents are organized in small lipid/protein domains, known as "lipid rafts." Membrane rafts are a specific type of lipid domains enriched in sphingolipids, including gangliosides (GSLs), sphingomyelin, and cholesterol, that result from tight hydrophobic interactions among these molecules, leading to the spontaneous formation of aggregates that separate from glycerophospholipids in cell membranes (Simons *et al.*, 1997). Because of their lipid composition, the physical state of these domains is supposed to be similar to a liquid-ordered (Lo) phase, structurally and dynamically distinct from the rest of the lipid bilayer, which is, in turn, assumed to be in a liquid-disordered phase (Ipsen *et al.*, 1987). On the basis of the peculiar biophysical and biochemical properties of lipid rafts, a number of possible interactions with various subcellular structures have been suggested. It is well known that these lipid microdomains may function to concentrate or segregate different proteins to form a glycosignaling domain. Several studies described a role for rafts in a variety of cellular processes, such as cell signaling, lipid and protein sorting, membrane organization and trafficking, immune response, and cell death (Malorni *et al.*, 2007). Indeed, a large variety of proteins have been detected in these microdomains, including tyrosine kinase receptors (EGF-R), G proteins, Src-like tyrosine kinases (lck, lyn, fyn), protein kinase c isozymes, glycosylphosphatidylinositol (GPI)-anchored proteins, adhesion molecules, and the death–inducing signaling complex. Thus, a general function of lipid rafts in signal transduction may be to allow the lateral segregation of proteins within the plasma membrane, providing a mechanism for the compartmentalization of signaling components, concentrating certain components in lipid rafts, including those of importance in apoptosis, and excluding others (Simons *et al.*, 1997). Several studies indicate that lipid rafts could play a role in the apoptotic cascade. Although they are detected in various cell types, their role has been studied mostly in lymphocytes undergoing apoptosis induced by CD95/Fas. In particular, it was proposed that rafts are the primary site of action of the enzyme sphingomyelinase, which catalyzes the hydrolysis of sphingomyelin, resulting in ceramide formation; ceramides deriving from sphingomyelin hydrolysis are essential mediators of apoptotic signals originating from CD95/Fas (Grassmé *et al.*, 2001). Moreover, a role for gangliosides as structural components of the multimolecular signaling complex involved in CD95/Fas receptor-mediated apoptotic pathway has been reported (Garofalo *et al.*, 2003).

Although lipid rafts are considered ubiquitous constituents of plasma membrane, recent lines of evidence also indicated that they are present on organelles, including the Golgi apparatus and a subcompartment of the endoplasmic reticulum. Raft-like microdomains have also been detected on mitochondrial membranes after CD95/Fas triggering (Garofalo *et al.*, 2005). Formation of a multimolecular complex that includes VDAC-1, Bcl-2 family, and fission proteins, for example, h-Fis, has been demonstrated. Thus, the complex may represent preferential sites where some key reactions can take place and can be catalyzed, leading to either survival or death of T cells.

Different methods for studying lipid rafts in cell apoptosis are reported in Table 6.1.

Table 6.1 Methods used for studying lipid rafts in apoptosis

Apoptosis	Electron microscopy
	TUNEL
	Annexin V
	Activity of caspases
	DNA laddering
	Mitochondrial membrane potential
Lipid rafts	Sucrose gradient analysis
	Coimmunoprecipitation
	FRET
	Single particle tracking
	Fluorescence correlation spectroscopy
	Scanning confocal microscopy
	Electron microscopy

2. THE CONTRIBUTION OF BIOCHEMICAL APPROACHES

In the last years, numerous different techniques have been considered for studying the association of apoptosis-related proteins with lipid rafts. The use of nonionic detergent extraction to generate low-density detergent-resistant membranes (DRMs) has had a major role in implicating rafts in cellular functions (Brown *et al.*, 2000). Although this treatment may disrupt lipid–lipid interactions, a minor fraction of cell membranes is preserved and can be isolated as DRMs. Because detergent extraction may also disrupt several lipid–protein interactions, only a few proteins, interacting strongly with highly ordered domains, retain their association with lipids and are recovered in DRMs (Schuck *et al.*, 2003). Essentially, all of these proteins have a saturated hydrocarbon chain modification, as a membrane anchor inserted into the raft domain of the outer plasma membrane leaflet and many transmembrane proteins require palmitoylation for their association to lipid rafts. Accordingly, isolation of DRMs represents a valuable tool for the analysis of raft-associated proteins involved in cell apoptosis and a useful starting point for defining membrane subdomains and/or protein–ganglioside interaction. Applying a variety of detergents may affect lipid–protein interactions (Schuck *et al.*, 2003). Specific protein–lipid interactions within rafts have been studied preferentially by coimmmunoprecipitation experiments (Garofalo *et al.*, 2002, 2003) or by fluorescence resonance energy transfer (FRET) (Edidin, 2003). In addition, many new approaches for detecting heterogeneity in cell membranes have emerged that rely on the distinct diffusion characteristics or enhanced proximity between raft components. Single particle tracking (SPT) has enabled us to measure the diffusion characteristics of GPI-anchored proteins. Fluorescence correlation spectroscopy, which combines different evaluations of biophysical properties of the plasma membrane, can contribute to better analyze the dynamics of raft components in living cells.

Some of these approaches are of widespread use in laboratory practice (mainly coimmunoprecipitation experiments), whereas others are specifically employed under certain experimental conditions. Coimmunoprecipitation has become the tool of choice for analyzing the dynamics of raft components and their functional interaction with proteins involved in cell apoptosis (Fig. 6.1).

2.1. Coimmunoprecipitation

The starting material can be (i) a cell pellet or (ii) the DRM fraction, containing lipid rafts, isolated by sucrose density-gradient ultracentrifugation after Triton X-100 (TX-100) solubilization at 4°, according to Iwabuchi *et al.* (2000).

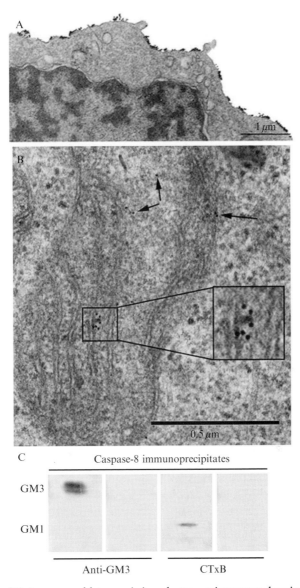

Figure 6.1 (A) Immunogold transmission electron microscopy showing the typical clustered distribution of gold particles, corresponding to ganglioside-enriched micro-domains, on cell plasma membrane of a T cell. (B) Immunogold labeling of GD3 in CEM cells induced to apoptosis by treatment with anti-CD95/Fas. Cells were prepared following the standard embedding procedure. Ultrathin sections, picked up on gold grids, were labeled using an anti-GD3 MoAb (1:30) as the primary antibody and, subsequently, with an antimouse IgM-10 nm gold conjugated (1:10). Note the distribution of GD3 molecules, which appear specifically localized on mitochondrial membranes. (Inset) The well-arranged distribution of gold particles at higher magnification.

Detergents

Although TX-100 is the most widely used detergent for the purification of lipid rafts, other detergents have been used for this purpose, including Brij96, Brij98, CHAPS, Triton X-14, and Lubrol WX (Schuck *et al.*, 2003). The use of detergents to solubilize the cell membrane leads to the generation of small micelles and large particles that can be separated by centrifugation at low temperature (4°) for 17 to 18 h at 200,000g in a 5 to 35% linear discontinous gradient of sucrose (Iwabuchi *et al.*, 2000). Before starting solubilization, make sure all lysing/solutions are prechilled on ice, as it has been also shown that the amount of lipids in the plasma membrane raft is increased at low temperature.

2.2. Protocol

2.2.1. Cultured cells

Both adherent cell lines and cells growing in suspension can be examined.

2.2.2. Reagents

- Lysis buffer: 10 mM Tris-HCl, pH 8.0, 150 mM NaCl, 1% Nonidet P-40, 1 mM phenylmethylsulfonyl fluoride, 10 mg of leupeptin/ml, 1 mM sodium orthovanadate
- Protein A/G acrylic or magnetic beads

2.3. Coimmunoprecipitation from cell pellet

1. Collect the culture supernatant, centrifuge at 300g for 10 min, and wash twice with ice-cold phosphate-buffered saline (PBS).
2. After removing the supernatant, resuspend the pellet in 1 ml ice-cold lysis buffer.
3. Transfer the lysate to a 1-ml homogenization tube and set on ice for at least 20 to 30 min.
4. Homogenize the samples using a tight-fitting Dounce pestle (20 strokes), maintaining the tube on ice at all times.
5. Transfer the lysate to a 1.5-ml tube and centrifuge for 5 min at 1300g to remove nuclei and large cellular debris.
6. Determine the protein concentration in the supernatant fraction using the Bio-Rad kit (Bio-Rad, Richmond, CA) with bovine serum albumin (BSA) as standard.

(C) T cells, undergoing apoptosis induced by CD95/Fas, were immunoprecipitated with anticaspase-8. The immunoprecipitates were analyzed for the presence of ganglioside molecules by TLC immunostaining using anti-GM3 MoAb or cholera toxin, B subunit (CTxB) (from Garofalo *et al.*, 2003).

7. Mix the supernatant fraction with protein A/G acrylic beads (Sigma Chemical Co., St. Louis, MO), stirring by a rotary shaker for 2 h at 4°to preclear nonspecific binding.

8. Centrifuge the lysate for 1 min at 500g at 4°.

9. Collect the supernatant.

10. Incubate with specific antibodies for specific binding with proteins, stirring by a rotary shaker for 2 h at 4°.

11. Add protein A/G acrylic beads for 1 h.

12. Centrifuge the sample for 1 min at 500g at 4°.

13. Remove the supernatant. Resuspend the pellet in 0.5 ml of H_2O and transfer it into a 15-ml tube for ganglioside extraction.

2.4. Coimmunoprecipitation from DRM fractions

Detergent-resistant membranes, obtained as described (Iwabuchi *et al.*, 2000), are subjected to coimmunoprecipitation following the procedure starting from step 7.

2.5. Ganglioside analysis in immunoprecipitate samples

Immunoprecipitation samples are subjected to GSL analysis by the high-performance thin-layer chromatography (HPTLC) technique. HPTLC has become the tool of choice for the identification and quantification of GSLs in small amounts of sample extracted from the plasma membrane of T cells, based on the fact that the HPTLC plate has a high-resolution property for glycosphingolipid separation. Of importance, the detection of glycosphingolipids after HPTLC separation is achieved by using specific colorimetric reagents or specific anti-GSLs monoclonal antibodies, followed by peroxidase-conjugated secondary antibodies. The latter represents an important method used to identify with high sensitivity small amounts (10 ng) of protein-associated ganglioside.

2.6. Reagents

1. HPTLC aluminium-backed silica gel 60 (20 × 20) plates (Merck, Darmstadt, Germany)

2. Solvent system for TLC (chloroform:methanol:0.25% aqueous KCl (5:4:1) (v:v:v)

3. Resorcinol-HCl reagent. Resorcinol (200 mg) in 20 ml of distilled water is mixed with 80 ml of concentrated HCl and 0.25 ml of 0.1 M $CuSO_4$

4. 0.5% (w/v) solution of poly(isobutyl methacrylate) beads (Polyscience, Warrington, PA) dissolved in hexane (PIM solution)

5. 0.1% (w/v) BSA/PBS

6. Monoclonal antibodies against glycosphingolipids

7. Secondary antibody, horseradish peroxidase-conjugated antimouse immunoglobulins
8. Enhanced chemiluminescence (ECL) Western blotting detection reagents (Amersham Life Science Ltd., Buckinghamshire, UK)

2.7. Procedure

Gangliosides from immunoprecipitates are extracted according to the method of Svennerholm *et al.* (1980), with minor modifications.

2.7.1. Ganglioside extraction

1. Extract the immunoprecipitate in chloroform:methanol:water (4:8:3) (v:v:v) and centrifuge twice at 1500g for 30 min at room temperature.
2. Collect the supernatant and add water, resulting in a final chloroform: methanol:water ratio of 1:2:1.4 (v:v:v).
3. Centrifuge the gradient at 1500g for 30 min at room temperature.
4. Collect the upper phase containing polar glycosphingolipids.
5. Purify the upper phase of salts and low molecular weight contaminants using Bond elute C18 columns (Superchrom, Milan, Italy). According to Williams and McCluer, 1980.

2.7.2. TLC Immunostaining

1. Resuspend the ganglioside extract in 10 μl of chloroform:methanol 2:1 (v:v) and separate on HPTLC aluminum-backed plates.
2. After HPTLC, dry the plate thoroughly with a hair dryer to remove the organic solvent.
3. Soak gently the dried plate for 90 s in 0.5% PIM solution in a glass box of appropriate size.
4. Remove the plate from the PIM solution and air dry.
5. Incubate the plate with a buffer containing 5% BSA/PBS for 30 min at room temperature to block nonspecific absorption of antibodies.
6. Remove the blocking solution and wash the plate gently for 10 min with 0.5% BSA/PBS with two changes of the buffer.
7. Add the first antibody (anti-GSLs) diluted with 0.1% BSA/PBS for 1 h at room temperature.
8. Wash the plate with 0.5% BSA/PBS with three changes of the buffer and then incubate it with the peroxidase-labeled secondary antibody solution for 1 h at room temperature.
9. Wash with 0.5% BSA/PBS in the same way.
10. Using ECL, visualize the positive reaction.

2.8. Isolation of mitochondria

2.8.1. Reagents and buffers

Mitochondria isolation buffer (MIB): 220 mM mannitol, 68 mM sucrose, 10 mM KCl, 1 mM EDTA, 1 mM EGTA, 10 mM HEPES-KOH (pH 7.4), 0.1% (w/v) BSA, supplemented with a cocktail of protease inhibitors

2.8.2. Equipment

Glass Dounce homogenizer (15 ml) with a tight Teflon pestle
Glass Dounce homogenizer (2 ml) with a glass pestle (B type)

2.9. Procedure

Preparation of cytosolic extracts

Cytosolic extracts are obtained according to Bossy-Wetzel et al. (2000). Cells are harvested by centrifugation at 200g for 10 min at 4°, washed twice with cold PBS, pH 7.4, and resuspended in 3 volumes of cold cytosolic extraction buffer. After incubation on ice for 30 min, cells are disrupted by homogenization with a glass Dounce homogenizer and a tight glass pestle, applying 50 strokes. Nuclei and unaffected cells are removed by centrifugation at 800g for 10 min at 4° in an Eppendorf centrifuge. Supernatants are transferred to new Eppendorf tubes and centrifuged further at 22,000g for 30 min. Supernatants are then used for detergent solubilization.

2.10. Detergent solubilization of isolated mitochondria

Reagents and buffers

Extraction buffer: 25 mM HEPES, pH 7.5, 0.15 M NaCl, 1% Triton X-100, and 100 U/ml kallikrein/aprotinin
Solubilization buffer: 50 mM Tris-HCl, pH 8.8, 5 mM EDTA, and 1% SDS

2.11. Procedure

Supernatants containing isolated mitochondria are detergent solubilized according to Skibbens et al. (1989). Briefly, mitochondria are lysed with 1 ml of extraction buffer for 20 min on ice. Lysates are collected and centrifuged for 2 min in a Brinkmann microfuge at 10,000g at 4°. Supernatants, containing Triton X-100 soluble material, are collected; pellets are centrifuged a second time (30 s) in order to remove the remaining soluble material. Pellets are then solubilized in 100 μl of solubilization buffer. DNA is sheared by passage through a 22-gauge needle. Both Triton X-100-soluble and -insoluble material are then analyzed by Western blot for specific proteins.

2.12. Coimmunoprecipitation from isolated mitochondria

Isolated mitochondria are subjected to coimmunoprecipitation following the procedure reported earlier.

3. The Contribution of Morphological Analyses

Numerous different morphological techniques have been considered for studying the complex sequence of structural modifications of lipid rafts during cell apoptosis. These are mainly qualitative analyses that can be carried out by means of light (LM) and electron microscopy (EM) techniques. Morphometric analyses, essentially performed by LM, can also be carried out. The use of these different technical approaches can provide information on the surface or intracellular localization of lipid rafts (Fig. 6.1).

3.1. Light microscopy

The aim of immunocytochemistry is to localize antigens by labeling them with specific antibodies. Analyzing lipid raft dynamics in cell apoptosis, antiganglioside antibodies may be considered a good tool. However, we recommend that biochemical analysis be used to confirm that a particular ganglioside is present in the cell or tissue under test.

3.1.1. Reagents

Available antiganglioside antibodies are summarized by Schwarz *et al.* (2000).

3.1.2. Fixation

The major problem to be taken into consideration when using antiganglioside antibodies is the choice of fixative (Schwarz *et al.*, 2000). For detection of lipid rafts on the cell surface, incubation with antibodies prior to fixation (prefixation) with 4% formaldehyde for 30 min at room temperature yields reproducible patterns of immunofluorescence. Alternatively, in order to detect raft-like microdomains associated with intracellular organelles (i.e., mitochondrion), lymphocytic cells may be postfixed with 4% paraformaldehyde in PBS for 30 min at room temperature and then permeabilized with 0.5% Triton X-100 in PBS for 5 min at room temperature (Garofalo *et al.*, 2005).

3.2. Protocol

1. Soak in Hanks' balanced salt solution.
2. Incubate with anti–GSLs MoAb for 1 h (times and temperatures of incubation with primary antibodies are empirical and need to be determined for each antibody and cell type) at 4°.

3. Wash three times in PBS, pH 7.4.
4. Fixate with 4% paraformaldehyde in PBS for 30 min (times and temperatures of incubation with primary antibodies are empirical and need to be determined for each antibody and cell type) at room temperature.
5. Wash three times in PBS, pH 7.4.
6. Stain with FITC-conjugated secondary antibody for 45 min (times and temperatures of incubation with primary antibodies are empirical and need to be determined for each antibody and cell type) at 4°.
7. Wash three times in PBS, pH 7.4.
8. Counterstain with Texas red-conjugated antiprotein antibody.
9. Wash three times in PBS, pH 7.4.
10. Mount in 0.1 M Tris-HCl, pH 9.2, containing 60% glycerol (v:v).
11. Acquire images through a confocal laser-scanning microscope.

The green (FITC) and red (Texas red) fluorophores are excited simultaneously at 488 and 518 nm and are observed by two different detectors. Before image acquisition, samples are scanned at different filter conditions to choose a setting that reduces the overlap of emission spectra with a maximal signal-to-noise ratio. Then, acquired images are processed by subtracting a scaled version of green from red series, and vice versa. The scale factor (bleed-through factor) is determined by scanning single-stained samples in a dual fluorescence scanning configuration. Samples are counterstained with Texas red fluorophore, which reduces fluorescence overlapping greatly. Serial optical sections are assembled in depth-coding mode. Acquisition and processing are carried out using appropriate confocal software.

3.3. Postembedding electron microscopy

There are two most common procedures that allow one to visualize cellular antigens with transmission electron microscopy: preembedding and postembedding techniques. Preembedding is used to label surface antigens only. It consists of an antigen–antibody reaction on living cells, that is, before the embedding and sectioning procedure. In contrast, the postembedding technique should be used when an internal antigen would be labeled. It consists of an antigen-antibody reaction in ultrathin sections, that is, after the embedding and sectioning procedure.

For both of these methods, subcellular antigens recognized by primary antibodies can be localized and visualized with a transmission electron microscope (TEM) using gold-conjugated secondary antibodies to provide areas of high electron density.

Colloidal gold can be coupled either to protein A, a protein from *Staphylococcus* cell walls that binds the Fc portion of immunoglobulins, or to a secondary antibody.

Preservation of excellent structural details and, at the same time, maintenance of antigenic sites reactivity are the hardest but primary concerns for immune gold labeling. In order to preserve antigenicity, conventional electron microscopy procedure should be modified. In fact, few antigens survive to the routine fixation and embedding procedures. The tendency is to use a gentler aldehyde fixation, combined with low temperature embedding, that has been shown to better preserve the antigenicity of biological samples (Gonzalez *et al.*, 1997; Scala *et al.*, 1992) but, to some extent, this procedure compromises the ultrastructural features of the cells.

3.4. Fixation

The reactivity and the stability of diverse antigens differ so widely from each other that no standardized method of fixation can be used for all EM immunocytochemistry experiments.

Initially, it would be routinely worthwhile to assess the preparation of the sample. In fact, it is important to check if the antigen to be labeled is resistant to the fixation procedures and if the sample ultrastructural features could be preserved. Milder fixation regimens normally required to retain antigenicity can actually result in the loss of ultrastructural details.

Depending on the chemical nature of such an antigen, successful immunolabeling may be obtained with routine preparations.

Unfortunately, for most antigens the standard procedure results in a complete, or at least severely impaired, loss of antigenicity. This means that a range of preparative procedures, varying in their degree of fixation strength, is recommended to optimize the compromise between ultrastructural preservation and immunolabeling efficiency.

In our experience, proteins are affected more profoundly by the standard embedding procedures, whereas lipid antigens are more likely to be preserved (Garofalo *et al.*, 2005).

3.5. Embedding medium

Biological specimens are embedded in resin in order to allow thin sectioning. Many different resin formulas are currently available and all these mixtures have been used for postembedding labeling (Bogers *et al.*, 1996; Brorson, 1998). For immunocytochemistry purposes, acrylic resins that polymerize under ultraviolet light (UV) at 4° are preferred. Epoxy resins are exposed at 65° for polymerization with the possible denaturing effects of heat. Furthermore, acrylic resins have low viscosity and are hydrophilic, thus enhancing the subsequent immunolabeling on the sections.

3.6. Embedding procedure

1. Fixation: antigens are quite commonly glutaraldehyde sensitive. Hence the variation of the concentration of this fixative has to be established experimentally for every antigen under consideration. Split the sample into three lots and fix as follows.
   Sample 1: 2.5% glutaraldehyde in sodium cacodylate buffer (pH 7.2; 0.1 M)
   Sample 2: 1% glutaraldehyde in sodium cacodylate buffer (pH 7.2; 0.1 M)
   Sample 3: 4% paraformaldehyde + 0.01% or 0.25% glutaraldehyde in sodium cacodylate buffer (pH 7.2; 0.1 M) for 20 min at room temperature or overnight at 4°
2. Wash in 0.1 M sodium cacodylate buffer three to four times for 5 min each.
3. Postfixation for sample 1 only: postfix in 1% OsO_4 in sodium cacodylate buffer for 30 to 60 min and carefully wash in sodium cacodylate buffer three to four times for 5 min each.

   Although most antigens are osmium tetroxide sensitive, some of them may resist this mild osmium fixation, thus improving preservation of ultrastructural details, especially membranes and membranous structures.

4. Dehydrate in graded series of ethanol solutions (50, 70, 95, and 100% twice) for 10 min each.
5. Resin embedding:
   - Embed sample 1, which is OsO_4 treated, in Epoxy resin–absolute ethanol mixtures (1:2; 1:1; 2:1; absolute) at room temperature and proceed as for the standard embedding procedure.
   - Embed the other samples in acrylic resin as follows: withdraw absolute ethanol and add resin. The time and conditions of each acrylic resin are different; see the data sheet of the chosen resin. For example, Unicryl resin is directly added as absolute: leave the sample to infiltrate with resin for 4 h with at least two changes of resin and leave overnight at 4°.
6. Polymerization:
   - Polymerize sample 1 at 65° for 48 h.
   - Place Unicryl samples in *closed* gelatin capsules with fresh resin. Leave them to polymerize by suspension in a rack held 35 cm above a UV light lamp for 72 h.

3.7. Postembedding procedure

Sections of about 80 nm are picked up on inert 200 mesh grids, such as gold or nickel ones. For all incubations, grids are placed on 30-μl droplets on Parafilm within a petri dish. Buffer rinses are carried out by floating grids on 50-μl droplets. Take great care not to wet the opposite grid face while transferring grids from one solution to another.

1. For epoxy resin sections only, etching with 0.5% sodium metaperiodate for 5 to 10 min at room temperature is required to allow exposure of antigenic sites.
2. To avoid nonspecific background labeling, a blocking step in PBS containing 1% BSA as blocking agent for 20 to 30 min at room temperature is recommended.
3. Incubate in specific primary antibody for 1 h at room temperature or overnight at 4° in a humidified chamber. The latter is preferable, as it appears to produce lower levels of nonspecific binding.

To determine the optimal conditions for labeling with minimum background, the concentration of the primary antibody should be determined experimentally by serial dilution in PBS containing 1% BSA.

Polyclonal antibody dilutions are usually 1:10, 1:100, 1:1000, and 1:10,000.
Monoclonal antibody dilutions are normally between 1:5 and 1:30.
Concurrent incubations should be performed in order to confirm the specific labeling of the sections:

- Omit the primary antibody and replace with PBS containing 1% BSA for checking the secondary antibody.
- Replace the primary antibody with normal nonimmune serum, obtained from the same animal, for checking the primary antibody.

4. Wash in PBS containing 1% BSA several times for 5 min each.
5. Incubate with gold–conjugated secondary antibody diluted 1:10 to 1:50 in PBS containing 1% BSA for 1 h at room temperature.
The usual range of gold probe size for TEM is 5 to 20 nm. Probes 5 nm in size have the advantage of improving spatial resolution over the tissue. However, because of their size, they present some disadvantage in the observation at the EM.
6. Wash in PBS containing 1% BSA several times for 5 min each.
7. Fix grids in 2.5% glutaraldehyde in 0.1 M sodium cacodylate buffer for 15 min at room temperature.
8. Wash grids twice in distilled water for 5 min each.
9. Staining is recommended for acrylic resin sections (5 min in a saturated solution of aqueous uranyl acetate and 1 min in lead citrate). Staining is instead facultative for epoxy resin sections. To avoid insoluble precipitates of lead carbonate over the sections, lead citrate staining should be carried out in a covered petri dish containing sodium hydroxide pellets.

REFERENCES

Bogers, J. J., Nibbeling, H. A., Deelderm, A. M., and van Marck, E. A. (1996). Quantitative and morphological aspects of Unicryl versus Lowicryl K4M embedding in immunoelectron microscopic studies. *J. Histochem. Cytochem.* **44**, 43–48.

Bossy-Wetzel, E., and Green, D. R. (2000). Assay for cytochrome c release from mitochondria during apoptosis. *Methods Enzymol.* **322**, 235–242.

Brorson, S. H. (1998). Antigen detection on resin sections and methods for improving the immunogold labeling by manipulating the resin. *Histol. Histopathol.* **13**, 275–281.

Brown, D. A., and London, E. (2000). Structure and function of sphingolipid and cholesterol-rich membrane rafts. *J. Biol. Chem.* **275**, 17221–17224.

Edidin, M. (2003). The state of lipid rafts: From model membranes to cells. *Annu. Rev. Biophys. Biomol. Struct.* **32**, 257–283.

Garofalo, T., Giammarioli, A. M., Misasi, R., Tinari, A., Manganelli, V., Gambardella, L., Pavan, A., Malorni, W., and Sorice, M. (2005). Lipid microdomains contribute to apoptosis-associated modifications of mitochondria in T cells. *Cell Death Differ.* **12**, 1378–1389.

Garofalo, T., Lenti, L., Longo, A., Misasi, R., Mattei, V., Pontieri, G. M., Pavan, A., and Sorice, M. (2002). Association of GM3 with Zap-70 induced by T cell activation in plasma membrane microdomains: GM3 as a marker of microdomains in human lymphocytes. *J. Biol. Chem.* **227**, 11233–11238.

Garofalo, T., Misasi, R., Mattei, V., Giammarioli, A. M., Malorni, W., Pontieri, G. M., Pavan, A., and Sorice, M. (2003). Association of the death-inducing signaling complex with microdomains after triggering through CD95/Fas: Evidence for caspase-8-ganglioside interaction in T cells. *J. Biol. Chem.* **278**, 8309–8315.

Gonzalez Santander, R., Martinez Cuadrado, G., Gonzalez-Santander Martinez, M., Monteagudo, M., Martinez Alonso, F. J., and Toledo Lobo, M. V. (1997). The use of different fixatives and hydrophilic embedding media (Historesin and Unicryl) for the study of embryonic tissues. *Microsc. Res. Tech.* **36**, 151–158.

Grassmé, H., Jekle, A., Riehle, A., Schwarz, H., Berger, J., Sandhoff, K., Kolesnick, R., and Gulbins, E. (2001). CD95 signaling via ceramide-rich membrane rafts. *J. Biol. Chem.* **276**, 20589–20596.

Ipsen, J. H., Karlstrom, G., Mouritsen, O. G., Wennerstrom, H., and Zuckermann, M. J. (1987). Phase equilibria in the phosphatidylcholine-cholesterol system. *Biochim. Biophys. Acta* **905**, 162–172.

Iwabuchi, K., Handa, K., and Hakomori, S. I. (2000). Separation of glycosphingolipid-enriched microdomains from caveolar membrane characterized by presence of caveolin. *Methods Enzymol.* **312**, 488–494.

Malorni, W., Giammarioli, A., Garofalo, T., and Sorice, M. (2007). Dynamics of lipid raft components during lymphocyte apoptosis: The paradigmatic role of GD3. *Apoptosis* **12**, 941–949.

Scala, C., Cenacchi, G., Ferrari, C., Pasquinelli, G., Preda, P., and Manara, G. C. (1992). A new acrylic resin formulation: A useful tool for histological, ultrastructural, and immunocytochemical investigations. *J. Histochem. Cytochem.* **40**, 1799–1804.

Schuck, S., Honsho, M., Ekroos, K., Shevchenko, A., and Simons, K. (2003). Resistance of cell membranes to different detergents. *Proc. Natl. Acad. Sci. USA.* **100**, 5795–5800.

Schwarz, A., and Futerman, A. H. (2000). Immunolocalization of gangliosides by light microscopy using anti-ganglioside antibodies. *Methods Enzymol.* **312**, 179–187.

Simons, K., and Ikonen, E. (1997). Functional rafts in cell membranes. *Nature* **387**, 569–572.

Skibbens, J. E., Roth, M. G., and Matlin, K. S. (1989). Differential extractability of influenza virus hemagglutinin during intracellular transport in polarized epithelial cells and nonpolar fibroblasts. *J. Cell Biol.* **108,** 821–832.

Svennerholm, L., and Fredman, P. A. (1980). A procedure for the quantitative isolation of brain gangliosides. *Biochim. Biophys. Acta* **617,** 97–109.

Williams, M. A., and McCluer, R. H. (1980). The use of Sep-Pak C18 cartridges during the isolation of gangliosides. *J. Neurochem.* **35,** 266–269.

CHAPTER SEVEN

APOPTOSOME ASSEMBLY

Yigong Shi

Contents

Abstract

Apoptosome refers to the multimeric protein complex that mediates activation of an initiator caspase at the onset of apoptosis. This chapter describes the assembly of three related apoptosomes from mammals, fruit flies, and worms. The assembly of the mammalian apoptosome, which is responsible for the activation of caspase-9, involves Apaf-1 and requires cytochrome c and ATP/dATP binding. Assembly of the apoptosome in *Drosophila melanogaster*, which activates caspase-9 homologue Dronc, involves the Apaf-1 homologue known as Dark/Hac-1/Dapaf-1. In *Caenorhabditis elegans*, assembly of the CED-4 apoptosome requires EGL-1-mediated dissociation of CED-9 (a Bcl-2 homologue) from the CED-4–CED-9 complex and subsequent oligomerization of CED-4. Recent biochemical and structural investigation revealed insights into the assembly and function of the various apoptosomes.

Department of Molecular Biology, Lewis Thomas Laboratory, Princeton University, Princeton, New Jersey

Methods in Enzymology, Volume 442

ISSN 0076-6879, DOI: 10.1016/S0076-6879(08)01407-9

1. INTRODUCTION

Apoptosis, the prevalent form of programmed cell death, plays a central role in the development and homeostasis of all multicellular organisms (Danial and Korsmeyer, 2004; Horvitz, 2003; Rathmell and Thompson, 2002; Wang, 2001). A molecular hallmark of apoptosis is the activation of caspases—specific proteases that execute cell death through the cleavage of multiple protein substrates.

Apoptotic caspases are classified into two general classes: initiator (or apical) and effector (or executioner). The onset of apoptosis requires a cascade of sequential activation of initiator and effector caspases (Budihardjo et al., 1999; Degterev et al., 2003; Fuentes-Prior and Salvesen, 2004; Riedl and Shi, 2004; Thornberry and Lazebnik, 1998). The critical involvement of caspase in apoptosis was discovered in the nematode *Caenorhabditis elegans*, in which the indispensable gene *ced-3* was found to encode a cysteine protease (Xue et al., 1996; Yuan et al., 1993) and to closely resemble the mammalian caspase, interleukin 1β-converting enzyme (or caspase-1), the first cloned caspase (Cerreti et al., 1992; Thornberry et al., 1992). More than a dozen distinct mammalian caspases have been identified, with 11 from the human genome (Shi, 2002b). At least seven mammalian caspases, including four initiator and three effector caspases, are known to play important roles in apoptosis. Caspase-9 is an important initiator caspase in mammalian cells and is activated by cleavage (Fig. 7.1) in response to several forms of intrinsic cell death stimuli. The functional homologues of caspase-9 are Dronc in

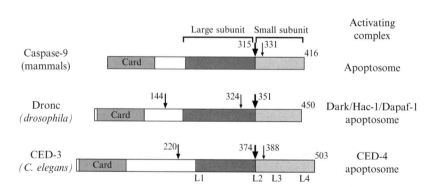

Figure 7.1 Three homologous initiator caspases from mammals, *Drosophila*, and *C. elegans*. Caspase-9, Dronc, and CED-3 are drawn to scale. The position of the first intrachain cleavage (between the large and small subunits) is highlighted by a large arrow, whereas additional cleavages are represented by medium and small arrows. The CARD domains are indicated.

Drosophila melanogaster and CED-3 in *C. elegans* (Fig. 7.1). CED-3 is the only known apoptotic caspase in *C. elegans*, likely fulfilling a dual role of both initiator and effector caspase.

All caspases are synthesized in cells as catalytically inactive zymogens. The activation of an effector caspase, such as caspase-3, is executed by an initiator caspase, such as caspase-9, through proteolytic cleavage after a specific internal Asp residue to separate the large and small subunits of the mature caspase (Thornberry and Lazebnik, 1998). Following intrachain cleavage, the catalytic activity of an effector caspase is increased by several orders of magnitude (Salvesen and Dixit, 1999). Once activated, the effector caspases are responsible for the proteolytic degradation of a broad spectrum of cellular targets that ultimately lead to cell death (Thornberry and Lazebnik, 1998).

In contrast to effector caspases, activation of an initiator caspase invariably requires an apoptosome complex and its regulation is poorly understood. Although intrachain cleavage is essential for the activation of effector caspases, it has only a modest effect on the catalytic activity of some initiator caspases, such as caspase-9, and may not be required for their activation (Srinivasula *et al.*, 2001; Stennicke *et al.*, 1999). The fully processed, free caspase-9 exhibits a low level of catalytic activity, comparable to that of the unprocessed caspase-9 zymogen.

The primary component of the mammalian apoptosome is Apaf-1, which oligomerizes in the presence of cytochrome *c* and ATP/dATP. The Apaf-1 apoptosome is responsible for the activation of caspase-9 (Cain *et al.*, 1999; Hu *et al.*, 1999; Rodriguez and Lazebnik, 1999; Saleh *et al.*, 1999; Zou *et al.*, 1999). In *Drosophila*, the activation of Dronc requires an octameric protein complex of Dark (Rodriguez *et al.*, 1999), also known as Hac-1(Zhou *et al.*, 1999) or Dapaf-1(Kanuka *et al.*, 1999), a homologue of Apaf-1. In *C. elegans*, the activation of CED-3 caspase zymogen is facilitated by the CED-4 complex (Chinnaiyan *et al.*, 1997; Irmler *et al.*, 1997; Seshagiri and Miller, 1997; Wu *et al.*, 1997; Yang *et al.*, 1998; Yuan and Horvitz, 1992), which exhibits significant sequence homology to Apaf-1. In each case, the initiator caspase is recruited into and activated within the described adaptor protein complex involving Apaf-1, Dark, or CED-4. These adaptor protein complexes are oligomeric and are generally referred to as apoptosomes.

There are other forms of apoptosomes. For example, the activation of caspase-2 and caspase-8 in mammalian cells depends on the PIDDosome (Tinel and Tschopp, 2004) and the death-inducing signaling complex (DISC) (Kischkel *et al.*, 1995), respectively. However, neither PIDDosome nor DISC has been completely reconstituted *in vitro* using purified recombinant proteins. This chapter focuses its discussion on the classic apoptosomes involving Apaf-1 and its homologues in *Drosophila* and *C. elegans*.

2. Apoptosome for Caspase-9 Activation

2.1. History

Caspase-9 represents the most thoroughly characterized initiator caspase and its activation is mediated by the apoptosome, a multimeric complex involving Apaf-1, cytochrome *c*, and the cofactor dATP/ATP.

Cytochrome *c* had been known as a crucial molecule for energy production in mitochondria for over half a century. It is generally viewed as a molecule supporting life; thus the discovery of a crucial role for cytochrome *c* in apoptosis, from Xiaodong Wang and colleagues, was a shock to the field and it took years for some cell death researchers to genuinely accept the finding. Cytochrome *c* was first found to be an important cofactor for the activation of caspase-3 (Liu *et al.*, 1996). Subsequently, the cellular receptor for cytochrome *c* was cloned and shown to be a novel protein, named Apaf-1 for apoptotic protease–activating factor 1 (Zou *et al.*, 1997). The oncoprotein Bcl-2 was shown to block apoptosis in part by preventing the release of cytochrome *c* from mitochondria (Yang *et al.*, 1997). These observations support the concept that cytochrome *c* plays an essential role in apoptosis. Finally, the concept was solidified by the identification of cytochrome *c* and dATP-dependent formation of an Apaf-1/caspase-9 complex (Li *et al.*, 1997), which subsequently activates the effector caspases, caspase-3 and -7.

Subsequent characterization revealed that, in the presence of dATP or ATP, cytochrome *c* and Apaf-1 assemble into a ≈1.4 MDa complex, dubbed the "apoptosome" (Cain *et al.*, 1999; Hu *et al.*, 1999; Rodriguez and Lazebnik, 1999; Saleh *et al.*, 1999; Zou *et al.*, 1999) (Fig. 7.2A). The CARD domain of Apaf-1 in the apoptosome interacts with the CARD domain of procaspase-9, resulting in the recruitment and subsequent activation of procaspase-9 zymogen (Li *et al.*, 1997; Qin *et al.*, 1999). Surprisingly, caspase-9 bound to the apoptosome exhibits a catalytic activity that is three orders of magnitude higher than that of the isolated caspase-9, prompting the concept of a holoenzyme (Rodriguez and Lazebnik, 1999). The primary function of the apoptosome appears to be allosteric regulation of the catalytic activity of caspase-9. Supporting this notion, procaspase-9 zymogen possesses a basal level of activity in the absence of the activation cleavage (Stennicke *et al.*, 1999) and this activity can be increased to the same level as the cleaved caspase-9 by the apoptosome (Srinivasula *et al.*, 2001).

2.2. Domains, structure, and assembly

Apaf-1, the central component of apoptosome, contains three distinct domains, an N-terminal CARD, an expanded nucleotide-binding domain, and 13 WD40 repeats at its carboxy-terminal half. The CARD is

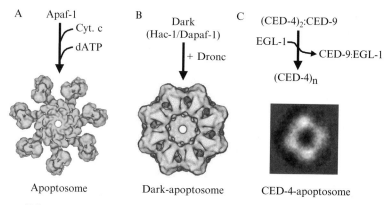

Figure 7.2 Assembly of apoptosomes in mammals, *Drosophila*, and *C. elegans*. (A) Assembly of the apoptosome in mammalian cells requires cytochrome *c* and dATP or ATP. The apoptosome consists of seven molecules of Apaf-1 bound to cytochrome *c*. (B) Assembly of the Dark apoptosome in *Drosophila* does not require cytochrome *c*. The assembly is triggered *in vitro* through incubation with either dATP or Dronc zymogen. (C) Assembly of the CED-4 apoptosome is initiated by displacement of CED-9 by EGL-1 from the CED-4–CED-9 complex. The freed CED-4 dimer further oligomerizes to form the CED-4 apoptosome.

responsible for the recruitment of caspase-9 through interactions with the CARD of caspase-9 (Li *et al.*, 1997; Qin *et al.*, 1999). Together, the CARD and the nucleotide-binding domains are responsible for the oligomerization of Apaf-1 in the presence of cytochrome *c* and dATP. The WD40 repeats are thought to interact with cytochrome *c*, because removal of this domain in Apaf-1 resulted in constitutive binding and activation of caspase-9 (Hu *et al.*, 1998; Srinivasula *et al.*, 1998). There are four distinct Apaf-1 splicing variants, but all variants contain these three essential domains. The longest form of these variants, containing 1248 amino acids, is thought to be responsible for the proapoptotic activity of Apaf-1.

Apaf-1 exists in an inactive conformation in normal cells and is activated through binding to cytochrome *c* and dATP during apoptosis (Li *et al.*, 1997). How does Apaf-1 maintain itself in an inactive state prior to dATP/ATP binding? The answer to this question was revealed by the crystal structure of a WD40-deleted Apaf-1 (Riedl *et al.*, 2005). In the structure, five distinct domains, CARD, three-layered α/β domain, helical domain I, winged-helix domain, and helical domain II, pack closely against each other through extensive interdomain interactions (Riedl *et al.*, 2005). These interactions result in the partial burial of the caspase-9-binding interface. Unexpectedly, the bound nucleotide is ADP, which is deeply buried and serves as an organizing center to strengthen interactions among these four adjoining domains (Riedl *et al.*, 2005). Structural analysis suggests that binding of nucleotide may induce significant conformational changes in

Apaf-1 and that these conformational changes may drive the formation of the caspase-9-activating apoptosome. Thus this structure also provides a plausible explanation to the question why ATP/dATP binding is required for the activation of Apaf-1.

How does the activated Apaf-1 molecule assemble into an apoptosome? The structure of the apoptosome, determined by electron cryomicroscopy (cryo-EM), reveals a wheel-shaped architecture with sevenfold symmetry (Acehan et al., 2002; Yu et al., 2005) (Fig. 7.2A). The CARD and the expanded nucleotide-binding domain of Apaf-1 form the central hub; the WD40 repeats constitute the extended spokes (Fig. 7.2A). Knowledge of the crystal structure of the autoinhibited Apaf-1 at an atomic resolution allowed assignment of the individual domains in the EM map (Yu et al., 2005), although these features await further confirmation at an atomic resolution. Despite the relatively low resolution, the cryo-EM structure confirms the structural involvement of cytochrome c in formation of the apoptosome. Docking of procaspase-9 to this apoptosome resulted in a dome-shaped structure in the center (Acehan et al., 2002). Limited by the low resolution, the apoptosome structure does not allow assignment of atomic interactions. Consequently, the molecular underpinnings of caspase-9 activation remain to be elucidated.

The kinetics and regulation of apoptosome assembly are fairly complex and are only beginning to be appreciated. The bound nucleotide in Apaf-1 is likely to undergo dynamic changes during apoptosome assembly. Cytochrome c binding to the monomeric Apaf-1 was reported to induce hydrolysis of dATP to dADP, which was subsequently replaced by exogenous dATP (Kim et al., 2005). This observation led to the hypothesis that hydrolysis of dATP and subsequent exchange of dADP by dATP were two required steps for apoptosome assembly (Kim et al., 2005). In another study, the monomeric Apaf-1 was found to contain predominantly ADP, not ATP or dATP (Bao et al., 2007).

Apoptosome assembly is likely regulated by other cellular factors, such as nucleotide exchange factors and/or ATPase activating proteins, to impact on apoptosome assembly, although no specific factor has been identified. The oncoprotein prothymosin-α was shown to negatively regulate caspase-9 activation by inhibiting apoptosome formation (Jiang et al., 2003). The small molecule α-(trichloromethyl)-4-pyridineethanol (PETCM) relieved prothymosin-α inhibition and allowed apoptosome formation, although how PETCM accomplished this task remains unclear (Jiang et al., 2003).

Physiological concentrations of calcium ion have been shown to inhibit the assembly of apoptosome through blocking the exchange of ADP/dADP for ATP/dATP (Bao et al., 2007). Calcium was found to bind to the autoinhibited Apaf-1 directly and to stabilize its ADP-bound conformation.

Calcium is critical to many cellular processes, including cell proliferation and apoptosis (Berridge et al., 2003; Parekh and Penner, 1997; Santella, 1998). The switch between cell proliferation and death depends on the spatial and temporal organization of calcium entry and concentration (Fanelli et al., 1999; Lang et al., 2005; Lang et al., 2000). The finding that calcium might regulate the Apaf-1-dependent apoptotic pathway by interfering with the assembly of apoptosome has important implications for the role of calcium homeostasis on cell death.

2.3. Function of apoptosome

The function of an assembled apoptosome is to activate caspase-9. However, the underlying molecular mechanisms by which caspase-9 is activated by apoptosome have remained incompletely understood and somewhat controversial.

The prevailing model for initiator caspase activation is induced proximity, which states that the initiator caspases autoprocess themselves when brought into close proximity of each other (Salvesen and Dixit, 1999). A refinement of the induced proximity model is the proximity-driven dimerization model (Boatright and Salvesen, 2003). Based on this model, the heptameric Apaf-1 apoptosome may recruit multiple copies of inactive procaspase-9 into close proximity of one another. The high concentrations of procaspase-9 monomers in the apoptosome are thought to favor dimerization and hence activation (Acehan et al., 2002; Renatus et al., 2001). This model is consistent with the observed second-order activation of caspase-9 by a miniapoptosome (Pop et al., 2006). In addition, a fusion protein between a dimeric leucine zipper and caspase-9 led to significant enhancement of its catalytic activity, which suggests that the dimerization of caspase-9 might be sufficient for its activation (Yin et al., 2006). In contrast to these studies, an engineered caspase-9, which exists as a constitutive homodimer in solution, exhibited a much lower level of catalytic activity compared to the apoptosome-activated caspase-9 (Chao et al., 2005). This observation was taken to imply an induced conformation model for the apoptosome-mediated activation of caspase-9, in which caspase-9 binding to the apoptosome was thought to result in an altered active site conformation and consequent activation of caspase-9 (Chao et al., 2005).

The proximity-driven dimerization model and the induced conformation model are not always mutually exclusive. For example, the essence of proximity-driven dimerization for an initiator caspase is to orient the active site conformation for more efficient substrate binding and catalysis (Fuentes-Prior and Salvesen, 2004). Nonetheless, definitive proof to either of these models, which likely entails structure of the apoptosome holoenzyme at an atomic resolution, remains at large.

2.4. Protocol for assembly and detection of apoptosome

There are several published protocols for the assembly of apoptosome (Cain et al., 1999; Hu et al., 1999; Rodriguez and Lazebnik, 1999; Saleh et al., 1999; Zou et al., 1999). An outline is given here, but the original papers should be consulted and the techniques adapted as required. In each case, cytochrome c and a nucleotide triphosphate are required to incubate with the Apaf-1 protein. In addition to dATP and ATP, several other nucleotide analogs have also been shown to support the assembly of apoptosome (Genini et al., 2000; Leoni et al., 1998). Reconstitution of apoptosome in vitro requires three essential components: recombinant Apaf-1 protein, cytochrome c, and dATP or ATP. The full-length Apaf-1 was expressed in baculovirus-infected insect cells and affinity purified (Cain et al., 1999; Hu et al., 1999; Rodriguez and Lazebnik, 1999; Saleh et al., 1999; Zou et al., 1999). Reconstitution of the miniapoptosome, involving a WD40-deleted variant of Apaf-1 (residues 1–591), does not require cytochrome c but still requires dATP/ATP (Riedl et al., 2005).

The protocols described here have been reported previously (Bao et al., 2007). For the assembly of apoptosome, the full-length Apaf-1 was incubated with 100 μM cytochrome c in the presence of 1 mM dATP at 4° overnight or at 22° for 10 min. For the assembly of miniapoptosome, Apaf-1 (residues 1–591) was incubated with 1 mM dATP at 4° overnight or at 22° for 10 min. The assembly of apoptosome was considerably faster at 22° than at 4°. However, the assembled apoptosome appears to exhibit a low solubility at 22° and tends to precipitate out of solution. Therefore, the recommended temperature for the assembly of apoptosome is 4° overnight if the starting concentration of Apaf-1 is more than 0.1 mg/ml. For lower concentrations of Apaf-1, the condition of 22° for 10 min may be used.

The formation of apoptosome is usually detected by gel filtration chromatography, in which the elution volume for Apaf-1 will be significantly smaller upon apoptosome assembly. Apaf-1 can be detected by Coomassie blue staining, in the case of abundant protein, or by Apaf-1-specific polyclonal or monoclonal antibody, in the case of trace amount of Apaf-1. The functional detection of apoptosome assembly relies on caspase-9 activity, which usually employs a fluorogenic peptide substrate. In one published report (Bao et al., 2007), the assay was performed at 25° in a buffer containing 20 mM HEPES (pH 7.0), 100 mM NaCl, and 5 mM dithiothreitol. To a final volume of 200 μl, a final concentration of 2 μM Apaf-1, 1 mM dATP, 10 μM cytochrome c, and 0.2 μM procaspase-9 was added for a 10-min incubation at 22°. Then the caspase-9 substrate LEHD-AMC was added to the reaction at a final concentration of 200 μM. Conversion of the fluorogenic substrate was measured in a Hitachi F2500 fluorescence spectrophotometer using an excitation wavelength of 380 nm and an emission wavelength of 440 nm. When Apaf-1 (residues 1–591) was assayed, cytochrome c was omitted from the reaction.

3. DARK APOPTOSOME FOR DRONC ACTIVATION

Dronc is the functional homologue of the mammalian initiator caspase-9 in *Drosophila* and is required for programmed cell death during the normal development of fruit flies (Chew *et al.*, 2004; Daish *et al.*, 2004; Dorstyn *et al.*, 1999). One important downstream target of Dronc is the effector caspase Drice, a homologue of mammalian caspase-3. The activation of Dronc in *Drosophila* cells requires Dark (also known as Hac-1 or Dapaf-1) (Kanuka *et al.*, 1999; Rodriguez *et al.*, 1999; Zhou *et al.*, 1999), the functional homologue of mammalian Apaf-1. However, in contrast to caspase-9, the CARD domain of Dronc is removed in the mature Dronc caspase in *Drosophila* cells (Muro *et al.*, 2004), suggesting a mode of activation different from that of caspase-9.

In contrast to the Apaf-1 apoptosome, cytochrome c was not required for assembly of the Dark apoptosome and, when added, cytochrome c did not bind to the Dark apoptosome (Yu *et al.*, 2006) (Fig. 7.2B). Dark was shown to assemble into an apoptosome in the presence of dATP (Yu *et al.*, 2006). However, it is unclear whether the availability of dATP/ATP to Dark is regulated in cells, and if so, how? The recombinant, full–length Dark protein exists as a monomer in solution (Yu *et al.*, 2006). In the absence of added dATP/ATP, incubation of the full–length Dark protein with Dronc zymogen results in the immediate formation of an apoptosome complex, within which the Dronc zymogen is activated (Shi, unpublished data). This observation suggests that the Dark apoptosome may be assembled upon encounter with the Dronc zymogen.

The cryo-EM structure of the Dark apoptosome revealed two wheel-shaped particles assembled face to face, each involving eight molecules of Dark (Yu *et al.*, 2006). Structural analysis of the Dark apoptosome using relevant crystal structures showed that, despite some apparent differences, the Dark apoptosome and the Apaf-1 apoptosome share a number of important features. For example, placement of the CARD domain in the apoptosome appears to be highly conserved. This analysis also suggested that a single wheel-shaped particle may represent the functional Dark apoptosome in *Drosophila* (Fig. 7.2B). It is important to note that, although both Dark and Apaf-1 belong to the AAA+ ATPases, the proposed models of oligomerization for Dark and Apaf-1 are quite different from that of other

AAA+ ATPases (Diemand and Lupas, 2006). The structure of the Dark apoptosome at an atomic resolution is needed to resolve this discrepancy.

How does the Dark apoptosome mediate activation of the Dronc zymogen? Available evidence suggests that, in contrast to the Apaf-1 apoptosome, the Dark apoptosome merely functions as a scaffold to bind Dronc and facilitates its maturation through autocatalytic cleavages. Unlike caspase-9, the CARD domain of Dronc is removed from the mature Dronc caspase (Yan *et al.*, 2006a) (Fig. 7.1). Consequently, the mature Dronc caspase is not associated with the Dark apoptosome (Shi, unpublished data). Similar to effector caspases, the mature Dronc caspase exhibits a drastically enhanced catalytic activity *in vitro* compared to the Dronc zymogen (Yan *et al.*, 2006a). The molecular basis for the differential activities was attributed to the observation that the mature Dronc formed a homodimer, whereas the uncleaved Dronc zymogen existed exclusively as a monomer (Yan *et al.*, 2006a). Thus the autocatalytic cleavage in Dronc induces its stable dimerization, which presumably allows the two adjacent monomers to mutually stabilize their active sites, leading to activation. Crystal structure of a CARD-deleted Dronc zymogen revealed an unproductive conformation at the active site, which is consistent with the observation that the zymogen remains catalytically inactive (Yan *et al.*, 2006a).

4. CED-4 Apoptosome for CED-3 Activation

4.1. Pathway, structure, and assembly

Genetic analysis in *C. elegans* led to the identification of four genes, *egl-1*, *ced-9*, *ced-4*, and *ced-3*, that control the death of 131 somatic cells during hermaphrodite development (Horvitz, 1999, 2003). Death of these somatic cells was caused by the CED-3 caspase. Similar to caspase-9 and Dronc, CED-3 is synthesized as an inactive zymogen in cells and must undergo an activation process that is mediated by the adaptor molecule CED-4 (Chinnaiyan *et al.*, 1997; Irmler *et al.*, 1997; Seshagiri and Miller, 1997; Wu *et al.*, 1997; Yang *et al.*, 1998; Yuan and Horvitz, 1992). In normal cells, the proapoptotic protein CED-4 is sequestered by the mitochondria-bound protein CED-9 (Chen *et al.*, 2000; Chinnaiyan *et al.*, 1997; Hengartner and Horvitz, 1994; James *et al.*, 1997; Spector *et al.*, 1997; Wu *et al.*, 1997), thus unable to activate CED-3. At the onset of apoptosis, the inhibitory CED-4/CED-9 interaction is disrupted by the proapoptotic protein EGL-1 (Conradt and Horvitz, 1998; del Peso *et al.*, 1998, 2000; Parrish *et al.*, 2000), which is activated transcriptionally in cells destined to die. The released CED-4 is thought to undergo homo-oligomerization, which then facilitates the activation of CED-3 (Yang *et al.*, 1998) (Fig. 7.2C).

Recent structural investigation revealed that one molecule of CED-9 interacts with an asymmetric dimer of CED-4 but only specifically recognizes one of the two CED-4 molecules through an extensive interface dominated by hydrogen bonds (Yan et al., 2005, 2006b). These specific interactions prevent CED-4 from activating CED-3. EGL-1 binding induces significant conformational changes in CED-9 that result in the dissociation of CED-9 from the CED-4 dimer (Yan et al., 2005). The freed CED-4 dimer further oligomerizes to form a CED-4 apoptosome, which facilitates the autoactivation of CED-3 (Fig. 7.2C). How CED-4 assembles into an oligomer and exactly how the CED-4 apoptosome facilitates CED-3 autoactivation await further investigation.

4.2. Protocol for assembly and detection of CED-4 apoptosome

The CED-4 protein has not been expressed successfully in a soluble and functional form as a recombinant protein. Consequently, assembly of the CED-4 apoptosome relies on the displacement of CED-9 from the CED-4–CED-9 complex by the recombinant EGL-1 protein (Yan et al., 2005). The technique summarized here is an outline of the steps used to prepare a recombinant CED-4–CED-9 complex. The researcher should be familiar with these primary techniques of molecular biology.

Preparation of a soluble, recombinant CED-4–CED-9 complex was also a technical challenge that had plagued us and other laboratories for years. We eventually succeeded in isolation of the CED-4–CED-9 complex using a coexpression strategy at 15° or lower, which presumably facilitates protein folding. The transmembrane segment-deleted CED-9 (residues 1–251) contains a His6 tag at its amino terminus and the full-length CED-4 (residues 1–549) is untagged. The CED-4–CED-9 complex is affinity purified using a standard protocol for Ni-NTA resin. This complex contains two copies of CED-4 and a single molecule of CED-9. The CED-4–CED-9 complex is incubated with an excess amount of recombinant EGL-1 protein (residues 1–91). The freed CED-4 apoptosome is subsequently isolated by gel filtration.

The formation of CED-4 apoptosome is usually detected by gel filtration chromatography, in which the elution volume for CED-4 will be significantly smaller compared to that of the CED-4–CED-9 complex. CED-4 can be detected by Coomassie blue staining, in the case of abundant protein, or by CED-4-specific polyclonal or monoclonal antibody, in the case of trace amounts of CED-4. Functional detection of CED-4 apoptosome assembly relies on a CED-3 activation assay. In one published report (Yan et al., 2005), the WT full-length CED-3 was in vitro translated using the TNT T7 quick-coupled transcription and translation system (Promega) at 30° for 25 min. The CED-4 apoptosome was added to the translation product and incubated at 30° for another 25 min before samples were taken out for SDS-PAGE analysis and visualized by autoradiography. The result shows that CED-4 facilitated the autocatalytic cleavage of CED-3 greatly (Yan *et al.*, 2005).

5. CONCLUSION

Our understanding of apoptosome assembly is far from complete. At present, the apoptosome complex involving Apaf-1, Dark, or CED-4 has been reconstituted successfully using homogeneous recombinant proteins. The *in vitro* reconstitution of specific apoptosomes represents the first essential step in understanding the structure and mechanism of the apoptosomes. Kinetics and regulation of apoptosome assembly are only beginning to be investigated. The differential role of ATP/dATP in the assembly of various apoptosomes has yet to be scrutinized. The structure of apoptosomes at an atomic resolution is expected to reveal critical insights into the assembly and mechanisms of the various apoptosomes.

ACKNOWLEDGMENTS

This work was supported by grants from the NIH (CA095218, CA090269, and GM072633) and Princeton University. Y.S. is the Warner-Lambert Parke-Davis Professor of Molecular Biology. The author thanks members of his laboratory for discussion.

REFERENCES

Acehan, D., Jiang, X., Morgan, D. G., Heuser, J. E., Wang, X., and Akey, C. W. (2002). Three-dimensional structure of the apoptosome: Implications for assembly, procaspase-9 binding and activation. *Mol. Cell* **9**, 423–432.

Bao, Q., Lu, W., Rabinowitz, J. D., and Shi, Y. (2007). Calcium blocks formation of apoptosome by preventing nucleotide exchange in Apaf-1. *Mol. Cell* **25**, 181–192.

Berridge, M. J., Bootman, M. D., and Roderick, H. L. (2003). Calcium signalling: Dynamics, homeostasis and remodeling. *Nat. Rev. Mol. Cell Biol.* **4**.

Boatright, K. M., and Salvesen, G. S. (2003). Mechanisms of caspase activation. *Curr. Opin. Cell Biol.* **15**, 725–731.

Budihardjo, I., Oliver, H., Lutter, M., Luo, X., and Wang, X. (1999). Biochemical pathways of caspase activation during apoptosis. *Annu. Rev. Cell Dev. Biol.* **15**, 269–290.

Cain, K., Brown, D. G., Langlais, C., and Cohen, G. M. (1999). Caspase activation involves the formation of the aposome, a large (≈700 kDa) caspase-activating complex. *J. Biol. Chem.* **274**, 22686–22692.

Cerreti, D. P., Kozlosky, C. J., Mosley, B., Nelson, N., Van Ness, K., Greenstreet, T. A., March, C. J., Kronheim, S. R., Druck, T., Cannizzaro, L. A., *et al.* (1992). Molecular cloning of the interleukin-1 beta converting enzyme. *Science* **256**, 97–100.

Chao, Y., Shiozaki, E. N., Srinivasula, S. M., Rigotti, D. J., Fairman, R., and Shi, Y. (2005). Engineering a dimeric caspase-9: A re-evaluation of the induced proximity model for caspase activation. *PLoS Biol.* **3,** e183.

Chen, F., Hersh, B. M., Conradt, B., Zhou, Z., Riemer, D., Gruenbaum, Y., and Horvitz, H. R. (2000). Translocation of *C. elegans* CED-4 to nuclear membranes during programmed cell death. *Science* **287,** 1485–1489.

Chew, S. K., Akdemir, F., Chen, P., Lu, W. J., Mills, K., Daish, T., Kumar, S., Rodriguez, A., and Abrams, J. M. (2004). The apical caspase dronc governs programmed and unprogrammed cell death in Drosophila. *Dev. Cell* **7,** 897–907.

Chinnaiyan, A. M., O'Rourke, K., Lane, B. R., and Dixit, V. M. (1997). Interaction of CED-4 with CED-3 and CED-9: A molecular framework for cell death. *Science* **275,** 1122–1126.

Conradt, B., and Horvitz, H. R. (1998). The *C. elegans* protein EGL-1 is required for programmed cell death and interacts with the Bcl-2-like protein CED-9. *Cell* **93,** 519–529.

Daish, T. J., Mills, K., and Kumar, S. (2004). Drosophila caspase DRONC is required for specific developmental cell death pathways and stress-induced apoptosis. *Dev. Cell* **7,** 909–915.

Danial, N. N., and Korsmeyer, S. J. (2004). Cell death: Critical control points. *Cell* **116,** 205–219.

Degterev, A., Boyce, M., and Yuan, J. (2003). A decade of caspases. *Oncogene* **22,** 8543–8567.

del Peso, L., Gonzalez, V. M., Inohara, N., Ellis, R. E., and Nunez, G. (2000). Disruption of the CED-9.CED-4 complex by EGL-1 is a critical step for programmed cell death in Caenorhabditis elegans. *J. Biol. Chem.* **275,** 27205–27211.

del Peso, L., Gonzalez, V. M., and Nunez, G. (1998). Caenorhabditis elegans EGL-1 disrupts the interaction of CED-9 with CED-4 and promotes CED-3 activation. *J. Biol. Chem.* **273,** 33495–33500.

Diemand, A. V., and Lupas, A. N. (2006). Modeling AAA+ ring complexes from monomeric structures. *J. Struct. Biol.* **156,** 230–243.

Dorstyn, L., Colussi, P. A., Quinn, L. M., Richardson, H., and Kumar, S. (1999). DRONC, an ecdysone-inducible *Drosophila* caspase. *Proc. Natl. Acad. Sci. USA* **96,** 4307–4312.

Fanelli, C., Coppola, S., Barone, R., Colussi, C., Gualandi, G., Volpe, P., and Ghibelli, L. (1999). Magnetic fields increase cell survival by inhibiting apoptosis via modulation of calcium influx. *FASEB J.* **13,** 95–102.

Fuentes-Prior, P., and Salvesen, G. S. (2004). The protein structures that shape caspase activity, specificity, activation and inhibition. *Biochem. J.* **384,** 201–232.

Genini, D., Budihardjo, I., Plunkett, W., Wang, X., Carrera, C. J., Cottam, H. B., Carson, D. A., and Leoni, L. M. (2000). Nucleotide requirements for the *in vitro* activation of the apoptosis protein-activating factor-1-mediated caspase pathway. *J. Biol. Chem.* **275,** 29–34.

Hengartner, M. O., and Horvitz, H. R. (1994). *C. elegans* cell survival gene ced-9 encodes a functional homolog of the mammalian proto-oncogene bcl-2. *Cell* **76,** 665–676.

Horvitz, H. R. (1999). Genetic control of programmed cell death in the nematode *Caenorhabditis elegans*. *Cancer Res.* **59,** 1701–1706.

Horvitz, H. R. (2003). Worms, life, and death (Nobel lecture). *Chembiochem* **4,** 697–711.

Hu, Y., Benedict, M. A., Ding, L., and Nunez, G. (1999). Role of cytochrome c and dATP/ATP hydrolysis in Apaf-1-mediated caspase-9 activation and apoptosis. *EMBO J.* **18,** 3586–3595.

Hu, Y., Ding, L., Spencer, D. M., and Nunez, G. (1998). WD-40 repeat region regulates apaf-1 self-association and procaspase-9 activation. *J. Biol. Chem.* **273,** 33489–33494.

Irmler, M., Hofmann, K., Vaux, D., and Tschopp, J. (1997). Direct physical interaction between the *Caenorhabditis* elegans 'death proteins' CED-3 and CED-4. *FEBS Lett.* **406,** 189–190.

James, C., Gschmeissner, S., Fraser, A., and Evan, G. I. (1997). CED-4 induces chromatin condensation in *Schizosaccharomyces pombe* and is inhibited by direct physical association with CED-9. *Curr. Biol.* **7,** 246–252.

Jiang, X., Kim, H. E., Shu, H., Zhao, Y., Zhang, H., Kofron, J., Donnelly, J., Burns, D., Ng, S. C., Rosenberg, S., and Wang, X. (2003). Distinctive roles of PHAP proteins and prothymosin-alpha in a death regulatory pathway. *Science* **299,** 223–226.

Kanuka, H., Sawamoto, K., Inohara, N., Matsuno, K., Okano, H., and Miura, M. (1999). Control of the cell death pathway by Dapaf-1, a Drosophila Apaf-1/CED-4-related caspase activator. *Mol. Cell* **4,** 757–769.

Kim, H. E., Du, F., Fang, M., and Wang, X. (2005). Formation of apoptosome is initiated by cytochrome c-induced dATP hydrolysis and subsequent nucleotide exchange on Apaf-1. *Proc. Natl. Acad. Sci. USA* **102,** 17545–17550.

Kischkel, F. C., Hellbardt, S., Behrmann, I., Germer, M., Pawlita, M., Krammer, P. H., and Peter, M. E. (1995). Cytotoxicity-dependent APO-1 (Fas/CD95)-associated proteins form a death-inducing signaling complex (DISC) with the receptor. *EMBO J.* **14,** 5579–5588.

Lang, F., Foller, M., Lang, K. S., Lang, P. A., Ritter, M., Gulbins, E., Vereninov, A., and Huber, S. M. (2005). Ion channels in cell proliferation and apoptotic cell death. *J. Membr. Biol.* **205,** 147–157.

Lang, F., Ritter, M., Gamper, N., Huber, S., Fillon, S., Tanneur, V., Lepple-Wienhues, A., Szabo, I., and Gulbins, E. (2000). Cell volume in the regulation of cell proliferation and apoptotic cell death. *Cell. Physiol. Biochem.* **10,** 417–428.

Leoni, L. M., Chao, Q., Cottam, H. B., Genini, D., Rosenbach, M., Carrera, C. J., Budihardjo, I., Wang, X., and Carson, D. A. (1998). Induction of an apoptotic program in cell-free extracts by 2-chloro-2'-deoxyadenosine 5'-triphosphate and cytochrome c. *Proc. Natl. Acad. Sci. USA* **95,** 9567–9571.

Li, P., Nijhawan, D., Budihardjo, I., Srinivasula, S. M., Ahmad, M., Alnemri, E. S., and Wang, X. (1997). Cytochrome c and dATP-dependent formation of Apaf-1/caspase-9 complex initiates an apoptotic protease cascade. *Cell* **91,** 479–489.

Liu, X., Kim, C. N., Yang, J., Jemmerson, R., and Wang, X. (1996). Induction of Apoptosis program in cell-free extracts: Requirement for dATP and cytochrome c. *Cell* **86,** 147–157.

Muro, I., Monser, K., and Clem, R. J. (2004). Mechanism of Dronc activation in *Drosophila* cells. *J. Cell Sci.* **117,** 5035–5041.

Parekh, A. B., and Penner, R. (1997). Store depletion and calcium influx. *Physiol. Rev.* **77,** 901–930.

Parrish, J., Metters, H., Chen, L., and Xue, D. (2000). Demonstration of the in vivo interaction of key cell death regulators by structure-based design of second-site suppressors. *Proc. Natl. Acad. Sci. USA* **97,** 11916–11921.

Pop, C., Timmer, J., Sperandio, S., and Salvesen, G. S. (2006). The apoptosome activates caspase-9 by dimerization. *Mol. Cell* **22,** 269–275.

Qin, H., Srinivasula, S. M., Wu, G., Fernandes-Alnemri, T., Alnemri, E. S., and Shi, Y. (1999). Structural basis of procaspase-9 recruitment by the apoptotic protease-activating factor 1. *Nature* **399,** 547–555.

Rathmell, J. C., and Thompson, C. B. (2002). Pathways of apoptosis in lymphocyte development, homeostasis, and disease. *Cell* **109,** S97–S107.

Renatus, M., Stennicke, H. R., Scott, F. L., Liddington, R. C., and Salvesen, G. S. (2001). Dimer formation drives the activation of the cell death protease caspase 9. *Proc. Natl. Acad. Sci. USA* **98,** 14250–14255.

Riedl, S. J., Li, W., Chao, Y., Schwarzenbacher, R., and Shi, Y. (2005). Structure of the apoptotic protease activating factor 1 bound to ADP. *Nature* **434**, 926–933.

Riedl, S. J., and Shi, Y. (2004). Molecular mechanisms of caspase regulation during apoptosis. *Nat. Rev. Mol. Cell Biol.* **5**, 897–907.

Rodriguez, A., Oliver, H., Zou, H., Chen, P., Wang, X., and Abrams, J. M. (1999). Dark is a Drosophila homologue of Apaf-1/CED-4 and functions in an evolutionarily conserved death pathway. *Nat. Cell Biol.* **1**, 272–279.

Rodriguez, J., and Lazebnik, Y. (1999). Caspase-9 and Apaf-1 form an active holoenzyme. *Genes Dev.* **13**, 3179–3184.

Saleh, A., Srinivasula, S. M., Acharya, S., Fishel, R., and Alnemri, E. S. (1999). Cytochrome c and dATP-mediated oligomerization of Apaf-1 is a prerequisite for procaspase-9 activation. *J. Biol. Chem.* **274**, 17941–17945.

Salvesen, G. S., and Dixit, V. M. (1999). Caspase activation: The induced-proximity model. *Proc. Natl. Acad. Sci. USA* **96**, 10964–10967.

Santella, L. (1998). The role of calcium in the cell cycle: Facts and hypotheses. *Biochem. Biophys. Res. Commun.* **244**, 317–324.

Seshagiri, S., and Miller, L. K. (1997). Caenorhabditis elegans CED-4 stimulates CED-3 processing and CED-3-induced apoptosis. *Curr. Biol.* **7**, 455–460.

Shi, Y. (2002a). Apoptosome: The cellular engine for the activation of caspase-9. *Structure* **10**, 285–288.

Shi, Y. (2002b). Mechanisms of caspase inhibition and activation during apoptosis. *Mol. Cell* **9**, 459–470.

Shi, Y. (2004). Caspase activation: Revisiting the induced proximity model. *Cell* **117**, 855–858.

Spector, M. S., Desnoyers, S., Hoeppner, D. J., and Hengartner, M. O. (1997). Interaction between the *C. elegans* cell-death regulators CED-9 and CED-4. *Nature* **385**, 653–656.

Srinivasula, S. M., Ahmad, M., Fernandes-Alnemri, T., and Alnemri, E. S. (1998). Autoactivation of procaspase-9 by Apaf-1-mediated oligomerization. *Mol. Cell* **1**, 949–957.

Srinivasula, S. M., Saleh, A., Hedge, R., Datta, P., Shiozaki, E., Chai, J., Robbins, P. D., Fernandes-Alnemri, T., Shi, Y., and Alnemri, E. S. (2001). A conserved XIAP-interaction motif in caspase-9 and Smac/DIABLO mediates opposing effects on caspase activity and apoptosis. *Nature* **409**, 112–116.

Stennicke, H. R., Deveraux, Q. L., Humke, E. W., Reed, J. C., Dixit, V. M., and Salvesen, G. S. (1999). Caspase-9 can be activated without proteolytic processing. *J. Biol. Chem.* **274**, 8359–8362.

Thornberry, N. A., Bull, H. G., Calaycay, J. R., Chapman, K. T., Howard, A. D., Kostura, M. J., Miller, D. K., Molineaux, S. M., Weidner, J. R., Aunins, J., Elliston, K. O., Ayala, J. M., *et al.* (1992). A novel heterodimeric cysteine protease is required for interleukin-1 beta processing in monocytes. *Nature* **356**, 768–774.

Thornberry, N. A., and Lazebnik, Y. (1998). Caspases: Enemies within. *Science* **281**, 1312–1316.

Tinel, A., and Tschopp, J. (2004). The PIDDosome, a protein complex implicated in activation of caspase-2 in response to genotoxic stress. *Science* **304**, 843–846.

Wang, X. (2001). The expanding role of mitochondria in apoptosis. *Genes Dev.* **15**, 2922–2933.

Wu, D., Wallen, H. D., and Nunez, G. (1997). Interaction and regulation of subcellular localization of CED-4 by CED-9. *Science* **275**, 1126–1129.

Xue, D., Shaham, S., and Horvitz, H. R. (1996). The *Caenorhabditis elegans* cell-death protein CED-3 is a cysteine protease with substrate specificities similar to those of the human CPP32 protease. *Genes Dev* **10**, 1073–1083.

Yan, N., Chai, J., Lee, E. S., Gu, L., Liu, Q., He, J., Wu, J. W., Kokel, D., Li, H., Hao, Q., Xue, D., and Shi, Y. (2005). Structure of the CED-4-CED-9 complex provides insights into programmed cell death in *Caenorhabditis elegans*. *Nature* **437,** 831–837.

Yan, N., Huh, J. R., Schirf, V., Demeler, B., Hay, B. A., and Shi, Y. (2006a). Structure and activation mechanism of the *Drosophila* initiator caspase Dronc. *J. Biol. Chem.* **281,** 8667–8674.

Yan, N., Xu, Y., and Shi, Y. (2006b). 2:1 Stoichiometry of the CED-4-CED-9 complex and the tetrameric CED-4: Insights into the regulation of CED-3 activation. *Cell Cycle* **5,** 31–34.

Yang, J., Liu, X., Bhalla, K., Kim, C. N., Ibrado, A. M., Cai, J., Peng, T.-I., Jones, D. P., and Wang, X. (1997). Prevention of apoptosis by Bcl-2: Release of cytochrome c blocked. *Science* **275,** 1129–1132.

Yang, X., Chang, H. Y., and Baltimore, D. (1998). Essential role of CED-4 oligomerization in CED-3 activation and apoptosis. *Science* **281,** 1355–1357.

Yin, Q., Park, H. H., Chung, J. Y., Lin, S. C., Lo, Y. C., da Graca, L. S., Jiang, X., and Wu, H. (2006). Caspase-9 holoenzyme is a specific and optimal procaspase-3 processing machine. *Mol. Cell* **22,** 259–268.

Yu, X., Acehan, D., Menetret, J. F., Booth, C. R., Ludtke, S. J., Riedl, S. J., Shi, Y., Wang, X., and Akey, C. W. (2005). A structure of the human apoptosome at 12.8 A resolution provides insights into this cell death platform. *Structure* **13,** 1725–1735.

Yu, X., Wang, L., Acehan, D., Wang, X., and Akey, C. (2006). Three-dimensional structure of a double apoptosome formed by the *Drosophila* Apaf-1 related killer. *J. Mol. Biol.* **355,** 577–589.

Yuan, J., and Horvitz, H. R. (1992). The *Caenorhabditis elegans* cell death gene ced-4 encodes a novel protein and is expressed during the period of extensive programmed cell death. *Development* **116,** 309–320.

Yuan, J., Shaham, S., Ledoux, S., Ellis, H. M., and Horvitz, H. R. (1993). The *C. elegans* cell death gene Ced-3 encodes a protein similar to mammalian interleukin-1 beta-converting enzyme. *Cell* **75,** 641–652.

Zhou, L., Song, Z., Tittel, J., and Steller, H. (1999). HAC-1, a *Drosophila* homolog of Apaf-1 and CED-4 functions in developmental and radiation-induced apoptosis. *Mol. Cell* **4,** 745–755.

Zou, H., Henzel, W. J., Liu, X., Lutschg, A., and Wang, X. (1997). Apaf-1, a human protein homologous to *C. elegans* CED-4, participates in cytochrome c–dependent activation of caspase-3. *Cell* **90,** 405–413.

Zou, H., Li, Y., Liu, X., and Wang, X. (1999). An APAF-1-cytochrome c multimeric complex is a functional apoptosome that activates procaspase-9. *J. Biol. Chem.* **274,** 11549–11556.

CHAPTER EIGHT

Caspases: Determination of Their Activities in Apoptotic Cells

Alena Vaculova *and* Boris Zhivotovsky

Contents

Abstract

Caspases, a family of cysteine proteases, were identified as major regulators of apoptotic cell death. These enzymes are involved not only in initiation but also in an execution phase of apoptosis by cleaving more than 400 substrates. This cleavage mediates a majority of the typical biochemical and morphological changes in apoptotic cells, such as cell shrinkage, chromatin condensation, DNA fragmentation, and plasma membrane blebbing. In addition to their role in cell death, caspases fulfill various nonapoptotic functions in living cells. Thus, detection of caspase activation/activity is essential for the understanding of different biological processes. It can be used as a biochemical marker for apoptosis induced by diverse stimuli. This chapter describes a set of methods available for characterization and/or quantification of caspase activation, including immunoblotting, cleavage of synthetic substrates, affinity labeling, flow cytometry and different microscopic techniques. In addition, several

Institute of Environmental Medicine, Division of Toxicology, Karolinska Institute, Stockholm, Sweden

Methods in Enzymology, Volume 442
ISSN 0076-6879, DOI: 10.1016/S0076-6879(08)01408-0

methods discussing the attempts for *in vivo* analyzing of caspase activity are
included. Advantages and disadvantages of each method are compared.

1. INTRODUCTION

Caspases, cysteinyl aspartate proteinases, are a family of the evolutionary
conserved enzymes that cleave their substrates following an aspartate residue.
To date, at least 14 mammalian caspases have been identified. Although
caspases have been recognized to play an important role in apoptotic cell
death, it has became clear that they are also involved in other important
biological processes, such as inflammation, cell differentiation, proliferation
and cell cycle regulation, cell division, and fusion (Droin *et al.*, 2008; Lamkanfi
et al., 2007; Nhan *et al.*, 2006). During apoptosis, caspases are responsible for
the cleavage of specific substrates, being involved in propagation of typical
apoptosis-related biochemical and morphological changes (Fig. 8.1). Therefore, the detection of activated caspases is considered to be a significant marker
for this mode of cell death.

Caspases are synthesized as inactive precursors (zymogens, procaspases).
The procaspase molecule (32–57 kDa) contains several domains: N-terminal prodomain and large (17–21 kDa) and small (10–13 kDa) subunits.

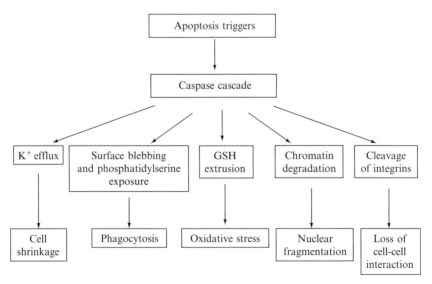

Figure 8.1 Selected consequences of caspase actions during apoptosis. Apoptosis-
inducing compounds trigger activation of the caspase cascade, which leads to the
cleavage of target proteins and results in various biochemical and morphological mani-
festations of cell death.

In response to apoptotic stimulus, proteolytic processing of procaspase at the specific aspartic acid residue site between large and small subunits occurs. For enzyme maturation, formation of a heterotetramer consisting of two heterodimers (large and small subunits) derived from two precursor molecules is required. Based on the length of a prodomain, caspases are divided into two classes: those containing long (more than 100 amino acids) or short (less than 30 amino acids) prodomains. The presence of the long prodomain is characteristic for initiator caspases, while effector caspases possess the short one. Specific motifs important for the interaction of initiator caspases with adaptor proteins are localized within the long prodomain. They enable the attachment of the procaspase molecule to the death signaling complex and its autocatalytic activation. The death effector domain has been found in procaspase-8 and -10, whereas procaspase-2 or -9 has been shown to contain the caspase recruitment domain. However, effector caspases lacking the long prodomain require cleavage by initiator caspases for their activation (Fig. 8.2) (Kumar, 2007; Zhivotovsky, 2003).

The strong preference for cleavage of the peptide bond C-terminal to aspartic acid at the so-called P1 site is a unique characteristic of caspases compared to other enzymes (except for granzyme B). Caspases recognize at least four amino acids (tetrapeptide) in their substrates, P1 to P4 (amino acids N-terminal to the cleavage site). Apart from P1, the P4 site is also a crucial determinant of caspase specificity (Thornberry et al., 1997). Based on the tetrapeptide sequence specificity, these enzymes can be divided into

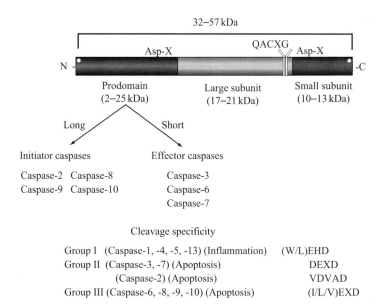

Figure 8.2 The structural and functional organization of caspases.

three groups. The optimal recognition motif for the members of the group I is (W/L)EHD (caspase-1, -4, -5, -13). The members of the group II (e.g., caspase-3, -7) prefer DEXD, and caspases of group III cleave after (I/L/V) EXD (caspase-6, -8, -9, -10). For caspase-2, the amino acid in the P5 position is also important for efficient cleavage, and its cleavage specificity is closely related to the effector caspases (-3, -7). Importantly, similarly to other initiator caspases, caspase-2 contains a long prodomain on N termini. Therefore, the combination of these features makes caspase-2 unique among other caspases (Fig. 8.2) (Baliga et al., 2004; Zhivotovsky and Orrenius, 2005).

Almost all procaspases are located in the cytosol, and procaspase-2 is also constitutively present in nuclei (Colussi et al., 1998; Zhivotovsky et al., 1999). Significant differences between localization of inactive procaspases and active caspase molecules have also been reported (Zhivotovsky et al., 1999). In apoptotic cells, caspases are located or associated with various intracellular compartments, including cytosol, mitochondria, endoplasmic reticulum, or Golgi apparatus. The intracellular translocation of active caspases can be critical for the development of the apoptotic process. For example, a deficiency in the relocalization of active caspase-3 into the nucleus, associated with the impaired subsequent proteolytic cleavage of specific nuclear proteins, has been shown to be responsible for the resistance of non-small cell lung cancer cells to undergo apoptosis (Joseph et al., 2001).

As inappropriate activation of caspases could have fatal consequences for cell fate, caspase activity needs to be tightly regulated at different levels of intracellular signaling pathways. Two main pathways leading to caspase activation, extrinsic and the intrinsic, have been identified (Scaffidi et al., 1999). Although they can operate separately, the cross talk between these pathways has been described that might lead to amplification of the apoptotic response. Signaling through extrinsic, death receptor (DR)-mediated pathway begins with the ligation of DRs, followed by formation of the death-inducing signaling complex (DISC) and caspase-8 activation. In so-called "type I cells," the active caspase-8 activates procaspase-3 directly, which is followed by the execution of apoptosis. In "type II cells," active caspase-8 cleaves the cytoplasmic member of the Bcl-2 family proteins Bid, and truncated Bid is then responsible for translocation of the apoptotic signal to mitochondria. This leads to a release of proapoptotic factors from the intermembrane space of mitochondria, such as cytochrome c, formation of the apoptosome complex (cytochrome c, dATP, apoptotic protease-activating factor-1, Apaf-1, and procaspase-9), and caspase-9 activation. Active caspase-9 then cleaves effector caspases. Importantly, noncaspase proteases, such as granzyme B, may also play a role in apoptosis regulation, being involved in caspase activation (Metkar et al., 2003) and cleavage of several caspase substrates (e.g., PARP, Bid) (Casciola-Rosen et al., 2007; Froelich et al., 1996).

Different protein factors might inhibit caspase activity. Some of them are known to be produced by viruses in order to prolong the life of host cells and enable a sufficient replication of the virus. These include cytokine response modifier A, viral inhibitor of apoptosis proteins (vIAP), or viral FLICE inhibitory protein (vFLIP). Cellular homologues of these proteins (e.g., cIAPs, cFLIP) are important endogenous regulators of caspase activity in mammalian cells. It has been documented that cFLIP is a competitive inhibitor of caspase-8 at the level of DISC and a key regulator of the DR-mediated extrinsic apoptotic pathway; cIAPs can directly physically interact with caspases (e.g., caspase-3, -7, -9) and inhibit their activity upstream or downstream of mitochondria. Their function can be opposed by Smac/DIABLO or Omi/HtrA, proapoptotic molecules containing IAP-binding motifs, which are released from mitochondria following an apoptotic stimulus. Other factors involved in the regulation of caspase activity include several Bcl-2 family proteins, heat shock proteins, nitric oxide, hydrogen peroxide, oxidative stress, or Akt (Rupinder et al., 2007; Turk and Stoka, 2007).

To date, the growing list of caspase substrates involves at least 400 different protein molecules. While some of them are considered to be common targets for caspase-mediated cleavage in all cells, cleavage of the others may be cell type specific. Although the significance of cleavage of some caspase substrates has been identified, the functional consequences of cleavage of many of these substrates still remain to be elucidated. Moreover, the association of cleavage of these substrates with other apoptotic events requires further investigation. Caspase substrates include proteins that fulfill different functions inside the cell, being involved in cytoskeleton organization (e.g., cytokeratins, lamin, fodrin, actin, vimentin), DNA replication, repair or degradation [e.g., DNA polymerase epsilon, poly(ADP-ribose)polymerase (PARP), inhibitor of caspase-activated DNase, respectively], cell cycle regulation (e.g., retinoblastoma protein pRb), cell adhesion (e.g., some cadherins, catenins), regulation of transcription (e.g., CREB, SRF, STAT1), translation (e.g., DAP5, some eIFs), kinase signal transduction (e.g., PKC, MEK), and many others (Fischer et al., 2003; Timmer and Salvesen, 2007).

Determination of caspase activity and detection of caspase cleavage products can be performed using different methodological approaches. These include detection of caspase processing (electrophoresis, immunoblotting), analysis of enzymatic activity using labeled synthetic caspase substrates or inhibitors (e.g., spectrophotometry, fluorimetry, chemiluminescence techniques, ELISA), or detection of the target caspase substrate cleavage (e.g., immunoblotting). This chapter describes the methods (for in vitro as well as in vivo applications) used at present, together with their modifications for fluorescence or confocal microscopy, and flow cytometry. In addition to the background information and experimental protocol details, advantages and disadvantages of these techniques are provided, followed by recommendations and warnings on some critical parameters.

2. METHODS FOR ANALYSIS OF CASPASE PROCESSING, ACTIVATION, AND SUBSTRATE CLEAVAGE

2.1. Processing of procaspases (Western blotting)

As mentioned earlier, caspases are initially synthesized as inactive zymogen precursors (procaspases), which can be activated rapidly upon enzymatic processing at specific sites containing an aspartic acid. Procaspase processing with the formation of fragments related to small and large subunits can be detected by immunoblotting approach and specific antibodies. A wide variety of different types of antibodies (and also of different quality) are available commercially. The most commonly used antibodies recognize epitopes that are present within the procaspase molecule and/or its cleavage fragments, enabling the detection of caspase precursors and/or its cleavage (processed) products (Fig. 8.3). Moreover, antibodies against neoepitopes that are uncovered at the C terminus during particular caspase cleavage have also been generated, recognizing only cleaved caspases, and not original procaspase molecules. Except for Western blotting, caspase processing can also be utilized by different other methods (see later).

Briefly, after the separation by SDS-PAGE, cellular proteins are transferred from the gel to a membrane (usually PVDF). The membrane is then blocked and probed with a specific primary antibody recognizing the procaspase and/or its cleavage products. The antibody binding (and thus presence of the appropriate protein) can finally be visualized using a horseradish peroxidase (HRP)-conjugated secondary antibody and an enhanced chemiluminescence (ECL) kit. Immunodetection can be repeated several times using the same membrane after its stripping and blocking (for details, see the protocol given later) to identify processing of more than one caspase. This method can be performed using different experimental models—cell cultures, tissue sections, and biopsies. Moreover, not only the whole cell lysates, but also protein extracts isolated from different subcellular fractions can be investigated. The latter approach can provide very useful information about intracellular localization of the processed/unprocessed (active/inactive)

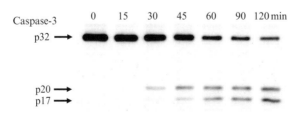

Figure 8.3 Time course (0–120 min) of procaspase-3 processing in Jurkat cells treated with anti-CD95 antibody (250 ng/ml, clone CH-11, MBL, Japan) detected by Western blotting according to the protocol provided in the text.

caspases. Alternatively, caspase processing can be detected using purified or recombinant caspase proteins.

2.1.1. Preparation of samples

5X Sample (Laemmli) buffer [625 μl 1 M Tris-HCl, pH 6.8, 1 ml glycerol, 2 ml 10% sodium dodecyl sulfate (SDS), 0.5 ml 0.5% (w/v) bromophenol blue in H_2O, 0.5 ml 2-mercaptoethanol, add water to 10 ml, store at 4 °C]

1. Place 1 × 10^6 cells in a tube and centrifuge for 5 min at 300 g, 4 °C. Add phosphate-buffered saline (PBS), centrifuge again, and discard supernatant.
2. Two methods may be used to obtain proteins from cells for immuno-blotting: (a) resuspend the exact number of intact cells directly in sample buffer or (b) mix an equal amount of proteins extracted from cells with sample buffer.
   a. Resuspend pellet (1 × 10^6 cells) in 100 μl of sample buffer and incubate for 5 min in a boiling water bath. Let cool.
   b. Determine the protein concentration in the cell lysates using the appropriate kit (e.g., BCA protein assay kit, Pierce), adjust the protein extracts to the equal concentration, add the sample buffer, and boil. Let cool.

If samples become highly viscous as a consequence of the release of chromosomal DNA, shear the DNA either by sonication or by repeated passage through a 23-gauge needle.

2.1.2. Gel electrophoresis

2.1.2.1. Preparation of SDS-polyacrylamide minigels   Minigels are useful because they require shorter run times and significantly smaller quantities of reagents are sufficient (the latter is especially helpful in immunoassays).

29% (w/v) acrylamide/1% (w/v) bisacrylamide stock (e.g., Bio-Rad)

1.5 M Tris-HCl, pH 8.8 (dissolve 54.5 g of Tris base in 150 ml of water, adjust to pH 8.8 with 10 M HCl, add water to a final volume of 300 ml, and store at 4 °C)

0.5 M Tris-HCl, pH 6.8 (dissolve 18 g of Tris base in 150 ml of distilled water, adjust to pH 6.8 with 10 M HCl, add water to a final volume of 300 ml, and store at 4 °C)

10% (w/v) ammonium persulfate (APS; e.g., Bio-Rad, 100 mg in 1 ml of distilled water, prepare just before gel casting)

10% (w/v) SDS (dissolve 10 g of SDS in 90 ml of distilled water with gentle stirring and then bring to 100 ml)

N,N,N',N'-Tetramethylethylenediamine (TEMED; e.g., Bio-Rad)

3. Assemble the glass plates for the Mini-Protean electrophoresis II system (Bio-Rad) according to the manufacturer's instructions and check for leakage.

4. Prepare separating gel (0.375 M Tris-HCl, pH 8.8). Because the molecular masses of procaspases vary between 30 and 55 kDa, a 12% gel is recommended. To make one set (two gels) for Mini-Protean II with 1.5-mm-thick spacers, combine the following:

    6.7 ml H_2O

    5 ml 1.5 M Tris-HCl, pH 8.8

    200 μl 10% SDS

    8 ml 29% acrylamide/1% bisacrylamide stock

    100 μl 10% APS

    10 μl TEMED

5. Pour the separating gel into the gap between the glass plates. Leave sufficient space for the stacking gel to be added later (the length of the comb teeth plus 5 mm). Using a pipette, carefully overlay the gel with distilled water, using a steady, even rate of delivery to prevent mixing.

6. Allow gel to polymerize 45 to 60 min. Rinse off the overlay water completely.

7. Prepare the stacking gel (4% gel, 0.125 M Tris-HCl, pH 6.8). For two gels (one set), combine the following:

    6.1 ml H_2O

    2.5 ml 0.5 M Tris-HCl, pH 6.8

    100 μl 10% SDS

    1.3 ml 29% acrylamide/1% bisacrylamide stock

    50 μl 10% APS

    10 μl TEMED

8. Pour stacking gel solution directly on top of the polymerized separating gel. Immediately insert a clean Teflon comb into the stacking gel solution, being careful to avoid trapping air bubbles. Top up with more stacking gel. Allow gel to polymerize 20 to 30 min.

2.1.2.2. Set up and run gel electrophoresis

10× electrode (running) buffer (dissolve 90 g of Tris base, 432 g glycine, and 30 g SDS in 3 liters of distilled water, pH 8.3; for one electrophoretic run, mix 50 ml of 10× stock with 450 ml of water)

9. Release the gel holder from the casting stand. Assemble the Mini-Protean II electrophoresis system (Bio-Rad) or equivalent equipment according to the manufacturer's instructions and fill it with 1× electrode buffer.

10. Place the glass plates with the gel into the cell and add 1× electrode buffer.

11. Remove the comb by pulling it straight up slowly and gently.
12. Load samples (20 to 30 μl) into wells under the electrode buffer with Prot/Elec Tips (Bio-Rad). Load one well with 5 μl of low-range prestained SDS-PAGE markers.
13. Attach the electrophoresis apparatus to an electric power supply. Run the gel at constant 130 V in a 4 to 8 °C cold room until the bromophenol blue reaches the bottom of the resolving gel ($\approx$1.5 h).
14. After electrophoresis is complete, turn off the power supply and disconnect the electrical leads.

2.1.3. Immunoblotting of proteins

Transfer buffer (14.4 g glycine, 3 g Tris, 150 ml methanol, adjust with distilled water to 1 liter)

15. Approximately 20 min prior to the completion of SDS-PAGE, prepare the PVDF (Millipore) membrane—prewet it in methanol for 3 s, wash in distilled water, and equilibrate in transfer buffer. Prepare the Whatman papers of appropriate size.
16. Remove the lid of the electrophoretic system and pour off the electrode buffer.
17. Remove glass plate sandwich. Push one of the spacers of the sandwich out to the side of the plates without removing it. Gently twist the spacer so that the upper glass plate pulls away from the gel.
18. Mark gel orientation by cutting a right corner from the bottom of the gel and then immerse the gel in transfer buffer.
19. Assemble the sandwich for electroblotting. It is important to wear gloves when handling gel, filter paper, and membranes. All the components wet carefully, avoid any air bubbles that would interfere with a proper transfer and assemble them in the following order: a pad, filter papers, the gel, PVDF membrane, the filter papers, a pad. Remember that the proteins plus SDS are negatively charged and will migrate to the anode.
20. Place the sandwich cassette into the tank. Fill in the tank with transfer buffer (4 °C).
21. Connect to the electrodes (check the orientation and the correct direction of transfer).
22. Transfer proteins from the gel onto a PVDF membrane by electrotransfer. For blotting, a Mini Trans-Blot Module and Bio-Ice cooling unit can be used, and perform the transfer at a constant 100 V for 2 h in a cold room. Control for transfer efficiency is made by checking the transfer of prestained SDS-PAGE markers.

2.1.4. Immunodetection

HSB 4× (high salt Tris buffer; dissolve 48.4 g Tris base and 233.8 g NaCl in
  1.5 liters of distilled water, adjust pH to 7.5 with 10 M HCl, and add
  water to 2 liters)
HSBT (high salt Tris buffer with Tween; mix 500 ml of 4×HSB and
  1500 ml of distilled water; add 1 ml Tween 20 and mix gently)
LSB (low salt Tris buffer; dissolve 12.1 g Tris base and 18 g NaCl in 1.5
  liters of distilled water, adjust pH to 7.5 with 10 M HCl, and add water to
  2 liters)
Stripping buffer (add 12.5 ml of 1 M Tris-HCl, pH 6.8, 0.77 ml 2-mercap-
  toethanol, and 20 ml 10% SDS to 76 ml of distilled water and mix;
  prepare immediately before use)
Antibodies: primary and secondary
Primary antibodies:
  Caspase-1 (Calbiochem)
  Caspase-2 (BD Biosciences)
  Caspase-3 (BD Biosciences, Cell Signaling Technology)
  Caspase-6 (Cell Signaling Technology)
  Caspase-7 (BD Biosciences)
  Caspase-8 (BD Biosciences, Cell Signaling Technology)
  Caspase-9 (Calbiochem, Cell Signaling Technology)
  Caspase-10 (Calbiochem)
  β-Actin (Sigma)
  Glyceraldehyde-3-phosphate dehydrogenase (Trevigen)
Secondary antibodies:
HRP-conjugated antimouse-IgG, antirabbit-IgG, antigoat-IgG (Pierce,
  Amersham)

23. Transfer the membrane into 1× HSB containing 5% dry nonfat milk
    and 0.05% sodium azide. Keep at least 1 h on a rocker in a 4 to 8 °C cold
    room.
24. Wash membrane three times, 10 min each, in HSBT on a rocker at
    room temperature.
25. Put membrane into a plastic bag, box, or tube and add desired primary
    antibody in appropriate dilution (usually 1:100 to 1:5000) in 1× HSB
    containing 1% bovine serum albumin (BSA) and 0.05% sodium azide.
    Incubate 1 to 2 h on a rocker at room temperature.
26. Wash three times, 10 min each, with HSBT on a rocker at room
    temperature.
27. Add peroxidase-conjugated secondary antibody in appropriate dilution
    (usually 1:5000 to 1:10,000) in 1× HSBT containing 1% BSA or milk.
    Incubate for 1 h on a rocker at room temperature.

28. Wash three times, 10 min each, with HSBT on a rocker at room temperature.
29. Wash two times with LSB on a rocker at room temperature.
30. Develop membrane with ECL (ECL Western blotting detection system, Amersham Pharmacia Biotech) according to the manufacturer's instructions.
31. Cover membrane with plastic wrap and immediately expose to X-ray film (e.g., Fujifilm) from 30 s to 5 min in a X-ray cassette.
32. In order to evaluate the intensity of the bands detected on films, scan the film using a densitometer and proceed with densitometric analysis. To minimize the nonlinearities in the system response, always make sure to perform adequate calibration.

Following ECL detection, it is possible to reprobe the membrane with other antibodies against different caspases or other proteins of interest (see later). Moreover, markers such as glyceraldehyde-3-phosphate dehydrogenase or β-actin, the protein expression of which remains to be relatively constant, should also be detected. The values of optical density of the analyzed bands of target proteins are then related to the respective values of the marker protein. This provides a useful tool for comparison between individual samples and for verification of the total protein amount on the membrane.

33. To repeat the immunodetection procedure for the other protein of interest, submerge the membrane in stripping buffer and incubate for 30 min at 50 °C with agitation.
34. Wash the membrane with water and repeat steps 23 to 32.

The quality of the antibody used is crucial for the final outcome. Weak affinity for the protein or high antibody cross-reactivity is the most common problem encountered. Proper testing of the antibody, suitable positive control, and verification of the molecular weight of the detected bands are highly recommended. A common event during these experiments is that the decrease in procaspase level and an increase in amount of its cleavage products observed after immunodetection are not equally proportional, with the latter parameter being less evident. This can be caused by different affinity of the antibody for the procaspase and its fragments or by degradation of the unstable cleavage products. Thus, the proper timing and design of the experiment might substantially influence the obtained results.

However, it is not a general rule that all caspases require cleavage for their activation. As an example, caspase-9 has been shown to be activated without proteolytic processing (Stennicke et al., 1999). Allosteric activation of procaspase-9, accompanied by the conformational changes of its molecule, mediated by cytosolic factors, such as Apaf-1, has been demonstrated (Rodriguez and Lazebnik, 1999). However, processing but not activation of some caspases may be required to exert their particular functions. Typically, processed caspase-2 has been shown to induce mitochondria-mediated apoptosis independently of its enzymatic activity (Robertson *et al.*, 2004). Therefore, simultaneous detection of both procaspase processing/cleavage and caspase activity (see later) is highly recommended in order to obtain convincing and fully informative results.

2.2. Caspase activity (cleavage of selective peptide substrates)

Caspase activity assay is a relatively easy, fast, and convenient tool for the detection of caspase activation in cells undergoing apoptosis. This quantitative and sensitive method utilizes various substrates that are recognized and cleaved by appropriate caspases. At present, a wide variety of different synthetic low molecular weight caspase substrates containing appropriate peptide (tetrapeptide or pentapeptide) sequences with caspase cleavage sites are provided by a number of suppliers. The peptides can be conjugated with fluorogenic [7-amino-4-methylcoumarin (AMC); 7-amino-4-trifluoro-methylcoumarin; rhodamine-110] or chromogenic (*p*-nitroaniline) groups, which are released when the substrate is cleaved by a particular caspase. This is accompanied by a rise of a signal (fluorescence or absorbance), which intensity (measured by fluorometer or spectrophotometer, respectively) is proportional to the amount of cleaved substrate and depends on the extent of caspase activity. The caspase activity assay measured by the cleavage of fluoro-/chromogen-conjugated synthetic caspase substrates can be performed using both intact cells and cellular extracts. A detailed protocol describing the caspase activity assay using fluorescent caspase substrates is provided here.

Buffer, concentrated stock [100 mM HEPES, e.g., Roche, 10% (w/v) sucrose, 0.1% (w/v) CHAPS (3-[{3-cholamidopropyl}dimethylammonio]-1-propanesulfonate, e.g., Sigma]. Do not adjust pH. Divide into aliquots and store in 40-ml Falcon tubes or equivalent. Store up to 6 months at −20 °C or 2 to 3 weeks at 4 °C.

1× Substrate buffer (1 × SB): For each 96-well plate to be processed, add 25 μl of 1 M dithiothreitol (DTT; 5 mM final) and 5 μl of 0.1% Nonidet P-40 (NP-40; see recipe; 10^{-4} final) to 5 ml of concentrated stock, and adjust pH to 7.25. This buffer can be kept for several hours at room temperature.

1 M DTT: Add 309 mg DTT (Roche) to 2 ml of 0.01 M sodium acetate, pH 5.2. Filter sterilize, dispense into 50-μl aliquots, and store at −20 °C

0.1% (v/v) NP-40: Add 1 μl of NP-40 (Roche) to 1 ml water, vortex, and keep at 4 °C

Caspase substrate stock solutions: Prepare 200 mM stock solutions of each substrate to be tested in sterile ultrapure water. Dispense into 2-μl aliquots and store at −20 °C. All caspase substrates are from Enzyme Systems Products.

1. Place duplicate samples containing 2 to 3×10^6 cells in microcentrifuge tubes and centrifuge for 5 min at 1000 g, 4 °C.
2. Discard supernatant. Wash pellet with PBS and centrifuge for 5 min at 1000 g, 4 °C.
3. Discard supernatant. Resuspend each pellet in 25 μl PBS (check pH, which should be neutral) and transfer into individual wells of a 96-well culture plate. The plate should be kept very cold (either by keeping it for some time at −20 °C or by having it "floating" on liquid nitrogen). If necessary, plates with cells can be stored for several days at −80 °C.
4. Set up the computer control for the Fluoroscan II (LabSystems) plate reader. Set temperature to 37 °C. This takes about 3 to 4 min. Several computer programs are available for acquisition and data processing. The authors recommend Genesis II Windows-based microplate software (LabSystems and Life Sciences).
5. Immediately prior to use, prepare complete 1× SB by adding 1 μl of a 200 mM stock solution of the substrate of interest (final concentration 50 μM) to 4 ml of 1× SB.
6. Dispense 50 μl of complete SB with substrate to each test well of cells at room temperature, and place plate immediately on the Fluoroscan II. The final concentration of substrate in the wells is 33.3 μM.
7. Read samples. The maximum absorption for AMC is 354 nm. Fluorometric detection for AMC cleaved from peptide is at excitation 380 nm and emission 460 nm.
8. The results obtained have to be recalculated to the amount of total proteins, or the amount of cells present in individual samples. Therefore, simultaneous determination of either protein concentration or cell numbers in the samples is essential.

Fluorescent substrates provide higher sensitivity in comparison with chromogenic ones, and caspase activity can be detected even if only 5 to 10% of apoptotic cells are present in a total number of one million cells per sample. However, in some cases, the caspase activity assay using a fluorescence signal can suffer from a limited sensitivity due to an increased background caused by residual fluorescence of the fluorophores conjugated to peptide, cell autofluorescence, or spectral overlap of fluorescence of the substrate and product.

A highly sensitive strategy for the caspase activity assay, providing an extremely low background and a high signal-to-noise ratio, has been developed (Liu et al., 2005; O'Brien et al., 2005). This assay uses peptide-conjugated aminoluciferin (as a luminescent caspase substrate) and a thermostable luciferase. In the absence of active caspase, the caspase substrate does not act as a substrate for luciferase (as it is blocked effectively) and produces no light. However, when cleaved by a particular caspase, aminoluciferin is released and participates in a luminescence reaction, which is accompanied by the light generation. The luminescence signal is proportional to the extent of caspase activity present in the particular sample. Moreover, in the assay, cleavage of the luminescent substrate by caspase and luciferase oxidation can be coupled within one step; thus, the stability of the luminescence signal is achieved for several hours.

The caspase assay is sensitive to many factors that could interfere with the activity of the enzyme. Among them, salt concentration (a high salt concentration blocks caspase activity), pH (preferentially between 6.5 and 8), buffer composition, presence of stabilizing agents, additional inhibitors preserving caspase activity (e.g., proteasome inhibitors), and reducing conditions (protecting the cysteine residues contained in the active site) are of a great importance. Moreover, appropriate saturating substrate concentrations, as well as sufficient amounts of the cells or the protein (when using recombinant or purified endogenous caspase), are necessary in order to be able to correctly quantify the amounts of the active enzyme present in the sample.

The kinetics of caspase activation may differ depending on the cell and tissue type, type and concentration of an apoptosis inducer, and its exposure time. Furthermore, the heterogeneity and cell cycle distribution of the cells in the population may also be responsible for the fact that individual cells undergo apoptosis at different time points (varying between several hours or days). It is therefore important to monitor the changes in caspase activities over the appropriate time period, and empirically determine the interval when the activity peaks. This will help minimize wrong conclusions caused by the poor experimental design. Furthermore, correlation of the caspase activation with the other apoptotic events occurring in the cell is also very important for the description and proper understanding of the complex process of cell death.

Each caspase can be responsible for the cleavage of multiple target proteins and caspases may have overlapping substrate specificities. Using commercially available short peptide-based substrates, the promiscuity of caspases for different cleavage motifs has been demonstrated (for details, see Fig. 8.4) (McStay et al., 2008). Thus, the caspase activity assay may not always accurately distinguish between various caspases, especially those within the same group. Moreover, in certain cell types, underestimation of caspase activation using caspase activity assays has been demonstrated. One of the possibilities of how to explain this observation could be sequestration of the active caspase subunits together with their cleavage products into large spherical cytoplasmic inclusions (MacFarlane et al., 2000), making them less available for the substrate delivered in the caspase assay.

2.3. Affinity labeling and synthetic caspase inhibitors

The active enzymatic sites of the caspases are accessible not only to the substrates, but also to the inhibitors. Synthetic caspase inhibitors are small, cell-permeable molecules that bind to the caspase active sites in a reversible

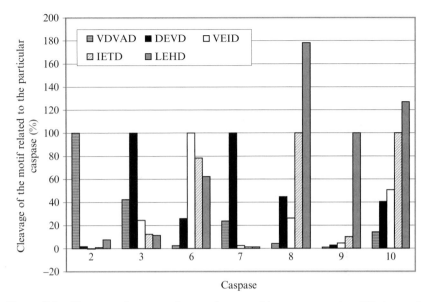

Figure 8.4 Cleavage of caspase substrates by recombinant caspases (modified according to McStay *et al.*, 2008). Substrate cleavage (%) is related to the cleavage of a particular motif reported for an individual caspase.

or irreversible manner. The binding is mediated by the presence of functional groups such as aldehyde (CHO), fluoromethylketone (FMK), chloromethylketone, or fluoroacyloxymethylketone. Inhibitors containing the CHO group are reversible, whereas others, forming the covalent adducts with the cysteine in the active site, are irreversible. The irreversible binding of active caspase is especially used for affinity labeling. The specificity of the binding is provided by the appropriate amino acid sequences (tetra- or pentapeptides), corresponding to those found in particular endogenous caspase substrates. The tripeptide moiety is typical for a pancaspase inhibitor, as it can recognize most active caspases present in the cell.

The caspase inhibitors are extensively used and well-established reagents for the evaluation of caspase activation as well as the general study of apoptosis. The role of caspases in the apoptotic process can be confirmed indirectly when these inhibitors, applied simultaneously with the apoptosis inductor, are responsible for inhibition or attenuation of downstream intracellular apoptotic events. Caspase activation also can be monitored directly using biotin- or fluorochrome-labeled inhibitors of caspases.

Various irreversible caspase inhibitors tagged with biotin as an affinity label have been synthesized. The conjugation with biotin allows the detection of active caspases based on the strong affinity of avidin for biotin. Biotin can be attached to the inhibitor molecule via linkers of various lengths. Originally, inhibitors containing short linkers were available, and biotin was present in a close proximity to the enzyme active site. For the detection (immunoblotting, visualization of biotin using streptavidin conjugated with HRP, and subsequent ECL kit) of the biotinylated inhibitor-labeled caspases, protein denaturation was an essential step. Recently, biotinylated caspase inhibitors with an extended linker allowing the affinity purification of the nondenatured enzyme from the cellular extracts have been introduced (Henzing et al., 2006).

Another method enabling the detection of caspase activity using biotin-labeled inhibitors is the enzyme-linked immunosorbent assay (ELISA). In this assay, a specific caspase antibody (recognizing both active and inactive caspase molecule) is precoated onto a microplate. Cellular extracts preincubated with the biotinylated caspase inhibitor (reacting only with active caspase) are then pipetted into the wells and total caspase is captured by the immobilized antibody. After washing out the unbound molecules, HRP-conjugated strep-tavidin (recognizing the biotin) is added to the wells. The appropriate HRP substrate is then added, and the intensity of the color product acquired is proportional to the detected amount of the active caspase.

2.4. Measurement of caspase activation by fluorescence/ confocal microscopy or flow cytometry in intact cells

2.4.1. Antibody-based detection of caspase activities

For the detection of processed (active) caspases in intact cells, antibodies against active caspase fragments are widely used tools. Application of the antibodies must be preceded by cell permeabilization. Once inside the cells, the anti-bodies recognize a neoepitope, which is exclusively exposed after specific cleavage of the procaspase molecule, but do not react with the full-length precursor. The antibody is either conjugated directly with fluorochrome or detected indirectly by the fluorochrome-conjugated secondary antibody. After removing the unbound substances, the fluorescence signal is detected by flow cytometry, fluorescence microscopy, or laser scanning cytometry.

2.4.2. Detection of caspase activity using synthetic substrates

Fluorescent caspase substrates containing two fluorochrome molecules have been introduced. The fluorochromes are attached to both the N and the C terminus of substrate molecules and are coupled together via the short peptide (containing specific caspase cleavage sequences). In this conformation, fluo-rescence is quenched. However, after cleavage of the peptide by the appropri-ate active caspase, fluorochromes are released, and this process is accompanied by an increase of fluorescence (green or red). The fluorescent signal (propor-tional to the amount of active caspase) can be detected using fluorescence microscopy, flow cytometry, or laser scanning cytometry. Various types of these fluorescent substrates are available commercially, including PhiPhiLux (for the detection of caspase-3 activity) or different CaspaLux substrates (for caspase-1, -6, -8, -9) (OncoImmunin, Inc., Calbiochem). There are several advantages of using these substrates. Because these substrates are cell permeable, they can be applied directly to the living intact cells (without any permeabilization or fixation), enabling evaluation of the caspase activity within a single cell. Moreover, simultaneous detection of the other apoptotic parameters can be performed using a particular sample.

Another approach for monitoring caspase activity in intact cells takes advantage of using fusion protein constructs composed of tandem fluorescence protein variants connected by a short peptide linker harboring at least one caspase cleavage site. Because of a very short distance between the two proteins, intensive direct fluorescent resonance energy transfer (FRET) occurs between donor and acceptor molecules, which share overlapping emission and excitation spectra, respectively. However, in the presence of active caspase, the linker is cleaved, which is accompanied by a decrease in FRET as a result of the physical separation of the two fluorescent protein moieties (Harpur et al., 2001). Initially, protein constructs composed of the enhanced cyan fluorescence protein as a donor and enhanced yellow fluorescence protein (EYFP) as an acceptor were used for the detection of caspase-3 activity (Luo et al., 2001; Rehm et al., 2002). However, these systems have been shown to be sensitive to changes in pH and chloride ion concentration, which may also occur during apoptosis. Under these conditions, signals that are not related to caspase activation could also be detected by the constructs, which would lead to misinterpretation of the final results, especially in in vivo experiments. Therefore, a FRET system resistant to pH and chloride concentration changes has been developed by Takemoto et al. (2003). The authors used a new variant of EYFP called Venus, whose absorption efficiency is significantly less sensitive to the H^+ or Cl^- concentration compared to EYFP (Takemoto et al., 2003). The resulting indicator for caspase activation, SCAT, has been used successfully for the detection of spatiotemporal activation of caspase-3 (SCAT3) or caspase-9 (SCAT9) in living cells. Moreover, the authors have also shown the reliability of this system in vivo (Takemoto et al., 2007). The measurement of FRET signals can be performed using fluorometry (Tawa et al., 2001) or fluorescence microscopy (Rehm et al., 2002). The flow cytometric FRET method for the analysis of caspase activity in living cells has also been introduced, facilitating performing the experiments in a high-throughput manner. Moreover, FRET-positive cells can be subsequently sorted and then used for additional biochemical or cell biology experiments.

2.4.3. Detection of caspase activity using synthetic inhibitors

The fluorochrome-labeled inhibitors of caspases (FLICA) provide a rapid, sensitive, relatively specific and ready-to-use tool for the evaluation of caspase activity (Bedner et al., 2000). The further important benefit of this method is the possibility of assessing the early caspase activation in individual cells, which also enables the estimation of the frequency of apoptosis within the cell population. FLICAs are carboxyfluorescein (FAM)-, sulforhodamine B (Sr)-, or fluorescein isothiocyanate (FITC)-labeled ligands that bind covalently (with 1:1 stochiometry) to active caspase enzyme sites (Darzynkiewicz et al., 2002). After cellular uptake and irreversible binding of the FLICA ligands to the active caspases, cells are washed in the appropriate buffer (see later) in order to eliminate the unbound fraction of FLICA molecules, and the fluorescence signal related to individual cells can be detected by fluorescence microscopy, flow cytometry, or laser scanning cytometry. An additional advantage of this method is that the simultaneous detection of other cell death parameters (e.g., plasma membrane integrity, phosphatidyl serine translocation, DNA fragmentation) can be performed within the single measurement (Bedner et al., 2000; Smolewski et al., 2001).

The following sample protocol is used for suspension cells (according to Immunocytochemistry Technologies, LLC).

1. Culture cells up to 1×10^6 cells/ml.
2. Induce apoptosis following the protocol and create positive and negative controls.
3. Reconstitute the reagent (FLICA, Immunocytochemistry Technologies, LLC) with 50 μl of dimethyl sulfoxide (DMSO) to form the stock concentrate (which can be frozen for future use).
4. Dilute the stock concentrate with 200 μl $1\times$ PBS to form the working solution.
5. Add $\approx$10 μl of the working solution directly to a 300- to 500-μl aliquot of the cell culture for labeling.
6. Incubate 1 to 4 h.
7. Wash and spin cells twice, or let incubate for 1 h with fresh medium.
8. If desired, label cells with Hoechst stain or propidium iodide or 7-AAD.
9. If desired, fix cells.
10. Analyze data using a fluorescence microscope, plate reader, or flow cytometer.

Unfortunately, the specificity of caspase inhibitors (labeled as well as unlabeled) is not absolute. As mentioned earlier, the tripeptide-containing inhibitors are not exclusively specific to any individual caspase and interact with almost all of them, but in the case of the caspase-2 only with a low affinity (Ekert et al., 1999). In addition, although individual tetrapeptide containing inhibitors should be expected to be relatively specific, they have been shown to interact simultaneously with several caspases. This is especially true in the case of the inhibitor containing the DEVD (caspase-3-like) sequence. In MCF-7 cells, which are known for the lack of caspase-3 protein, the strong fluorescence signal for FAM-DEVD-FMK has been detected, confirming that in addition to caspase-3, other caspases (caspase-7) were also bound with this FLICA ligand (Smolewski et al., 2001). The intracellular concentration of FLICA is also a very important factor in influencing the propensity of the FLICA ligand for the particular caspase molecule (Pozarowski et al., 2003).

It is important to consider that in addition to their ability to identify the particular active caspases, FLICAs can directly block caspase activity and, finally, inhibit apoptosis. This could significantly influence the reproducibility of the results, particularly when FLICAs are used for live nonfixed cells and for in situ experiments. However, cell fixation has been shown to induce the conformational changes of the inactive procaspase molecules, resulting in better accessibility of their active sites for the FLICA ligands and a false positive signal for caspase activity. Furthermore, the parts of the FLICA molecules containing the fluorochrome have been shown to interact with the newly uncovered hydrophobic epitopes of intracellular proteins of the apoptotic cells. This can contribute to a slight increase in the fluorescence signal and overestimation of the final results (Smolewski et al., 2001).

2.5. Measurement of caspase activation in vivo

2.5.1. Using synthetic caspase inhibitors

FLIVO is an injectable cell-permeable fluorescent probe acting as a pancaspase binding inhibitor, which is used for in vivo apoptosis quantification in live animals. It complements the aforementioned in vitro apoptosis detection by FLICA. FLIVO is composed of three parts: peptide inhibitor sequence targeted by all active caspases—VAD, fluoromethylketone (FMK) group enabling the formation of a covalent bond with active caspases, and fluorescent label: red—sulphorhodamine (SR, excitation/emission—565/600 nm) or green—carboxyfluorescein (FAM, 492/520). The choice of the fluorescence group depends on the other probes/fluorochrome-conjugated antibodies used simultaneously for dual-staining apoptotic studies in the animal model.

Once injected into the animal (e.g., mouse, rat, chicken), FLIVO circulates through the body, enters the cells, and binds covalently to the active caspases present inside the apoptotic cells, marking them red or green. The unbound reagent diffuses out of the cells and is removed by the circulation system of the animal. Tissues can be examined either directly using a window chamber system by confocal microscopy or, after sacrificing the animal and preparing tissue samples by fluorescence microscopy, by fluorometry or flow cytometry. Tissue sections obtained from the animal can also be fixed or frozen and stored (protected from light) for subsequent analysis for several months.

The following protocol is relatively easy and fast (according to Immunochemistry Technologies, LLC).

1. Dissolve the lyophilized content (usually enough for six mice) of one vial with 50 μl of DMSO.

2. Dilute the 10× injection buffer in dH2O (1:10).
3. Sterilize the injection buffer using a syringe filter (0.2 μm).
4. Add 200 μl of the sterile injection buffer to the reagent in the vial.
5. Inject 40 μl intravenously into the tail vein of each mouse. The total volume injected needs to be tested empirically and depends on the size of the animal and the frequency of apoptosis within its body.
6. Allow the reagent to circulate in the mouse; a 30-min interval is recommended.
7. Detect fluorescence (for methods see earlier discussion).

2.5.2. Using synthetic caspase substrates

Another approach for monitoring caspase activity *in vivo* is based on the hybrid recombinant reporter molecules consisting of the luciferase (Luc) reporter domain, estrogen receptor (ER) regulatory domain(s), and specific caspase cleavage site(s) inserted between them (Laxman *et al.*, 2002). When expressed in nonapoptotic mammalian cells, the level of luciferase activity is very low, as ER acts as its effective silencer. However, in the presence of active caspases, separation of the two domains (Luc and ER) occurs, which is accompanied by the restoration of Luc activity. Luc activation (a rapid increase in photon counts) can be detected using bioluminescence imaging (magnetic resonance), providing a noninvasive and dynamic approach for the evaluation of caspase activity in living animals. Several versions of the fusion proteins have been constructed. Of them, the dual ER reporter molecule ER–DEVD–Luc–DEVD–ER has been shown to be the most effective (Laxman *et al.*, 2002). The motif of the cleavage site can also be modified/substituted depending on the targeting of the particular caspases, which provides a unique opportunity to study the role of clinically important caspases and therapeutic modulations of their activities in intact animals.

2.6. Detection of the cleavage of caspase substrates

As mentioned in Section 1, there are multiple target caspase substrates inside the cells. Detection of cleavage of several of them is commonly used as a sensitive biochemical marker for apoptosis evaluation. This section provides an example of two of them, cleavage of Poly(ADP)ribose polymerase (PARP) and cytokeratin 18 (CK18).

2.6.1. PARP cleavage

Poly(ADP)ribose polymerase is a nuclear protein involved in DNA repair. During apoptosis, it is specifically cleaved by effector caspases with generation of the fragments of 89 and 24 kDa. Using specific antibodies against the particular epitopes of the PARP molecule, full length as well as cleaved proteins have been routinely detected by Western blotting. Development of the fluorochrome-conjugated antibodies that recognize only the specific

PARP cleavage product (89 kDa) enabled the fluorescence microscopy-based detections and the applications also in immunocytochemistry and immunohistochemistry. In addition, antibodies against the cleavage form of PARP are also available for FACS analysis (Li and Darzynkiewicz, 2000).

2.6.2. Cytokeratin 18 cleavage

Cytokeratin 18, a type I intermembrane filament protein, is an important component of the cytoskeleton of epithelial cells. During apoptosis, it is cleaved by caspase-3, -6, -7, and -9, which is accompanied by the appearance of a specific neoepitope in CK18 molecule (Caulin et al., 1997; Leers et al., 1999). This neoepitope is recognized exclusively by monoclonal antibodies (M30 Cytodeath). Antibody-based detection of CK18 cleavage is a very sensitive tool for apoptosis evaluation in cells and tissues from the very early until well-advanced stages. The CK18 neoepitope is known to be conserved in human, mouse, rat, and rabbit. However, the use of CK18 cleavage as the apoptotic marker is restricted only to normal epithelial cells and epithelium-derived tumors, being inapplicable, for example, in lymphoid and neuronal cells.

Specific unconjugated as well as conjugated monoclonal anti-CK18 neoepitope antibodies are available (e.g., Peviva, Roche). Unconjugated primary antibodies require an additional secondary antibody conjugated with either HRP (Western blotting) or fluorochrome (immunohistochem-istry, flow cytometry). However, for immunofluorescence and flow cyto-metry applications, the use of the anticleaved CK18 antibody conjugated directly to the fluorochrome is recommended, as it does not require the antimouse/rat/rabbit-IgG-fluorochrome secondary antibody. This can minimize potential problems with background fluorescence when analyzing samples of mouse/rat/rabbit origin. Moreover, the one step-procedure using the monoclonal antibody conjugated directly to fluorochrome (FITC) is faster and more convenient. As an example, a detailed protocol for flow cytometry detection of CK18 cleavage is provided.

1. Wash cells (one million) with PBS.
2. Fix the cells in ice-cold pure methanol (0.5 ml) at $-20\,°C$ for 30 min.
3. Wash cells in PBS twice.
4. Remove PBS.
5. Incubate the cells with 100 μl of M30 Cytodeath-FITC antibody work-ing solution (incubation buffer: PBS containing 1% BSA, final concen-tration of antibody 1 μg/ml) for 30 to 60 min at 15 to 25 $°C$ in the dark.
6. Wash cells with PBS twice.
7. Add 0.3 ml PBS, protect samples from light, and measure the FITC fluorescence (FL1-H) with a flow cytometer, evaluating the percentage of cells positive for CK18 cleavage (increased fluorescence compared to control) (Fig. 8.5).

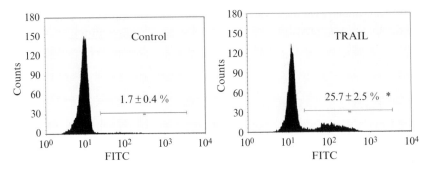

Figure 8.5 CK18 cleavage in HT-29 human colon cancer cells untreated (control) or treated with TRAIL (100 ng/ml) for 4 h, stained by M30 Cytodeath-FITC antibody, and measured using flow cytometry (FACS Calibur, Becton Dickinson) according to the protocol provided in the text. Results are means $\pm$ SEM of three independent experiments, Tukey test, $P > 0.05$.

Alternatively, the level of CK18 can also be detected using the ELISA-based method (M30 Apoptosense ELISA, Peviva). This assay is suitable for *in vitro*-cultured cells (cell lysates) as well as tumor extracts or patient blood/serum. The method is an example of a solid-phase sandwich enzyme immunoassay. Wells are coated with a monoclonal antibody that captures all CK18 molecules present in the sample. Another HRP-conjugated antibody added to the reaction mix specifically recognizes only the CK18 fragments containing the neoepitope. Unbound components are removed using washing buffer. Then, the appropriate HRP substrate is added to enable the visualization of the reaction. The amount of a color product is measured spectrophotometrically and is proportional to the amount of the specifically cleaved CK18. Appropriate standards and controls need to be processed simultaneously in order to quantify the results obtained. As the CK18 cleavage assay is only applicable for cells of an epithelial origin and not to lymphoid cells, CK18 is a very suitable marker for monitoring the apoptosis of carcinoma cells present in patient blood, especially following the anticancer treatment (Kramer *et al.*, 2004). This method may represent a very important and useful tool for evaluation of the tumor response to therapy.

3. CONCLUSION

The detection of caspase activation is considered to be an important biochemical marker of apoptosis. Different methods/kits for analysis of the processing and/or activation of caspases are available and provided by various companies. The choice of the method depends on several factors, including the experimental system used, cell type, and previous experience. However, because activation of caspases has been detected under nonapoptotic conditions and also caspase-independent cell death exists, it is highly recommended to perform some other apoptosis detection methods in parallel with the particular caspase activity assay in order to properly estimate and characterize the apoptotic cell death.

ACKNOWLEDGMENTS

Work in the authors' laboratory is supported by grants from the Swedish (3829-B04-09XAC) and Stockholm (061491) Cancer Societies, The Swedish Childhood Cancer Foundation (07/005), The Swedish Research Council (K2006-31X-02471-39-3), The Wenner-Gren Foundation, and the EC-FP-6 (Oncodeath and Chemores) and EC-FP-7 (APO-SYS) programs. The authors express their gratitude to Professor Sten Orrenius for permanent support.

REFERENCES

Baliga, B. C., Read, S. H., and Kumar, S. (2004). The biochemical mechanism of caspase-2 activation. *Cell Death Differ.* **11**, 1234–1241.

Bedner, E., Smolewski, P., Amstad, P., and Darzynkiewicz, Z. (2000). Activation of caspases measured in situ by binding of fluorochrome-labeled inhibitors of caspases (FLICA): Correlation with DNA fragmentation. *Exp. Cell Res.* **259**, 308–313.

Casciola-Rosen, L., Garcia-Calvo, M., Bull, H. G., Becker, J. W., Hines, T., Thornberry, N. A., and Rosen, A. (2007). Mouse and human granzyme B have distinct tetrapeptide specificities and abilities to recruit the bid pathway. *J. Biol. Chem.* **282**, 4545–4552.

Caulin, C., Salvesen, G. S., and Oshima, R. G. (1997). Caspase cleavage of keratin 18 and reorganization of intermediate filaments during epithelial cell apoptosis. *J. Cell Biol.* **138**, 1379–1394.

Colussi, P. A., Harvey, N. L., and Kumar, S. (1998). Prodomain-dependent nuclear localization of the caspase-2 (Nedd2) precursor: A novel function for a caspase prodomain. *J. Biol. Chem.* **273**, 24535–24542.

Darzynkiewicz, Z., Bedner, E., Smolewski, P., Lee, B. W., and Johnson, G. L. (2002). Detection of caspases activation in situ by fluorochrome-labeled inhibitors of caspases (FLICA). *Methods Mol. Biol.* **203**, 289–299.

Droin, N., Cathelin, S., Jacquel, A., Guery, L., Garrido, C., Fontenay, M., Hermine, O., and Solary, E. (2008). A role for caspases in the differentiation of erythroid cells and macrophages. *Biochimie* **90**, 416–422.

Ekert, P. G., Silke, J., and Vaux, D. L. (1999). Caspase inhibitors. *Cell Death Differ.* **6**, 1081–1086.

Fischer, U., Janicke, R. U., and Schulze-Osthoff, K. (2003). Many cuts to ruin: A comprehensive update of caspase substrates. *Cell Death Differ.* **10**, 76–100.

Froelich, C. J., Hanna, W. L., Poirier, G. G., Duriez, P. J., D'Amours, D., Salvesen, G. S., Alnemri, E. S., Earnshaw, W. C., and Shah, G. M. (1996). Granzyme B/perforin-mediated apoptosis of Jurkat cells results in cleavage of poly(ADP-ribose) polymerase to the 89-kDa apoptotic fragment and less abundant 64-kDa fragment. *Biochem. Biophys. Res. Commun.* **227**, 658–665.

Harpur, A. G., Wouters, F. S., and Bastiaens, P. I. (2001). Imaging FRET between spectrally similar GFP molecules in single cells. *Nat. Biotechnol.* **19,** 167–169.

Henzing, A. J., Dodson, H., Reid, J. M., Kaufmann, S. H., Baxter, R. L., and Earnshaw, W. C. (2006). Synthesis of novel caspase inhibitors for characterization of the active caspase proteome *in vitro* and in vivo. *J. Med. Chem.* **49,** 7636–7645.

Joseph, B., Ekedahl, J., Lewensohn, R., Marchetti, P., Formstecher, P., and Zhivotovsky, B. (2001). Defective caspase-3 relocalization in non-small cell lung carcinoma. *Oncogene* **20,** 2877–2888.

Kramer, G., Erdal, H., Mertens, H. J., Nap, M., Mauermann, J., Steiner, G., Marberger, M., Biven, K., Shoshan, M. C., and Linder, S. (2004). Differentiation between cell death modes using measurements of different soluble forms of extracellular cytokeratin 18. *Cancer Res.* **64,** 1751–1756.

Kumar, S. (2007). Caspase function in programmed cell death. *Cell Death Differ.* **14,** 32–43.

Lamkanfi, M., Festjens, N., Declercq, W., Vanden Berghe, T., and Vandenabeele, P. (2007). Caspases in cell survival, proliferation and differentiation. *Cell Death Differ.* **14,** 44–55.

Laxman, B., Hall, D. E., Bhojani, M. S., Hamstra, D. A., Chenevert, T. L., Ross, B. D., and Rehemtulla, A. (2002). Noninvasive real-time imaging of apoptosis. *Proc. Natl. Acad. Sci. USA* **99,** 16551–16555.

Leers, M. P., Kolgen, W., Bjorklund, V., Bergman, T., Tribbick, G., Persson, B., Bjorklund, P., Ramaekers, F. C., Bjorklund, B., Nap, M., Jornvall, H., and Schutte, B. (1999). Immunocytochemical detection and mapping of a cytokeratin 18 neo-epitope exposed during early apoptosis. *J. Pathol.* **187,** 567–572.

Li, X., and Darzynkiewicz, Z. (2000). Cleavage of Poly(ADP-ribose) polymerase measured in situ in individual cells: Relationship to DNA fragmentation and cell cycle position during apoptosis. *Exp. Cell Res.* **255,** 125–132.

Liu, J. J., Wang, W., Dicker, D. T., and El-Deiry, W. S. (2005). Bioluminescent imaging of TRAIL-induced apoptosis through detection of caspase activation following cleavage of DEVD-aminoluciferin. *Cancer Biol. Ther.* **4,** 885–892.

Luo, K. Q., Yu, V. C., Pu, Y., and Chang, D. C. (2001). Application of the fluorescence resonance energy transfer method for studying the dynamics of caspase-3 activation during UV-induced apoptosis in living HeLa cells. *Biochem. Biophys. Res. Commun.* **283,** 1054–1060.

MacFarlane, M., Merrison, W., Dinsdale, D., and Cohen, G. M. (2000). Active caspases and cleaved cytokeratins are sequestered into cytoplasmic inclusions in TRAIL-induced apoptosis. *J. Cell Biol.* **148,** 1239–1254.

McStay, G. P., Salvesen, G. S., and Green, D. R. (2008). Overlapping cleavage motif selectivity of caspases: Implications for analysis of apoptotic pathways. *Cell Death Differ.* **15,** 322–331.

Metkar, S. S., Wang, B., Ebbs, M. L., Kim, J. H., Lee, Y. J., Raja, S. M., and Froelich, C. J. (2003). Granzyme B activates procaspase-3 which signals a mitochondrial amplification loop for maximal apoptosis. *J. Cell Biol.* **160,** 875–885.

Nhan, T. Q., Liles, W. C., and Schwartz, S. M. (2006). Physiological functions of caspases beyond cell death. *Am. J. Pathol.* **169,** 729–737.

O'Brien, M. A., Daily, W. J., Hesselberth, P. E., Moravec, R. A., Scurria, M. A., Klaubert, D. H., Bulleit, R. F., and Wood, K. V. (2005). Homogeneous, bioluminescent protease assays: Caspase-3 as a model. *J. Biomol. Screen* **10,** 137–148.

Pozarowski, P., Huang, X., Halicka, D. H., Lee, B., Johnson, G., and Darzynkiewicz, Z. (2003). Interactions of fluorochrome-labeled caspase inhibitors with apoptotic cells: A caution in data interpretation. *Cytometry A* **55,** 50–60.

Rehm, M., Dussmann, H., Janicke, R. U., Tavare, J. M., Kogel, D., and Prehn, J. H. (2002). Single-cell fluorescence resonance energy transfer analysis demonstrates that caspase activation during apoptosis is a rapid process: Role of caspase-3. *J. Biol. Chem.* **277**, 24506–24514.

Robertson, J. D., Gogvadze, V., Kropotov, A., Vakifahmetoglu, H., Zhivotovsky, B., and Orrenius, S. (2004). Processed caspase-2 can induce mitochondria-mediated apoptosis independently of its enzymatic activity. *EMBO Rep.* **5**, 643–648.

Rodriguez, J., and Lazebnik, Y. (1999). Caspase-9 and APAF-1 form an active holoenzyme. *Genes Dev.* **13**, 3179–3184.

Rupinder, S. K., Gurpreet, A. K., and Manjeet, S. (2007). Cell suicide and caspases. *Vasc. Pharmacol.* **46**, 383–393.

Scaffidi, C., Schmitz, I., Zha, J., Korsmeyer, S. J., Krammer, P. H., and Peter, M. E. (1999). Differential modulation of apoptosis sensitivity in CD95 type I and type II cells. *J. Biol. Chem.* **274**, 22532–22538.

Smolewski, P., Bedner, E., Du, L., Hsieh, T. C., Wu, J. M., Phelps, D. J., and Darzynkiewicz, Z. (2001). Detection of caspases activation by fluorochrome-labeled inhibitors: Multiparameter analysis by laser scanning cytometry. *Cytometry* **44**, 73–82.

Stennicke, H. R., Deveraux, Q. L., Humke, E. W., Reed, J. C., Dixit, V. M., and Salvesen, G. S. (1999). Caspase-9 can be activated without proteolytic processing. *J. Biol. Chem.* **274**, 8359–8362.

Takemoto, K., Kuranaga, E., Tonoki, A., Nagai, T., Miyawaki, A., and Miura, M. (2007). Local initiation of caspase activation in Drosophila salivary gland programmed cell death in vivo. *Proc. Natl. Acad. Sci. USA* **104**, 13367–13372.

Takemoto, K., Nagai, T., Miyawaki, A., and Miura, M. (2003). Spatio-temporal activation of caspase revealed by indicator that is insensitive to environmental effects. *J. Cell Biol.* **160**, 235–243.

Tawa, P., Tam, J., Cassady, R., Nicholson, D. W., and Xanthoudakis, S. (2001). Quantitative analysis of fluorescent caspase substrate cleavage in intact cells and identification of novel inhibitors of apoptosis. *Cell Death Differ.* **8**, 30–37.

Thornberry, N. A., Rano, T. A., Peterson, E. P., Rasper, D. M., Timkey, T., Garcia-Calvo, M., Houtzager, V. M., Nordstrom, P. A., Roy, S., Vaillancourt, J. P., Chapman, K. T., and Nicholson, D. W. (1997). A combinatorial approach defines specificities of members of the caspase family and granzyme B: Functional relationships established for key mediators of apoptosis. *J. Biol. Chem.* **272**, 17907–17911.

Timmer, J. C., and Salvesen, G. S. (2007). Caspase substrates. *Cell Death Differ.* **14**, 66–72.

Turk, B., and Stoka, V. (2007). Protease signalling in cell death: Caspases versus cysteine cathepsins. *FEBS Lett.* **581**, 2761–2767.

Zhivotovsky, B. (2003). Caspases: The enzymes of death. *Essays Biochem.* **39**, 25–40.

Zhivotovsky, B., and Orrenius, S. (2005). Caspase-2 function in response to DNA damage. *Biochem. Biophys. Res. Commun.* **331**, 859–867.

Zhivotovsky, B., Samali, A., Gahm, A., and Orrenius, S. (1999). Caspases: Their intracellular localization and translocation during apoptosis. *Cell Death Differ.* **6**, 644–651.

CHAPTER NINE

Lysosomes in Apoptosis

Saška Ivanova, Urška Repnik, Lea Bojič, Ana Petelin, Vito Turk, *and* Boris Turk

Contents

Abstract

Lysosomes are specialized organelles for protein recycling and as such are involved in the terminal steps of autophagy. However, it has become evident that lysosomes also play an important role in the progression of apoptosis. This latter function seems to be dependent on lysosomal proteases, which need to be released into the cytosol for apoptosis to be efficient. Among the lysosomal proteases, the most abundant are the cysteine cathepsins and the aspartic protease cathepsin D, which seem to be the major apoptosis mediators. This chapter reviews the methods used to study lysosomes and lysosomal proteases.

Department of Biochemistry, Molecular and Structural Biology, J. Stefan Institute, Ljubljana, Slovenia

Methods in Enzymology, Volume 442
ISSN 0076-6879, DOI: 10.1016/S0076-6879(08)01409-2

1. INTRODUCTION

For many years, caspases (cysteinyl aspartate-specific proteses) have been thought to be the major proteases involved in the execution of the apoptotic program (Green and Kroemer, 1998; Salvesen and Dixit, 1997). However, recent findings suggest that proteases from the endosomal/lysosomal system, which in addition to proteases contain a number of other hydrolases, are also involved in this process, especially in pathological conditions (Guicciardi *et al.*, 2004; Turk *et al.*, 2002). The endosomal/lysosomal system is the place for the degradation of endocytosed and autophagocytosed cellular material, and endosomes/lysosomes have been for a long time erroneously considered as stable structures only involved in necrotic cell death, where the lysosomal proteases contribute to the degradation of cellular components in a rather nonspecific fashion (de Duve, 1969; Leist and Jäättelä, 2001; Turk *et al.*, 2002). Accumulating evidence suggests that partial lysosomal membrane permeabilization (LMP) with consequent release of lysosomal proteases into the cytosol may initiate and execute the apoptotic program in several models of apoptosis (Guicciardi *et al.*, 2004; Leist and Jäättelä, 2001; Stoka *et al.*, 2007; Turk *et al.*, 2002). The release of proteases into the cytosol seems to be a key event, although the precise mechanisms of LMP and the role of lysosomal enzymes in apoptosis are still largely unknown. This chapter describes methods used frequently for studying lysosomes and lysosomal proteases, in particular cysteine cathepsins, in apoptosis.

2. LYSOSOMAL MEMBRANE PERMEABILIZATION INDUCTION

Lysosomal membrane permeabilization induction is a critical step in the lysosomal pathway. Despite the fact that LMP was observed upon action of a number of stimuli, including death receptor engagement, oxidative stress, ultraviolet exposure, and lyosomotropic detergents (Table 9.1), the molecular mechanisms of LMP are only poorly understood. Perhaps the best characterized is the mechanism of lysosomotropic compounds (detergents), such as L-leucyl-L-leucine methyl ester (LeuLeuOMe), O-methyl-serine dodecylamide hydrochloride (MSDH), and C1311 (Burger *et al.*, 1999; Li *et al.*, 2000; Thiele and Lipsky, 1990, 1992). LeuLeuOMe, which is taken up through receptor-mediated endocytosis, becomes protonated within the acidic lysosomal compartments, where it accumulates. This results in a blockade of proton pumps and an increase in lysosomal pH followed by cathepsin C-mediated LeuLeuOMe polymerization within the

Table 9.1   Inducers of LMP and apoptosis

Inducer[a]	Amount/ concentration	References
Hydroxychloroquine	200 μM	Boya et al., 2003b
Chloroquine	30 μg/ml	Boya et al., 2003a
MSDH	<50 μM	Li et al., 2000; Zhao et al., 2003
LeuLeuOMe	0.1–3 mM	Cirman et al., 2004; Droga-Mazovec et al., 2008; Thiele and Lipsky, 1992, Uchimoto et al., 1999
Siramesine	0.1–15 μM	Ostenfeld et al., 2005
Vincristine	100–400 nM	Groth-Pedersen et al., 2007
Staurosporine	100–350 nM	Bidere et al., 2003; Johansson et al., 2003
αFas	0.1–0.5 μg/ml	Bojič et al., 2007; Brunk and Svensson, 1999; Wattiaux et al., 2007
Tumor necrosis factor-α	0.1–30 ng/ml	Fehrenbacher et al., 2004; Foghsgaard et al., 2001; Guiccardi et al., 2000, 2001
TRAIL	5 ng/ml	Guicciardi et al., 2007; Nagaraj et al., 2006; Werneburg et al., 2007
Hydrogen peroxide	0.1–30 μM	Antunes et al., 2001; Olejnicka et al., 1999; Yin et al., 2005
UVA/B	/	Bivik et al., 2006, 2007

[a] An incomplete list of LMP inducers is shown. The concentrations shown are approximate to give an orientation of the quantity of the reagent needed, but they have to be optimized for each cell line used.

lysosomes. Finally, polymerized LeuLeuOMe acts as a detergent within lysosomes, resulting in permeabilization of the lysosomal membrane (Thiele and Lipsky, 1990, 1992). MSDH and C1311 work in a similar fashion, that is, upon diffusion into acidic vacuolar compartment, they become protonated and accumulate in the lysosomes via "proton trapping." Finally, they work as detergents in their protonated form (Burger et al., 1999; Li et al., 1999). Also the classical antimalarial compounds chloroquine and hydroxychloroquine, which are weak amines, work in a similar fashion through accumulation in the lysosomes (Boya et al., 2003a,b).

Table 9.1 summarizes a number of cell death inducers also known to trigger LMP. However, it should be noted that the role of LMP is unclear in most of these pathways, with the exception of cell death induced by the lysosomotropic compounds (see earlier discussion), which clearly trigger lysosome-dependent cell death. In order to evaluate the role of lysosomes in certain pathway, one has to follow the kinetics of LMP in comparison with other cell death parameters, as LMP may be observed very late in the pathway, thereby being just a bystander event or eventually serving as an amplification loop.

3. METHODS USED FOR THE DETECTION OF LYSOSOME INTEGRITY

In order to be able to detect LMP, one has to be able to monitor the integrity of lysosomes, which gets compromised as a consequence of LMP. Most of the methods are based on the difference in pH between the lysosomes and the surrounding milieu. Lysosomes have namely an average pH of 4.6 to 5.0 due to the proton-pumping vacuolar (V) ATPases, which produce a proton gradient over the lysosomal membrane (Luzio *et al.*, 2007; Mellman *et al.*, 1986), whereas their surroundings have a pH value around neutral.

4. STAINING FOR ACID COMPARTMENTS/VESICLES

The simplest way to assess the integrity of lysosomes is to use lysosomotropic fluorescent dyes. Lysosomotropic compounds are weak bases that become charged after attracting one or more protons. Because the protonated form is no longer membrane permeable, they remain trapped in the acidic compartments. The most specific lysosomotropic dyes have pK_a values around 6, which implies that they will become selectively significantly more protonated in the lysosomes with a pH around 5, but not in the cytosol or compartments with a pH close to neutral (Kurz *et al.*, 2007).

Lysotracker (a trademark of Molecular Probes) and acridine orange (AO) are used frequently to label acidic compartments such as lysosomes, as well as secretory vesicles, because the procedure is quite simple. However, they should be used at a low concentration to reduce background staining. Lysotracker is available in several colors, and the most suitable product should be chosen based on the analytical method used (flow cytometry, fluorescence microscopy) and the scope of the experiment. For flow cytometry analysis, Lysotracker green is the most suitable because of its strong absorption at 488 nm. Both Lysotracker and AO have pH–independent emission spectra. However, AO is a metachromatic fluorochrome and its emission spectrum depends on the AO concentration. When activated by blue light (including 488 nm), AO will give a distinct red fluorescence (peak at about 650 nm) at high lysosomal concentrations and a weak green fluorescence (peak between 530 and 550 nm) at low cytosolic and nuclear concentrations (Antunes *et al.*, 2001; Millot *et al.*, 1997). Acidic vesicles should therefore appear red.

To assess the disruption of lysosomes, cells are usually stained after treatment to determine the acidity of the compartments. Under fluorescence microscope, untreated control cells should contain bright lysosomes, whereas cells with damaged lysosomes should only display uniform background fluorescence. In our experience, Lysotracker is extremely photosensitive. According to Molecular Probes, Lysotracker Red DND-99 is the only one suitable for aldehyde fixation and subsequent antifade mounting. In flow cytometry analysis, untreated control cells should have increased mean fluorescence intensity (MFI). However, it should be emphasized that with flow cytometry analysis, only the percentage of cells with damaged lysosomes can be evaluated and not the percentage of damaged lysosomes within one cell. Although fluorescence microscopy offers the possibility of qualitative determination whether the lysosomes of one cell are damaged completely or partially, a detailed and accurate quantitative analysis has not been reported as yet.

4.1. Reagents

Stock solution of 1 m*M* Lysotracker (Molecular Probes) should be prepared in dimethyl sulfoxide (DMSO), whereas a stock solution of AO at 1 mg/ml should be prepared in dH$_2$O.

4.2. Assay procedure

1. To stain suspension cells, dilute LysoTracker to a final concentration of 35 to 75 n*M*, depending on the cells or product used, using the growth medium. The same procedure can be applied to AO, except that the final concentration should be 2.5 to 5 μg/ml. The final concentration should be kept as low as possible to reduce potential artifacts due to dye overloading.

2. All cells should be stained after the treatment. For suspension cells, centrifuge cells, aspirate the supernatant, and resuspend the cells in probe-containing medium. For adherent cells, remove the medium and add probe-containing medium. Alternatively, if detached cells are to be included in the analysis, add medium containing twofold the final concentration of a probe directly to cell cultures without removing the medium containing the detached cells. With Lysotracker green DND-26, incubate cells under growing conditions for 5 to 10 min, for 30 min to 2 h with other Lysotracker reagents, and for 15 min with AO, respectively.

3. For suspension cells, repellet the cells by centrifugation and then resuspend them in fresh medium for microscopy analysis and in Hank's buffered salt

solution (HBSS) for flow cytometry analysis. For adherent cells, replace the probe-containing medium with fresh medium for microscopy analysis. Alternatively, for flow cytometry analysis, trypsinize the cells, dilute them with HBSS, centrifuge, and finally resuspend the pellet in HBSS.

4. Use a sample of untreated cells to adjust the sensitivity of the side scatter and forward scatter detectors and make sure that all the cells are displayed in the plot. A linear scale is generally recommended for a scatter plot unless cells differ in scatter properties a lot and exceed a linear scale range. Set the gate around the cells excluding debris. Adjust the sensitivity of the fluorescence detector (FL1 for Lysotracker green and FL3 for AO) in such a way that the untreated cells will have the fluorescence intensity between $10e^2$ and $10e^3$ a.u. on a logarithmic scale. Cells with increased lysosomal pH should have weaker fluorescence. Data can be presented in a histogram overlay. Valuable readouts include percentage of cells with weaker/stronger fluorescence and MFI or peak channel values of these subpopulations. Out of five different Lysotracker products available, two, green and yellow, absorb strongly at 488 nm and are therefore suitable for analysis by argon laser flow cytometers. Lysotracker green has the advantage of a narrow emission spectrum with little spillover into other detectors. It is detected in the FL1 channel and enables simultaneous detection of at least propidium iodide staining (detection in FL3 channel). AO displays green as well as red fluorescence and is therefore less suitable to be used along with other fluorophores.

5. Validation of Lysosomal Membrane Permeabilization

As an increase in lysosomal pH might not necessarily be a consequence of LMP, but might also be a consequence of specific effects of agents on V-ATPases or of an increased proton leakage, it is important to use an alternative method to verify LMP. Currently, it is still unclear whether LMP results in the selective release of certain proteases/proteins or in a simultaneous release of all proteins from the lysosomal lumen (Kroemer and Jäättelä, 2005). A selective release depending on the size of the proteins has been suggested based on experiments using fluorescein isothiocyanate (FITC)-labeled dextran particles of increasing size, which were loaded into lysosomes by endocytosis (Bidère *et al.*, 2003). However, there has not been a plausible explanation for this model at the molecular level (Kroemer and Jäättelä, 2005). Early alterations in lysosomal stability can be analyzed by staining the cells with AO before the treatment. Because of its metachromatic emission spectrum, rupture of AO-loaded lysosomes can be monitored as an increase in the cytoplasmic diffuse green fluorescence with a concomitant decrease in the lysosomal red fluorescence (Antunes *et al.*, 2001). An early molecular marker for lysosome integrity was to measure the activity of the lysosomal enzyme β-hexosaminidase (Mehta *et al.*, 1996), which is very large and is also present in the lysosomal membrane-bound form. In order to avoid possible artifacts, use of lysosomal proteases as the cytosolic marker is preferred now because of several reasons: (i) they are much smaller then β-hexosaminidase, (ii) they are often associated with cell death progression, and (iii) they are easy to assay.

6. Preparation of Cytosolic Extract

A critical step in assaying the cytosolic activity of cathepsins is preparation of the cytosolic extract. In order to measure the increase in cytosolic activity of the cathepsins after triggering apoptosis, cytosolic samples with minimal lysosomal contamination should be prepared. Several methods have been described; perhaps the easiest to use is the digitonin extraction method, which follows (Fogshgaard *et al.*, 2001).

6.1. Digitonin extraction method

Digitonin is a natural glycoside, which effectively solubilize lipids, and can serve as a very efficient detergent to permeabilize cell membranes. The digitonin concentration should be optimized for each cell line. It is important to work quickly and to keep all the reagents and cells on ice throughout the procedure.

Assay procedure

1. Grow 10^6 cells in 1 ml of appropriate medium.
2. Prepare serial dilutions of digitonin (1–200 μg/ml) in digitonin extraction buffer ($\times$ μg/ml digitonin, see later, 250 mM sucrose, 20 mM HEPES, pH 7.5, 10 mM KCl, 1.5 mM MgCl$_2$, 1 mM EDTA, 1 mM EGTA).
3. Collect cells by centrifugation (3 min, 3000g, 4°) in 1.5-ml Eppendorf tube and add 300 μl of ice-cold digitonin extraction buffer immediately to the cell pellet on ice.
4. Vortex tubes subsequently for 5 s and then keep them on ice for an additional 10 min and shake continuously.
5. Centrifuge (1 min, 14,000g, 4°) and remove the supernatant quickly. Measure lactate dehydrogenase (LDH) activity, cathepsin activity (see later), and protein concentration using a Bradford (Bio-Rad) or similar assay. Assays are recommended to be performed in small volumes (100–200 μl) in 96-well microtiter plates. The assay for LDH activity is started by adding 100 μl of buffer containing Tris-HCl, pH 7.3, 10 mM

sodium pyruvate, and 0.5 mM β-NADH to the sample. Oxidation of NADH is then monitored for 20 min at 37° using a microplate reader at 340 nm (Caruso *et al.*, 2006).

The enzymatic activity of LDH is used as a marker of cellular membrane permeability, whereas cathepsin activity is used as a marker of lysosomal leakage. The appropriate digitonin concentration is the one where the activity of LDH starts to increase, the protein extraction based on Bradford assay is nearly complete, and the cathepsin activity is still very low. There is usually a narrow window of digitonin concentration, which can be used without significantly affecting the integrity of lysosomes.

6.2. Preparation of the cytosolic extract with a needle

Another method can be used for the preparation of organelle-free cytosolic extracts (Ellerby *et al.*, 1997). An advantage of this method is that it does not require any detergent, whereas a disadvantage is that it requires heavy instrumentation (ultracentrifuge). In addition, the method requires a lot of skill and is therefore less recommended for beginners.

6.2.1. Assay procedure

1. Grow 10^6 cells/ml in a 10-cm plate in the appropriate medium.
2. Wash cells twice (5 min, 3000g, 4°) in isotonic buffer (200 mM mannitol, 70 mM sucrose, 1 mM EGTA, 10 mM HEPES).
3. Homogenize cells by passaging them through a needle. The number of passages (typically 15–25 passages) and the size of the needle should be optimized for each cell line.
4. Remove nuclei and unbroken cells by centrifugation at 500 g for 5 min and by two consecutive centrifugation steps, the first at 10,000g for 5 min and the second at 100,000g for 1 h.
5. Transfer the supernatant to a fresh Eppendorf tube and determine the protein concentration using the Bradford (Bio-Rad) or a similar assay.
6. Control experiments and assays for organelle markers should always be performed, as it is very easy to damage the organelles.

7. Lysosomal Cathepsins

Cathepsins are one of the largest groups of enzymes within the lysosomes. The most abundant are the cysteine cathepsins, which comprise a group of 11 related enzymes in human (B, C, F, H, K, L, O, S, V, W, and X), and the aspartic protease cathepsin D (Turk *et al.*, 2000, 2001, 2002). Because cathepsins B, L, and D are the most abundant among the lysosomal proteases, they are used most often as markers. The most common approaches are either immunoblot detection in different subcellular fractions or detection of lysosomal protease activity in the cytosolic fractions of cell lysates or in whole cells. Alternatively, colocalization studies with two-color immunofluorescence staining can be used, although the latter method is used less frequently.

All the cathepsins are optimally active at acidic pH; however, cysteine cathepsins are also highly active at neutral pH, although their lifetime under these conditions is limited as a consequence of irreversible unfolding. The most unstable is cathepsin L, which under these conditions (i.e., neutral pH and 37°) is active only for a minute or so in comparison with cathepsin S, which is stable for hours (Turk et al., 1995, 2002). Acidification of cytosol during apoptosis, however, may prolong their lifetime and enables them to cleave cytosolic substrates (Lagadic-Gossmann et al., 2004; Turk et al., 1995). Currently, with the exception of Bid and some other antiapoptotic members of the Bcl2 family of proteins (Blomgran et al., 2007; Cirman et al., 2004; Droga-Mazovec et al., 2008; Reiners et al., 2002; Stoka et al., 2001), almost no other characteristic physiological substrates have been identified during apoptosis. As they are relatively unspecific proteases, they generally degrade their substrates, Bid being an exception, thereby making this approach not very suitable. However, because of their relative abundance, it is possible to detect their activity using small synthetic substrates.

8. Synthetic Cathepsin Substrates

A number of different synthetic di- or tripeptides are available commercially from various suppliers (Bachem, Bubendorf, Switzerland; Calbiochem, San Diego, CA). The substrates are N-blocked peptides containing a reporter group at the C terminus, which is released by the active protease. The most commonly used reporter group for fluorimetric detection of cathepsin activity is 7-amino-4-methylcoumarin (AMC; excitation at 370–380 nm and emission at 460 nm), although 7-amino-4-trifluoromethylcoumarin (AFC; excitation at 405 nm and emission at 500 nm) can also be used. AFC can also be quantified colorimetrically at 380 nm, although the assay is less sensitive than the fluorimetric one. Chromogenic substrates based on the p-nitroanilide (pNA) reporter group are also used and can be recorded at 405 to 410 nm. In general, fluorimetric assays are substantially more sensitive and are therefore particularly useful in assaying low cathepsin concentrations, such as found in the cytosolic extracts. However, these substrates are not very selective and therefore will be hydrolyzed by different cathepsins present in the cytosolic extract (see Table 9.2).

The substrate recommended to be used first is Z-Phe-Arg-AMC, which can be hydrolyzed by all cathepsin endopeptidases. Some warning is perhaps

Table 9.2 Synthetic substrates of cysteine cathepsins[a]

Substrate	Cathepsin	k_{cat}/K_m values $[M^{-1}s^{-1}]$
Phe Arg	B, K, L, S, F, V	0.9×10^5–5.1×10^6 (S ≈ V ≈ K < B << L; data for F vary between 1.0 and 56.0×10^5, depending on the expression system used)
Arg Arg	B	≈1.1×10^5
Phe Val Arg	S	2.4×10^5
Val Val Arg	S	8.1×10^5
Leu Arg	K, S, L	2.4×10^5–2×10^6 (S ≈K << L)

[a] A compilation of the core peptide sequences of some commercially available substrates together with the cathepsins, which cleave them. The blocking group is generally benzyloxycarbonyl- (Z-), and reporter groups are most often pNA or AMC, although AFC can be also used. Data are given for the AMC substrates (Brömme *et al.*, 1993, 1996, 1999; Fonovic *et al.*, 2004; Wang *et al.*, 1998; an excellent review on cathepsin substrates is given by Kirschke *et al.*, 1995). However, values should only be taken as approximate, as the experiments were done at pH 5.5 to 6.0 and at 25°.

meaningful here as the substrate can be also hydrolyzed to some extent by trypsin-like proteases. The other commonly used substrate, which allows relatively selective measurement of cathepsin B activity, is Z-Arg-Arg-AMC. Other substrates provide somewhat higher selectivity (e.g., Z-Phe-Val-Arg-AMC or Z-Val-Val-Arg-AMC are much better substrates for cathepsin S), but they are not completely specific.

The most often used substrate for cathepsin D is denatured or FITC-hemoglobin assayed at pH 3.5 to 3.7 (Conner, 2004; De Lumen and Tappel, 1970). However, it should be noted that cysteine cathepsins may also degrade hemoglobin under the same conditions. There are also some synthetic octapeptide fluorogenic or chromogenic peptidic substrates available, such as Lys-Pro-Ile-Glu-Phe*Nph-Arg-Leu (*point of cleavage; Nph, *p*-nitrophenylalanine; Beyer and Dunn, 1996). An important note is also that at pH >5.0 cathepsin D at least *in vitro* exhibits little or no proteolytic activity due to deprotonation of both catalytic Asp residues, so its catalytic function under these conditions may be questionable.

8.1. Measurement of cysteine cathepsin activity in cytosolic fractions

8.1.2. Reagents

Stock solutions: prepare stock solution of appropriate substrate (see Table 9.2) in DMSO at 10 mM for AMC substrates. Stock solutions are stable for at least 2 to 3 months at −20°.

Buffer: 100 mM phosphate buffer containing 1 mM EDTA and 1 mM dithiothreitol, pH 6.0.

8.1.2. Assay procedure

The assay is designed to be performed in a small volume (100 μl) in a 96-well microtiter plate.

1. Add 50 to 100 μg of proteins from cytosolic extract into a well and fill with dH$_2$O to 40 μl.
2. Add 50 μl of phosphate buffer and incubate for 15 min at 37°.
3. Prior to use, dilute the stock solution of appropriate substrate in DMSO to 400 μM (50-fold) and add phosphate buffer in a 1:1 ratio (v/v) to get a 200 μM working solution. Add 10 μl of the working substrate solution to the sample to the final substrate concentration of 20 μM, which is sufficient to allow detection in the linear range of substrate consumption. Care should be taken not to exceed the 5% (v/v) DMSO total, as higher concentrations diminish cathepsin activity.
4. Measure kinetics of generation of the free fluorescent reporter group at 37° (although it is also possible to perform the experiment at lower temperature such as 25°). The initial rate of hydrolysis is determined from the observed progress curve. It is important to ensure that rates are calculated from the linear portion of the progress curve. If substrate depletion occurs too rapidly to allow the rate measurement, dilute the cytosolic extract and repeat the assay.

Calbiochem has also developed InnoZyme cathepsin B, L, and S activity assay kits, which are designed to measure enzyme activity in cell lysates. This method is similar to the aforementioned and is also based on a fluorogenic substrate hydrolysis in a 96-well plate. For the detailed protocol, see the Calbiochem Web page. However, one should be aware that all substrates used are semispecific and can be also cleaved by other cathepsins and sometimes even other proteases.

8.2. Detection of cathepsin activity in living cells

In addition to measuring the cathepsin activity in the cytosolic extract, one can measure the activity of cathepsins in living cells. ICT's Magic Red substrate-based kits are available commercially for measuring cathepsin B [MR-(RR)$_2$], K (MR-(LR)$_2$), and L (MR-(FR)$_2$] activity from Immuno-chemistry Technologies. These kits allow quick visualization of intracellular cathepsin activity in living cells. ICT's Magic Red detection kits utilize cresyl violet as the fluorophore. Following enzymatic cleavage at one or both arginine (R) amide linkage sites, the mono- and nonsubstituted cresyl violet fluorophores generate red fluorescence (emission wavelength >610 nm) when excited at 540 to 590 nm. Intracellular cresyl violet substrate hydrolysis can be monitored by the accumulation of red fluorescent product within various organelles using a fluorescence microscope or microtiter plate fluorimeter. The detailed protocol is available on http://www.immunochemistry.com/

CathepsinKits.htm. An improvement of the aforementioned relatively non-specific substrates are various cell-permeable, activity-based probes, which can be much more selective and are becoming more and more popular (Blum *et al.*, 2005; Watzke *et al.*, 2007).

8.2.1. Cathepsin inhibitors

In order to validate whether cathepsins are also actively involved in apoptosis, one can block their activity using small molecule inhibitors. There are several cathepsin inhibitors available on the market, such as L-*trans*-epoxysuccinyl (OEt)-Leu-3-methylbutylamide (E-64) (Peptide Research Institute, Osaka, Japan), Z-Phe-Ala-fmk (Sigma), and *N*-(L-trans-propylcarbamoyloxirane-2-carbonyl)-L-isoleucyl-L-prolin (CA-074) (Peptide Research Institute, Osaka, Japan) (for more, see Shaw, 1990). Whereas the first two are broad-spectrum cathepsin inhibitors, the latter is specific for cathepsin B. In addition to cysteine cathepsins, E-64 also inhibits calpains, so care should be taken to exclude calpains using a more selective calpain inhibitor such as the inhibitory calpastatin-derived peptide (Bachem, Bubendorf, Switzerland). For cellular studies, cell-permeable esterified forms of inhibitors should be used (E-64d and CA-074Me). However, CA-074Me is not completely specific for cathepsin B as it was found to also bind cathepsin S and some other as yet unidentified proteins (Bogyo *et al.*, 2000). Stock solutions of the inhibitors can be made in DMSO, which are stable for several months at $-20°$. A 1 h incubation time at 10 to 20 μM final concentration in the medium is sufficient to completely inhibit all the cathepsins in cells (Rozman-Pungerčar *et al.*, 2003). Higher concentrations (≥ 100 μM) of both inhibitors are cytotoxic and should not be used.

In addition, one should use caution when using typical "caspase-specific" fluoromethyl- or chloromethyketone inhibitors such as *N*-benzyloxycarbonyl-Val-Ala-Asp(OMe) fluoromethylketone (Z-VAD-fmk), Z-DEVD-fmk/cmk, or Y-VAD-fmk/cmk in apoptosis evaluation, as they can nonspecifically block cathepsins in living cells at higher concentrations (50–100 μM, depending on cells used) and *in vitro* (also at low concentrations) (Rozman-Pungerčar *et al.*, 2003). Usually, 10 to 15 μM Z-VAD-fmk, which is the most often used inhibitor, is sufficient to abolish any caspase activity, but does not block the cathepsins. However, it is advisable to perform a control experiment and measure the remaining cathepsin activity in the whole cell extract in the presence and absence of Z-VAD-fmk as described earlier using Z-Phe-Arg-AMC as a substrate. If the activity of cathepsins is not reduced by more than 5 to 10% (within experimental error), the inhibitor concentration is low enough to proceed with the experiment.

As cathepsin D is an aspartic protease, it can be inhibited by pepstatin A, a general inhibitor of aspartic proteases. In addition to cathepsin D, the inhibitor will only inhibit cathepsin E, which is the only other intracellular aspartic protease. Similarly as with other inhibitors, high concentrations (>100 μM) should be avoided.

9. CLEAVAGE OF PROTEIN SUBSTRATES

Although only a very small number of intracellular cathepsin substrates not linked with intralysosomal protein degradation have been described so far, it is advisable to check *in vitro* whether an individual cathepsin can actually cleave the potential substrate identified in a cellular assay in order to validate it. Recombinant or purified cathepsins can be obtained commercially (Sigma-Aldrich; Biomol International, R&D Systems) or prepared as recombinant proteins (Brömme *et al.*, 2004). The experimental design of such experiments has been described excellently earlier in this series for caspases (Stennicke and Salvesen, 2000). The only differences are in buffer composition and in the order of adding various components to the reaction mixture. As cysteine cathepsins are prone to irreversible pH-induced inactivation at neutral pH, they should be kept in their normal storage buffers (100 mM phosphate buffer, pH 6.0, for cathepsins B and C; 100 mM phosphate buffer, pH 6.8, for cathepsin H; 100 mM phosphate buffer, pH 7.0, for cathepsin S; 50 mM acetate buffer, pH 5.0, for cathepsins X, L, and K; and 50–200 mM acetate buffer, pH 4.2, for cathepsin D) containing 1 mM EDTA. Therefore, cathepsins should be added last in a very small volume to the reaction mixture, consisting of 100 mM phosphate buffer pH 7.2, containing 2 mM dithiothreitol (final concentration) and 1 mM EDTA, and the protein substrate (either as recombinant/purified protein or translated protein). Incubation time can vary between 15 min to 1 h, and then the reaction mixture can be treated as described (Stennicke and Salvesen, 2000). However, because cathepsins generally degrade their substrates, no specific degradation products are expected to be observed and the analysis will be essentially qualitative. Anyhow, exceptions such as Bid cleavage generating a defined tBid fragment can also be found (Cirman *et al.*, 2004; Stoka *et al.*, 2001).

Prior to enzymatic assays, it is advisable to active site titrate the enzymes. General procedures how to titrate proteases have been described very elegantly earlier in this series (Bieth, 1995; Stennicke and Salvesen, 2000) and for cysteine cathepsins in this series (Barrett and Kirschke, 1981) and elsewhere (Turk *et al.*, 1993), so the readers are encouraged to follow these procedures. Titration is performed with E-64, which inhibits all cysteine cathepsins, including cathepsin X. The preferred substrate for cysteine cathepsins is Z-Phe-Arg-pNA, which is a poorer cathepsin substrate than Z-Phe-Arg-AMC, as one can avoid substrate consumption in this way (Turk *et al.*, 1993). One should not forget that the protease concentration should be higher in this type of experiments ($\geq$100 nM). For cathepsin D, pepstatin A is used.

REFERENCES

Antunes, F., Cadenas, E., and Brunk, U. T. (2001). Apoptosis induced by exposure to a low steady-state concentration of H_2O_2 is a consequence of lysosomal rupture. *Biochem. J.* **356**, 549–555.

Barrett, A. J., and Kirschke, H. (1981). Cathepsin B, cathepsin H, and cathepsin L. *Methods Enzymol.* **80**, 535–561.

Beyer, B. M., and Dunn, B. M. (1996). Self-activation of recombinant human lysosomal procathepsin D at a newly engineered cleavage junction, "short" pseudocathepsin D. *J. Biol. Chem.* **271**, 15590–15596.

Bidere, N., Lorenzo, H. K., Carmona, S., Laforge, M., Harper, F., Dumont, C., and Senik, A. (2003). Cathepsin D triggers Bax activation, resulting in selective apoptosis-inducing factor (AIF) relocation in T lymphocytes entering the early commitment phase to apoptosis. *J. Biol. Chem.* **278**, 31401–31411.

Bieth, J. G. (1995). Theoretical and practical aspects of protease inhibition kinetics. *Methods Enzymol.* **248**, 59–84.

Bivik, C., Rosdahl, I., and Ollinger, K. (2007). Hsp70 protects against UVB induced apoptosis by preventing release of cathepsins and cytochrome c in human melanocytes. *Carcinogenesis.* **28**, 537–544.

Bivik, C. A., Larsson, P. K., Kågedal, K. M., Rosdahl, I. K., and Ollinger, K. M. (2006). UVA/B-induced apoptosis in human melanocytes involves translocation of cathepsins and Bcl-2 family members. *J. Invest. Dermatol.* **126**, 1119–1127.

Blomgran, R., Zheng, L., and Stendahl, O. (2007). Cathepsin-cleaved Bid promotes apoptosis in human neutrophils via oxidative stress-induced lysosomal membrane permeabilization. *J. Leukocyte Biol.* **81**, 1213–1223.

Blum, G., Mullins, S. R., Keren, K., Fonovic, M., Jedeszko, C., Rice, M. J., Sloane, B. F., and Bogyo, M. (2005). Dynamic imaging of protease activity with fluorescently quenched activity-based probes. *Nat. Chem. Biol.* **1**, 203–209.

Bogyo, M., Verhelst, S., Bellingard-Dubouchaud, V., Toba, S., and Greenbaum, D. (2000). Selective targeting of lysosomal cysteine proteases with radiolabeled electrophilic substrate analogs. *Chem. Biol.* **7**, 27–38.

Bojič, L., Petelin, A., Stoka, V., Reinheckel, T., Peters, C., Turk, V., and Turk, B. (2007). Cysteine cathepsins are not involved in Fas/CD95 signalling in primary skin fibroblasts. *FEBS Lett.* **581**, 5185–5190.

Boya, P., Andreau, K., Poncet, D., Zamzami, N., Perfettini, J. L., Metivier, D., Ojcius, D. M., Jaattela, M., and Kroemer, G. (2003a). Lysosomal membrane permeabilization induces cell death in a mitochondrion-dependent fashion. *J. Exp. Med.* **197**, 1323–1334.

Boya, P., Gonzalez-Polo, R. A., Poncet, D., Andreau, K., Vieira, H. L., Roumier, T., Perfettini, J. L., and Kroemer, G. (2003b). Mitochondrial membrane permeabilization is a critical step of lysosome-initiated apoptosis induced by hydroxychloroquine. *Oncogene.* **22**, 3927–3936.

Brömme, D., Bonneau, P. R., Lachance, P., Wiederanders, B., Kirschke, H., Peters, C., Thomas, D. Y., Storer, A. C., and Vernet, T. (1993). Functional expression of human cathepsin S in *Saccharomyces cerevisiae*: Purification and characterization of the recombinant enzyme. *J. Biol. Chem.* **268**, 4832–4838.

Brömme, D., Li, Z., Barnes, M., and Mehler, E. (1999). Human cathepsin V functional expression, tissue distribution, electrostatic surface potential, enzymatic characterization, and chromosomal localization. *Biochemistry.* **38**, 2377–2385.

Brömme, D., Nallaseth, F. S., and Turk, B. (2004). Production and activation of recombinant papain-like cysteine proteases. *Methods.* **32**, 199–206.

Brömme, D., Okamoto, K., Wang, B., and Biroc, S. (1996). Human cathepsin O_2, a matrix protein degrading cysteine protease expressed in osteoclasts: Functional expression of human cathepsin O_2 in *Spodoptera frugiperda* and characterization of the enzyme. *J. Biol. Chem.* **271**, 2126–2132.

Brunk, U. T., and Svensson, I. (1999). Oxidative stress, growth factor starvation and Fas activation may all cause apoptosis through lysosomal leak. *Redox Rep.* **4**, 3–11.

Burger, A. M., Jenkins, T. C., Double, J. A., and Bibby, M. C. (1999). Cellular uptake, cytotoxicity and DNA-binding studies of the novel imidazoacridinone antineoplastic agent C1311. *Br. J. Cancer.* **81**, 367–375.

Caruso, J. A., Mathieu, P. A., Joiakim, A., Zhang, H., and Reiners, J. J., Jr. (2006). Aryl hydrocarbon receptor modulation of tumor necrosis factor-alpha-induced apoptosis and lysosomal disruption in a hepatoma model that is caspase-8-independent. *J. Biol. Chem.* **281**, 10954–10967.

Cirman, T., Orešić, K., Mazovec, G. D., Turk, V., Reed, J. C., Myers, R. M., Salvesen, G. S., and Turk, B. (2004). Selective disruption of lysosomes in HeLa cells triggers apoptosis mediated by cleavage of Bid by multiple papain-like lysosomal cathepsins. *J. Biol. Chem.* **279**, 3578–3587.

Conner, G. E. (2004). Cathepsin D. *In* "Handbook of Proteolytic Enzymes" (A. J. Barrett, N. D. Rawlings, and J. F. Woessner, eds.). Academic Press, London.

de Duve, C. (1969). *In* "Lysosomes in Biology and Pathology" (J. T. Dingle and H. B. Fell, eds.), Vol. 1, pp. 3–40. North-Holland, Amsterdam.

De Lumen, B. O., and Tappel, A. L. (1970). Fluorescein-hemoglobin as a substrate for cathepsin D and other proteases. *Anal. BioChem.* **36**, 22–29.

Droga Mazovec, G., Bojič, L., Petelin, A., Ivanova, S., Romih, R., Repnik, U., Salvesen, G. S., Turk, V., and Turk, B. (2008). Cysteine cathepsins trigger caspase-dependent cell death through cleavage of Bid and antiapoptotic Bcl-2 homologues. *J. Biol. Chem.,* DOI 10.1074/jbc.M802513200.

Ellerby, H. M., Martin, S. J., Ellerby, L. M., Naiem, S. S., Rabizadeh, S., Salvesen, G. S., Casiano, C. A., Cashman, N. R., Green, D. R., and Bredesen, D. E. (1997). Establishment of a cell-free system of neuronal apoptosis: Comparison of premitochondrial, mitochondrial, and postmitochondrial phases. *J. Neurosci.* **17**, 6165–6178.

Fehrenbacher, N., Gyrd-Hansen, M., Poulsen, B., Felbor, U., Kallunki, T., Boes, M., Weber, E., Leist, M., and Jäättelä, M. (2004). Sensitization to the lysosomal cell death pathway upon immortalization and transformation. *Cancer Res.* **64**, 5301–5310.

Foghsgaard, L., Wissing, D., Mauch, D., Lademann, U., Bastholm, L., Boes, M., Elling, F., Leist, M., and Jäättelä, M. (2001). Cathepsin B acts as a dominant execution protease in tumor cell apoptosis induced by tumor necrosis factor. *J. Cell Biol.* **153**, 999–1010.

Fonovic, M., Brömme, D., Turk, V., and Turk, B. (2004). Human cathepsin F: Expression in baculovirus system, characterization and inhibition by protein inhibitors. *Biol. Chem.* **385**, 505–509.

Green, D. R., and Kroemer, G. (1998). The central executioner of apoptosis: Mitochondria or caspases? *Trends Cell Biol.* **8**, 267–271.

Groth-Pedersen, L., Ostenfeld, M. S., Høyer-Hansen, M., Nylandsted, J., and Jäättelä, M. (2007). Vincristine induces dramatic lysosomal changes and sensitizes cancer cells to lysosome-destabilizing siramesine. *Cancer Res.* **67**, 2217–2225.

Guicciardi, M. E., Bronk, S. F., Werneburg, N. W., and Gores, G. J. (2007). cFLIPL prevents TRAIL-induced apoptosis of hepatocellular carcinoma cells by inhibiting the lysosomal pathway of apoptosis. *Am. J. Physiol. Gastrointest. Liver Physiol.* **292**, G1337-1346.

Guicciardi, M. E., Deussing, J., Miyoshi, H., Bronk, S. F., Svingen, P. A., Peters, C., Kaufmann, S. H., and Gores, G. J. (2000). Cathepsin B contributes to TNF-alpha-mediated hepatocyte apoptosis by promoting mitochondrial release of cytochrome c. *J. Clin. Invest.* **106**, 1127–1137.

Guicciardi, M. E., Leist, M., and Gores, G. J. (2004). Lysosomes in cell death. *Oncogene.* **23,** 2881–2890.

Guicciardi, M. E., Miyoshi, H., Bronk, S. F., and Gores, G. J. (2001). Cathepsin B knockout mice are resistant to tumor necrosis factor-alpha-mediated hepatocyte apoptosis and liver injury: Implications for therapeutic applications. *Am. J. Pathol.* **159,** 2045–2054.

Johansson, A. C., Steen, H., Ollinger, K., and Roberg, K. (2003). Cathepsin D mediates cytochrome c release and caspase activation in human fibroblast apoptosis induced by staurosporine. *Cell Death Differ.* **10,** 1253–1259.

Kirschke, H., Barrett, A. J., and Rawlings, N. D. (1995). Lysosomal cysteine proteases. *In* "Protein Profile" (P. Sheterline, ed.), Vol. 2, pp. 1587–1643. Academic Press, London.

Kroemer, G., and Jäättelä, M. (2005). Lysosomes and autophagy in cell death control. *Nat. Rev. Cancer.* **5,** 886–897.

Kurz, T., Terman, A., and Brunk, U. T. (2007). Autophagy, aging and apoptosis: The role of oxidative stress and lysosomal iron. *Arch. Biochem. Biophys.* **462,** 220–230.

Lagadic-Gossmann, D., Huc, L., and Lecureur, V. (2004). Alterations of intracellular pH homeostasis in apoptosis: Origins and roles. *Cell Death Differ.* **11,** 953–961.

Leist, M., and Jäättelä, M. (2001). Four deaths and a funeral: From caspases to alternative mechanisms. *Nat. Rev. Mol. Cell Biol.* **2,** 589–598.

Li, W., Yuan, X., Nordgren, G., Dalen, H., Dubowchik, G. M., Firestone, R. A., and Brunk, U. T. (2000). Induction of cell death by the lysosomotropic detergent MSDH. *FEBS Lett.* **470,** 35–39.

Luzio, J. P., Pryor, P. R., and Bright, N. A. (2007). Lysosomes: Fusion and function. *Nat. Rev. Mol. Cell Biol.* **8,** 622–632.

Mehta, D. P., Ichikawa, M., Salimath, P. V., Etchison, J. R., Haak, R., Manzi, A., and Freeze, H. H. (1996). A lysosomal cysteine proteinase from *Dictyostelium discoideum* contains N-acetylglucosamine-1-phosphate bound to serine but not mannose-6-phosphate on N-linked oligosaccharides. *J. Biol. Chem.* **271,** 10897–10903.

Mellman, I., Fuchs, R., and Helenius, A. (1986). Acidification of the endocytic and exocytic pathways. *Annu. Rev. BioChem.* **55,** 663–700.

Millot, C., Millot, J. M., Morjani, H., Desplaces, A., and Manfait, M. (1997). Characterization of acidic vesicles in multidrug-resistant and sensitive cancer cells by acridine orange staining and confocal microspectrofluorometry. *J. Histochem. Cytochem.* **45,** 1255–1264.

Nagaraj, N. S., Vigneswaran, N., and Zacharias, W. (2006). Cathepsin B mediates TRAIL-induced apoptosis in oral cancer cells. *J. Cancer Res. Clin. Oncol.* **132,** 171–183.

Olejnicka, B. T., Andersson, A., Tyrberg, B., Dalen, H., and Brunk, U. T. (1999). Beta-cells, oxidative stress, lysosomal stability, and apoptotic/necrotic cell death. *Antioxid. Redox Signal.* **1,** 305–315.

Ostenfeld, M. S., Fehrenbacher, N., Høyer-Hansen, M., Thomsen, C., Farkas, T., and Jäättelä, M. (2005). Effective tumor cell death by sigma-2 receptor ligand siramesine involves lysosomal leakage and oxidative stress. *Cancer Res.* **65,** 8975–8983.

Reiners, J. J., Jr., Caruso, J. A., Mathieu, P., Chelladurai, B., Yin, X. M., and Kessel, D. (2002). Release of cytochrome c and activation of pro-caspase-9 following lysosomal photodamage involves Bid cleavage. *Cell Death Differ.* **9,** 934–944.

Rozman-Pungerčar, J., Kopitar Jerala, N., Bogyo, M., Turk, D., Vasiljeva, O., Stefe, I., Vandenabeele, P., Brömme, D., Puizdar, V., Fonovic, M., Trstenjak-Prebanda, M., Dolenc, I., *et al.* (2003). Inhibition of papain-like cysteine proteases and legumain by "caspase-specific" inhibitors: When reaction mechanism is more important than specificity. *Cell Death Differ.* **10,** 881–888.

Salvesen, G. S., and Dixit, V. M. (1997). Caspases: Intracellular signaling by proteolysis. *Cell.* **91,** 443–446.

Shaw, E. (1990). Cysteinyl proteinases and their selective inactivation. *Adv. Enzymol. Relat. Areas Mol. Biol.* **63**, 271–347.

Stennicke, H. R., and Salvesen, G. S. (2000). Caspase assays. *Methods Enzymol.* **322**, 91–100.

Stoka, V., Turk, B., Schendel, S. L., Kim, T. H., Cirman, T., Snipas, S. J., Ellerby, L. M., Bredesen, D., Freeze, H., Abrahamson, M., Bromme, D., Krajewski, S., Reed, J. C., Yin, X. M., Turk, V., and Salvesen, G. S. (2001). Lysosomal protease pathways to apoptosis: Cleavage of Bid, not pro-caspases, is the most likely route. *J. Biol. Chem.* **276**, 3149–3157.

Stoka, V., Turk, V., and Turk, B. (2007). Lysosomal cysteine cathepsins: Signaling pathways in apoptosis. *Biol. Chem.* **388**, 555–560.

Thièle, D. L., and Lipsky, P. E. (1990). Mechanism of L-leucyl-L-leucine methyl ester-mediated killing of cytotoxic lymphocytes: Dependence on a lysosomal thiol protease, dipeptidyl peptidase I, that is enriched in these cells. *Proc. Natl. Acad. Sci. USA* **87**, 83–87.

Thièle, D. L., and Lipsky, P. E. (1992). Apoptosis is induced in cells with cytolytic potential by L-leucyl-L-leucine methyl ester. *J. Immunol.* **148**, 3950–3957.

Turk, B., Bieth, J. G., Björk, I., Dolenc, I., Turk, D., Cimerman, N., Kos, J., Čolić, A., Stoka, V., and Turk, V. (1995). Regulation of the activity of lysosomal cysteine proteinases by pH-induced inactivation and/or endogenous protein inhibitors, cystatins. *Biol. Chem. Hoppe Seyler.* **376**, 225–230.

Turk, B., Krizaj, I., Kralj, B., Dolenc, I., Popovic, T., Bieth, J. G., and Turk, V. (1993). Bovine stefin C: A new member of the stefin family. *J. Biol. Chem.* **268**, 7323–7329.

Turk, B., Stoka, V., Rozman-Pungerčar, J., Cirman, T., Droga-Mazovec, G., Oresic, K., and Turk, V. (2002). Apoptotic pathways: Involvement of lysosomal proteases. *Biol. Chem.* **383**, 1035–1044.

Turk, B., Turk, D., and Turk, V. (2000). Lysosomal cysteine proteases: More than scavengers. *Biochim. Biophys. Acta.* **1477**, 98–111.

Turk, V., Turk, B., and Turk, D. (2001). Lysosomal cysteine proteases: Facts and opportunities. *EMBO J.* **20**, 4629–4633.

Uchimoto, T., Nohara, H., Kamehara, R., Iwamura, M., Watanabe, N., and Kobayashi, Y. (1999). Mechanism of apoptosis induced by a lysosomotropic agent, L-leucyl-L-leucine methyl ester. *Apoptosis.* **4**, 357–362.

Wang, B., Shi, G. P., Yao, P. M., Li, Z., Chapman, H. A., and Brömme, D. (1998). Human cathepsin F. *J. Biol. Chem.* **273**, 32000–32008.

Wattiaux, R., Wattiaux-de Coninck, S., Thirion, J., Gasingirwa, M. C., and Jadot, M. (2007). Lysosomes and Fas-mediated liver cell death. *Biochem. J.* **403**, 89–95.

Watzke, A., Kosec, G., Kindermann, M., Jeske, V., Nestler, H. P., Turk, V., Turk, B., and Wendt, K. U. (2008). Selective activity-based probes for cysteine cathepsins. Angew Chem. Int. Ed. Engl **47**, 406–409.

Werneburg, N. W., Guicciardi, M. E., Bronk, S. F., Kaufman, S. H., and Gores, G. J. (2007). Tumor necrosis factor-related apoptosis-inducing ligand activates a lysosomal pathway of apoptosis that is regulated by Bcl-2 proteins. *J. Biol. Chem.,* **282**, 28960–28970.

Yin, L., Stearns, R., and González-Flecha, B. (2005). Lysosomal and mitochondrial pathways in H_2O_2-induced apoptosis of alveolar type II cells. *J. Cell. BioChem.* **94**, 433–445.

Zhao, M., Antunes, F., Eaton, J. W., and Brunk, U. T. (2003). Lysosomal enzymes promote mitochondrial oxidant production, cytochrome c release and apoptosis. *Eur. J. BioChem.* **270**, 3778–3786.

CHAPTER TEN

MORE THAN TWO SIDES OF A COIN?

HOW TO DETECT THE MULTIPLE ACTIVITIES OF TYPE 2 TRANSGLUTAMINASE

Carlo Rodolfo,* Laura Falasca,[†] Giuseppina Di Giacomo,*
Pier Giorgio Mastroberardino,* *and* Mauro Piacentini*,[†]

Contents

Abstract

Programmed cell death (PCD) by apoptosis has been widely characterized as a process in which the expression and protein activation of a gene must be regulated in a very precise way in order to achieve the elimination of the dying cell without disturbing the neighborhoods. One of the first genes observed to be induced during the onset of PCD is the one coding for type 2 transglutaminase (TG2). Since the late 1990s, the unveiling of different new properties and enzymatic activities suggested the involvement of TG2 in a variety of cellular processes other than PCD and rendered the study of this protein more and more complicated.

* Department of Biology, University of Rome Tor Vergata, Rome, Italy
[†] INMI-IRCCS "L. Spallanzani," Rome, Italy

Methods in Enzymology, Volume 442
ISSN 0076-6879, DOI: 10.1016/S0076-6879(08)01410-9

1. INTRODUCTION

"Tissue" or type 2 transglutaminase (TG2) is the most peculiar member of a large family of transamidating acyltransferases called trans-glutaminases (Lorand and Graham, 2003). In the beginning, TG2 was characterized as a Ca^{2+}-dependent cross-linking enzyme, able to catalyze posttranslational modifications of proteins at the level of glutamine and lysine residues [for a complete review, see Fesus and Piacentini (2002)]. When the cell death program takes place, in both physiological or pathological settings [i.e., mammary gland regression, HIV and HCV infection (Amendola *et al.*, 1996; Nardacci *et al.*, 2003; Nemes *et al.*, 1996)], TG2 expression increases and the Ca^{2+}-dependent activation of its cross-linking activity leads to formation of the insoluble protein polymers called apoptotic bodies. These observations gave rise to the hypothesis that TG2's transamidating activity would have been an ideal mechanism for the cell to prevent the release of the intracellular material during the later phases of apoptosis (Fesus and Piacentini, 2002; Fesus *et al.*, 1989; Piacentini *et al.*, 2005).

Since the mid to late 1990s, the unveiling of new activities exerted by TG2 shifted the enzyme from simply a cross-linking agent to a multifunc-tional player, involved in all cellular activities (Fesus and Piacentini, 2002; Lorand and Graham, 2003). In 1994 Nakaoka and co-workers identified TG2 as the α subunit of the α_1-adrenergic receptor, suggesting that TG2 might behave as a G-protein and be involved in cellular signaling (Nakaoka *et al.*, 1994). Since then, many research groups have focused their interest on the possibility that the two mutually exclusive activities of the protein may be exerted not only at different cellular districts but also in diverse condi-tions and may have different effects on cell survival (Antonyak *et al.*, 2006; Mehta *et al.*, 2006; Piacentini *et al.*, 1991; Verma and Mehta, 2007). In the cytosol, TG2 might interact with cytoskeleton protein components and then contribute to rearrangements and stabilization of the cytoskeleton under stressful conditions. When TG2 is secreted outside of the cell, interactions with integrins and fibronectin stabilize the extracellular matrix (Facchiano *et al.*, 2006; Yuan *et al.*, 2007).

The contribution of TG2 to a multitude of cellular processes involves the enzyme in both neurodegenerative (i.e., Huntington's disease and Parkinson's disease) and metabolic (i.e., coeliac) diseases and points to it as a possible target for some therapies. The aim of this chapter is to describe experimental procedures that allow detection of the TG2 enzyme and activities under normal conditions as well as upon induction of programmed cell death. In addition, this chapter also reports on a method used to evaluate the effect of TG2 on the phagocytosis of apoptotic cells.

2. TG2 ACTIVITY ASSAYS

The first enzymatic activity characterized for TG2 was the transamidating one. This activity could be detected by means of the classical protocol, based on measurement of the incorporation of radiolabeled putrescine into dimethylcasein (DMC), or by means of the newest one, based on the incorporation of 5-(biotinamido)pentylamine (BPA, EZ-link, Pierce) into protein substrates. In addition, the use of BPA allows researchers to perform *in situ* analysis of TG2 activity as well as purification of TG2 protein substrates in a very simple way.

2.1. Radiometric assay

The TG2-catalyzed incorporation of radiolabeled putrescine into DMC is measured by means of the following procedure.

- In a final volume of 0.1 ml, add 1 mg/ml DMC, 0.2 mCi [1,4-^{14}C or ^{3}H] putrescine dihydrochloride (GE Healthcare), 0.5 mM putrescine (Sigma), 200 mM Tris-HCl (pH 8.0), 10 mM dithiothreitol, and 10 mM CaCl$_2$.
- Prewarm the reaction mixture for 10 min at 37 °C prior to the addition of TG2, either purified protein, from *in vitro* transcription–translation reaction (TnT, Promega) or commercial enzyme (N-Zyme or R&D) or 100 μg of total protein extracts, from any cell line or tissue, in Tris-HCl, pH 8.0, to start the reaction.
- Terminate the enzyme reaction, after 20 min incubation at 37 °C, by the addition of 1 ml of ice-cold 50% trichloroacetic acid (TCA).
- Place the tubes in ice for at least 30 min.
- Centrifuge at 10,000g for 15 min at 4 °C to collect the precipitated proteins.
- Wash twice with 0.1 ml of 10% TCA, using the same centrifugation conditions.
- Dissolve the final pellet with 0.1 ml of 1 M NaOH, combine with 4 ml of Aquasol (NEN), and perform liquid scintillation counting.

The measured activity can be expressed as picomoles of putrescine incorporated into protein per hour per milligram of protein.

2.2. *In situ* TG2 activity assay

In situ TG activity can be quantified by determining the incorporation of BPA into protein substrates using a microplate assay with neutravidin–horseradish peroxidase (HRP) (Zhang *et al.*, 1998).

- Incubate cells for 1 h with 3 to 5 mM BPA.

- Treat the cells in order to induce TG2 activation (i.e., staurosporine) and then incorporate BPA into protein substrates.
- Harvest cells in homogenate buffer (50 mM Tris-HCl, pH 7.5, 150 mM NaCl, and 1 mM EDTA).
- Load 2 μg of total protein into each well of a 96-well microtiter plate.
- Add 50 μl of homogenate buffer and incubate the plates overnight at 4 °C.
- Add to each well 200 μl of 5% bovine serum albumin (BSA) and 0.01% Tween 20 in borate saline buffer (100 mM boric acid, 20 mM sodium borate, and 80 mM NaCl, pH 8.5) and then incubate for 1 h at 37 °C.
- Rinse each well once with 1% BSA and 0.01% Tween 20 in borate saline buffer.
- Add to each well 100 μl of HRP-conjugated streptavidin (1:1000) in 1% BSA and 0.01% Tween 20 in borate saline buffer and incubate at 22 °C for 2 h.
- Rinse the wells four times with 1% BSA and 0.01% Tween 20 in borate saline buffer.
- Add 200 μl of substrate solution (0.4 mg of o-phenylenediamine dihydrochloride/ml of 0.05 mM sodium citrate phosphate buffer, pH 5.0) to each well and incubate for 20 min at 22 °C.
- Stop the reactions by the addition of 50 μl of 3 mM HCl to each well.
- The amount of proteins into which BPA had been incorporated is estimated by measuring the absorbance at 492 nm on a microplate spectrophotometer.

An alternative method used to detect TG2 activity on total protein extracts or *in situ* on frozen tissue sections, using a GFP derivative as substrate, has been reported by Furutani *et al.* (2001).

2.3. Purification of biotinylated protein substrates

The labeling procedure described earlier could be used also for the purification of the TG2 protein substrates (Lesort *et al.*, 1998; Rodolfo *et al.*, 2004).

- Harvest the labeled cells and wash in ice-cold phosphate-buffered saline (PBS).
- Disrupt the cells with CHAPS-IP buffer (10 mM HEPES, pH 7.4, 150 mM NaCl, 150 mM KCl, 1% CHAPS) for 30 min on ice.
- Add 100 μl of streptavidin paramagnetic particles (MagneSphere, Promega) or avidin resin to 100 μg of total protein lysate.
- Incubate for 2 h on a rotating wheel at 4 °C.
- Wash the beads (or the avidin resin) six times with CHAPS-IP buffer.
- Solubilize the proteins attached to the beads by direct boiling into protein sample buffer or elute from the avidin resin by incubation with 5 mM biotin.
- Analyze the eluted protein by Western blotting or mass spectrometry.

3. GTPASE ASSAY

The other historically characterized enzymatic activity is the ability to bind and hydrolyze GTP in order to function as a G-protein. This activity could be detected in a very simple way, but only for the purified protein, as follows.

- In a final volume of 50 μl mix 1 μCi (γ-^{32}P)GTP (3000 Ci/mmol), 5 μM GTP, 6 mM MgCl$_2$ 1 mM EDTA, and 50 mM Tris-HCl, pH 7.5.
- Add purified or *in vitro*-transcripted/translated TG2 at different concentrations.
- Incubate the reaction for 30 min at 37 °C.
- Stop the reaction by the addition of 750 μl ice-chilled 5% activated charcoal in 50 mM NaH$_2$PO$_4$.
- Centrifuge at 10,000 g for 5 min at 4 °C.
- Mix 400 μl of the supernatant with 3.5 ml of scintillation fluid and measure the release of ^{32}Pi in a scintillation counter.

4. TG2 KNOCKOUT MICE AS MODEL SYSTEM

TG2 knockout mice were generated some years ago in order to assess the role of TG2 "*in vivo*" in physiological as well as pathological conditions (Bailey *et al.*, 2004; De Laurenzi and Melino, 2001). Unfortunately, since the beginning it has become evident that the study of this model would have been very difficult. In fact, the analysis of tissues slides, as well as protein extracts, has to be considered not fully reliable, due to the lack of specificity of the commercially available anti-TG2 antibodies (Fig. 10.1). Figure 10.1A shows a characteristic pattern of bands obtained by probing a filter with different commercially available antibodies. It is clearly visible that some of the antibodies recognized a nonspecific signal very close to that of the real TG2. Although with little difficulty, the real TG2 corresponding signal it is even detectable, if the running conditions of the gel are appropriate. However, when the same antibodies are used in techniques such as immunohistochemistry or immunofluorescence on tissue slides or cultured cells, the nonspecific signal is clearly visible in TG2$^{-/-}$ mice, rendering this kind of approach completely unreliable (Fig. 10.1B).

It appears quite clear that scientists must pay a lot of attention in the evaluation of results obtained by means of the different anti-TG2 antibodies and must provide always the necessary controls when dealing with such a subject.

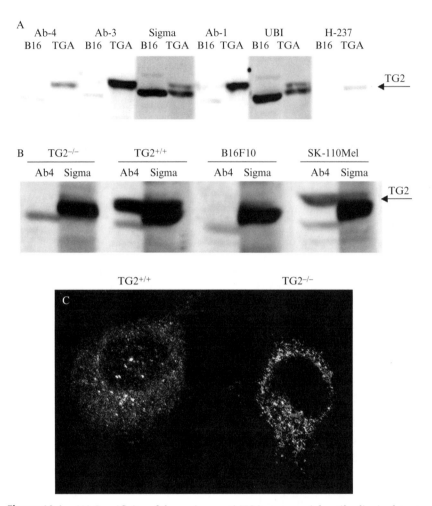

Figure 10.1 (A) Specificity of the various anti–TG2 commercial antibodies in detecting TG2 expression in B16, murine melanoma cells, and TGA, human neuroblastoma cells overexpressing TG2. Ab-1, Ab-3, and Ab-4 are antibodies with different specificities from Neomarkers, Sigma and UBI are the same antibody produced by two different companies, and H-237 is from Santa Cruz. (B) Comparison of the ability of "murine-specific" commercial antibodies to recognize TG2 in TG2$^{+/+}$, TG2$^{-/-}$, B16F10 murine melanoma, and SK-110 Mel human melanoma protein extracts. It appears clear that, for Western blotting detection of TG2, the most reliable antibodies for human cell extracts are the Neomarkers, whereas for murine protein extracts the most reliable is Neomarker AB4. (C) Immunofluorescence staining of TG2 in peritoneal macrophages from TG2$^{+/+}$ and TG2$^{-/-}$ mice using the Ab-4 antibody from Neomarkers. Confocal images show intense staining of the cell cytoplasm for both genotypes, indicating that there is no specific staining of TG2. Original magnification: 63×. It appears quite clear that the antibody recognizes a lot of the nonspecific signal, rendering impossible the detection of mouse TG2 by this approach.

5. IMMUNOCYTOCHEMISTRY AND IMMUNOFLUORESCENCE DETECTION OF TG2

The evaluation of TG2 expression under different conditions (i.e., apoptosis induction, starvation, virus infection) could be performed easily in human-derived cultured cells, while it is quite difficult when dealing with murine cell lines or tissues. In fact, as depicted in Fig. 10.1 the available commercial antibodies are not so useful in detecting TG2 in either Western blot or immunocytochemistry.

- Wash cells cultured on glass chamber slides or coverslips once with ice-cold PBS and fix for 20 to 30 min at room temperature with 4% paraformaldehyde/0.19% picric acid.
- After fixing, rinse three times with ice-cold PBS.
- Permeabilize with 0.1% SDS (or TX-100) in PBS for 10 min at room temperature.
- Block for 20 min at room temperature in 10% fetal bovine serum (FBS) (with or without 5% skim milk) in PBS and 0.1% SDS (or TX-100).
- Wash twice with 0.1% SDS (or TX-100) in PBS.
- Add primary anti-TG2 antibody, AB-1 Neomarkers diluted at 2 μg/ml, in 10% FBS (with or without 5% skim milk) in PBS and 0.1% SDS (or TX-100) and incubate for 1 h at room temperature (or overnight at 4 °C).
- Wash three times with PBS and incubate for 30 min–1 h at room temperature with the appropriate Alexa Fluor-conjugated secondary antibody diluted 1/1000 in 10% FBS in PBS and 0.1% SDS (or TX-100).
- Wash three times with PBS, remembering to counterstain nuclei with DAPI or Hoechst during the second wash (incubate for 10–20 min at room temperature).
- Cover slides with appropriate mounting medium and analyze under an epifluorescence or confocal microscope.

This technique allows one to analyze the difference of TG2 expression in a cell population as well as changes in intracellular TG2 distribution. However, to analyze the localization of TG2 at the ultrastructural level, we use immunogold labeling and electron microscopy on tissue sections. The standard fixation and embedding procedures for electron microscopy, which guarantee preservation of the ultrastructure of the samples, do not preserve the reactivity and stability of antigens. Thus we use a gentler aldehyde fixation combined with a low temperature embedding protocol. The best results are obtained when samples (human liver biopsies) are fixed in 2% paraformaldehyde/0.2% glutaraldehyde, dehydrated in ethanol, and then embedded in LR White resin.

Ultrathin sections are mounted on nickel grids coated with plastic film (e.g., Formvar) to improve the stability of the sections and immunolabeled by floating the grids on the surface of reagent droplets and washed in between by floating on the surface of a large volume of buffer or water.

- To inactivate residual aldehyde groups after aldehyde fixation, incubate the grids with 0.05 M glycine in PBS for 15 min.
- Transfer the grid to 5% BSA, 0.1% CWFS gelatin, and 5% FBS in PBS and incubate for 30 min.
- Incubate the grid with primary anti–TG2 antibody, AB-1 Neomarkers (CUB 7402) diluted at 8 μg/ml for 1 h at room temperature (4 °C overnight using 4 μg/ml antibody dilution; longer incubations with higher dilutions produce more specific labeling).
- Wash three times for 5 min in PBS.
- Incubate with the gold conjugate secondary antibody, diluted between 1:10 and 1:100 in PBS buffer for 1 h.
- Transfer the grid to a series of droplets of PBS buffer (5× 2 min) to remove unbound gold conjugate.
- Stain embedded sections in 5% uranyl acetate and wash in distilled water (5× 2 minutes) before examining under an electron microscope.

Gold-conjugated secondary antibodies can be purchased in various sizes. For single labeling we suggest using goat antimouse IgG conjugated to 15-nm gold particles (Bio-Cell, Cardiff, UK) (Fig. 10.2).

6. *In Vitro* Phagocytosis Assays

Macrophages, obtained by a PBS wash of the TG2$^{-/-}$ and TG2$^{+/+}$ mice peritoneal cavity and enriched by adherence selection for 1 h at 37 °C, are cultured in RPMI 1640 medium, supplemented with 10% FBS, 2 mM glutamine, 100 units/ml penicillin, and 100 mg/ml streptomycin at 37 °C and 5% CO_2.

Macrophages are very sensitive to external stimuli and/or environmental changes. For this reason it is highly recommended to allow them to recover their basal condition culture for at least 48 h before use. In addition, in order to avoid needing to detach cells by means of aggressive mechanical or chemical methods, it is important to plate the cells directly into the most suitable conditions, according to the kind of analysis to be performed.

- For light microscopy analysis, plate 5 × 10^5 cells/well into 2-well chamber slides.
- For scanning electron microscopy, plate 1 × 10^6 cells/well onto round coverslips contained in 24-well plates.
- For cytokine assays, plate 3 × 10^6 cells/well into 6-well plates.

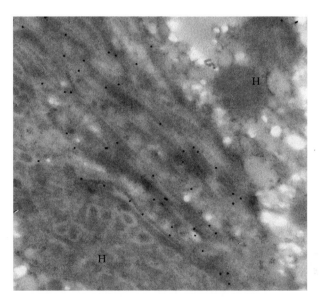

Figure 10.2 Immunogold localization of TG2 in HCV-infected livers (early stage of hepatic injury). TG2 (black dots) is present in the dilated spaces between the lateral domains of neighboring hepatocytes (H), and, as shown by high magnification observations, anti-TG2 conjugated-gold particles are associated with the presence of fibril bundles. Original magnification: 40,000×.

In order to study apoptotic cell uptake, a variety of different targets can be used; however, human lymphocytes could be considered the most convenient target, as they can be easily induced to die.

Peripheral blood lymphocytes are isolated by differential centrifugation using Ficoll-Paque, washed, and plated in RPMI 1640 medium supplemented with 10% FBS, 2 mM glutamine, 100 units/ml penicillin, and 100 mg/ml streptomycin at 37 °C and 5% CO_2. After 1 h, nonadherent cells are collected and, after morphological examination to determine that the majority of them are lymphocytes, the cells are cultured for 24 h prior to use.

Apoptotic cell death is induced either by incubation with cycloheximide (1 μM, 12 h) or by exposure to ultraviolet irradiation at 254 nm for 10 min, followed by 3 h of culture.

The percentage of apoptotic cells is quantified by flow cytometry analysis using annexin V and propidium iodide (PI) staining. Under these conditions, cells will be ≤5% PI positive and 50 to 80% annexin V positive.

For the *in vitro* phagocyte interaction assays, apoptotic lymphocytes are washed twice, suspended in fresh medium, and then added to the macrophage monolayer (3:1) in chamber slides, and the cells are allowed to interact at 37 °C.

After an appropriate period of time, cells are counterstained with hematoxylin–eosin, the uptake of apoptotic cells is determined by counting

400 macrophages in light microscopy, and results are expressed as the percentage of macrophages containing apoptotic bodies. This method requires one to be very careful in order to discriminate between internalized and surface-bound particles.

7. TG2 Acts as a Protein Disulfide Isomerase (PDI)

Evaluation of the extent to which the different activities of TG2 contribute to a specific experimental setting may not be an easy task when referring to kinase and PDI activities. This latter activity has been reported previously *in vitro*, and only recently has there been published evidence suggesting that, in the presence of low calcium levels, TG2 is able to catalyze the formation of disulfide bonds (Hasegawa *et al.*, 2003; Mastroberardino *et al.*, 2006).

To assess whether TG2 may be responsible for the formation of disulfide bonds *in vivo*, any of the hypothetical substrates may be analyzed by a modified Western blot procedure. It is important to remember that the action of TG2 as a PDI could be revealed by the comparison of SDS-PAGE migration on protein extracts belonging from both $TG2^{+/+}$ and $TG2^{-/-}$ mice tissues or siRNA silenced cells under reducing and nonreducing conditions. Under reducing conditions, the levels of the putative PDI substrate should be comparable in both $TG2^{+/+}$ and $TG2^{-/-}$ tissues/cells. However, when TG2-dependent disulfide bonds are preserved under non-reducing conditions, samples belonging to tissues/cells lacking TG2 exhibit a drastic reduction in the levels of the high molecular weight products (see Mastroberardino et al., 2006). This result clearly indicates a direct *in vivo* correlation between the formation of disulfide bonds and the presence of TG2. By means of this approach, we demonstrated that TG2–PDI activity is involved in the folding of some mitochondrial proteins as well as in the correct assembly of mitochondrial respiratory complexes. Interestingly, while the ATP synthase β chain is a substrate for PDI activity, the α chain is not modified, thus indicating a high degree of specificity in the system (Mastroberardino *et al.*, 2006).

8. Conclusions

The methods reported here are well-known procedures used to detect classical TG2 activities, which have been paralleled by some new enzymatic activities during the past few years. There are reports of the ability of TG2 to perform as a kinase and as a PDI. Assays that allow determining these activities are classical assays for kinase and PDI and can be retrieved from Hasegawa *et al.* (2003) and Mishra and Murphy (2004).

ACKNOWLEDGMENTS

The author express their gratitude to Associazione Italiana Ricerca sul Cancro; MiUR-PRIN 2006 to C.R and M.P., Telethon Grant N. GGP06254 to M.P. for their support.

REFERENCES

Amendola, A., Gougeon, M. L., Poccia, F., Bondurand, A., Fesus, L., and Piacentini, M. (1996). Induction of "tissue" transglutaminase in HIV pathogenesis: Evidence for high rate of apoptosis of CD4+ T lymphocytes and accessory cells in lymphoid tissues. *Proc. Natl. Acad. Sci. USA* **93**, 11057–11062.

Antonyak, M. A., Jansen, J. M., Miller, A. M., Ly, T. K., Endo, M., and Cerione, R. A. (2006). Two isoforms of tissue transglutaminase mediate opposing cellular fates. *Proc. Natl. Acad. Sci. USA* **103**, 18609–18614.

Bailey, C. D., Graham, R. M., Nanda, N., Davies, P. J., and Johnson, G. V. (2004). Validity of mouse models for the study of tissue transglutaminase in neurodegenerative diseases. *Mol. Cell. Neurosci.* **25**, 493–503.

De Laurenzi, V., and Melino, G. (2001). Gene disruption of tissue transglutaminase. *Mol. Cell. Biol.* **21**, 148–155.

Facchiano, F., Facchiano, A., and Facchiano, A. M. (2006). The role of transglutaminase-2 and its substrates in human diseases. *Front. Biosci.* **11**, 1758–1773.

Fesus, L., and Piacentini, M. (2002). Transglutaminase 2: An enigmatic enzyme with diverse functions. *Trends Biochem. Sci.* **27**, 534–539.

Fesus, L., Thomazy, V., Autuori, F., Ceru, M. P., Tarcsa, E., and Piacentini, M. (1989). Apoptotic hepatocytes become insoluble in detergents and chaotropic agents as a result of transglutaminase action. *FEBS Lett.* **245**, 150–154.

Furutani, Y., Kato, A., Notoya, M., Ghoneim, M. A., and Hirose, S. (2001). A simple assay and histochemical localization of transglutaminase activity using a derivative of green fluorescent protein as substrate. *J. Histochem. Cytochem.* **49**, 247–258.

Hasegawa, G., Suwa, M., Ichikawa, Y., Ohtsuka, T., Kumagai, S., Kikuchi, M., Sato, Y., and Saito, Y. (2003). A novel function of tissue-type transglutaminase: Protein disulphide isomerase. *Biochem. J.* **373**, 793–803.

Lesort, M., Attanavanich, K., Zhang, J., and Johnson, G. V. (1998). Distinct nuclear localization and activity of tissue transglutaminase. *J. Biol. Chem.* **273**, 11991–11994.

Lorand, L., and Graham, R. M. (2003). Transglutaminases: Crosslinking enzymes with pleiotropic functions. *Nat. Rev. Mol. Cell Biol.* **4**, 140–156.

Mastroberardino, P. G., Farrace, M. G., Viti, I., Pavone, F., Fimia, G. M., Melino, G., Rodolfo, C., and Piacentini, M. (2006). "Tissue" transglutaminase contributes to the formation of disulphide bridges in proteins of mitochondrial respiratory complexes. *Biochim. Biophys. Acta* **1757**, 1357–1365.

Mehta, K., Fok, J. Y., and Mangala, L. S. (2006). Tissue transglutaminase: From biological glue to cell survival cues. *Front. Biosci.* **11**, 173–185.

Mishra, S., and Murphy, L. J. (2004). Tissue transglutaminase has intrinsic kinase activity: Identification of transglutaminase 2 as an insulin-like growth factor-binding protein-3 kinase. *J. Biol. Chem.* **279**, 23863–23868.

Nakaoka, H., Perez, D. M., Baek, K. J., Das, T., Husain, A., Misono, K., Im, M. J., and Graham, R. M. (1994). Gh: a GTP-binding protein with transglutaminase activity and receptor signaling function. *Science* **264,** 1593–1596.

Nardacci, R., Ciccosanti, F., Falasca, L., Lo Iacono, O., Amendola, A., Antonucci, G., and Piacentini, M. (2003). Tissue transglutaminase in HCV infection. *Cell Death Differ.* **10** (Suppl 1), S79–S80.

Nemes, Z., Jr., Friis, R. R., Aeschlimann, D., Saurer, S., Paulsson, M., and Fésüs, L. (1996). Expression and activation of tissue transglutaminase in apoptotic cells of involuting rodent mammary tissue. *Eur. J. Cell Biol.* **70,** 125–133.

Piacentini, M., Amendola, A., Ciccosanti, F., Falasca, L., Farrace, M. G., Mastroberardino, P. G., Nardacci, R., Oliverio, S., Piredda, L., Rodolfo, C., and Autuori, F. (2005). Type 2 transglutaminase and cell death. *Prog. Exp. Tumor Res.* **38,** 58–74.

Piacentini, M., Fesus, L., Farrace, M. G., Ghibelli, L., Piredda, L., and Melino, G. (1991). The expression of "tissue" transglutaminase in two human cancer cell lines is related with the programmed cell death (apoptosis). *Eur. J. Cell Biol.* **54,** 246–254.

Rodolfo, C., Mormone, E., Matarrese, P., Ciccosanti, F., Farrace, M. G., Garofano, E., Piredda, L., Fimia, G. M., Malorni, W., and Piacentini, M. (2004). Tissue transglutaminase is a multifunctional BH3-only protein. *J. Biol. Chem.* **279,** 54783–54792.

Verma, A., and Mehta, K. (2007). Tissue transglutaminase-mediated chemoresistance in cancer cells. *Drug Resist. Updat.* **10,** 144–151.

Yuan, L., Siegel, M., Choi, K., Khosla, C., Miller, C. R., Jackson, E. N., Piwnica-Worms, D., and Rich, K. M. (2007). Transglutaminase 2 inhibitor, KCC009, disrupts fibronectin assembly in the extracellular matrix and sensitizes orthotopic glioblastomas to chemotherapy. *Oncogene* **26,** 2563–2573.

Zhang, J., Lesort, M., Guttmann, R. P., and Johnson, G. V. (1998). Modulation of the in situ activity of tissue transglutaminase by calcium and GTP. *J. Biol. Chem.* **273,** 2288–2295.

CHAPTER ELEVEN

Granzymes and Cell Death

Denis Martinvalet,* Jerome Thiery,* *and* Dipanjan Chowdhury[†]

Contents

* Immune Disease Institute and Department of Pediatrics, Harvard Medical School, Boston, Massachusetts
† Dana Farber Cancer Institute and Department of Radiation Oncology, Harvard Medical School, Boston, Massachusetts

Methods in Enzymology, Volume 442
ISSN 0076-6879, DOI: 10.1016/S0076-6879(08)01411-0

Abstract

Granzymes are cell death-inducing serine proteases released from cytotoxic granules of cytotoxic T lymphocytes and natural killer cells during granule exocytosis in response to viral infection or against transformed cells marked for elimination. A critical cofactor for the granule exocytosis pathway is perforin, which mediates the entry of granzymes into target cells, where they cleave specific substrates that initiate DNA fragmentation and apoptosis. One of the biggest challenges in studying the biology of granzymes has been the functional redundancy of granzymes in animal models making an *in vitro* experimental system essential. This chapter discusses methods to study granzyme function *in vitro* under physiologically relevant experimental conditions.

1. INTRODUCTION

Cytotoxic T lymphocytes (CTL) and natural killer (NK) cells eliminate transformed tumor cells and virally infected cells, primarily using the granule exocytosis pathway (Chowdhury and Lieberman, 2008; Melief and Kast, 1992). This involves the release of the pore-forming protein perforin (PFN), which delivers granzyme serine proteases to the cytosol of target cells. Both perforin and the granzymes are stored in cytotoxic granules and are released into the immunological synapse upon triggering by engagement with a target cell (Pipkin and Lieberman, 2007). Granzyme A and granzyme B (GzmA, GzmB) are the most abundant granzymes. GzmB, which cleaves after aspartic acid residues in many of the same substrates as caspases, has been the most extensively studied. However, work from several laboratories has begun to elucidate the roles of GzmA and other (so-called orphan) granzymes in cell death pathways. GzmB promotes caspase-dependent and -independent apoptosis, whereas GzmA, the most abundant protease in CTLs and NK cells, triggers caspase-independent apoptosis (Lieberman, 2003). The GzmB (and caspase) mitochondrial pathway leads to reactive oxygen species (ROS) generation, dissipation of $\Delta\Psi m$, and mitochondrial outer membrane permeabilization. All these events trigger the release of cytochrome c and other proapoptotic molecules from the mitochondrial intermembrane space. Treatment of target cells with superoxide scavengers that neutralize ROS completely blocks cell death by CTLs expressing all granzymes, underlining the importance of mitochondrial damage in granzyme/perforin-mediated apoptosis (Martinvalet *et al.*, 2005). DNA damage by GzmB is mediated primarily by activation of the caspase-activated DNase (CAD) following proteolytic cleavage of its inhibitor ICAD either directly by human GzmB or indirectly by executioner caspases, such as caspase-3 (Lord *et al.*, 2003).

Cell death induced by GzmA and perforin is rapid. Target cells have all the morphological features of apoptosis, and chromatin condensation and nuclear fragmentation can be readily seen within a few hours (Lieberman and Fan, 2003). However, GzmA does not activate the caspases or induce cleavage of most key caspase pathway substrates, such as bid or ICAD. Cells that are resistant to caspase-mediated cell death, including cells that over-express bcl-2, are sensitive to the apoptotic effects of GzmA (Martinvalet *et al.*, 2005). This may be important for immune defense against cancers and viruses that have devised strategies for evading caspase-mediated apoptosis. A key feature of GzmA-mediated cell death is the induction of single-stranded DNA nicks. These nicks produce multikilobase fragments, which are not detected by conventional assays that measure apoptotic DNA damage (Fan *et al.*, 2003; Shresta *et al.*, 1999).

GzmA-, GzmB-, and perforin-deficient mice were generated a decade ago, and early studies have been reviewed extensively (Russell and Ley, 2002). Although perforin-deficient mice are severely immunodeficient and compromised in their ability to defend against viruses and tumors, mice deficient in any granzyme have only subtle differences compared with wild-type animals. Although only one molecule (perforin) effectively delivers the granzymes into target cells, each of the granzymes can trigger cell death. These experiments highlight the functional redundancy of the granzymes. Therefore, to understand the molecular mechanism of granzyme function it is important to develop *in vitro* assays to measure cell death induced by individual granzymes. The biggest hurdle in doing these experiments is purifying enzymatically active granzymes and perforin. This chapter first discusses the purification procedure of GzmB, GzmA, and perforin. The final section of the chapter focuses on assaying granzyme- and perforin-mediated cell death.

2. Purification of Rat Perforin from RNK-16 NK-like Leukemia Cells

This method is based on the use of the rat LGL leukemia cell line RNK-16. Cells are expanded in the rat peritoneal cavity and then PFN is purified from isolated cytotoxic granules.

2.1. Growth of RNK-16 cells in rats

Cytotoxic granules are isolated from rat RNK-16 NK-like leukemia cells, serially passaged *in vivo* as ascites in Fischer F344 rats.

Protocol

F344 rats (200–225 g) are injected intraperitoneally with 1.5 ml of pristane (2,5,10,14-tetramethyl-pentadecane) (Sigma, St. Louis, MO) 4 to 7 days before RNK-16 cells injection to prepare peritoneal cavity. About 5 to 10 × 10^7 RNK-16 cells are injected into each rat. Cells are washed twice in Hank's balanced salt solution (HBSS) and resuspended in 2 ml per rat of HBSS for intraperitoneal injection. After 11 days postintraperitoneal injection of RNK-16 cells, rats are monitored every day for ascites production. Most of the time, rats are sacrificed after 12 to 15 days postintraperitoneal injection by CO_2 inhalation and RNK-16 cells are harvested immediately. Typically, the yield of RNK-16 cells per rat can range from 1 to 10 × 10^9. Aliquots of RNK-16 cells that appear healthy and are found in a high yield of cells without blood contamination are used for reinjection or frozen immediately for future use.

2.2. Cytotoxic granules isolation from RNK-16 cells

2.2.1. Solutions

All solutions must be cold prior to use.

HHH buffer: HBSS containing 10 mM HEPES, pH 7.2, and 100 U/ml heparin (Sigma-Aldrich, Louis, MO) (make fresh)

Relaxation buffer: Distilled H_2O containing 10 mM KCl, 3.5 mM MgCl$_2$, 10 mM PIPES, pH 6.8, 1.25 mM EGTA, pH 8.0, and 1 mM ATP (relaxation buffer can be prepared in advance without ATP)

40% adjusted Percoll: relaxation buffer containing 40% Percoll (w/v) (Sigma), 250 mM sterile sucrose, 10 mM HEPES, pH 7.2, and 7 mM HCl (make fresh)

Solubilization buffer: Distilled H_2O containing 1 M NaCl, 20 mM Na-acetate, pH 4.5, and 2 mM EDTA, pH 8.0 (keep at 4°)

2.2.2. Protocol

1. Harvest RNK-16 cells from rats using HHH buffer (CO_2 asphyxiation and bilateral thoracotomy 5 min after CO_2 exposure are used for euthanasia before harvesting the cells). To harvest cells, pin the rat on its back to a pad and slit the skin. Then perform a small incision into the rat abdomen and inject 10 ml of HHH buffer with a 10-ml pipette to wash the peritoneal cavity, mix slowly, and withdraw. Keep all harvested cells on ice until the end of the procedure. Expect to use 100 to 150 ml of HHH buffer to extract all RNK-16 cells. Then count the cells and keep aliquots for reinjection or storage at −80° for future use. Centrifuge at 200g for 5 min and remove the supernatant (carefully take the top layer first, which is pristane, and then remove the HHH buffer). Wash the cells three times

with cold HBBS (keep at 4° at all times) and then resuspend the cells at 10^8/ml in relaxation buffer.

2. Prechill cavitation/disruption bombs (Parr Instrument Company, Moline, IL) at 4° overnight before the procedure. Place cells in cavitation bombs in a 50-ml tube (or Erlenmeyer if sample is bigger than 50 ml) surrounded by ice and place a stir bar in the tube. Slowly pressurize to 450 psi with N_2 and stir at 4° for 25 min. Then gradually release pressure and harvest the contents into a new tube on ice. If pressure drops below 250 psi, release remaining pressure quickly through the intake valve and collect remaining sample in the original tube.

3. Remove nuclei by differential centrifugation at 4°. To do this, spin the sample at 400g for 10 min and retain the supernatant. Wash the pellet twice in equal volumes of relaxation buffer, spin at 400g for 5 min, and add supernatants to the original supernatant. Spin pooled supernatants at 400g for 5 min. The final supernatant (PNS) should be free of nuclei and whole cells.

4. Concentration of cytotoxic granules and removal of cytoplasmic contamination (optional). Centrifuge PNS at 15,000g for 10 min to pellet granules and heavy organelles and resuspend the pellet in relaxation buffer with a 20-gauge spinal needle (Popper & Son's Inc., New Hyde Park, NY).

5. Fractionate the cell lysate and collect granules. To do this, perform a 20-ml continuous gradient of 40% adjusted Percoll (4°) in a 26-ml rigid polycarbonate tube (Beckman Instruments Inc., Palo Alto, CA) by centrifuging at 20,000 rpm for 10 min. Then carefully layer 5 ml/tube of PNS above the 20 ml of preformed gradient of 40% adjusted Percoll. Centrifuge for 35 min at 20,000 rpm.

6. Extract cytotoxic granules with a 20-gauge spinal needle from the bottom of the gradient by harvesting the bottom 5 ml from each tube (granules should form a visible layer). Pool the granules fractions and then pellet Percoll by centrifugation overnight at 22,000 rpm (granules form a thin layer above the Percoll pellet).

7. Remove supernatant and resuspend granule layer in solubilization buffer with a 20-gauge spinal needle (2 ml per 10^9 original RNK-16 cells) and store at −80°. After resuspension of the granules, the suspension can be stored indefinitely at −80° before further PFN purification. The best results to disrupt the granules membranes come from solubilization with at least one freeze/thaw cycle overnight at −80°.

2.3. Purification of rat perforin

For perforin isolation, the immobilized metal affinity chromatography (IMAC) procedure is used (Winkler *et al.*, 1996).

2.3.1. Materials/reagents/solutions

Fast protein liquid chromatography (FPLC) system (Bio-Rad, Hercules, CA) and columns (Pharmacia Biotechnology, Piscataway, NJ)

Dynamic loop

IMAC resin charged with cobalt: TALON Superflow metal affinity resin (Clontech, Mountain View, CA)

10-ml desalting columns: Econo-Pac 10DG disposable chromatography columns (Bio-Rad)

Ultrafiltration concentrator: Amicon Ultra centrifugal filter device with low-binding Ultracel-50K membranes (Millipore, Billerica, MA)

Equilibration buffer (buffer A): distilled H_2O containing 1 M NaCl, 20 mM HEPES, pH 7.2, and 10% (w/v) betaine

Elution buffer (buffer B): equilibration buffer containing 1 M imidazole, pH 7.5

2.3.2. Protocol

All buffers and suspension are kept on ice.

1. Keep the frozen tube of solubilized granules for 2 h on ice, keeping the suspension mixed by inverting the tube every 15 to 20 min, then centrifuge the suspension at 15,000g for 10 min, carefully decant the supernatant, and filter through a 0.45-μm low-protein binding syringe filter (Millipore) to remove membrane fragments (if the filter is blocked, change the filter carefully).

2. During step 1, equilibrate desalting columns with 25 ml of cold equilibration buffer (1 column/2.5 ml of supernatant).

3. Run 2.5 ml of supernatant through each desalting column, discard the flow through, add 3.5 ml of equilibration buffer, retain the flow through, and combine all samples for the FPLC step. Filter one more time through a 0.45-μm low-protein binding syringe filter. *It is very important to note that at this point PFN is no longer in a chelated buffer and can be inactivated by traces of Ca^{2+} and that the FPLC step must be realized as quickly as possible.*

4. Prepare a 5-ml column of IMAC TALON Superflow resin, equilibrate it with 10 column volumes (CVs) of equilibration buffer, load sample into a dynamic loop, and attach to a FPLC work station. The addition of EGTA into each fraction collector tube is performed before FPLC run (for a 1 mM final concentration).

5. Run the following FPLC program with buffer A (equilibration buffer) and buffer B (elution buffer): wash the column with 100% buffer A at 1.5 ml/min for 1 CV, inject sample at 1 ml/min, wash the column with a linear gradient of 1 to 2% buffer B at 1 ml/min for 5 CVs, and then elute with a linear gradient of 2 to 40% buffer B at 1.5 ml/min for 7 CVs and 40 to 100% buffer B at 1.5 ml/min for 5 CVs. Monitor elute fractions

continuously at 280 nm. With this method, two peaks absorbing at 280 nm are observed, the first one containing both granzymes and proteoglycan and the second one perforin (Fig.11.1A).

6. Collect 10 to 20 μl of each fraction and screen for hemolytic activity by a hemoglobin release assay (see next section). Then, collect all hemolytic positive fractions (containing perforin) and concentrate with an ultra-centrifugation concentrator to a final volume of 200 μl in the presence of 20 μl fatty acid-free bovine serum albumin (BSA).

7. Aliquot purified active perforin and store at −80°. Determine protein concentration by the BCA assay (Pierce, Rockford, IL) using BSA for calibration. SDS-PAGE and Western blot are used to control quality and purity of the purified perforin (Fig. 11.1B). Analyze a sample by electrophoresis on a 12% polyacrylamide-SDS gel. Visualize protein bands using silver staining procedure (Invitrogen, Carlsbad, CA). For Western blot, transfer proteins to a nitrocellulose membrane and detect perforin using rabbit antiperforin (Cell Signaling, Danvers, MA).

2.4. Perforin activity detection

2.4.1. Hemoglobin release assay

Sheep red blood cells (Rockland Immunochemicals, Gilbertsville, PA) are washed three times with cold HBSS and resuspended at 1% (v/v) in HBSS containing 4 mM CaCl$_2$. One hundred microliters of red blood cells solution is transferred to a 96-well V-bottom microplate, and 10 to 20 μl of each FPLC fractions is added. The plate is incubated at room temperature for 20 min. The plate is then centrifuged for 5 min at 400g, and 50 μl of cell-free supernatant is transferred from each well to a 96-well flat-bottom microplate. The hemoglobin release into supernatant is detected with an ELISA reader at 412 nm. The percentage release is calculated against the maximal hemolysis determined by adding 0.01% (v/v) saponin (Sigma) to the red blood cells as [(experimental hemolysis − spontaneous hemolysis)/ (maximal hemolysis − spontaneous hemolysis)] × 100.

2.4.2. Perforin loading and detection by fluorescence microscopy

Cells are grown on collagen-coated coverslips (Sigma–Aldrich) and treated with a sublytic concentration of PFN. The sublytic PFN dose is determined independently for each cell line, PFN preparation, and experiment as the concentration required to induce 5 to 20% trypan blue or propidium iodide uptake (Shi et al., 1992). Cells are equilibrated in cell buffer (HBSS with 10 mM HEPES, pH 7.5, 4 mM CaCl$_2$, 0.4% BSA), and PFN, diluted in PFN buffer (HBSS, 10 mM HEPES, pH 7.5), is added at the required dose. Cells are then fixed for 20 min in PBS containing 2% paraformaldehyde, washed, and incubated in PBS containing 50 mM NH$_4$Cl for 20 min.

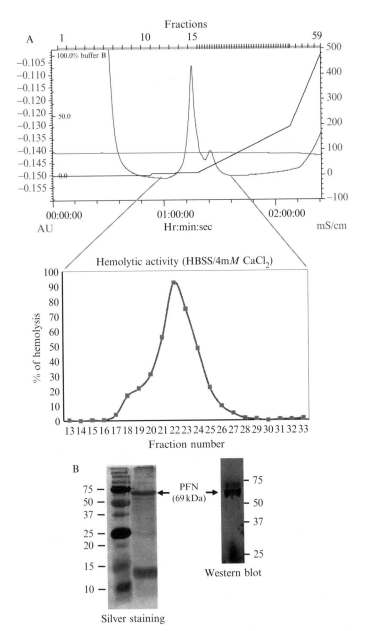

Figure 11.1 Purification of perforin and measuring hemolytic activity. (A) The profile of eluted protein from the FPLC is shown (top). Fractions and corresponding hemolytic activity (bottom) show that the highest activity corresponds to the second absorption peak at 280 nm. (B) A sample is analyzed after concentration by electrophoresis on a 12% polyacrylamide-SDS gel. Protein bands are visualized using the silver staining procedure (left). Molecular weight marker proteins are indicated in the left lane of the gel. For immunoblot (right), proteins are transferred to a nitrocellulose membrane and a 69-kDa perforin band is detected.

Cells are then washed with PBS and permeabilized for 5 min in permeabilization buffer (0.2% Triton X-100 in PBS). After two washes in PBS, coverslips are placed in blocking solution [10% fetal calf serum (FCS) in PBS] for 30 min, washed once in PBS, and incubated for 1 h at room temperature with primary antibody (anti-PFN #3693, Cell Signaling Technology Inc., Danvers, MA) in incubation buffer (PBS/0.05% Triton X-100). Cells are then washed three times with incubation buffer and incubated 1 h at room temperature with Alexa-Fluor-conjugated donkey antirabbit secondary antibody (Molecular Probe, Carlsbad, CA) in incubation buffer containing 5% normal donkey serum. Cells are then washed three times in PBS and mounted in Vectashield mounting medium containing 4′,6-diamidino-2-phenylindole (DAPI) (Vector Laboratories, Burlingame, CA) before analysis with a fluorescence microscope (Fig.11.2).

3. Expression and Purification of Recombinant Granzyme B in a Baculovirus System

This method is based on the use of the baculovirus-infected *Spodoptera frugiperda* (Sf) cell-9 and the BTI-TN-5B1-4 clone of *Trichoplusia* commonly referred to as High Five (H5). The cDNA encoding the full-length mouse GzmB was amplified previously by reverse transcription and polymerase chain reaction (RT-PCR) from IL-3-derived BalB/c mouse bone morrow-derived mast cell as described previously (Xia et al., 1998).

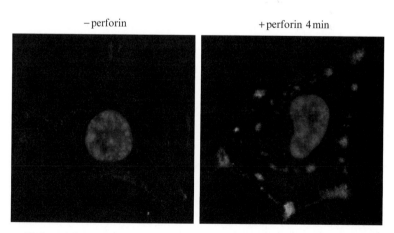

−perforin   +perforin 4 min

Figure 11.2 Perforin detection by fluorescence microscopy. Within 4 min of treatment with sublytic PFN, HeLa cells show large intracellular and membrane-bound vesicles that stain for PFN (right). Perforin is stained using the primary antibody, rabbit polyclonal antiperforin, and the secondary antibody, donkey antirabbit-Alexa 488. Nuclei are stained with DAPI.

3.1. Material/reagent/solution

Insect cell incubator (Wheaton Science Products, Millville, NJ), spinner flask (Bellco Glass Inc, Vineland, NJ), and Sf9 and H5 cells (Invitrogen) are grown in EX-CELL 420 and EX-CELL 405 media, respectively (JRH Bioscience, Inc., Lenexa, KS). The FPLC system (Bio-Rad), Prep/Scale TFF cartridges (Millipore, Bedford, MA), LP-1 pump (Amicon, Pineville, NC), and anti-FLAG M2 agarose, 5,5'-dithiobis 2-notrobenzoic acid (DTNB, i.e., Ellman's reagent)(Sigma-Aldrich), BCA protein assay kit (Pierce Biochemical, Rockford, IL), and granzyme B substrate [Boc-Ala-Ala-Asp-Bzl (AAD), MP Biomedical, Solon, OH].

3.2. Growth of Sf9 and H5 cells

Sf9 and H5 cells are maintained and amplified at 27° in serum-free insect medium in spinner flasks under agitation at 80 to 90 rpm.

3.3. Virus production

Sf9 cells (2×10^6), grown in monolayer, are cotransfected with 4 μg of modified pVL1393 plasmid (Invitrogen), encoding an FLAG-tagged GzmB cDNA plus 1 μg of linearized wild-type AcMNPV baculovirus DNA using cationic liposome (Xia *et al.*, 1998). After 5 days of incubation at 27°, the culture is spun at 1500 rpm for 10 min, and the supernatant containing the recombinant baculovirus (viral stock) is stored at 4° until used for virus amplification or protein expression. Viral stock is diluted 10^3-, 10^5-, and 10^7-fold and recombinant baculovirus clones are identified by plaque assay (King and Possee 1992).

3.4. Virus amplification

Sf9 cells at 0.8 to 1×10^6 cells/ml are cultured in 400 ml serum-free insect medium in a 1-liter spinner flask.

The virus stock is vortexed vigorously before infecting Sf9 cells at a multiplicity of infection (MOI = ratio of infectious virus particles to cells) of 0.25 to 0.5.

The culture is maintained for 6 to 7 days, keeping the cell density under 10^6 cells/ml by adding fresh medium.

After 6 to 7 days, 90% of cells should die. The supernatant is harvested by spinning the culture at 1500 rpm for 5 min.

The virus concentration should be determined by plaque assay.

Virus stock protected from light can be kept at 4° for at least 4 weeks.

3.5. GzmB expression using H5 cells

H5 cells grown in suspension in spinner flasks at a concentration of $\approx 10^6$ cells/ml are infected at an MOI of 10. Forty-eight hours later, the supernatant is harvested and concentrated using the Prep/Scale TFF cartridges.

3.6. GzmB purification

The concentrated supernatant (100 ml) is loaded onto an anti-FLAG M2 affinity column using a FPLC system. The FPLC program is run with buffer A (50 mM Tris-HCl, 150 mM NaCl, pH 7.4) and buffer B (0.1 M glycine-HCl, pH 3.5). The column is washed with 100% buffer A for 5 CVs at 1 ml/min, and the column is eluted with a step gradient 100% buffer B for 2 CVs. Eluted fractions are monitored continuously at a wavelength of 280 nm. The 2-ml fractions are collected in tube containing 50 ml of 1 M Tris-HCl, pH 8, to neutralize the sample. The protein concentration is measured with a BCA protein assay kit.

3.7. GzmB activity assay

This protocol is for a 96-well plate format. Five microliters of fraction is added to 200 μl AAD buffer (50 mM Tris HCl, pH 7.5, 0.2 mM AAD, 0.2 mM DTNB) in each well of a 96-well plate and is incubated for 5 min, and the plate is read at OD_{405} in a microplate reader.

4. EXPRESSION AND PURIFICATION OF RECOMBINANT HUMAN GRANZYME A IN *ESCHERICHIA COLI*

Expression of active GzmA in any cells is not possible because of its toxicity. To circumvent this problem, Beresford and colleagues (1997) expressed an inactive form of the protein that was activated after purification. Human GzmA cDNA is PCR amplified with Vent polymerase (New England Biolabs, Ipswich, MA) from reverse-transcribed mRNA isolated from mitogen-activated peripheral blood cells. Primers are constructed from the human GzmA sequence to contain *Bam*HI and *Eco*RI restriction sites. The PCR product is directionally ligated into pGBT9 (Clonetech, Mountain View, CA). The granzyme insertion is excised and modified by PCR amplification with primers encoding an enterokinase site 5′ of the predicted first amino acid of the active enzyme and *Bam*HI and *Xho*I restriction sites for insertion into pet26b (Novagen, Gibbstown, NJ). This construct is available from Addgene (Cambridge, MA).

4.1. Material/reagent/solution

Fast protein liquid chromatography system (Bio-Rad), mono-S columns (Pharmacia Biotechnology), IMAC Sepharose 6 fast flow (GE Healthcare, Piscataway, NJ), sonicator (Branson sonifier 450), dialysis tubing 10 kDa MWCO (Spectrum Laboratory, Greensboro, NC) and ultrafiltration (10 kDa MWCO) Centricon Concentrator (Fisher Scientific, Pittsburgh, PA).

Escherichia coli strain BL21 DE3 (Novagen, Gibbstown, NJ), LB medium, LB agar, isopropyl β-D-1-thiogalactopyranoside (IPTG), enterokinase, kanamycin, and Coomasie blue (Sigma-Aldrich). Note that similar results are obtained using S-Sepharose from GE Healthcare.

4.2. Solutions

$8\times$ binding buffer (160 mM Tris-HCl, pH 7.9, 4 M NaCl, 40 mM imidazole), $8\times$ wash buffer (160 mM Tris-HCl, pH 7.9, 4 M NaCl, 480 mM imidazole), $8\times$ elution buffer (80 mM Tris-HCl, pH 7.9, 4 M imidazole, 2 M NaCl), $8\times$ charge buffer (400 mM NiSO$_4$)

Enterokinase buffer (50 mM Tris-HCl, pH 7.5, 1 mM CaCl$_2$, 200 mM NaCl)

S-column buffer A (50 mM bis-Tris, pH 5.8, 500 mM NaCl), S-column buffer B (50 mM bis-Tris, pH 5.8, 1 M NaCl)

4.3. Expression of GzmA in *E. coli*

4.3.1. Transformation of Bl21DE3 cells

1. Add 1 μg of plasmid to cells.
2. Place on ice for 30 min.
3. Heat shock for 40 s at 42°.
4. Place on ice for 2 min.
5. Add 500 μl of room temperature SOC medium (supplied with competent cells).
6. Shake at 200 to 250 rpm for 1 h at 37°.
7. Plate cells on 50 μg/ml kanamycin plates: 50 μl on first plate, 100 μl on second, and 250 μl on the last one.
8. Incubate overnight at 37°.

4.3.2. Replication of plate colonies on IPTG-coated plate

1. To make an IPTG plate, spread 125 μl of 200 mM IPTG on LB plate. Let it dry and then grow replica plate overnight at 37°.
2. Mark colonies that disappear on IPTG plate. Use these colonies for GzmA production.

4.3.3. Growth of 8 liters of GzmA expressing BL21DE3

1. Set four flasks with 50 g of LB broth and 2 liters of water and autoclave for 30 min.
2. Cool at room temperature.
3. Add kanamycin to a final concentration of 50 μg/ml.
4. From each 2-liter broth take 50 ml to make a seed culture.
5. Seed each 50-ml seed culture with one single colony from step 2 in Section 4.3.2.
6. Grow seed cultures overnight at 37°.
7. Use seed culture to seed the large culture, one seed culture for each 2-liter bottle.
8. Grow large culture until OD_{600} reaches 0.7.
9. Induce the large culture with IPTG at a final concentration of 250 mg/ml.
10. Incubate for 4 h at 37° with agitation at 200 to 250 rpm. Harvest bacteria by spinning culture at 5000 rpm for 15 min.

4.4. Purification of GzmA protein

1. Resuspend bacteria in 240 ml of cold 1× binding buffer containing 0.1% NP-40.
2. Put 40 ml of lysate in polypropylene tubes and sonicate each tube on ice for 3 min total with a cycle of a 30-s pulse and 10 s off at level 4.
3. Spin the sample for 1 h at 30,000g at 4°.
4. Filter the supernatant with a 0.45-μm filter.
5. Load sample on Ni^{2+} IMAC column with a flow rate of 1 ml/min.
6. Run the FPLC program with buffer A (wash buffer) and buffer B (elution buffer): Wash the column with 100% buffer A for 20 CVs at 1 ml/min, elute column with a linear gradient up to 100% buffer B for 2 CVs, and wash column with an additional 2 CVs of buffer B. Monitor eluted fractions continuously at wavelength 280 nm. Collect 2-ml fractions starting from the beginning of the elution step until the end of the program.
7. Screen fractions by analyzing an aliquot of every other fraction on a nondenaturing SDS gel and stain with Coomassie blue.
8. GzmA appears predominantly as a 50-kDa band (dimer).
9. Combine positive fractions, avoiding the fraction containing the GzmA monomer (25-kDa band), add 1 unit of enterokinase per liter of initial culture, and dialyze against enterokinase buffer overnight at room temperature with stirring.
10. Monitor the efficiency of enterokinase cleavage by analyzing an aliquot of untreated GzmA side by side with the enterokinase-treated sample on a nondenaturing SDS gel and stain with Coomassie blue.

11. The processed form of GzmA has a lower molecular weight than the unprocessed form.
12. Dialyze the GzmA preparation against S-column buffer A for at least 3 h with agitation.
13. Spin the sample at 1500 rpm to get rid of the precipitated material (contaminant proteins).
14. Filter sample with a 0.45-μm filter.
15. Load sample on S-column at 0.5 ml/min equilibrated with S-column buffer A.
16. Run FPLC program: Run the FPLC program with buffer A (S-column buffer A) and buffer B (S-column buffer B). Wash the column with 100% buffer A for 10 CVs at 1 ml/min, elute column with a linear gradient up to 100% buffer B for 2 CVs, and wash column with an additional 2 CVs of buffer B. Monitor eluted fractions continuously at 280 nm wavelength. Collect 2-ml fractions starting from the beginning of the elution step until the end of the program. One sharp peak corresponding to active GzmA is obtained at 0.7 to 0.8 mM NaCl.

4.5. Measuring GzmA activity

4.5.1. Fluorometric assay

This protocol is for a 96-well plate format. This activity assay is a fast way to screen GzmA-positive fractions; however, GzmA activity should be further tested in a cleavage assay (described later) and cell death experiment (next section). Five microliters of fraction is added to 200 μl BLT buffer (50 mM Tris-HCl, pH 7.5, 0.2 mM BLT, 0.2 mM DTNB) in each well of a 96-well plate. Incubate for 5 min and measure in a microplate reader at OD$_{405}$. Wells with buffer but no purified fractions serve as negative control, and as determined empirically by earlier work (Henkart *et al.*, 1987), fractions with an OD higher than 0.2 are positive for GzmA activity. These fractions are pooled and concentrated using the Centricon MWCO 10 kDa.

4.5.2. Cleavage assay

The protein SET was identified as a bona fide substrate of GzmA in the Lieberman laboratory (Beresford *et al.*, 1997). We use the cleavage efficiency of recombinant SET protein as a measure of GzmA activity (Fig. 11.3). Two micrograms of recombinant human SET is incubated with various amounts of purified GzmA for 3 h in 50 μl final volume of PBS at 37°. Ten microliters of each reaction is resolved on a 12.5% SDS gel and immunoblotted with rabbit polyclonal anti-SET antibody (1:2000). Almost all the SET protein is cleaved at 0.25 μM GzmA.

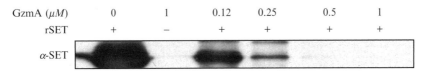

GzmA (μM)	0	1	0.12	0.25	0.5	1
rSET	+	–	+	+	+	+

α-SET

Figure 11.3 GzmA cleavage activity. Indicated amounts of purified GzmA were incubated with 2 μg of recombinant human SET. The reaction mix was analyzed by immunoblot using the rabbit polyclonal anti-SET antibody. It is noteworthy that the antibody does not cross-react with GzmA.

5. Granzyme-Mediated Apoptosis Assay

The exact mechanism by which granzymes enter their target cells *in vivo* is not clear but a sublethal dose of perforin is both necessary and sufficient for this process *in vitro*. Therefore, the first step is to determine the appropriate dose of perforin that does not kill cells via necrosis but is enough to facilitate granzyme-mediated apoptosis.

5.1. Perforin titration

1. Plate 10^4 adherent cells (HeLa)/well on a flat-bottom 96-well plate the night before, or 5×10^4 suspension cells (K562) are taken in a 96-well V-bottom plate.
2. Wash cells in buffer C (HBSS with 10 mM HEPES, pH 7.5, 2 mM CaCl$_2$, 0.4% BSA) and resuspend in 30 μl of the buffer.
3. Serially dilute purified perforin in buffer P (HBSS with 10 mM HEPES, pH 7.5); the diluted form at this stage is at a concentration of 4×.
4. Add 15 μl of buffer P to cells (this is to keep conditions identical to the granzyme-mediated cell death assay described later) and also add 15 μl of the perforin diluted in step 3.
5. Incubate at 37° for 15 min.
6. Measure cell death by counting trypan blue-positive cells, or propidium iodide staining (analyzed by flow cytometry).
7. Perforin causing approximately 10 to 15% cell death is considered a sublytic dose.

5.2. Granzyme A- and B-mediated cell death assay

5.2.1. Chromium release assay

1. Harvest 10^6 HeLa or K562 target cells from cultures, resuspend in 1 ml of RPMI/10% FCS, and then label with 100 μCi [^{51}Cr] Na$_2$CrO$_4$ for 1 h at 37°, followed by three washes with HBSS.

2. Resuspend cells in buffer C. Distribute target cells in triplicate (30 μl/ well) in V-bottom microtiter plates.
3. Dilute perforin in buffer P to 4× of the sublethal dose determined in the earlier step.
4. Dilute GzmA or GzmB to 4× of the final desired concentration also with buffer P. The amount of granzyme required to kill cells varies, depending both on the target cell and on the quality of the purified Gzm.
5. Mix the diluted Gzm and perforin together and add to the cells. Please include controls such as "untreated cells," "perforin alone," and "Gzm alone," substituting the reagent with buffer P.
6. Incubate plates for 1 to 4 h at 37°, 5% CO_2.
7. After a short centrifugation (2 min at 500g), collect 50 μl of supernatants and measure ^{51}Cr release in a TopCount counter. Determine spontaneous and maximum release by incubating target cells in buffer P and in a 1% NP-40 solution, respectively. Calculate the percentage of lysis as follows: [(experimental ^{51}Cr release − spontaneous ^{51}Cr release)/(maximum ^{51}Cr release − spontaneous ^{51}Cr release)] ×100.

5.2.2. Annexin/propidium iodide (PI) staining assay

1. Plate 10^4 adherent cells (HeLa)/well on a flat-bottom 96-well plate the night before, or 5×10^4 suspension cells (K562) are taken in a 96-well V-bottom plate.

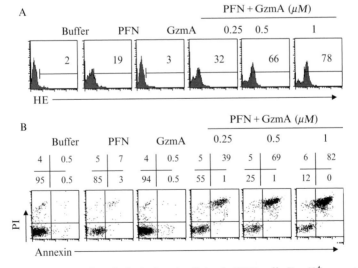

Figure 11.4 GzmA- and perforin-induced cell death. K562 cells (5×10^4) were treated with buffer alone, sublytic dose of PFN alone, GzmA alone, or PFN plus GzmA for 1 h at 37°. ROS was measured by staining with 2 μM of HE (A), and cell death was followed by annexin + PI staining (B). PFN and GzmA treatment induces dose-dependent ROS production and cell death.

2. Wash cells in buffer C and resuspend in 30 μl of the buffer.
3. Follow steps 3, 4, 5, and 6 from the previous section.
7. Add 60 μl of PI (10 μg/ml) and FITC-conjugated annexin (1:25, BD Pharmingen, Franklin Lakes, NJ) and incubate at room temperature in the dark for 20 min. Alternatively, monitor the ROS production as an indicator of cell death by staining cells with 2 μM of hydroethidine (HE) (Martinvalet *et al.*, 2005).
8. Analyze by flow cytometry (Fig. 11.4).

6. Conclusion

Cytotoxic T lymphocyte- and NK-cell mediated apoptosis has been studied at the cellular level for many decades, but only recently have we started to investigate the molecular aspects of this process. This chapter provided protocols necessary to reconstitute this cell death pathway *in vitro* by purifying granzymes and perforin and then by performing cell death assays on different target cells. The ability to quantitate the efficiency of cell death induced by granzymes/perforin on target cells with different geno-types will allow us to investigate the importance of different proteins in the granzyme-mediated cell death pathways.

ACKNOWLEDGMENT

We would like to thank Judy Lieberman (Harvard Medical School) for providing unpub-lished protocols and data.

REFERENCES

Beresford, P. J., Kam, C. M., Powers, J. C., and Lieberman, J. (1997). Recombinant human granzyme A binds to two putative HLA-associated proteins and cleaves one of them. *Proc. Natl. Acad. Sci. USA* **94,** 9285–9290.

Chowdhury, D., and Lieberman, J. (2008). Death by a thousand cuts: Granzyme pathways of programmed cell death. *Annu. Rev. Immunol.* **26,** 389–420.

Fan, Z., Beresford, P. J., Oh, D. Y., Zhang, D., and Lieberman, J. (2003). Tumor suppressor NM23-H1 is a granzyme A-activated DNase during CTL-mediated apoptosis, and the nucleosome assembly protein SET is its inhibitor. *Cell* **112,** 659–672.

Henkart, P. A., Berrebi, G. A., Takayama, H., Munger, W. E., and Sitkovsky, M. V. (1987). Biochemical and functional properties of serine esterases in acidic cytoplasmic granules of cytotoxic T lymphocytes. *J. Immunol.* **139,** 2398–2405.

King, L. A., and Possee, R. D. (1992). "The Baculovirus Expression System: A Laboratory Guide." London: Chapman & Hall, London.

Lieberman, J. (2003). The ABCs of granule-mediated cytotoxicity: New weapons in the arsenal. *Nat. Rev. Immunol.* **3,** 361–370.

Lieberman, J., and Fan, Z. (2003). Nuclear war: the granzyme A-bomb. *Curr. Opin. Immunol.* **15,** 553–559.

Lord, S. J., Rajotte, R. V., Korbutt, G. S., and Bleackley, R. C. (2003). Granzyme B: A natural born killer. *Immunol. Rev.* **193,** 31–38.

Martinvalet, D., Zhu, P., and Lieberman, J. (2005). Granzyme A induces caspase-independent mitochondrial damage, a required first step for apoptosis. *Immunity* **22,** 355–370.

Melief, C. J., and Kast, W. M. (1992). Lessons from T cell responses to virus induced tumours for cancer eradication in general. *Cancer Surv.* **13,** 81–99.

Pipkin, M. E., and Lieberman, J. (2007). Delivering the kiss of death: Progress on understanding how perforin works. *Curr. Opin. Immunol.* **19,** 301–308.

Russell, J. H., and Ley, T. J. (2002). Lymphocyte-mediated cytotoxicity. *Annu. Rev. Immunol.* **20,** 323–370.

Shi, L., Kam, C. M., Powers, J. C., Aebersold, R., and Greenberg, A. H. (1992). Purification of three cytotoxic lymphocyte granule serine proteases that induce apoptosis through distinct substrate and target cell interactions. *J. Exp. Med.* **176,** 1521–1529.

Shresta, S., Graubert, T. A., Thomas, D. A., Raptis, S. Z., and Ley, T. J. (1999). Granzyme A initiates an alternative pathway for granule-mediated apoptosis. *Immunity* **10,** 595–605.

Winkler, U., Pickett, T. M., and Hudig, D. (1996). Fractionation of perforin and granzymes by immobilized metal affinity chromatography (IMAC). *J. Immunol. Methods* **191,** 11–20.

Xia, Z., Kam, C. M., Huang, C., Powers, J. C., Mandle, R. J., Stevens, R. L., and Lieberman, J. (1998). Expression and purification of enzymatically active recombinant granzyme B in a baculovirus system. *Biochem. Biophys. Res. Commun.* **243,** 384–389.

CHAPTER TWELVE

Investigation of the Proapoptotic Bcl-2 Family Member Bid on the Crossroad of the DNA Damage Response and Apoptosis

Sandra S. Zinkel

Contents

Medicine, Vanderbilt University School of Medicine, Nashville, Tennessee

Methods in Enzymology, Volume 442
ISSN 0076-6879, DOI: 10.1016/S0076-6879(08)01412-2

Abstract

The BCL-2 family of apoptotic proteins encompasses key regulators proximal to irreversible cell damage. BID, a "BH3-only" proapoptotic family member, plays a critical role in connecting death signals through surface death receptors such as Fas and tumor necrosis factor-α to the core apoptotic pathway at the mitochondria. BID is activated downstream of death receptors by caspase-8 cleavage and N-myristoylation to target mitochondria where it activates BAX, BAK, and the downstream apoptotic pathway. In addition to its role in apoptosis, a role has been uncovered for BID in regulating the DNA damage-induced intra-S phase checkpoint that does not require its death-promoting BH3 domain. Following DNA damage, BID is found in the nucleus where it is phosphorylated by ATM and plays a role in the intra-S phase checkpoint. This checkpoint role is dependent on ATM-mediated phosphorylation at position 78. Thus, BID has two distinct and separable functions: an apoptotic function mediated by caspase cleavage and its BH3 domain and a cell cycle/DNA repair function mediated by phosphorylation by the DNA damage kinase ATM. Studies indicate that the pro-death activity of BID is inhibited by phosphorylation. Taken together, these findings suggest interaction between the two functions of BID. An area of intense research pursuit is determining what dictates how cells respond to DNA damage. Some cells arrest the cell cycle, whereas others undergo apoptosis. We hypothesize that BID acts at the interface between the DNA damage response and apoptosis, in position to signal a cell either to undergo cell cycle arrest and initiate DNA repair or to undergo apoptosis. This chapter describes the techniques used to characterize the role of BID in apoptosis and the DNA damage response.

1. INTRODUCTION

The prototype family member BCL2 was identified at the chromosomal breakpoint of the t(14;18) translocation found in 75% of follicular lymphomas and established a third class of oncogenes in which resistance to cell death allows accrual of excess cells, potentially facilitating accumulation of the additional mutations linked with tumor progression (Bakhshi *et al.*, 1985; Cleary and Sklar, 1985; Tsujimoto *et al.*, 1985). Members of the family possess up to four conserved α-helical domains, designated BH1, BH2, BH3, and BH4 (Adams and Cory, 1998; Kelekar and Thompson, 1998). Mutagenesis studies of BCL-2 indicate that the conserved domains are necessary for the interaction with proapoptotic members and for the inhibition of cell death (Yin *et al.*, 1994).

Both pro- and antiapoptotic family members have been identified. The multidomain proapoptotic proteins BAX and BAK serve as a critical gateway in the intrinsic pathway of apoptosis, operating both at mitochondria and endoplasmic reticulum (Wei *et al.*, 2001; Zong *et al.*, 2003). Cells doubly deficient for both BAX and BAK are resistant to multiple death stimuli. The subset of proteins that contain homology only in the death-directing BH3 domain (BH3-only) link upstream death signals to the checkpoint of BCL-2 multidomain members. An emerging paradigm in the field is that these BH3-only proteins may play additional roles, embedded in essential processes within the cell.

Proapoptotic BID is unique among BH3-only BCL2 family members in interconnecting death receptors to the mitochondrial amplification loop of the intrinsic apoptotic pathway. BID was cloned through interaction with BCL2 and BAX (Wang *et al.*, 1996) and was purified biochemically as a protein mediating cytochrome *c* release from mitochondria following the activation of death receptors (Luo *et al.*, 1998). *In vitro* studies of mitochondria and recombinant truncated BID indicate that it activates the multidomain BCL2 family members BAX or BAK, resulting in allosteric conformational change and release of cytochrome *c* (Desagher *et al.*, 1999; Wei *et al.*, 2000).

The role of *Bid* in normal development and cellular homeostasis has been characterized using mice in which *Bid* has been disrupted. *Bid*-deficient mice are viable and execute developmental cell death normally (Yin *et al.*, 1999). When challenged with the agonistic anti-fas antibody, *Bid*-deficient mice are resistant to the hepatocellular apoptosis that kills wild-type mice, indicating a critical role for BID in this Fas-signaled death. Aging *Bid*-deficient mice spontaneously develop a myeloproliferative disorder with elevated absolute neutrophil counts, and over time, the mice progress to a fatal clonal disorder resembling chronic myelomonocytic leukemia (Zinkel *et al.*, 2003). Myeloid progenitors from *Bid*-deficient mice exhibit resistance to death receptor-induced apoptosis and demonstrate a competitive advantage *in vivo*. These studies indicate an essential role for BID in maintaining myeloid homeostasis and in suppressing leukemogenesis.

In addition to its role in apoptosis, we have uncovered a role for BID in regulating the DNA damage-induced intra-S phase checkpoint that does not require its death-promoting BH3 domain (Kamer *et al.*, 2005; Zinkel *et al.*, 2005)(Fig. 12.1). Following DNA damage, BID is found in the

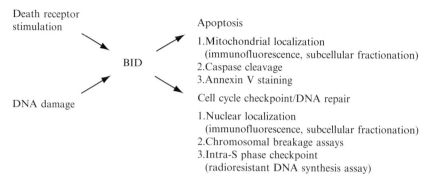

Figure 12.1 Model for the dual function of BID. Following death receptor stimulation, BID initiates a proapoptotic program at mitochondria. After DNA damage, BID is phosphorylated in the nucleus and plays a role in cell cycle checkpoint control.

nucleus, where it is phosphorylated by ATM and plays a role in the intra-S phase checkpoint. This checkpoint role is dependent on ATM-mediated phosphorylation at position 78. Thus, BID has two distinct and separable functions: an apoptotic function mediated by caspase cleavage and its BH3 domain and a cell cycle/DNA repair function mediated by phosphorylation by the DNA damage kinase ATM. Studies indicate that the pro-death activity of BID is inhibited by phosphorylation (Desagher *et al.*, 2001). Taken together, these findings suggest interaction between the two functions of BID.

Hematopoietic cells are particularly sensitive to low to moderate levels of genotoxic stress relative to other cell types, relying on apoptosis to prevent accumulation of mutations. In order to maintain homeostasis and prevent leukemogenesis following DNA damage, the pathways directing cell cycle checkpoints and apoptosis must be carefully balanced and coordinated. These cells therefore are particularly well suited to the study of the interface between the DNA damage response and apoptosis. Our laboratory has focused our effort on primary hematopoietic cells as well as immortalized myeloid progenitor cells.

To facilitate biochemical studies of the signals governing the interface between the DNA damage response and apoptosis, we use Hox 11 to immortalize myeloid progenitor cells. This technique has the advantage of immortalizing myeloid progenitor cells at a relatively uniform stage of development (Fig. 12.2), with cells predominantly in the promyelocyte to myelocyte stage of development by morphology. In addition, Hox11-immortalized cells display intact DNA damage-induced cell cycle checkpoints, as well as a relatively high percentage of cells in S phase, making them a useful system in which to study the interface between the DNA damage response and apoptosis (Hawley *et al.*, 1994; Zinkel *et al.*, 2005).

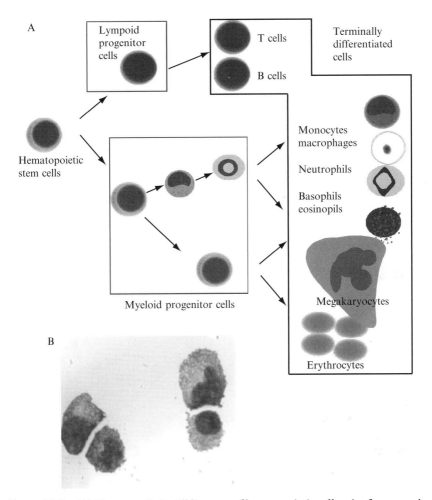

Figure 12.2 (A) Hematopoiesis. All lineages of hematopoietic cells arise from a puripotent stem cell. Progenitor cells then commit to the myeloid or lymphoid lineage (red boxes) prior to terminal differentiation (black box). (B) Cytospin of Hox11-immortalized myeloprogenitor cells. One thousand cells in 50 μl of PBS were centrifuged at 700 rpm for 7 min in a cytospin. Cells were allowed to dry and were then stained with May–Grunwald–Giemsa stain (Sigma). (See color insert.)

 ## 2. Isolation of Myeloid Precursor Cells from Mouse Bone Marrow Cells

2.1. Materials

Phosphate-buffered saline (PBS), culture medium: RPMI, 10% fetal calf serum (FCS), L-glutamine, and streptomycin/penicillin

Dissection instruments (two sets of forceps/scissors), 70- μm cell strainer (Fisher), 5-ml syringes, 20-gauge needles, 15- and 50-ml conical tubes, tissue culture flasks, tissue culture plates

Cytokines: Stem cell factor (SCF), granulocyte colony-stimulating factor (GCSF), interleukin 3 (IL-3), granulocyte–macrophage colony stimulating factor (GMCSF), Preprotech or R&D

Magnetic beads for lineage depletion [sheep anti-Rat Dynabeads (Dynal Biotech)]

Magnetic stand for Eppendorf tubes (Promega)

Metal stand for Miltenyi magnets (MACS multistand, Miltenyi)

Magnet for MS columns (MiniMACS separation unit, Miltenyi)

MS columns (one column per Sca1+ sample)

2.2. Buffers and media

PBS

RPMI10: RPMI + 10% FCS, L-glutamine, and 100 U/ml streptomycin/penicillin

DMEM20: DMEM + 20% FCS L-glutamine, and 100 U/ml streptomycin/penicillin

Myelocult 5300 (Stem Cell Technologies), L-glutamine, and 100 U/ml streptomycin/penicillin

Infection medium (IMDM): 20% FCS, 100 U/ml penicillin/streptomycin, 2 mM glutamine, 10 ng/ml IL-3, 20 ng/ml SCF, 10 ng/ml GMCSF, and 2 ng/ml GCSF

Red blood cell lysis buffer: 10 mM Tris-HCl, pH 7.2, 0.83% NH_4Cl

Staining buffer: 3% FCS in PBS

2.3. Methods

1. Euthanize mice according to the approved method at your institution.
2. To harvest bone marrow, excise the end of the femur using a single-edged razor blade.
3. Flush the bone marrow from femurs of 6- to 12-week-old mice using a 20-gauge needle and 5 ml of RPMI. Control wild-type mice should be strain, age, and sex matched.
4. Centrifuge bone marrow at 1200 rpm for 5 min and remove medium.
5. Lyse red blood cells by resuspending bone marrow cells in 2 ml of red blood cell lysis buffer for 3 min at room temperature.
6. Add 5 ml of RPMI to stop lysis.
7. Centrifuge bone marrow at 1200 rpm for 5 min and remove medium.
8. Proceed to purification of Lin⁻ cells.

3. Isolation of Myeloid Precursor Cells

Myeloid precursor cells (MPCs) are isolated by lineage depletion followed by positive selection of Sca-1$^+$ cells by magnetic beads (Miltenyi).

4. Purification of Lin⁻ Cells

1. Resuspend bone marrow cells at a concentration of 10 million per milliliter in staining buffer.
2. Add antibodies to lineage markers [1:100 dilution of biotin-conjugated anti-Gr-1, B220, Ter119; purified anti-CD3 (BD Biosciences)].
3. Incubate cells for 30 min on ice.
4. Wash cells three times in 10 ml of staining buffer.
5. Wash magnetic beads with 10 ml of staining buffer to wash out the azide [sheep antirat Dynabeads (Dynal Biotech)].
6. Add two magnetic beads per cell (concentration of beads is 4×10^8 beads/ ml) to bone marrow cells.
7. Incubate for 30 min on ice.
8. Using a magnetic separation stand (Promega), deplete differentiated bone marrow cells that express lineage markers. Place cells on a magnetic stand in a 1.5-ml Eppendorf tube for 1 min to allow cells bound to magnetic beads to adhere to the side of the tube.
9. Using a plugged Pasteur pipette, remove buffer and nonadherent cells and transfer to another Eppendorf tube.
10. Repeat step 9.

5. Sca-1 Positive Selection

1. Count lineage-depleted cells and resuspend in 0.5% FCS (Sca1 staining buffer) at 10 million cells per milliliter.
2. Incubate with 1/100 volume FITC–conjugated Sca-1 (Pharmingen) for 15 min on ice.
3. Wash three times with 10 ml of staining buffer.
4. Resuspend cells in 0.5% FCS (Sca1 staining buffer) at 100 million cells per milliliter.
5. Incubate cells with 1/10 volume anti–FITC conjugated to magnetic beads (Miltenyi) for 30 min at 4°.

6. Wash cells three times with 10 ml of staining buffer.
7. Resuspend cells in 1 ml staining buffer.

6. MAGNETIC SEPARATION

1. Place an MS column in a MiniMACS separation unit on the metal stand.
2. Wash with 500 μl of staining buffer.
3. Load cells onto the washed MS column (Milltenyi) by gravity.
4. Reload the flow through.
5. Wash the column three times with 1 ml of staining buffer.
6. Remove the column from the magnet and elute cells with 1 ml of Sca1 staining buffer.

7. CULTURE OF LINEAGE-DEPLETED, SCA1+ BONE MARROW CELLS

1. Centrifuge cells at 1200 rpm for 7 min.
2. Resuspend cells at 10 million cells per milliliter in DMEM20 or myelocult 5300 (Stem Cell Technologies) media supplemented with SCF (100 ng/ml).
3. Add GCSF (100 ng/ml) to the aforementioned culture after 1 day.
4. Grow NIH 3T3 hph-HOX11 retroviral producer cells (Hawley et al., 1994) to 90 to 100% confluence in 100-mm tissue culture dishes. One dish per culture is needed.
5. Irradiate retroviral producer cells with 3000 rad.
6. Change the medium and incubate 24 h before adding myeloid progenitor cells.
7. After 3 days, add myeloid progenitor cells to irradiated NIH 3T3 hph-HOX11 retroviral producer cells in 100-mm tissue culture dishes containing 10 ml of infection medium. Culture for 3 days at 37° in 5% CO_2 (Hawley et al., 1994).
8. Expand the cells growing in suspension in IMDM 20% FCS, 100 U/ml penicillin/streptomycin, and 2 mM glutamine with 10% conditioned medium from WEHI cells as a source of IL-3. The medium should be changed every third day to replenish growth factors and cytokines. These cells should initially be expanded slowly and maintained at a density of 1 million cells per milliliter. Cells have a tendency to terminally differentiate if maintained at a lower density. Cells should be expanded and aliquots of early passages should be frozen for experimental use.

8. WEHI-CONDITIONED MEDIUM (SOURCE OF IL-3) (WARNER *ET AL.*, 1969)

WEHI cells are grown in RPMI/10% FCS:

1. Split cells 1:2 every second day until cells reach a volume of 400 to 500 ml.
2. Remove 10 ml of cell culture in T75 to maintain actively growing cells.
3. Allow large culture of cells to grow for 5 to 7 days until medium is bright yellow.
4. Spin down cells in large centrifuge.
5. Filter supernatant using a 500-ml filter unit.
6. Aliquot into 50-ml conical tubes.

9. CHROMOSOMAL BREAKAGE ASSAYS

Bid $^{-/-}$ leukemias display increased genomic instability, as evidenced by chromosomal translocations and trisomies. To determine whether BID plays a role in maintaining genomic integrity, we evaluated chromosomal integrity following treatment with mitomycin c (Fig. 12.3).

Figure 12.3   Metaphase spread of mitomycin c–treated *Bid*$^{-/-}$ MPCs. Cells were treated with 100 n*M* mitomycin c for 24 h. The black arrow indicates a quadriradial, the blue arrow indicates a chromosome fragment, and the red arrow indicates a chromosome break. (See color insert.)

9.1. Materials

Mitomycin c (Sigma)
Colcemid (Gibco Karyo Max)
0.068 M KCl
Methanol: glacial acetic acid 3:1 (should be made up fresh)
Glass microscope slides (plain glass, not coated with adhesives) (Fisher)
Giemsa stain (Fisher)

9.2. Protocol

1. Treat cells with 100 μM mitomycin c for 24 h.
2. Arrest cells in metaphase with 0.1 μg/ml colcemid.
3. Incubate cells in 0.068 M KCl for 15 min.
4. Add 1/20 volume methanol.
5. Centrifuge cells at 1200 rpm for 7 min.
6. Resuspend cells in 5 ml of methanol:glacial acetic acid (3:1) and incubate for 10 min at room temperature. (The methanol:glacial acetic acid should be made fresh just before use.)
7. Centrifuge cells at 1200 rpm for 7 min.
8. Repeat steps 6 and 7 two additional times. During the third incubation, place cells on ice.
9. Following the last fixation, resuspend cells in 500 μl of methanol:glacial acetic acid (3:1).
10. Drop cells onto glass slides as follows. Drop approximately 30 μl of cells 12 to 15 inches onto a microscope slide (do not use coated slides) held at a 45° angle. We humidify our slides by placing over a 37° water bath for a few minutes prior to dropping.
11. When the drop begins to look grainy, invert slides and hold over a 37° water bath for 2 s (until the slide fogs).
12. Dry slides on a 55° heating block for 5 min.
13. Stain metaphase spreads for 6 to 8 min with Giemsa stain (Gibco).
14. Photograph 50 metaphases per sample and score for chromosomal damage per metaphase as follows: give each chromosome break a score of +1 and each triradial or quadriradial form a score of +2. Then calculate the number of chromosome breaks per metaphase spread.

10. RADIORESISTANT DNA SYNTHESIS

10.1. Materials

96-well round-bottomed tissue culture plates
[*methyl*-^{14}C]Thymidine (NEN Life Science Products, Inc.)

[*methyl*-^{3}H]Thymidine (NEN Life Science Products, Inc.)
10% (w/v) trichloroacetic acid
70% ethanol
95% ethanol
25-mm glass microfiber filters (Whatman GF/C)
Scintillation vials (6 ml polyethylene, Perkin Elmer)
Scintillation fluid (Aquasol, Perkin Elmer)
10-place filter manifold (Fisher-FH225V)
^{137}Cs irradiator
Scintillation counter

10.2. Methods

1. Label 1×10^5 cells in 200 μl of the appropriate medium with10nCi of [^{14}C]thymidine for 24h. This prelabeling provides an internal control for cell number by allowing normalization for total DNA content of samples.
2. Remove the medium containing [^{14}C]thymidine and replace with fresh medium.
3. Incubate the cells for another 24h.
4. Remove medium and replace with fresh medium.
5. Irradiate the cells in a ^{137}Cs irradiator. For each cell type, the dose of ionizing radiation should be titrated so that [^{3}H]thymidine incorporation is decreased by 50% following irradiation in the control cells. Generally, cells should be irradiated with between 2 and 10 Gy, although certain cell types may require higher or lower doses.
6. Incubate cells for 1 h.
7. Pulse label cells with 2.5 μCi of [^{3}H]thymidine/ml for 30min.
8. An additional set of control samples should be included that contain only ^{14}C to allow correction for channel crossover (see Data Analysis, below).
9. Harvest cells, wash twice with PBS, and transfer to Whatman filters that have been placed on the filter manifold.
10. Wash the filters with 5 ml of ice-cold 10% trichloroacetic acid, 5 ml 70% ethanol, and 5 ml 95% ethanol.
11. Air dry the filters and then place in a scintillation tube with 5 ml of Aquasol scintillation fluid.
12. Measure the amount of radioactivity in a liquid scintillation counter. The resulting ratios of ^{3}H counts per minute to ^{14}C counts per minute, corrected for those counts per minute that were the result of channel crossover, are a measure of DNA synthesis (Fig. 12.4).

11. DATA ANALYSIS

The principle of liquid scintillation counting involves conversion of the kinetic energy emitted by the decay of β particles into ultraviolet (UV) light, producing approximately 10 photons per keV of energy. The intensity

Figure 12.4 Radioresistant DNA synthesis. Cells were labeled with ^{14}C and treated with a DNA-damaging agent. After 1 h of incubation, cells were incubated with ^{3}H and analyzed in a liquid scintillation counter.

of the emitted light is proportional to the initial energy of the β particle. This UV light is detected by the photo cathode photomultiplier tube (PMT) of the liquid scintillation counter and is converted to an electrical pulse that is proportional to the number of photons. The PMT analyzer collects events according to the energy range or channel into which it falls. The energy emission spectra for ^{3}H (mean energy of emission equal to 18.6 keV) and ^{14}C (mean energy of emission equal to 156) overlap. It is therefore necessary to correct for the events that are the result of channel crossover. This is accomplished as follows.

The scintillation counter should be set to count two channels: channel 1—a low-energy range from 0 to 18.6 keV (^{3}H and some ^{14}C)—and channel 2—an intermediate-energy range from 18.6 to 156 keV (^{14}C).

The adjusted ^{3}H counts are equal to (channel 1 counts) − [(channel 2 counts) × (1-^{14}C fraction)/(^{14}C fraction)].

The adjusted ^{14}C counts are equal to the channel 2 counts/^{14}C fraction.

The ^{14}C fraction is calculated using the counts from the control sample, containing ^{14}C only, and is equal to (average channel 1 counts)/(average channel 1 counts + average channel 2 counts).

The rate of DNA synthesis is equal to the ratio of the adjusted ^{3}H to the adjusted ^{14}C counts.

12. SUBCELLULAR FRACTIONATION

The BCL-2 family of proteins represents a key regulatory checkpoint to apoptosis through mitochondria. Proapoptotic BID interconnects death receptor signaling to the core apoptotic machinery at mitochondria through interaction with the multidomain BCL-2 family members BAX and BAK. Following DNA damage, however, BID has been found in the nucleus, is phosphorylated by ATM and/or ATR, and displays increased radioresistant DNA synthesis, suggesting a role in the intra-S phase checkpoint. Subcellular localization thus represents a key mechanism for determining the role of BID in a given cellular environment. We have therefore undertaken subcellular fractionation and immunofluorescence studies to evaluate BID localization within the cell following DNA damage and death receptor signaling.

12.1. Materials

Benzonase nuclease (Novagen)
β-Glycerophosphate (Sigma)
Sodium orthovanadate (Sigma)
Sodium fluoride (Sigma)
Microcystin LR (Sigma)
Complete, EDTA-free protease inhibitor cocktail (Roche)
Hydroxyurea (Sigma)
Triton X-100 (10%)

12.2. Buffers

Buffer A: 10 mM Tris, pH 7.5, 1.5 mM MgCl$_2$, 10 mM KCl, and 0.5 mM dithiothreitol (DTT)
Buffer B: 20 mM Tris, pH 7.5, 20% glycerol, 420 mM NaCl, 1.5 mM MgCl$_2$, 0.2 mM EDTA,
1 mM DTT, and 10% glycerol
Buffer D: 20 mM Tris, pH 7.5, 100 mM KCl, 12.5 mM MgCl2, 0.1 mM EDTA, 1 mM DTT, and 10% glycerol

Add 10 mM β-glycerophosphate, 2 mM sodium orthovanadate, 10 mM NaF, 1/50 volume protease inhibitors, and 1 μM microcystin LR to buffers immediately prior to use.

12.3. Methods

1. Treat 30 million cells with DNA-damaging agent of choice. We use 1 mM HU (30 μl of a 1 M stock) for 2 h.
2. Spin down treated and untreated cells in a 50-ml conical tube.
3. Wash once with PBS.
4. Resuspend cells in 200 μl hypotonic buffer (buffer A).
5. Transfer to a chilled Eppendorf tube.
6. Add 1/10 volume of 1% Triton X-100 buffer dropwise with gentle mixing.
7. Spin at 800g for 10 min at 4° .
8. Remove supernatant and put in a chilled Eppendorf tube. This is the cytosolic fraction.
9. Wash the pellet (nuclei) twice in 200 μl buffer A.
10. Resuspend pellet in 100 μl buffer B.
11. Leave on ice for 30 min. Gently mix every 5 min.
12. Centrifuge at 10,000 rpm for 30 min.
13. Centrifuge for 15 min. at 10,000 rpm.

14. Remove the supernatant and place in a chilled Eppendorf tube. This is the soluble nuclear fraction.
15. If the protein is to be immunoprecipitated, dialyze against buffer D.
16. Resuspend the pellet in 100 μl buffer D.
17. Add 0.5 μl benzonase nuclease.
18. Incubate at 4° on a nutator for 30 min.
19. Centrifuge for 15 min at 10,000 rpm.
20. Remove the supernatant and place in a chilled Eppendorf tube. This is the chromatin fraction.

13. IMMUNOFLUORESCENCE STAINING ON ADHERENT CELLS

We generate epitope-tagged BID by introducing the cDNA for BID into the retroviral vectors pOZ-FH-N and pOZ-FH-C. These vectors are a derivative of the MMLV-based pOZ vector contructed by Bruce Howard and colleagues and modified by Nakatani and Ogryzko (2003) to generate BID that is tagged at either the N or the C terminus with FLAG and hemagglutinin (HA). These vectors contain a bicistronic transcriptional unit that allows expression of two proteins from a single transcript to allow tight coupling to the selectable marker, the interleukin-2α chain receptor (CD25). Expression is checked by transient expression in 293T cells followed by SDS polyacrylamide gel electrophoresis (PAGE) and immunoblot for BID protein, as well as FLAG and HA. Retroviral supernatants are generated by transient transfection of the ecotropic packaging cell line BOSC with the appropriate expression plasmid. Tagged BID is introduced into $Bid^{-/-}$ mouse embryonic fibroblasts (MEFS) by retroviral transduction, and transduced cells are stained with antihuman CD25 (Caltag) followed by sheep antimouse Dynal beads. Magnetic sorting collects positive cells. Levels of tagged BID protein are verified by Western blot of BID and compared to endogenous BID levels in wild-type MEFS.

We have introduced BID fused to a HA tag into fibroblasts by retroviral transduction. Cells expressing endogenous levels of BID are isolated by flow sorting, and BID expression levels are verified using SDS-PAGE followed by immunoblotting with anti-BID antibodies (R&D).

13.1. Materials

Autoclaved 22 × 22-mm glass coverslips #1 (Fisher)
Polylysine (Fisher)
DMEM10
Six-well tissue culture plates (Sarstedt)

Mitotracker red (Molecular Probes-Invitrogen)

Methanol:acetone, 3:1 (at −20°)

Normal goat serum (NGS; Invitrogen)

PBS (Gibco)

2- μm filters (Millipore)

Poly-L-lysine solution: 50 μg/ml in 10 mM Tris, pH 8.0 [0.02 g poly-L-lysine (Sigma) in 396 ml water and 4 ml 1 M Tris, pH 8.0]

Opti-MEM (Gibco)

Fugene 6 (Roche)

Polybrene 10 mg/ml in sterile water (Sigma)

PE-conjugated anti-human CD25 (Calbiochem)

0.05% trypsin (Gibco)

Vectashield mounting medium (Vector Labs, Inc.)

Alexa-fluor 488-conjugated anti-HA (Molecular Probes)

13.2. Retroviral transduction

1. Plate 293T cells on a 10-cm dish at a density of 1 million cells per plate the evening before transfection.
2. Remove the medium and replace with 5 ml of fresh DMEM.
3. Mix together 250 μl of opti-MEM, 10 μg of plasmid DNA containing FLAGHA-tagged BID, and 60 μl of Fugene 6.
4. Incubate at room temperature for 30 min; a precipitate should form.
5. Add dropwise to the 293T cells and mix gently.
6. Incubate at 37 °C, 5% CO_2, for 48 h.
7. After 24 h, plate Bid$^{-/-}$ MEFs on a 10-cm dish at a density of 1 million cells per plate.
8. After 48 h, remove the viral supernatant from the 293T cells.
9. Filter viral supernatant through a 0.4- μm filter to remove any cell debris or nonadherent 293T cells.
10. Dilute viral supernatant 1:2 with DMEM, and add polybrene to 4 μg/ml.
11. Remove media from BID$^{-/-}$ MEFs.
12. Add viral supernatant to MEFs.
13. Remove viral supernatant and replace with fresh DMEM after 24 h.

13.3. Isolation of virally transduced cells

1. After 48 h, remove media and wash once with 5 ml of sterile PBS.
2. Add 5 ml of 0.05% trypsin.
3. Incubate for 5 min at room temperature or until cells begin to lift off of the plate.
4. Add 5 ml of DMEM and pipette cells off of the plate.
5. Spin at 1200 rpm for 7 min.

6. Resuspend in 500 μl of 3% FBS.
7. Add 5 μl of PE-conjugated anti–CD25.
8. Incubate on ice in the dark (cover with foil) for 30 min.
9. Wash three times with 3% FBS.
10. Resuspend in 500 μl 0.5% FBS.
11. Sort for PE+ cells by FACS (we use the Vanderbilt flow sorting facility).
12. Alternatively, cells may be incubated with sheep antirat Dynal beads as used earlier for lineage depletion, and positive cells isolated using a magnet.

13.4. Pretreatment of coverslips for immunofluorescence

1. Work in a tissue culture hood.
2. Place six sterile coverslips into each well of a six-well tissue culture plate.
3. Add 2 ml of polylysine solution
4. Make sure that the entire coverslip is covered with solution; if not, add more polylysine solution.
5. Incubate at room temperature for 30 min.
6. Aspirate off the polylysine solution.
7. Wash three times with PBS.
8. If slides are not to be used immediately, leave the last wash of PBS on the slides and seal the plate with Parafilm.
9. Store at 4°.

13.5. Cell culture

1. Plate 2 to 3 ml of sorted fibroblasts (at a concentration of 1×10^4/ml in DMEM10) into each of six wells of a plate containing coverslips. If the coverslips are not covered completely, add additional medium.
2. Place in incubator for 24 h. Cells should be 50% confluent at the time of staining.
3. Treat cells with death stimulus.
4. Make up media with Mitotracker red.
   Final concentration: 75 nM
   Stock: 1 mM
   Add 0.9 μl per 12 ml medium
5. Remove medium from cells and replace with medium containing Mitotracker red.
6. Incubate 30 min.
7. Remove medium with Mitotracker red.
8. Wash three times with PBS.
9. Add 4 ml of methanol:acetone (3:1) and incubate at −20° for 10 min.
10. Wash three times with PBS.

13.6. Blocking

1. Make up blocking solution: 5% NGS in PBS.
2. Spin at 13,000 rpm for 2 min to remove particulate matter.
3. Block for 1 h at room temperature. Make certain that the slides remain covered with blocking solution for the entire incubation period.
4. Wash three times with PBS.

13.7. Primary antibody

1. Dilute the primary antibody in 5% NGS (anti-HA Alexa Fluor 488, 1:100)
2. Spin at 13,000 rpm for 2 min.
3. Apply to slides using a p200. We use 200 μl of antibody to cover a 22-mm^2 slide.
4. Incubate at room temperature for 1 h.
5. Remove the primary antibody. We save the primary antibody and reuse it one time.
6. Wash three times with PBS by rocking the plate gently for 10 min.
7. Blot the edges of the slide with a Kimwipe; do not let the slide dry completely.
8. Pipette 10 μl of Vectashield (Vector Labs) onto a microscope slide.
9. Invert stained coverslip onto the microscope slide.
10. Seal the coverslip using clear nail polish.
11. Staining may be visualized using a fluorescent microscope. We use a Nikon Eclipse E600 microscope.
12. Slides may be kept at 4° for up to 1 week.

14. ANNEXIN V STAINING

BID performs two distinct roles, an apoptotic role directed by caspase cleavage and interaction with other Bcl-2 family members through the BH3 domain and a cell cycle checkpoint/DNA repair role directed by phosphorylation by ATM and/or ATR. To study the apoptotic response of cells harboring wild-type BID or BID mutated in one of the aforementioned domains in response to DNA-damaging agents, we have used annexin V/PI staining. During the early stages of apoptosis, cells expose phosphatidylserine on the surface of their plasma membrane. Annexin V is a phospholipid-binding protein that, when conjugated to a fluorophore, has been used to detect exposed phosphatidylserine by flow cytometry. Propidium iodide (PI) is a dye that is excluded from living cells, but is incorporated into DNA when the cell membrane becomes permeable late in apoptosis (Fig. 12.5).

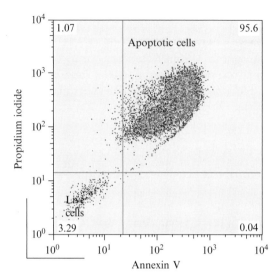

Figure 12.5 Flow cytometry analysis of annexin V/PI staining. One million MPCs were treated with etoposide for 24 h. Cells were stained with annexin V FITC and PI, and flow cytometry was performed on a BD Facscalibur flow cytometer.

14.1. Materials

3% FBS in PBS
Propidium iodide (Sigma) stock solution is 50 μg/ml in PBS
Annexin V FITC (Biovision, Inc.)
FACS tubes (BD Falcon
10 × staining buffer: 0.1 M M HEPES, pH 7.4; 1.4 M NaCl; 25 mM CaCl$_2$.
   Dilute to 1× prior to use.

14.2. Methods

1. Spin down 1 million cells.
2. Resuspend in 100 μl staining buffer + 0.5 μl annexin V FITC.
3. Incubate 30 minutes at room temperature in the dark (cover with foil).
4. Put into FACS tube.
5. Add 100 μl 3% FBS.
6. Immediately prior to analysis, add 4 μl of 50 μg/ml PI.

14.3. Controls

1. No annexin V
2. No PI

15. CONCLUSION

The assays described in this chapter have allowed us to define a dual role for BID in regulating cell death following death receptor stimulation and a cell cycle checkpoint/ DNA repair role following DNA damage. An area of intense research interest is determining what dictates how cells respond to DNA damage. Some cells arrest the cell cycle and initiate DNA repair, whereas others undergo apoptosis. BID, with its dual roles in the interface between these two pathways, is well positioned to play a key role in determining the fate of a cell following DNA damage.

ACKNOWLEDGMENTS

This work was supported by 1K08 CA098394, 1 R01 HL088347, a Kimmel Foundation Scholar award, and a G&P Foundation Scholar award to SSZ.

REFERENCES

Adams, J. M., and Cory, S. (1998). The Bcl-2 protein family: Arbiters of cell survival. *Science* **281**, 1322–1326.

Bakhshi, A., Jensen, J. P., Goldman, P., Wright, J. J., McBride, O. W., Epstein, A. L., and Korsmeyer, S. J. (1985). Cloning the chromosomal breakpoint of t(14;18) human lymphomas: Clustering around JH on chromosome 14 and near a transcriptional unit on 18. *Cell* **41**, 899–906.

Cleary, M. L., and Sklar, J. (1985). Nucleotide sequence of a t(14;18) chromosomal breakpoint in follicular lymphoma and demonstration of a breakpoint-cluster region near a transcriptionally active locus on chromosome 18. *Proc. Natl. Acad. Sci. USA* **82**, 7439–7443.

Desagher, S., Osen-Sand, A., Montessuit, S., Magnenat, E., Vilbois, F., Hochmann, A., Journot, L., Antonsson, B., and Martinou, J. C. (2001). Phosphorylation of bid by casein kinases I and II regulates its cleavage by caspase 8. *Mol. Cell* **8**, 601–611.

Desagher, S., Osen-Sand, A., Nichols, A., Eskes, R., Montessuit, S., Lauper, S., Maundrell, K., Antonsson, B., and Martinou, J. C. (1999). Bid-induced conformational change of Bax is responsible for mitochondrial cytochrome c release during apoptosis. *J. Cell Biol.* **144**, 891–901.

Hawley, R. G., Fong, A. Z., Lu, M., and Hawley, T. S. (1994). The HOX11 homeobox-containing gene of human leukemia immortalizes murine hematopoietic precursors. *Oncogene* **9**, 1–12.

Kamer, I., Sarig, R., Zaltsman, Y., Niv, H., Oberkovitz, G., Regev, L., Haimovich, G., Lerenthal, Y., Marcellus, R. C., and Gross, A. (2005). Proapoptotic BID is an ATM effector in the DNA-damage response. *Cell* **122**, 593–603.

Kelekar, A., and Thompson, C. B. (1998). Bcl-2-family proteins: The role of the BH3 domain in apoptosis. *Trends Cell Biol.* **8**, 324–330.

Luo, X., Budihardjo, I., Zou, H., Slaughter, C., and Wang, X. (1998). Bid, a Bcl2 interacting protein, mediates cytochrome c release from mitochondria in response to activation of cell surface death receptors. *Cell* **94**, 481–490.

Nakatani, Y., and Ogryzko, V. (2003). Immunoaffinity purification of mammalian protein complexes. *Methods Enzymol.* **370,** 430–444.

Tsujimoto, Y., Gorham, J., Cossman, J., Jaffe, E., and Croce, C. M. (1985). The t(14;18) chromosome translocations involved in B-cell neoplasms result from mistakes in VDJ joining. *Science* **229,** 1390–1393.

Wang, K., Yin, X. M., Chao, D. T., Milliman, C. L., and Korsmeyer, S. J. (1996). BID: A novel BH3 domain-only death agonist. *Genes Dev.* **10,** 2859–2869.

Warner, N. L., Moore, M. A., and Metcalf, D. (1969). A transplantable myelomonocytic leukemia in BALB-c mice: Cytology, karyotype, and muramidase content. *J. Natl. Cancer Inst.* **43,** 963–982.

Wei, M. C., Lindsten, T., Mootha, V. K., Weiler, S., Gross, A., Ashiya, M., Thompson, C. B., and Korsmeyer, S. J. (2000). tBID, a membrane-targeted death ligand, oligomerizes BAK to release cytochrome c. *Genes Dev.* **14,** 2060–2071.

Wei, M. C., Zong, W. X., Cheng, E. H., Lindsten, T., Panoutsakopoulou, V., Ross, A. J., Roth, K. A., MacGregor, G. R., Thompson, C. B., and Korsmeyer, S. J. (2001). Proapoptotic BAX and BAK: A requisite gateway to mitochondrial dysfunction and death. *Science* **292,** 727–730.

Yin, X. M., Oltvai, Z. N., and Korsmeyer, S. J. (1994). BH1 and BH2 domains of Bcl-2 are required for inhibition of apoptosis and heterodimerization with Bax. *Nature* **369,** 321–323.

Yin, X. M., Wang, K., Gross, A., Zhao, Y., Zinkel, S., Klocke, B., Roth, K. A., and Korsmeyer, S. J. (1999). Bid-deficient mice are resistant to Fas-induced hepatocellular apoptosis. *Nature* **400,** 886–891.

Zinkel, S. S., Hurov, K. E., Ong, C., Abtahi, F. M., Gross, A., and Korsmeyer, S. J. (2005). A role for proapoptotic BID in the DNA-damage response. *Cell* **122,** 579–591.

Zinkel, S. S., Ong, C. C., Ferguson, D. O., Iwasaki, H., Akashi, K., Bronson, R. T., Kutok, J. L., Alt, F. W., and Korsmeyer, S. J. (2003). Proapoptotic BID is required for myeloid homeostasis and tumor suppression. *Genes Dev.* **17,** 229–239.

Zong, W. X., Li, C., Hatzivassiliou, G., Lindsten, T., Yu, Q. C., Yuan, J., and Thompson, C. B. (2003). Bax and Bak can localize to the endoplasmic reticulum to initiate apoptosis. *J. Cell Biol.* **162,** 59–69.

CHAPTER THIRTEEN

ASSEMBLY, PURIFICATION, AND ASSAY OF THE ACTIVITY OF THE ASC PYROPTOSOME

Teresa Fernandes-Alnemri *and* Emad S. Alnemri

Contents

Department of Biochemistry and Molecular Biology, Center for Apoptosis Research, Kimmel Cancer Institute, Thomas Jefferson University, Philadelphia, Pennsylvania

Methods in Enzymology, Volume 442
ISSN 0076-6879, DOI: 10.1016/S0076-6879(08)01413-4

Abstract

Pyroptosis is an inflammatory form of cell death mediated by caspase-1. Until recently, little was known about the mechanism by which caspase-1 is specifically activated to induce pyroptosis. Using biochemical and time-lapse confocal bioimaging approaches, it has been shown that caspase-1 is activated during pyroptosis by a large supramolecular assembly termed the pyroptosome. Biochemical and mass spectroscopic analyses revealed that the pyroptosome assembly is an oligomer of dimers of the adaptor protein ASC. Only one distinct pyroptosome is formed in each cell when macrophages or monocytes are stimulated with proinflammatory stimuli, which rapidly recruits and activates caspase-1, resulting in pyroptosis. This chapter describes methods for real-time observation and recording of the pyroptosome assembly process in live THP-1 monocytes. It also describes biochemical methods for the assembly, purification, and assay of the ASC pyroptosome from the THP-1 cell line, which could be adapted for use with other cell lines containing ASC, such as primary mouse macrophages. Finally, it describes methods for the *in vitro* reconstitution of a functional ASC pyroptosome from the recombinant ASC protein produced in *Escherichia coli*.

1. INTRODUCTION

Caspase-1 is a critical element of innate immunity and the host defense against pathogenic infections (Dinarello, 1998). Caspase-1 cleaves the inactive pro-interleukin (IL)-1β and pro-IL-18 to produce the active proinflammatory cytokines IL-1β and IL-18, respectively. Activation of caspase-1 occurs in response to pathogenic infection or cellular stress and requires the assembly of intracellular molecular platforms, called inflammasomes (Drenth and van der Meer, 2006; Mariathasan, 2007; Ogura *et al.*, 2006; Petrilli *et al.*, 2005). A number of molecules involved in the assembly of different inflammasomes have been identified, including ICE-protease activating factor (Ipaf), Nalp1, cryopyrin/Nalp3, pyrin, and ASC (Agostini *et al.*, 2004; Delbridge and O'Riordan, 2006; Martinon and Tschopp, 2005; Poyet *et al.*, 2001; Srinivasula *et al.*, 2002; Ting *et al.*, 2006; Yu *et al.*, 2006). Genetic studies revealed that the Ipaf inflammasome is specifically activated by flagellin of intracellular pathogens such as *Salmonella typhimurium* and *Legionella pneumophila*, possibly via recognition of flagellin by the regulatory LRR domain of Ipaf (Amer *et al.*, 2006; Franchi *et al.*, 2006; Mariathasan *et al.*, 2004; Miao *et al.*, 2006). However, gene deletion studies in the mouse revealed that Nalp1 inflammasome is activated by anthrax lethal toxin (Boyden and Dietrich, 2006), whereas *in vitro* studies with reconstituted human Nalp1 inflammasome demonstrated that it can also be activated by the microbial product muramyl-dipeptide and ribonucleoside triphosphates

(Faustin *et al.*, 2007). Both the Ipaf and Nalp1 inflammasomes associate directly with inactive monomeric procaspase-1 via CARD–CARD interactions and prompt its dimerization and activation (Faustin *et al.*, 2007; Poyet *et al.*, 2001).

In contrast to Ipaf and Nalp1 inflammasomes, the cryopyrin/Nalp3 and pyrin inflammasomes activate caspase-1 (Agostini *et al.*, 2004; Yu *et al.*, 2006) with the help of the adaptor protein ASC (Srinivasula *et al.*, 2002). The specific stimuli that activate the pyrin inflammasome are not yet known, but studies suggest that engagement of pyrin by the cytoskeleton-associated protein PSTPIP1 leads to activation of the pyrin inflammasome (Yu *et al.*, 2007). The cryopyrin inflammasome, however, is activated in response to diverse signaling pathways triggered specifically by infection with intracellular bacteria *Listeria monocytogenes* and *Staphylococcus aureus*, TLR agonists plus potassium-depleting agents such as ATP, nigericin, or maitotoxin, danger signal monosodium urate, antiviral compounds R837 and R847, bacterial RNA, and viral double-stranded RNA (Kanneganti *et al.*, 2006a,b; Mariathasan *et al.*, 2006; Martinon *et al.*, 2006; Sutterwala *et al.*, 2006). How these diverse stimuli activate the cryopyrin inflammasome is still unclear.

The activation of caspase-1 by these inflammasomes leads to processing of pro-IL-1β and pro-IL-18, resulting in inflammation. However, in cases such as when macrophages or monocytes are infected with intracellular bacteria, the activation of caspase-1 can lead to an inflammatory form of cell death called pyroptosis (Cookson and Brennan, 2001; Fink and Cookson, 2005, 2006; Swanson and Molofsky, 2005). Studies reveal that extensive oligomerization of the adaptor protein ASC downstream of the inflammasomes, triggered by low intracellular ionic strength/potassium depletion, is largely responsible for directing caspase-1 activation toward cell death (Fernandes-Alnemri *et al.*, 2007). Using live cell confocal bioimaging techniques, we observed that treatment of THP1 monocytes/macrophages expressing an ASC-GFP fusion protein with lipopolysaccharides (LPS) or several other proinflammatory agents triggers formation of a supramolecular assembly of ASC that recruits and activates procaspase-1 (Fernandes-Alnemri *et al.*, 2007), leading to pyroptosis. Because of its direct involvement in caspase-1 activation and pyroptosis, we called this ASC supramolecular assembly the "pyroptosome." Biochemical studies revealed that the mature ASC pyroptosome is composed of oligomerized ASC dimers and caspase-1 and does not contain upstream inflammasome components such as cryopyrin, suggesting that inflammasomes might function as initial catalysts to induce the formation of ASC dimers, which then dissociate from the inflammasome and undergo additional oligomerization to form the large pyroptosome. This chapter describes protocols for live cell bioimaging of pyroptosome formation in THP-1 cells. It also describes protocols used for purification, reconstitution, and assay of the pyroptosome activity.

2. Methods

Studying the role of the adaptor protein ASC in caspase-1 activation requires working with cells of myeloid origin such as monocytes and macrophages. The THP-1 monocytic cell line is a useful model system that has been used extensively to study the role of caspase-1 and other upstream molecules in the cellular responses to diverse inflammatory stimuli. THP-1 are not easy to transfect with plasmids by standard procedures and require methods involving retroviral transduction to stably express exogenous genes in these cells. Therefore, to observe in real time how ASC activates caspase-1 in response to treatment of THP-1 monocytes with diverse stimuli, this chapter describes how to design an ASC-GFP fusion protein expression vector, transduce it stably into THP-1 cells, and finally visualize ASC oligomerization by time-lapse confocal microscopy.

3. Generation of Stable THP-1 Cells Expressing an ASC–Green Fluorescent Protein (GFP) Fusion Protein to Visualize ASC Oligomerization

Fusing ASC with a GFP allows visualization of the subcellular localization and behavior of ASC under different cellular conditions. Because the N-terminal pyrin domain (PYD) of ASC is important for its recruitment by upstream molecules such as pyrin or cryopyrin and for its self-dimerization, it is important to fuse the GFP tag to the C terminus of ASC. Tagging of ASC at the C terminus with the GFP does not interfere with its oligomerization or functionality. We found that MSCV retroviral expression vectors are suitable for the construction of the ASC–GFP fusion construct and for stable transduction into THP-1 cells.

3.1. Construction of the ASC–GFP retroviral expression construct

The procedure for construction of the retroviral vector used to make ASC–GFP stable THP-1 cells is outlined here.

1. Amplify the ASC coding region by polymerase chain reaction (PCR) using an ASC-specific start primer carrying an *Eco*RI restriction site, 5′-CGGGAATTCGATGGGGCGCGCGCGCGAC-3′, an end primer carrying an *Xho*I restriction site, 5′-CCGCTCGAGGCTCCGCTC-CAGGTCCTCC-3′, and an appropriate plasmid containing the full-length ASC cDNA (e.g., LIFESEQ1406906, Open Biosystems) as the template.

2. Clean up the resulting PCR product using a PCR purification kit (Qiagen or comparable product) and digest with *Eco*RI and *Xho*I restriction enzymes (as recommended by supplier).

3. Fractionate the digested PCR product by gel electrophoresis on a 2% low melting point agarose gel stained with ethidium bromide, excise the 588-bp band from the gel, and clean up using Qiagen's gel purification kit or comparable product. Determine the concentration of the purified fragment by spectroscopy.

4. Ligate the PCR fragment to an *Eco*RI- and *Sal*I-digested pEGFP-N1 plasmid (Clontech) to obtain the pEGFP-N1-ASC fusion plasmid. We used the rapid ligation kit supplied by Roche. This will result in an in-frame fusion of the ASC and GFP coding regions. Follow standard bacterial transformation and screening procedures. Confirm the reading frame of the ASC–GFP fusion cDNA by sequence analysis of the resulting pEGFP-N1-ASC plasmid clones. You can further confirm expression of the fusion construct by transient transfection into HEK293T cells and visualize the GFP fluorescence 24 h posttransfection.

5. Cut the pEGFP-N1-ASC plasmid with the *Not*I restriction enzyme and blunt the resulting linearized plasmid using standard methods. Then cut the linearized plasmid with *Bgl*II to excise the ASC–EGFP fusion cDNA.

6. Fractionate on a 1% low melting point agarose gel to separate the vector from the ASC–EGFP cDNA fragment, which can then be ligated into the *Hpa*I/*Bgl*II sites of pMSCVpuro plasmid (Clontech). This will result in directional cloning of the ASC–EGFP cDNA in the multiple cloning sites of pMSCVpuro under the control of the 5′LTR-Ψ+ promoter. Confirm the pMSCVpuro-ASCgfpN1 by sequence analysis. We recommend using the pMSCV 5′ and 3′ sequencing primers.

7. Test the resulting functionality of the pMSCVpuro-ASCgfpN1 plasmid by transfection into HEK293 cells using standard LipofectAMINE transfection (Invitrogen) procedures. Examine the transfected cells under a fluorescent microscope 24 h posttransfection. If the plasmid encodes the correct ASC–GFP fusion protein, you will be able to see the ASC–GFP fluorescence distributed evenly in the cytoplasm and nucleus of the transfected cells 24 to 36 h after transfection. Some cells will show intensely fluorescent large aggregates of ASC–GFP.

3.2. Generation of the ASC–GFP retrovirus and transduction of THP-1 cells

To generate the ASC-GFP retrovirus, it is necessary to use the amphotropic packaging cell line Phoenix-Ampho (G.P. Nolan's laboratory, Stanford University Medical Center, Stanford, CA. http://www.stanford.edu/group/nolan/retroviral_systems/retsys.html), which is a HEK293T cell line stably transfected with constructs capable of producing gag-pol, and the amphotropic envelope protein for the Moloney murine leukemia virus (MoMuLV). This cell line is available from Orbigen.

The following procedure describes how to generate the ASC–GFP retrovirus in the Phoenix-Ampho and how to use this retrovirus to stably transduce THP-1 to generate THP-1-ASC-GFP cells.

1. Seed Phoenix-Ampho cells (1×10^6 cells) in 100-mm dishes in complete DMEM/F12 culture medium. The next day, transfect cells with 1 to 1.5 μg pMSCVpuro-ASC-EGFP-N1 plasmid using the LipofectAMINE transfection method as recommended by the supplier.

2. Forty-eight hours after transfection, harvest the cells by trypsinization and sort GFP-expressing cells by fluorescence-activated cell sorting (FACS). Grow the fluorescent cells until confluent in T75 tissue culture flasks. There is no need for the use of puromycin for selection of stable cells, as FACS can achieve the same goal.

3. Repeat sorting two more times over a period of 4 weeks by FACS until more than 95% of the Phoenix cells are stably expressing the ASC–GFP fusion protein. The culture medium of these cells should now contain a high titer of the ASC–GFP encoding retrovirus. Safety precautions should be followed during generation and handling of these stable retrovirus producing cells.

4. To transduce THP-1 cells with the ASC–GFP encoding retrovirus, it is preferable to prepare the retrovirus in THP-1 culture medium (complete RPMI 1640 medium). To do that, seed the stable ASC–GFP Phoenix cells in complete DMEM/F12 culture medium at a density of 2×10^6 cells/100-mm dish. When the cells are 60 to 70% confluent (usually in 24 h after seeding), remove the culture medium and replace it with 10 ml of complete RPMI 1640 medium. Culture the cells in this medium for 24 h and then collect the culture medium, which should contain the retroviral particles. Add polybrene (Sigma) at a final concentration of 4 μg/ml to the retrovirus-containing medium and then filter through a 0.45-μm membrane filter. We use 0.45-μm GE cellulose acetate sterile syringe filters.

5. Pellet exponentially growing THP1 cells by a brief centrifugation at 1200 rpm for 5 min. Resuspend the cells in the filtered retrovirus-containing medium at a density of 5×10^5 cells/ml and then pipette 2 ml of the suspended cells into each well of a six-well plate. Centrifuge the six-well plate for 60 min at 2500rpm at 37° in an Eppendorf 5810 R tabletop centrifuge equipped with six-well plate adaptors. After centrifugation, plates must be placed back in a CO_2 incubator at 37° for 2 to 3 h. Collect cells from the six-well plate and transfer to a T175 flask. Add an equal amount of fresh THP-1 medium to the cells and allow cells to recover for 24 h. Pellet cells by centrifugation at 1200 rpm for 5 min and then resuspend in complete RPMI1640 medium. Split cells if necessary into two T175 flasks and allow to grow for 72 h before sorting by FACS.

6. The GFP expressing cells should be sorted by FACS three times over a period of 5 weeks until more than 95% of the cells are stable and GFP positive. If preferred, independent stable THP-1 populations can be generated with different levels of ASC–GFP expression by sorting the cells into separate pools of low, medium, or high fluorescence. Do not select stable cell clones using puromycin selection because this causes many cells to die and may result in clonal selection of cell death-resistant clones. We found that nonclonal selection by FACS is sufficient to produce stable THP-1 cells. The stable expression of ASC-GFP fusion protein in THP-1 cells must be verified by Western blotting using an ASC-specific antibody (MLB) and by additional functional assays as described later.

   Complete RPMI 1640: RPMI 1640 supplemented with 10 mM N-(2-hydroxyethyl) piperazine-N'-(2-ethanesulfonic acid), 1 mM sodium pyruvate, 1.5 g/L sodium bicarbonate, 2 mM L-glutamine, 55 μM β-mercaptoethanol, 10% fetal bovine serum, 200 μg·ml^{-1} penicillin, and 100 μg·ml^{-1} streptomycin sulfate.

   Complete DMEM/F12: DMEM/F12 medium supplemented with 10% fetal bovine serum, 200 μg·ml^{-1} penicillin, and 100 μg·ml^{-1} streptomycin sulfate.

3.3. Confocal time-lapse bioimaging of live THP-1 cells

Results have demonstrated that treatment of THP-1 cells with some pathogen or nonpathogen-derived proinflammatory agents induces rapid oligomerization of ASC followed by cell death (Fernandes-Alnemri *et al.*, 2007). The THP-1–ASC–GFP cells are ideal to study and observe in real time or by time-lapse imaging the effect of a broad range of stimuli on the behavior of ASC in live cells and the consequences of ASC oligomerization on cell viability. The following steps describe how to prepare THP-1–ASC–GFP cells for confocal time-lapse bioimaging.

1. Pellet exponentially growing THP-1–ASC–GFP cells by a brief centrifugation at 1200 rpm for 5 min. Resuspend the cells in fresh medium containing phorbol 12-myristate-13-acetate (PMA) (0.5 μM) at a density of 5 × 10^5 cells/ml. Pipette 1 ml of the suspended cells into 35-mm cover glass bottom culture dishes (MatTek Corp.) and allow cells to attach for 2 h. These culture dishes are specifically designed for live confocal bioimaging.

2. Remove the PMA containing medium and replace it with fresh medium without PMA. The cells are now ready to be treated with proinflammatory agents. Normally we leave the cells in fresh medium without PMA overnight, but shorter times can be used.

3. If nuclear and/or mitochondrial staining is desired, stain cells with Hoecht 33342 and/or Mitotracker red stains according to the manufacturer's instructions for 30 min before microscopy.

4. Place the culture dish in the microscope sample chamber. We use an LSM 510 META confocal microscope system (Carl Zeiss) controlled by Zeiss AIM software and equipped with a temperature- and CO_2-controlled sample chamber for live cell imaging, but comparable microscopes can be used. It is critical that the temperature of the sample be maintained between 37 and 38° and CO_2 be maintained at 5%. We found that temperatures below 37° reduce pyroptosome formation.

5. Stimulate the cells with crude LPS (1–5 μg/ml) or other proinflammatory stimuli. Ten minutes after stimulation, start recording images from the GFP, Hoecht, and Mitotracker fluorescence signals simultaneously every 17.5 s for an additional 30 min. In our assays we use the following settings: EGFP is excited with a 488-nm argon laser and its 505- to 545-nm emitted light is collected in a photomultiplier tube (PMT). Hoechst is excited with a 405-nm HeNe laser and its 420- to 480-nm emitted light is collected in a second PMT. Mitotracker red is excited with a 543-nm HeNe laser and its 545- to 700-nm emitted light is collected by the META PMT array. Transmitted laser light is collected to produce a differential interference microscopic image. The eight-bit images are converted into a RGB format, the median is filtered to eliminate noise, montaged or overlayed, and is converted into movie files at five frames per second using Metamorph 7.3 (Molecular Devices Corp.).

In unstimulated cells, ASC–GFP should be distributed evenly in the cytoplasm and nucleus. However, ASC begins to oligomerize and form single clusters (pyroptosomes) as early as 10 min after stimulation with crude LPS or other stimuli (Fernandes-Alnemri et al., 2007). The ASC pyroptosomes do not form simultaneously in all cells and some cells may not form the pyroptosome even after prolonged stimulation. Only one ASC pyroptosome will be formed in each cell (Fig. 13.1), and this clearly precedes morphological changes characteristic of pyroptosis. We observed no correlation between the intensity of the ASC-GFP and pyroptosome formation. Pyroptosome formation was seen equally in cells containing very faint or intense ASC–GFP fluorescence. For representative time-lapse movies of ASC pyroptosome formation in THP-1–ASC–GFP cells, see supplementary movies in Fernandes-Alnemri et al. (2007).

4. BIOCHEMICAL ASSAY OF ASC PYROPTOSOME FORMATION IN STIMULATED THP-1 OR PRIMARY MACROPHAGES

When working with cells that do not express ASC–GFP, it is not possible to visualize ASC pyroptosome formation by fluorescence microscopy. However, because of their large size (1–2 μm) it is feasible to separate

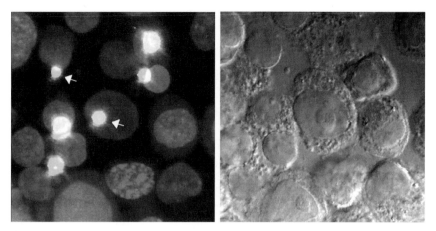

Figure 13.1 ASC–GFP pyroptosomes in LPS-treated THP-1–ASC–GFP cells. Cells were seeded in 35-mm cover glass bottom culture dishes, primed with PMA, and then treated with crude LPS for 60 min as described in the text. Cells were fixed, stained with DAPI, and then observed and photographed by fluorescence confocal microscopy (63× magnification). (Left) Note the large perinuclear bright fluorescent ASC pyroptosomes (indicated by arrows) formed in several cells. (Right) A phase-contrast confocal micrograph of the same cells in the left micrograph is shown.

the ASC pyroptosomes by conventional biochemical techniques from other cellular components for further biochemical analysis. The technique described here allows partial purification and detection of pyroptosome in stimulated THP-1 cells or primary macrophages by Western blot analysis.

1. In a 50-ml centrifuge tube prepare a 40-ml suspension of exponentially growing THP-1 cells in complete RPMI 1640 culture medium at a density of 1×10^6 cells/ml.

2. Treat cells with PMA (0.5 μM) and then transfer cells immediately into four 100-mm dishes (10 ml/dish). Allow cells to attach for 3 h, change medium with fresh medium, and leave cells in the tissue culture incubator overnight.

3. The next day, pretreat cells with zVAD (50 μM) for 30 min to prevent cell death, and then add LPS (1–10 μg/ml) to two of the plates and PBS to the remaining two plates (control) and incubate for 2 h.

4. Harvest cells from the LPS-treated plates and combine in one 50-ml centrifuge tube. Do the same for the control plates. Centrifuge the LPS-treated and control cells at 1200 rpm for 5 min, remove medium, resuspend the two cell pellets each in 1 ml PBS, and transfer to two 1.5-ml Eppendorf tubes. Pellet cells at 1500 rpm in an Eppendorf tabletop 5417 R refrigerated microcentrifuge. Pelleted cells may be frozen at −80° or proceed with purification of pyroptosomes.

5. Add 0.5 ml buffer A to each of the pellets and lyse cells on ice by syringing 30× using a 1-ml syringe with a 21-gauge needle.

6. Centrifuge the lysates at 1800 rpm for 8 min. Remove supernatants and transfer to an empty Eppendorf tube, taking care not to disturb the nuclear pellet. Remove a 30-μl sample and save for Western blot analysis.

7. Dilute the remaining supernatants with 2 volumes of buffer A and then filter the diluted supernatants with 5-μm Ultrafree-CL centrifugal filters (Millipore) at 2000g. This filtration step removes any contaminating nuclei and unlysed cells.

8. Collect the clarified lysates and dilute with 1 volume of CHAPS buffer to lyse organelles such as mitochondria and then centrifuge the lysates at 5000 rpm for 8 min in a tabletop Eppendorf 5417 R refrigerated microcentrifuge. Discard the supernatants and keep the pellets.

9. Resuspend each pellet in 1 ml CHAPS buffer by gentle pipetting up and down and then centrifuge the tubes at 5000 rpm for 8 min. Repeat this step two times to wash the pyroptosomes.

10. Resuspend pellets in 30 to 40 μl CHAPS buffer. For Western blot analysis, mix an equal amount of the resuspended pellets and 2× SDS sample buffer, boil, and then fractionate on a 12% SDS-polyacrylamide gel. Immunoblot the fractionated proteins with an anti-ASC antibody. A large amount of ASC should be seen in the pellet from the LPS-treated cells and a very small amount of ASC in the pellet from the control untreated cells. The presence of a small amount of ASC in the control pellet is normal and is due to nonspecific precipitation of some ASC with cellular contaminants during the centrifugation step.

4.1. Assaying the oligomeric nature of the ASC pyroptosome by chemical cross-linking

Our biochemical analyses revealed that ASC pyroptosomes are composed of oligomerized ASC dimers. To verify the oligomeric state of ASC in the pyroptosome preparation described earlier, we use chemical cross-linking reagents such as disuccinimidyl suberate (DSS, Pierce) as described in the following procedure.

1. To the resuspended pellets in step 8 above, add DSS to a final concentration of 2 mM and incubate at room temperature for 20 to 30 min. Quench the reaction by adding an equal amount of 2× SDS sample buffer, boil, and then fractionate on a 12% SDS-polyacrylamide gel. DSS should be prepared fresh each time in dimethyl sulfoxide according to the manufacturer's instructions (Pierce).

2. Immunoblot the fractionated proteins with an anti–ASC antibody. There should be two major bands of ASC: a ≈27-kDa band corresponding to monomeric ASC and a ≈54-kDa band corresponding to dimeric ASC in the pellet from the LPS-treated cells (Fernandes-Alnemri et al., 2007).

Very little or no dimeric ASC should be present in pellets from the control untreated cells.

Buffer A: 20 mM HEPES-KOH, pH 7.5, 10 mM KCl, 1.5 mM MgCl$_2$, 1 mM EDTA, 1 mM EGTA, and 320 mM sucrose. Add phenyl-methylsulfonyl fluoride (PMSF) (0.1 mM) and protease inhibitor cocktail (Roche) prepared fresh each time.

CHAPS buffer: 20 mM HEPES-KOH, pH 7.5, 5 mM MgCl$_2$, 0.5 mM EGTA, and 0.1% CHAPS. Add PMSF (0.1 mM) and protease inhibitor cocktail (Roche) prepared fresh each time.

5. *In Vitro* Assembly and Purification of ASC Pyroptosomes from THP-1 Cell Lysates

Early studies demonstrated that incubation of THP-1 lysates prepared in a hypotonic buffer at 30 to 37° leads to the activation of caspase-1 (Ayala *et al.*, 1994; Yamin *et al.*, 1996). However, the molecular mechanism responsible for this activation was not elucidated. Employing similar procedures, we have demonstrated that incubation of THP-1 lysates prepared in a hypotonic CHAPS lysis buffer at 30 to 37° causes rapid assembly of the ASC pyroptosome, which recruits procaspase-1, leading to its activation (Fernandes-Alnemri *et al.*, 2007). This section describes a simple procedure for the assembly and purification of functional ASC pyroptosome from THP-1 cell lysates. This procedure can be used to prepare ASC pyroptosomes either from ASC–GFP expressing THP-1 cells or from the parental THP-1 cells.

5.1. Assembly and purification of ASC pyroptosomes from ASC–GFP expressing THP-1 cells

1. Grow ASC–GFP–THP-1 cells until they reach a density of 1×10^6 cells/ml in tissue culture T175 flasks (30 ml/flask). We recommend growing at least 10 T175 flasks (3×10^8 cells) of ASC–GFP–THP-1 cells to make a large preparation of ASC pyroptosomes; however, this procedure can be scaled down as needed. Harvest THP-1 cells by centrifugation in an Eppendorf 5810 R tabletop centrifuge or equivalent at 1200 rpm in 50-ml centrifuge tubes. Remove medium and resuspend each pellet in 1 ml PBS and transfer to 1.5-ml Eppendorf tubes. Pellet cells at 1500 rpm in an Eppendorf tabletop 5417 R refrigerated micro-centrifuge or equivalent. Cell pellets can now be frozen at −80° or proceed with purification of the pyroptosomes. Cell pellets do not need to be grown and harvested all at the same time, but frozen cell pellets can be used from different batches of cells. We normally freeze THP-1 cells in packed pellet sizes of $\leq$150 μl.

2. Add 2.5 volumes of hypotonic CHAPS buffer to each pellet (e.g., for a 100-μl packed cell pellet, add 250 μl CHAPS lysis buffer) and lyse cells on ice by syringing 25×. Work rapidly to prevent formation of pyroptosomes before the centrifugation step. Centrifuge the lysates at 14,000 rpm for 8 min in a refrigerated Eppendorf centrifuge set at 4° to remove nuclei and organelles. After this centrifugation step, remove the supernatants and transfer to 1-ml polycarbonate ultracentrifuge tubes (Beckman) and centrifuge at 100,000g in a TLA-120.2 rotor or equivalent in a Beckman Optima TLX benchtop ultracentrifuge for 30 min to obtain S100 supernatants. Remove the S100 supernatants gently not to disturb the pellet and transfer to 1.5-ml Eppendorf tubes. The ultracentrifugation step enhances the formation of large pyroptosomes. However, smaller pyroptosomes can be assembled using supernatants directly after the 14,000 rpm step. All steps must be performed on ice to prevent pyroptosome formation during the lysate preparation.

3. To induce the assembly of the ASC pyroptosomes, incubate the S100 supernatants from step 2 at 37° for 30 to 40 min. Normally, pyroptosomes form within the first 10 to 15 min. If caspase-1 is to be trapped on the pyroptosomes, add zVAD-fmk (50 μM) to the supernatants before incubation at 37°. During the incubation period, remove 2 μl of the supernatants, place on a microscope slide, and observe the formation of pyroptosomes using a fluorescent microscope.

4. After the incubation period, spin the reaction tubes at 2500 rpm in an Eppendorf tabletop 5417 R microcentrifuge. Remove the supernatants and keep the pellets, which contain the pyroptosomes. The pyroptosomes can be seen in the pellets by shining an ultraviolet (UV) light on the pellet using a hand-held UV lamp.

5. To purify the pyroptosomes further, resuspend the pellets in 2 ml CHAPS buffer and keep on ice. Prepare two 50% Percoll cushions in two 1.5-ml Eppendorf tubes by mixing 100 μl Percoll (Sigma) with 100 μl CHAPS buffer in each tube. Carefully layer the 1 ml of resuspended pyroptosomes over each Percoll cushion and centrifuge the tubes at 14,000 rpm for 10 min at 4°. This centrifugation allows the pyroptosomes to settle in the Percoll cushion bellow the interface. The contaminants stay in the interface above the Percoll cushion.

6. After the centrifugation step, determine the position of the pyroptosomes in the Percoll cushion by shining a UV light on the tube and marking the position with a black marker. Remove and discard the buffer layer above the interface, remove the interface layer, which might contain some crude pyroptosomes, and save in a clean Eppendorf tube. Remove as much of the Percoll cushion above the marked pure pyroptosomes (Fig. 13.2). Dilute the pyroptosomes in the remaining Percoll cushion with 1 ml CHAPS buffer and then centrifuge at 14,000 rpm. Wash the pelleted pyroptosomes two times with CHAPS and then

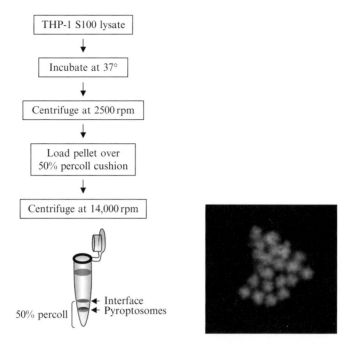

Figure 13.2 (Left) A schematic diagram of the *in vitro* pyroptosome assembly and purification steps. (Right) A confocal micrograph (63× magnification) of ASC–GFP pyroptosomes purified from THP-1–ASC–GFP cell lysates.

resuspend in 50 to 100 μl CHAPS buffer. The pyroptosomes can now be used for caspase-1 activation assays (see Fig. 13.3B) or stored at −80°.

7. If the interface material from step 6 contains a large amount of pyroptosomes (check under a microscope), resuspend it in CHAPS buffer, layer it over a 50% Percoll cushion, centrifuge at 14,000 rpm for 10 min at 4°, and continue as in step 6.

5.2. Assembly and purification of endogenous ASC pyroptosomes from THP-1 cells

Although the endogenous ASC pyroptosomes cannot be seen during the purification step, they can be purified using the exact procedure described earlier for ASC–GFP pyroptosomes. However, we found that labeling the ASC pyroptosomes with sulforhodamine (SR)-VAD-fmk (Apo Logix), which binds to and covalently labels activated caspase-1 on the pyroptosome, makes it easier to follow the pyroptosomes during the purification steps. To prepare ASC pyroptosomes from THP-1 cells, follow these steps.

1. Grow, harvest, and lyse THP-1 cells exactly as described in steps 1 and 2 for the THP–1-ASC–GFP cells.

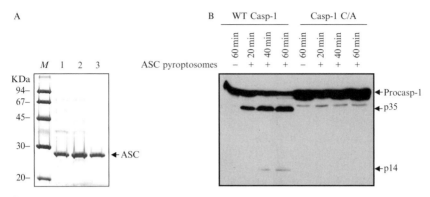

Figure 13.3 (A) Purification of recombinant ASC pyroptosomes from *E. coli* occlusion bodies. Lane 1, purified ASC occlusion bodies (1 μl) before 50% Percoll purification step. Lane 2, purified ASC occlusion bodies (1 μl) after 50% Percoll purification step. Lane 3, purified ASC pyroptosomes (5 μl) from bacterial occlusion bodies. Samples were fractionated on a 12.5% SDS-polyacrylamide gel and stained with Coomassie blue. (B) *In vitro* activation of caspase-1 by purified ASC pyroptosomes. ASC pyroptosomes were assembled from THP-1 S100 extracts *in vitro* and then purified as described in the text. The ASC pyroptosomes (75 ng) were incubated with S100 lysates from 293-caspase-1 (1st to 4th lanes) or 293-caspase-1-C285A (5th to 8th lanes) at 37° for the indicated periods of time. Samples were then fractionated by SDS-PAGE followed by Western blotting with the anti-Flag antibody. Note autoactivation of wild-type caspase-1 by the ASC pyroptosomes to generate the p35 and p14 fragments (1st to 4th lanes). No autoactivation can be seen with the active site C285A mutant procaspase-1 (5th to 8th lanes).

2. Add SR-VAD-fmk (1×) to the S100 supernatants and induce pyroptosome assembly at 37° as described earlier. After the incubation period, spin the reaction tubes at 2500 rpm. Discard the supernatants and then resuspend the pellets, which contain the pyroptosomes, in 100 μl CHAPS buffer. Take a 2-μl sample from the pyroptosome suspension and place on a microscope slide. Using a fluorescent microscope equipped with a rhodamine red filter, observe the pyroptosomes.
3. Follow steps 5 to 7 given earlier to further purify the pyroptosomes on Percoll cushions. However, as these pyroptosomes cannot be seen with a UV lamp, a microscope is needed to see the pyroptosomes in the Percoll cushion and the interface.

6. *IN VITRO* ASSEMBLY AND PURIFICATION OF RECOMBINANT ASC PYROPTOSOMES

Expression of ASC at a high level in bacteria results in oligomerization and aggregation of ASC in occlusion bodies. We have devised a procedure to purify, refold, and assemble ASC pyroptosomes using recombinant ASC expressed in *Escherichia coli*. This procedure involves isolation of ASC occlusion bodies from *E. coli*, denaturation in guanidine HCl, renaturation in a high salt buffer, and assembly of ASC pyroptosomes in a hypotonic CHAPS buffer.

6.1. Expression of ASC in bacteria

1. Transform DH5α *E. coli* with a pET-28-ASC plasmid or an equivalent construct. We found that placing a T7 or His6 tag at the N terminus of ASC enhances expression of the recombinant ASC protein in *E. coli*. We use a pET-28a-ASC plasmid in which the ASC open reading frame is cloned in the *EcoRI* and *XhoI* restriction sites of pET-28a(+) (Novagen) in frame with the His6-T7 tag. We also made another plasmid in which the His6 of the pET-28a-ASC plasmid is removed, leaving only the T7 tag at the N terminus of ASC. Both plasmids express large amounts of ASC in occlusion bodies.

2. Inoculate 100 ml LB medium containing kanamycin (0.025 mg/ml) with a single colony of *E. coli* cells containing the pET-28a-ASC plasmid and incubate overnight with shaking at 37°.

3. Inoculate 1 liter of LB medium containing kanamycin (0.025 mg/ml) antibiotic with 10 ml overnight culture. Incubate at 37° with shaking until A_{600} is 0.6 to 0.7.

4. Lower incubation temperature to 28°. Add isopropyl-β-D-thiogalactoside to a final concentration of 0.5 mM and incubate with shaking for a further 3 h.

5. Harvest cells by centrifugation at 6000g for 15 min at 4°. Resuspend cell pellets in PBS, aliquot in 1.5-ml Eppendorf tubes, and centrifuge tubes to obtain 100 μl packed pellets. Snap freeze and store at −80° until purification.

6.2. Purification of ASC occlusion bodies

1. To the 100-μl bacterial pellet in a 1.5-ml Eppendorf tube, add 1 ml CHAPS buffer containing protease inhibitor cocktail (Roche) and 0.1 mM PMSF, and resuspend by repeated pipetting up and down on ice.

2. Lyse cells by sonication on ice (25 bursts using Branson Sonifier 450 or equivalent equipped with a microtip for 1.5-ml Eppendorf tubes).

3. Centrifuge at 10,000 rpm for 5 min at 4° in an Eppendorf tabletop 5417 R refrigerated microcentrifuge or equivalent. Discard the supernatant and resuspend the pellet in 1 ml CHAPS buffer on ice by repeated pipetting until completely in suspension. SDS-PAGE analysis of a 1-μl aliquot of ASC occlusion bodies from this step should reveal a ≈90% pure ASC (Fig. 13.3A, lane 1).

4. Centrifuge the suspension at 8000 rpm for 5 min at 4°. Discard the supernatant and resuspend the pellet in 1 ml CHAPS buffer on ice by repeated pipetting until completely in suspension.
5. Prepare two 50% Percoll cushions in two 1.5-ml Eppendorf tubes by mixing 150 μl Percoll with 150 μl CHAPS buffer in each tube. Carefully layer 0.5 ml of the resuspended pellets over each Percoll cushion and centrifuge the tubes at 14,000 rpm for 10 min at 4°.
6. Remove and discard the buffer and interface layers and leave the white band at the bottom of the Percoll cushion. Dilute the white band in the Percoll cushion with 1 ml CHAPS buffer and then centrifuge at 12,000 rpm. Wash the pellet once with CHAPS buffer and then resuspend it in 250 μl CHAPS buffer. Make 50-μl aliquots and freeze at −80° or proceed with the purification of pyroptosomes. SDS-PAGE analysis of a 1-μl aliquot of ASC occlusion bodies from this step should reveal a ≈95% pure ASC (Fig. 13.3A, lane 2).

6.3. Assembly of ASC pyroptosomes from occlusion bodies

1. To a 50-μl aliquot of ASC occlusion bodies from step 6, add 50 μl of 8 M guanidine hydrochloride (GuHCl). Vortex until the solution becomes clear and then centrifuge at 14,000 rpm for 5 min to remove any insoluble material.
2. Transfer the 100-μl supernatant into a fresh 1.5-ml Eppendorf tube, dilute with 100 μl of CHAPS buffer containing 200 mM KCl, and vortex. Leave on ice for 5 min, add an equal volume (200 μl) of the same CHAPS/200 mM KCl buffer, and vortex. Leave on ice for another 5 min, add an equal volume (400 μl) of CHAPS/200 mM KCl buffer, and vortex. Leave on ice for another 5 min and then centrifuge at 14,000 rpm to remove any insoluble material. Transfer the supernatant to a fresh Eppendorf tube. The supernatant volume is now 800 μl and the concentration of GuHCl in the supernatant is 0.5 M.
3. Divide the supernatant in step 2 into eight 100-μl aliquots in fresh 1.5-ml Eppendorf tubes on ice, add 900 μl of CHAPS buffer without KCl gradually to each tube, 300 μl at a time, and vortex. Leave on ice for 5 min after each dilution. This procedure brings the concentration of GuHCl in each sample to 50 mM and the concentration of KCl to ~20 mM. This is important because pyroptosomes will not form in the presence of high concentrations of GuHCl or in the presence of KCl concentrations above 120 mM.
4. Incubate the diluted samples at 37° in a water bath for 30 to 40 min to induce pyroptosome formation.
5. After the incubation period, spin the pyroptosomes at 5000 rpm for 8 min. Discard the supernatants and keep the pellets, which contain the pyroptosomes. Wash the pyroptosomes once in CHAPS buffer

followed by centrifugation at 5000 rpm for 8 min. Spot a small sample of the pellet on a microscope slide and observe the pyroptosomes under a light microscope at 20 or 40× magnification.

6. To purify the pyroptosomes further, resuspend the pellets from the eight tubes in 1 ml CHAPS buffer and then layer the sample over a 50% Percoll cushion in a 1.5-ml Eppendorf tube as described earlier for pyroptosome purification from THP-1 cells. Centrifuge tube at 14,000 rpm for 8 min.

7. Remove and discard the buffer and interface layers. The pyroptosomes will fractionate as a distinct cloudy band within the Percoll cushion. Carefully collect this band in the Percoll cushion, dilute it with 1 ml CHAPS buffer, and then centrifuge at 10,000 rpm. Wash the pellet two times with CHAPS buffer and then resuspend it in 200 μl of CHAPS buffer. Divide into 20-μl aliquots and freeze at $-80°$. SDS-PAGE analysis of a 5-μl aliquot of ASC pyroptosomes from this step should reveal a $\approx$99% pure ASC (Fig. 13.3A, lane 3).

The concentration of the purified ASC pyroptosomes can be determined by fractionating a 5- to 10-μl sample of the purified pyroptosomes on an SDS-polyacrylamide gel followed by Coomassie blue staining as shown in Fig. 13.3A (lane 3). Measure the intensity of the ASC band by densitometry and calculate its concentration in the sample by comparing it to known concentrations of protein standards.

6.4. Assay of the activity of the pyroptosomes

For this assay, a source of procaspase-1 is needed. We do not recommend expressing procaspase-1 in bacteria because the yield of intact, soluble procaspase-1 from bacteria is very low and the protein is mostly processed. We recommend using 293T cells to express procaspase-1 because these cells do not express ASC. We have generated a stable 293T cell line expressing Flag-procaspase-1 (designated 293-caspase-1 cells) at concentrations comparable to those present in THP1 cells (Yu *et al.*, 2006). We find that the presence of the Flag tag at the N terminus of procaspase-1 actually improves its expression level in these cells. The following procedure describes a typical assay of pyroptosome activity using S100 extracts prepared from our stable 293-caspase-1 cell line.

1. Grow 293-caspase-1 cells in tissue culture T175 flasks until they become confluent. Harvest cells by trypsinization and then centrifuge at 1200 rpm to pellet cells. Resuspend pellets in PBS and then aliquot the resuspended cells into 1.5-ml Eppendorf tubes. Pellet cells by centrifugation at 1500 rpm to obtain $\approx$150-μl packed cell pellets and then store pellets at $-80°$ or proceed to step 2.

2. Lyse a 293-caspase-1 cell pellet in 3 volumes of CHAPS buffer on ice by syringing 25×. Centrifuge the lysates at 14,000 rpm for 8 min in an Eppendorf tabletop 5417 R refrigerated microcentrifuge or equivalent at 4° to remove nuclei and organelles. Transfer the supernatants to 1-ml polycarbonate ultracentrifuge tubes (Beckman) and centrifuge at 100,000g in a TLA-120.2 rotor or equivalent in a Beckman Optima TLX benchtop ultracentrifuge for 30 min to obtain S100 supernatants. Remove the S100 supernatants gently so as not to disturb the pellet and transfer to a 1.5-ml Eppendorf tube. The optimal total protein concentration in the extract is ≈10 μg/μl. Use the extract fresh or store as 100-μl aliquots at −80°.

3. To assay the ability of the purified pyroptosomes to activate procaspase-1, prepare 20-μl aliquots of 293-caspase-1 S100 extracts from step 2 in 1.5-ml Eppendorf tubes. Add increasing amounts of purified pyroptosomes (e.g., 0, 20, 40, 100, and 200 ng) to the S100 extracts and incubate at 37° for 60 min.

4. Stop the reaction by adding an equal volume of 2× SDS sample buffer to each reaction tube, boil, and then fractionate on a 12% SDS-polyacrylamide gel. Western blot the gel into a PVDF membrane and then immunoblot with an anti-Flag or anti-caspase-1 antibody to detect processing of caspase-1. If the pyroptosomes are functional, increased processing of caspase-1 with increasing amounts of pyroptosomes should be seen (Fig. 13.3B). No caspase-1 processing should be seen in samples that do not have pyroptosomes.

The pyroptosomes act as a platform to recruit and facilitate procaspase-1 dimerization. This process induces conformational changes in the caspase-1 dimer, resulting in its activation. Pyroptosome-induced caspase-1 processing depends on the catalytic activity of caspase-1. Therefore, no processing can be seen if a catalytically inactive procaspase-1 mutant, such as the active site mutant C285A, is used (Fig. 13.3B, 5th to 8th).

ACKNOWLEDGMENTS

The authors thank Jianghong Wu, Brian Miller, Margaret McCormick, and Wojciech Jankowski for their technical assistance. This work was supported by NIH Grants AG14357 and CA78890 (to E.S.A).

REFERENCES

Agostini, L., Martinon, F., Burns, K., McDermott, M. F., Hawkins, P. N., and Tschopp, J. (2004). NALP3 forms an IL-1beta-processing inflammasome with increased activity in Muckle-Wells autoinflammatory disorder. *Immunity* **20,** 319–325.

Amer, A., Franchi, L., Kanneganti, T. D., Body-Malapel, M., Ozoren, N., Brady, G., Meshinchi, S., Jagirdar, R., Gewirtz, A., Akira, S., and Nunez, G. (2006). Regulation of *Legionella* phagosome maturation and infection through flagellin and host Ipaf. *J. Biol. Chem.* **281**, 35217–35223.

Ayala, J. M., Yamin, T. T., Egger, L. A., Chin, J., Kostura, M. J., and Miller, D. K. (1994). IL-1 beta-converting enzyme is present in monocytic cells as an inactive 45-kDa precursor. *J. Immunol.* **153**, 2592–2599.

Boyden, E. D., and Dietrich, W. F. (2006). Nalp1b controls mouse macrophage susceptibility to anthrax lethal toxin. *Nat. Genet.* **38**, 240–244.

Cookson, B. T., and Brennan, M. A. (2001). Pro-inflammatory programmed cell death. *Trends Microbiol.* **9**, 113–114.

Delbridge, L. M., and O'Riordan, M. X. (2006). Innate recognition of intracellular bacteria. *Curr. Opin. Immunol.* **19**, 10–16.

Dinarello, C. A. (1998). Interleukin-1 beta, interleukin-18, and the interleukin-1 beta converting enzyme. *Ann. N. Y. Acad. Sci.* **856**, 1–11.

Drenth, J. P., and van der Meer, J. W. (2006). The inflammasome: A linebacker of innate defense. *N. Engl. J. Med.* **355**, 730–732.

Faustin, B., Lartigue, L., Bruey, J. M., Luciano, F., Sergienko, E., Bailly-Maitre, B., Volkmann, N., Hanein, D., Rouiller, I., and Reed, J. C. (2007). Reconstituted NALP1 inflammasome reveals two-step mechanism of caspase-1 activation. *Mol. Cell* **25**, 713–724.

Fernandes-Alnemri, T., Wu, J., Yu, J. W., Datta, P., Miller, B., Jankowski, W., Rosenberg, S., Zhang, J., and Alnemri, E. S. (2007). The pyroptosome: A supramolecular assembly of ASC dimers mediating inflammatory cell death via caspase-1 activation. *Cell Death Differ.* **14**, 1590–1604.

Fink, S. L., and Cookson, B. T. (2005). Apoptosis, pyroptosis, and necrosis: Mechanistic description of dead and dying eukaryotic cells. *Infect. Immun.* **73**, 1907–1916.

Fink, S. L., and Cookson, B. T. (2006). Caspase-1-dependent pore formation during pyroptosis leads to osmotic lysis of infected host macrophages. *Cell. Microbiol.* **8**, 1812–1825.

Franchi, L., Amer, A., Body-Malapel, M., Kanneganti, T. D., Ozoren, N., Jagirdar, R., Inohara, N., Vandenabeele, P., Bertin, J., Coyle, A., Grant, E. P., and Nunez, G. (2006). Cytosolic flagellin requires Ipaf for activation of caspase-1 and interleukin 1beta in salmonella-infected macrophages. *Nat. Immunol.* **7**, 576–582.

Kanneganti, T. D., Body-Malapel, M., Amer, A., Park, J. H., Whitfield, J., Taraporewala, Z. F., Miller, D., Patton, J. T., Inohara, N., and Nunez, G. (2006a). Critical role for cryopyrin/Nalp3 in activation of caspase-1 in response to viral infection and double-stranded RNA. *J. Biol. Chem.* **281**, 36560–36568.

Kanneganti, T. D., Ozoren, N., Body-Malapel, M., Amer, A., Park, J. H., Franchi, L., Whitfield, J., Barchet, W., Colonna, M., Vandenabeele, P., Bertin, J., Coyle, A., *et al.* (2006b). Bacterial RNA and small antiviral compounds activate caspase-1 through cryopyrin/Nalp3. *Nature* **440**, 233–236.

Mariathasan, S. (2007). ASC, Ipaf and Cryopyrin/Nalp3: Bona fide intracellular adapters of the caspase-1 inflammasome. *Microbes Infect.* **9**, 664–671.

Mariathasan, S., Newton, K., Monack, D. M., Vucic, D., French, D. M., Lee, W. P., Roose-Girma, M., Erickson, S., and Dixit, V. M. (2004). Differential activation of the inflammasome by caspase-1 adaptors ASC and Ipaf. *Nature* **430**, 213–218.

Mariathasan, S., Weiss, D. S., Newton, K., McBride, J., O'Rourke, K., Roose-Girma, M., Lee, W. P., Weinrauch, Y., Monack, D. M., and Dixit, V. M. (2006). Cryopyrin activates the inflammasome in response to toxins and ATP. *Nature* **440**, 228–232.

Martinon, F., Petrilli, V., Mayor, A., Tardivel, A., and Tschopp, J. (2006). Gout-associated uric acid crystals activate the NALP3 inflammasome. *Nature* **440**, 237–241.

Martinon, F., and Tschopp, J. (2005). NLRs join TLRs as innate sensors of pathogens. *Trends Immunol.* **26,** 447–454.

Miao, E. A., Alpuche-Aranda, C. M., Dors, M., Clark, A. E., Bader, M. W., Miller, S. I., and Aderem, A. (2006). Cytoplasmic flagellin activates caspase-1 and secretion of interleukin 1beta via Ipaf. *Nat. Immunol.* **7,** 569–575.

Ogura, Y., Sutterwala, F. S., and Flavell, R. A. (2006). The inflammasome: First line of the immune response to cell stress. *Cell* **126,** 659–662.

Petrilli, V., Papin, S., and Tschopp, J. (2005). The inflammasome. *Curr. Biol.* **15,** R581.

Poyet, J. L., Srinivasula, S. M., Tnani, M., Razmara, M., Fernandes-Alnemri, T., and Alnemri, E. S. (2001). Identification of Ipaf, a human caspase-1-activating protein related to Apaf-1. *J. Biol. Chem.* **276,** 28309–28313.

Srinivasula, S. M., Poyet, J. L., Razmara, M., Datta, P., Zhang, Z., and Alnemri, E. S. (2002). The PYRIN-CARD protein ASC is an activating adaptor for caspase-1. *J. Biol. Chem.* **277,** 21119–21122.

Sutterwala, F. S., Ogura, Y., Szczepanik, M., Lara-Tejero, M., Lichtenberger, G. S., Grant, E. P., Bertin, J., Coyle, A. J., Galan, J. E., Askenase, P. W., and Flavell, R. A. (2006). Critical role for NALP3/CIAS1/cryopyrin in innate and adaptive immunity through its regulation of caspase-1. *Immunity* **24,** 317–327.

Swanson, M. S., and Molofsky, A. B. (2005). Autophagy and inflammatory cell death, partners of innate immunity. *Autophagy* **1,** 174–176.

Ting, J. P., Kastner, D. L., and Hoffman, H. M. (2006). CATERPILLERs, pyrin and hereditary immunological disorders. *Nat. Rev. Immunol.* **6,** 183–195.

Yamin, T. T., Ayala, J. M., and Miller, D. K. (1996). Activation of the native 45-kDa precursor form of interleukin-1-converting enzyme. *J. Biol. Chem.* **271,** 13273–13282.

Yu, J. W., Fernandes-Alnemri, T., Datta, P., Wu, J., Juliana, C., Solorzano, L., McCormick, M., Zhang, Z., and Alnemri, E. S. (2007). Pyrin activates the ASC pyroptosome in response to engagement by autoinflammatory PSTPIP1 mutants. *Mol. Cell* **28,** 214–227.

Yu, J. W., Wu, J., Zhang, Z., Datta, P., Ibrahimi, I., Taniguchi, S., Sagara, J., Fernandes-Alnemri, T., and Alnemri, E. S. (2006). Cryopyrin and pyrin activate caspase-1, but not NF-kappaB, via ASC oligomerization. *Cell Death Differ.* **13,** 236–249.

CHAPTER FOURTEEN

NUCLEASES IN PROGRAMMED CELL DEATH

Kohki Kawane *and* Shigekazu Nagata

Contents

Abstract

DNA degradation is one of the hallmarks of programmed cell death, or apoptosis. Recent analyses of this process revealed that apoptotic DNA degradation is mediated by two independent mechanisms. First, the caspase-activated DNase (CAD) cell autonomously cleaves DNA into nucleosomal units in dying cells. Then, after the apoptotic cells are engulfed by macrophages, the fragmented DNA is further degraded by DNase II in the lysosomes of the macrophages. This chapter describes assay procedures for CAD and DNase II. It includes biochemical methods for quantifying DNase activity and cell culture systems to follow cell-autonomous and noncell-autonomous DNA degradation. These techniques

Department of Medical Chemistry, Graduate School of Medicine, Kyoto University, Yoshida, Sakyo-ku, Kyoto, Japan

Methods in Enzymology, Volume 442
ISSN 0076-6879, DOI: 10.1016/S0076-6879(08)01414-6

271

are useful for studying DNases that are involved in programmed cell death and for following the engulfment of apoptotic cells by phagocytes.

1. Introduction

Programmed cell death, or apoptosis, is a well-regulated cell death process that is critical for mammalian development and the maintenance of homeostasis in organisms (Jacobson *et al.*, 1997; Vaux and Korsmeyer, 1999). This process consists of two steps, killing and clearing. In the killing step, a cascade of caspases is activated by an intrinsic or extrinsic pathway and kills the cells by cleaving more than 300 cellular proteins (Danial and Korsmeyer, 2004; Fischer *et al.*, 2003). In the clearing step, macrophages and immature dendritic cells engulf the apoptotic cells by recognizing phosphatidylserine as an "eat me" signal, and they degrade all the components of the dead cells into amino acids and nucleotides for reuse (Henson *et al.*, 2001; Savill *et al.*, 2002).

In the killing step, chromosomal DNA is first cleaved into domain-sized fragments (50–100 kB) (Lagarkova *et al.*, 1995; Oberhammer *et al.*, 1993) and then into multimers of nucleosomal units (180 bp) (Wyllie, 1980). This type of DNA degradation is not observed in necrotic cell death and is regarded as one of the hallmarks of apoptotic cell death (Earnshaw, 1995). Digestion of the chromosomal DNA kills the cells quickly (Susin *et al.*, 2000), but this process itself is not required for cells to die (Sakahira *et al.*, 1998). Among the many nucleases proposed for apoptotic DNA degradation, caspase-activated DNase (CAD) was identified as the DNase responsible for apoptotic DNA degradation (Enari *et al.*, 1998). It is expressed ubiquitously, with predominant expression in thymocytes and lymphocytes that show clear apoptotic DNA fragmentation (Mukae *et al.*, 1998). Cells lacking CAD do not undergo apoptotic DNA fragmentation, indicating that CAD is solely responsible for this process (Kawane *et al.*, 2003; Nagase *et al.*, 2003).

The functional mouse CAD contains 344 amino acids and is synthesized with the help of its inhibitor (ICAD, inhibitor of CAD), which works as a specific chaperone for CAD and remains complexed with it (Sakahira and Nagata, 2002; Sakahira *et al.*, 2000). When cells receive apoptotic stimuli (death factors, anticancer drugs, or intrinsic signals), a cascade of caspases is activated, and caspase 3, downstream in the cascade, cleaves ICAD (McIlroy *et al.*, 1999). CAD, thus released from ICAD, forms a dimer that has a scissors-like structure (Woo *et al.*, 2004), attacks chromatin in the nucleus, and cleaves DNA at the spacer regions between nucleosomes, producing multimers of 180 bp. In cells lacking ICAD, CAD is not folded correctly and is degraded quickly (Nagase *et al.*, 2003; Sakahira and Nagata, 2002);

thus *ICAD*$^{-/-}$ cells behave like *CAD*$^{-/-}$ cells (Nagase *et al.*, 2003; Zhang *et al.*, 1998).

In addition to CAD, authentic DNase II, also called DNase IIα, and DNase II-like acid DNase (DLAD or DNase IIβ) are present in mammals (Evans and Aguilera, 2003). Mouse DNase II and DLAD are glycoproteins of 334 and 332 amino acids, respectively (Baker *et al.*, 1998; Shiokawa and Tanuma, 1999). Both are synthesized as a precursor carrying a signal sequence, transported into lysosomes, and exhibit DNase activity under acidic conditions (Nakahara *et al.*, 2007; Shiokawa and Tanuma, 1999). In the lysosomes, which also contain proteases and glycosidases, DNase II digests the DNA into nucleotides, destroying the chromatin structure. DNase II is expressed predominantly in macrophages and is responsible for degrading the DNA of engulfed apoptotic cells or nuclei expelled from erythroid precursors (Kawane *et al.*, 2001, 2003). *DNase II*$^{-/}$ mice contain many abnormal macrophages carrying undigested DNA throughout their bodies. These macrophages are activated to produce interferon β and tumor necrosis factor, leading to severe anemia or polyarthritis in these mice (Kawane *et al.*, 2006; Yoshida *et al.*, 2005). DLAD is specifically expressed in the fiber cells of the eye and is responsible for degrading the nuclear DNA during lens cell differentiation (Nishimoto *et al.*, 2003). *DLAD*$^{-/-}$ mice suffer from cataracts because the undegraded nuclear DNA accumulates in the lens fiber cells and blocks the light path.

The following sections describe biochemical methods for assaying CAD and DNase II. In addition, we describe an assay for the engulfment of apoptotic cells by macrophages using *CAD*$^{-/-}$ cells as prey.

2. Assays for CAD

CAD is a DNase with high specific activity. One nanogram of CAD protein is sufficient for detectable enzymatic activity (Sakahira *et al.*, 2001). Active CAD is heat labile, particularly at low protein concentrations, whereas ICAD and the ICAD–CAD complex are heat stable (Enari *et al.*, 1998). The latent form of CAD (CAD–ICAD complex) is present in healthy cells, and its DNase activity can be activated in cell lysates by treating with caspase 3 (Enari *et al.*, 1998; Halenbeck *et al.*, 1998). Apoptotic cells express active CAD, which can be distinguished from other DNases by its sensitivity to the inhibitory action of ICAD (Sakahira *et al.*, 1998). The cleavage of DNA by CAD produces fragments with predominantly flush ends carrying the 3'-hydroxyl group (Widlak *et al.*, 2001), which is a good substrate for terminal transferase. CAD requires Mg^{2+} but not Ca^{2+} for its DNase activity and works at neutral pH. CAD can be assayed by using naked plasmid DNA or isolated nuclei as a substrate, followed by agarose gel electrophoresis (Enari *et al.*, 1998). The digestion of nuclei with CAD produces a DNA ladder consisting of multimers of 180 bp (Fig. 14.1A), whereas the digestion of plasmid DNA with CAD yields fragmented DNA that appears as a smeary band on an agarose gel (Fig. 14.1B). Electrophoresis of the products on agarose gel is a simple and convenient way to detect the DNase activity of CAD, but it is not quantitative. To determine the DNase activity of CAD quantitatively, the number of $3'$ hydroxyl ends generated by the action of CAD can be determined by the incorporation of $[\alpha\text{-}^{32}P]$ ddATP by terminal transferase (Sakahira *et al.*, 2001)(Fig. 14.1C).

2.1. Materials and buffer

1. Plasmid DNA: any kind of plasmid DNA, preferably linearized DNA
2. Nuclei can be prepared from various tissues, including mouse liver (Enari *et al.*, 1995; Hughes and Cidlowski, 1997; Newmeyer and Wilson, 1991) and HeLa cells (Hughes and Cidlowski, 1997)
3. Caspase 3: recombinant caspase 3 can be produced in *Escherichia coli* as described (Kamens *et al.*, 1995). It is also available commercially from various companies (GE Healthcare Bioscience).
4. Terminal transferase from various companies (Takara Shuzo, Japan; Roche)
5. $[\alpha\text{-}^{32}P]$ddATP (specific activity, 0.1 μCi/pmol) from GE Healthcare
6. Buffer A ($2\times$): 20 mM HEPES-KOH (pH 7.2), 4 mM MgCl$_2$, 10 mM EGTA, 100 mM NaCl, 40%(v/v) glycerol, 20 mM dithiothreitol (DTT), 2 mg/ml bovine serum albumin (BSA), 2 mM (p-amidinophenyl)methanesulfonyl fluoride hydrochloride (p-APMSF)
7. TdT buffer ($5\times$): 125 mM Tris-HCl (pH 6.6), 12.5 mM CoCl$_2$, 1.0 M potassium cacodylate, and 1.25 mg/ml BSA

2.2. Preparation of CAD

2.2.1. Crude cell lysate

Cell lysates are prepared by the repeated freezing and thawing of cells treated with apoptotic stimuli such as Fas ligand or anticancer drugs as described (Enari *et al.*, 1995; Martin *et al.*, 1995). Lymphoid cells (mouse thymocytes, lymph node cells, or lymphoid cell lines) are better sources of CAD than fibroblasts or epithelial cells (mouse embryonic fibroblasts, hepatocytes, or HeLa cells) (Mukae *et al.*, 1998; Nagase *et al.*, 2003; Woo *et al.*, 2004).

2.2.2. Recombinant CAD

Recombinant CAD can be produced on a large scale in *E. coli* (Woo *et al.*, 2004), mammalian cells (293T or COS) (Enari *et al.*, 1998), or insect cells (Sf9) (Sakahira *et al.*, 1999) together with ICAD. It is recovered as a complex with ICAD. To prepare ICAD-free CAD, the CAD–ICAD complex is treated with caspase 3, and the CAD is separated from the cleaved ICAD by chromatography on an ion-exchange column. Recombinant CAD can also be synthesized in a cell-free system using an *in vitro* transcription–translation-coupled system supplemented with recombinant ICAD (Enari *et al.*, 1998).

2.3. Protocol for the CAD assay with agarose gel electrophoresis

1. Combine in a 1.5-ml microcentrifuge tube:
    a. 10 µl of twice concentrated buffer A
    b. 1.0 µl of 1 µg/ml plasmid DNA or 2×10^8/ml nuclei

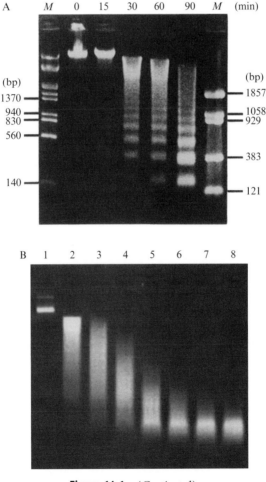

Figure 14.1 (*Continued*).

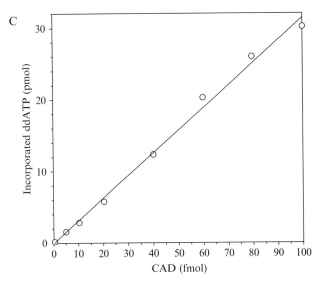

Figure 14.1 The DNase (CAD) activated in apoptosis. (A) DNA fragmentation using extracts from apoptotic cells. Mouse WR19L cells expressing Fas receptor (5×10^7 cells) were induced to undergo apoptosis by Fas engagement at 37° for the indicated periods, and cell extracts were prepared. Rat liver nuclei were incubated with the cell extracts (120 μg protein) at 37° for 60 min, and the chromosomal DNA was analyzed by agarose gel electrophoresis. Standard DNA fragments (lane M) are shown in base pairs. Reprinted from Enari *et al.* (1995). (B) DNA degradation by purified CAD. Plasmid DNA (1.0 μg) was incubated at 30° for 30 min with various amounts of CAD (lanes 1–8: 0, 5, 10, 20, 40, 60, 80, and 100 fmol), and the DNA was fractionated by electrophoresis on an agarose gel. Reprinted from Sakahira *et al.* (2001). (C) Quantitative assay for CAD. Plasmid DNA (1 μg) was incubated with the indicated amounts of CAD, and the number of generated free 3′-hydroxyl groups on DNA fragments was determined by measuring the incorporation of [α-^{32}P]ddATP in the presence of terminal transferase. Reprinted from Sakahira *et al.* (2001).

    c. Cell lysate (50–200 μg) from healthy or apoptotic cells, the purified ICAD–CAD complex (1.0–10 ng), or purified CAD (0.2–4.0 ng)

    d. 1.0 μl of 150 μg/ml recombinant caspase 3

    e. H$_2$O to a final volume of 20 μl

      Note: If the cell lysate from apoptotic cells or active CAD is used, caspase 3 is not necessary. To protect CAD from inactivation during the assay, the reaction mixture (buffer A) contains 20% glycerol. Because CAD contains 14 free thiol groups essential for its activity, the reaction mixture also contains a high concentration of DTT (10 mM) (Sakahira *et al.*, 2000).

2. Incubate for 0.5 to 2 h at 30°. *Note*: Because CAD is heat labile, we perform the assay at 30°.

3. Stop the reaction by adding 30 μl of 10 mM Tris-HCl (pH 8.0) buffer containing 1 mM EDTA.
4. Collect nuclei by centrifugation for 5 min at 10,000 rpm, resuspend them in 500 μl of 100 mM Tris-HCl (pH 8.5), 5 mM EDTA, 0.2 M NaCl, 0.2% SDS, and 0.2 mg/ml proteinase K, and incubate them at 37° for several hours.
5. Extract the DNA with an equal volume of phenol/chloroform mixture and recover it by ethanol precipitation.
6. Dissolve DNA in 20 μl of 10 mM Tris-HCl (pH 8.0) buffer containing 1 mM EDTA and 0.1 mg/ml RNase A and incubate it at 37° for 30 min.
7. Subject the recovered DNA (20–50%) to electrophoresis on a 1.5% agarose gel in Tris-borate-EDTA buffer in the presence of 0.5 μg/ml ethidium bromide.

Note: If plasmid DNA is used as the substrate, skip step 4.

2.4. Protocol for quantitative CAD assay

1. Incubate 1.0 μg of plasmid DNA with CAD prepared as described earlier (steps 1 and 2) in a final volume of 20 μl.
2. Stop the reaction by adding 1.0 μl of 200 mM Tris-HCl buffer (pH 7.5) containing 50 mM EDTA and inactivate the enzyme by heating the mixture for 15 min at 65°.
3. Remove aggregates by centrifugation at 15,000 rpm for 15 min at 4°.
4. Combine in a 1.5-ml microcentrifuge tube:
   a. 1.0 μl of the supernatant from step 3 containing the CAD-treated DNA fragments
   b. 1.0 μl of 12,500 units/ml terminal deoxynucleotide transferase
   c. 1.0 μl of 1.0 mCi/ml [α-^{32}P]ddATP
   d. 5.0 μl of five times concentrated TdT buffer
   e. H$_2$O to a final volume of 25 μl
5. Incubate the reaction mixture at 37° for 1 h.
6. Stop the reaction by adding 1.0 μl of 500 mM EDTA and heat the mixture at 65° for 10 min.
7. Add 1.0 ml of 5% (w/v) trichloroacetic acid (TCA) containing 50 μg/ml BSA and 50 μM ATP and leave the mixture on ice for 30 min to precipitate the DNA.
8. Using a filtration device connected to a water vacuum pump, apply the mixture to filters (Protran BA85, Whatman) that have been presoaked in 5% TCA.
9. Wash the filter twice with 5% TCA.
10. Measure the radioactivity retained on the filter using a scintillation counter.

3. ASSAY FOR DNASE II

The optimal pH for the DNase activity of DNase II is 4.8, whereas that for DLAD is 5.6 (Nakahara *et al.*, 2007). Neither DNase II nor DLAD requires a metal ion (Ca^{2+} or Mg^{2+}) for its DNase activity. They cleave DNA into fragments that carry 3′-phosphorylated ends, and the products are analyzed by electrophoresis on an agarose gel. Cell extracts prepared from *DNase II$^{-/-}$* tissues do not exhibit any DNase activity under acidic conditions (Fig. 14.2), indicating that DNase II is the only acid DNase present in lysosomes.

3.1. Materials and buffer

1. Plasmid DNA: Any kind of plasmid DNA, preferably linearized
2. Acid assay buffer (5×): 250 mM MOPS-NaOH buffer (pH 4.8) containing 50 mM EDTA for DNase II, and 250 mM MOPS-NaOH buffer (pH 5.9) containing 50 mM EDTA for DLAD

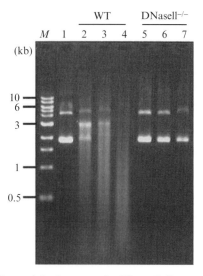

Figure 14.2 Acid DNase activity in mouse fetal liver. Cell extracts were prepared from the fetal liver of wild-type (lanes 2–4) or *DNase II$^{-/-}$* (lanes 5–7) embryos. Plasmid DNA (2 μg) was incubated at 37° for 60 min with the cell extracts (lanes 2 and 5, 6.0 μg; lanes 3 and 6, 12.0 μg; and lanes 4 and 7, 24.0 μg protein). After incubation, DNA was extracted and analyzed by agarose gel electrophoresis. Plasmid DNA incubated with buffer alone is shown in lane 1. Lane M shows DNA size markers, in kilobases. Reprinted with permission from Kawane *et al.* (2001).

3.2. Preparation of DNase II

3.2.1. Natural DNase II and DLAD

DNase II can be prepared from various tissues, including fetal liver, spleen, and bone marrow (Evans and Aguilera, 2003). In contrast, DLAD is expressed exclusively in the eye lens (Nishimoto *et al.*, 2003) and can be prepared from the eyes of newborn mice (Nakahara *et al.*, 2007). To extract DNase II and DLAD from lysosomes, tissues are homogenized in a high salt buffer (10 mM Tris-HCl, pH 7.5, 1.0 M NaCl, and 1.0 mM p-APMSF) (Kawane *et al.*, 2003). The homogenates are dialyzed against 10 mM Tris-HCl (pH 7.5) containing 1 mM EDTA.

3.2.2. Recombinant DNase II and DLAD

Recombinant DNase II and DLAD can be produced on a large scale from mammalian cells (293T or COS cells). Most of the DNase II expressed in 293T cells is secreted into the medium, whereas DLAD is retained in the cells (Nakahara *et al.*, 2007). DNase II and DLAD can be FLAG tagged at the C terminus and the recombinant proteins purified from the culture supernatant or cell lysates using an anti-FLAG mAb.

3.3. Protocol for the DNase II or DLAD assay

1. Combine in a 1.5-ml microcentrifuge tube:
   a. 60 μl of five times concentrated acid assay buffer
   b. 2.0 μl of 1.0 mg/ml plasmid DNA
   c. Cell lysates (10–300 μg) in less than 40 μl or purified DNase II or DLAD (5–100 ng)
   d. H$_2$O to a final volume of 300 μl
2. Incubate at 37° for 1 to 3 h.
3. Extract the DNA by successive treatment with an equal volume of phenol and then with the phenol/chloroform mixture.
4. Recover DNA by ethanol precipitation with 20 μg of glycogen as carrier.
5. Dissolve DNA in 50 μl of 10 mM Tris-HCl buffer (pH 7.5) containing 1.0 mM EDTA and 100 μg/ml RNase A and incubate at 37° for 20 min.
6. Use 12-μl aliquots for electrophoresis on a 1.5% agarose gel in Tris-borate-EDTA buffer.

4. ASSAYS FOR CELL-AUTONOMOUS APOPTOTIC DNA DEGRADATION

Chromatin attaches to the nuclear scaffold to form a series of 30- to 50-kb loops. During apoptotic cell death, the chromosomal DNA is first cleaved into large fragments (50–200 kB) and then is degraded into

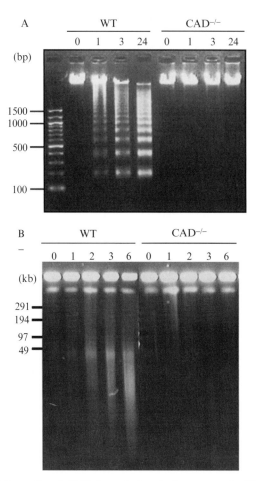

Figure 14.3 CAD-mediated DNA degradation during apoptosis. (A) DNA degradation into nucleosomal units. Thymocytes from wild-type (WT) or $CAD^{-/-}$ mice were incubated at 37° for the indicated periods of time with 10 μM dexamethasone, and their chromosomal DNA was analyzed by agarose gel electrophoresis. (B) Cleavage of chromosomal DNA into high molecular weight DNA. Thymocytes from WT or $CAD^{-/-}$ mice were incubated at 37° with 10 μM dexamethasone, encapsulated in a block of agarose, treated with proteinase K, and analyzed by pulsed-field gel electrophoresis. Reprinted from Kawane et al. (2003).

nucleosomal units (Lagarkova et al., 1995; Oberhammer et al., 1993). Cells lacking CAD show neither large-scale nor nucleosomal DNA degradation upon exposure to apoptotic stimuli (Kawane et al., 2003; Sakahira et al., 2001)(Fig. 14.3). Because the scaffold attachment regions are more sensitive to nuclease (Khodarev et al., 2000), the activated CAD first cleaves the DNA at these points, producing "large-scale chromatin fragmentation."

This unpacks the chromatin structure and allows CAD to gain access to the nucleosomes to cause nucleosomal DNA fragmentation.

DNA is sensitive to mechanical stress-induced shearing. For analysis of large-scale DNA degradation by pulsed-field electrophoresis, cells are first encapsulated in a block of agarose gel and treated *in situ* with protease to avoid artificial cleavage of the DNA during its preparation (Okada *et al.*, 2005). To analyze nucleosomal DNA fragmentation, the chromosomal DNA is isolated from the cells and separated by conventional agarose gel electrophoresis. DNA degradation can be analyzed using primary cells (thymocytes, lymphocytes, and embryonic fibroblasts) and cell lines that have been treated with various apoptotic stimuli. However, the apoptotic DNA fragmentation is seen most clearly with thymocytes and lymphocytes. Neither large-scale nor nucleosmal DNA degradation occurs in $CAD^{-/-}$ cells (Fig. 14.3), indicating that CAD is responsible for both processes.

4.1. Assay for large-scale DNA degradation

4.1.1. Materials and buffer

1. L buffer: 10 mM Tris-HCl buffer (pH 8.0) containing 20 mM NaCl and 100 mM EDTA
2. Proteinase K solution: 1 mg/ml proteinase K in L buffer containing 1% N-lauroylsarcosine
3. Low melting point agarose: ultrapure low melting point agarose from Invitrogen
4. Plug mold: disposable plug molds from GE Healthcare

4.1.2. Protocol

1. Wash cells (2×10^6) with phosphate-buffered saline (PBS) and suspend them in 100 μl of L buffer at room temperature.
2. To the cell suspension, add 100 μl of 1.6% low melting point agarose in L buffer that has been prewarmed to 42° and mix well.
3. Apply 100 μl of the mixture to the plug mold and leave it at 4° for 15 min.
4. Soak the gels in 2 ml of proteinase K solution and incubate twice at 50° for 24 h with gentle shaking, with a change to fresh proteinase K solution between the two incubations.
5. Wash the gels four times at room temperature with 2 ml of 10 mM Tris-HCl buffer (pH 8.0) containing 1 mM EDTA with gentle shaking.
6. Cut the gels in half and insert the pieces into the wells of a 1% agarose gel in twice diluted Tris-borate-EDTA buffer.
   Note: Set the sample block tightly against the bottom surface and the surface directed toward the gel. Fill the well with twice diluted Tris-borate-EDTA buffer.

7. Place the gel in a contour-clamped homogeneous electric field pulsed field system (GE Healthcare) and run the gel for 8 h under thermostatically controlled conditions (14°) using the parameters of 6.0 V/cm with a pulse ramp of 0.06 to 26.29 s.
8. Stain the gels with 0.5 μg/ml ethidium bromide and photograph them under ultraviolet illumination.

4.2. Assay for DNA degradation into nucleosomal units

4.2.1. Buffer

Lysis buffer: 100 mM Tris-HCl buffer (pH 8.5), 5 mM EDTA, 200 mM NaCl, 0.2% SDS, and 200 μg/ml proteinase K

4.2.2. Protocol

1. Wash the cells (5×10^5 cells) with PBS in a 1.5-ml microcentrifuge tube and suspend them in 0.2 ml of lysis buffer.
2. Incubate at 37° for 3 to 12 h. Add 20 μg of glycogen as carrier and precipitate the DNA with 0.2 ml of isopropanol.
3. Recover the DNA by centrifugation at 15,000 rpm for 15 min and dissolve it in 12 μl of 10 mM Tris-HCl buffer (pH 7.4) containing 1 mM EDTA and 100 μg/ml RNase A.
4. Incubate the mixture at 37° for 20 min.
5. Using aliquots (1.0–3.0 μl), run the DNA on 1.5% agarose gel in Tris-borate-EDTA buffer containing 0.5 μg/ml ethidium bromide.

5. ASSAY FOR NONCELL-AUTONOMOUS DNA DEGRADATION IN MACROPHAGES

$CAD^{-/-}$ mice do not show severe phenotype (Kawane et al., 2003) because the DNA of apoptotic cells is digested by DNase II in macrophages after they are engulfed. In definitive erythropoiesis, nuclei are expelled from erythroid cell precursors and are engulfed by macrophages present in the center of the erythroblastic islands. Their DNA is also degraded by the DNase II in macrophages (Kawane et al., 2001; Krieser et al., 2002). Thus, noncell-autonomous DNA degradation by macrophages can be assayed by incubating $CAD^{-/-}$ apoptotic cells or erythroid cell nuclei with macrophages. That is, the DNA of the $CAD^{-/-}$ apoptotic cells is cleaved in macrophages, which is detectable by TUNEL staining followed by flow cytometry (Fig.14.4A). If macrophages from $DNase II^{-/-}$ mice are used as

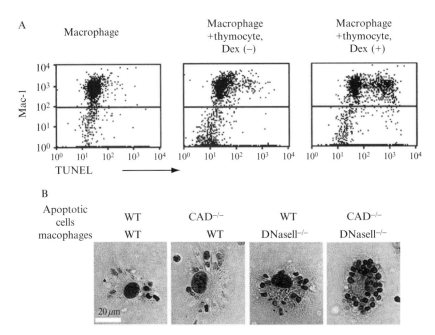

Figure 14.4 Degradation of the DNA of apoptotic cells in macrophages. (A) TUNEL assay of macrophages, showing DNA degradation within them. Thymocytes from CAD-deficient mice were treated with 10 μM dexamethasone (Dex). Thioglycollate-elicited mouse peritoneal macrophages were incubated at 37° for 90 min with healthy [Dex (-)] or apoptotic [Dex (+)] thymocytes. Adherent cells were stained with PE-conjugated Mac-1, followed by TUNEL staining, and analyzed by flow cytometry. The FACS profile of macrophages incubated without thymocytes is also shown. Reprinted from Hanayama *et al.* (2002). (B) Accumulation of undigested DNA in *DNase II*$^{-/-}$ macrophages, revealed by Feulgen staining. Macrophages were prepared from wild-type or *DNase II*$^{-/-}$ embryos. Thymocytes from wild-type or *CAD*$^{-/-}$ mice were induced to undergo apoptosis and were then added to wild-type or *DNase II*$^{-/-}$ macrophages. After incubation at 37° for 3 h, adherent cells were stained with Feulgen. If apoptotic *CAD*$^{-/-}$ thymocytes are used as prey for *DNase II*$^{-/-}$ macrophages, undigested DNA accumulate in lysosomes of macrophages. Reprinted from Kawane *et al.* (2003).

phagocytes, the macrophages accumulate a large amount of undegraded DNA, which can be stained by Feulgen or DAPI (Kawane *et al.*, 2003) (Fig. 14.4B). This system, noncell-autonomous DNA degradation of apoptotic cells in macrophages, has been used successfully to assay the engulfment of apoptotic cells (Hanayama *et al.*, 2002; Miyanishi *et al.*, 2007).

5.1. Materials and buffer

1. Monoclonal antibodies: rat anti-mouse FcγIII/II receptor and phycoerythrin (PE)-conjugated rat antimouse Mac-1 from BD PharMingen

2. Terminal deoxynucleotidyl transferase from Takara Shuzo, Kyoto
3. Fluorescein-12-labeled dUTP from Roche Diagnostics

5.1.1. Apoptotic cells

We prepare thymocytes from $CAD^{-/-}$ mice at the age of 4 to 10 weeks. Thymocytes are treated with 10 μM dexamethasone or Fas ligand for 4 h to induce apoptosis in about 60% of the cells, as judged by annexin V staining.

Note: We use dexamethasone or Fas ligand to induce apoptosis because these reagents induce apoptosis quickly and synchronously in thymocytes. Do not use cell preparations that contain more than 10% propidium iodide-positive necrotic cells.

5.1.2. Macrophages

Thioglycollate-elicited peritoneal macrophages and fetal thymic macrophages efficiently engulf apoptotic cells. To prepare thioglycollate-elicited peritoneal macrophages, 3% (w/v) Brewer's thioglycollate (Sigma) is injected intraperitoneally into C57BL/6 mice (12 weeks old), and peritoneal cells are collected 4 days after the injection, as described (Hanayama et al., 2002). Thymic macrophages are prepared from fetal thymus by collagenase and DNase I treatment as described (Kawane et al., 2003; Platt et al., 1996). Adherent cells are cultured for 3 weeks with macrophage colony-stimulating factor and used as thymic macrophages.

5.2. Assay by FACS

Protocol

1. Grow 2.5×10^5 macrophages in 48-well cell culture plates overnight at 37° in Dulbecco's modified Eagle's medium (DMEM) containing 10% fetal calf serum (FCS).
2. Add 1 to 2×10^6 apoptotic $CAD^{-/-}$ thymocytes to the wells and incubate them at 37° for 90 min to allow them to be engulfed.
3. Wash the macrophages with PBS, and detach the cells from the plate by incubating them at 37° in PBS containing 1 mM EDTA.
4. Incubate the macrophages at 4° for 10 min with 2.5 μg/ml rat antimouse FcγIII/II receptor in FACS staining buffer (PBS containing 2% FCS and 0.02% NaN$_3$).
5. Incubate the macrophages at 4° for 30 min with 4 μg/ml PE-rat anti-mouse Mac-1, fix with 1% paraformaldehyde in PBS, and incubate for 10 min with PBS containing 0.1% Triton X-100.
6. Suspend the cells in 50 μl of 100 mM cacodylate buffer (pH 7.2) containing 1 mM CoCl$_2$ and 0.01% BSA and incubate at 37° for 45 min with 100 units/ml terminal deoxynucleotidyl transferase and 2.5 μM fluorescein-12-labeled dUTP.

7. Analyze the cells by flow cytometry using a FACS Calibur (Becton-Dickinson).

5.3. Assay by feulgen staining and microscopic observation

Protocol

1. Grow 0.5 to 1.0×10^5 macrophage overnight at 37° in DMEM containing 10% FCS in eight-well Lab-Tek II chamber slides (Nalge Nunc) coated with 0.1% gelatin.
2. Add apoptotic thymocytes ($1.0–1.5 \times 10^6$ cells) to the macrophages and incubate at 37° for 1 to 3 h.
3. Wash the cells by dipping the slides into PBS, and fix the cells overnight at 4° with 4% formaldehyde in 85% methanol and 5% acetic acid.
4. Subject the cells to Feulgen staining using Schiff's solution (Merck).
5. Mount the slides with FluorSave (Calbiochem) and observe by light microscopy.

REFERENCES

Baker, K. P., Baron, W. F., Henzel, W. J., and Spencer, S. A. (1998). Molecular cloning and characterization of human and murine DNase II. *Gene* **215,** 281–289.

Danial, N. N., and Korsmeyer, S. J. (2004). Cell death: Critical control points. *Cell* **116,** 205–219.

Earnshaw, W. C. (1995). Nuclear changes in apoptosis. *Curr. Biol.* **7,** 337–343.

Enari, M., Hase, A., and Nagata, S. (1995). Apoptosis by a cytosolic extract from Fas-activated cells. *EMBO J.* **14,** 5201–5208.

Enari, M., Sakahira, H., Yokoyama, H., Okawa, H., Iwamatsu, A., and Nagata, S. (1998). A caspase-activated DNase that degrades DNA during apoptosis, and its inhibitor ICAD. *Nature* **391,** 43–50.

Evans, C. J., and Aguilera, R. J. (2003). DNase II: Genes, enzymes and function. *Gene* **322,** 1–15.

Fischer, U., Janicke, R. U., and Schulze-Osthoff, K. (2003). Many cuts to ruin: A comprehensive update of caspase substrates. *Cell Death Differ.* **10,** 76–100.

Halenbeck, R., MacDonald, H., Roulston, A., Chen, T. T., Conroy, L., and Williams, L. T. (1998). CPAN, a human nuclease regulated by the caspase-sensitive inhibitor DFF45. *Curr. Biol.* **8,** 537–540.

Hanayama, R., Tanaka, M., Miwa, K., Shinohara, A., Iwamatsu, A., and Nagata, S. (2002). Identification of a factor that links apoptotic cells to phagocytes. *Nature* **417,** 182–187.

Henson, P. M., Bratton, D. L., and Fadok, V. A. (2001). The phosphatidylserine receptor: A crucial molecular switch? *Nat. Rev. Mol. Cell Biol.* **2,** 627–633.

Hughes, F. M., Jr., and Cidlowski, J. A. (1997). Utilization of an *in vitro* assay to evaluate chromatin degradation by candidate apoptotic nucleases. *Cell Death Differ.* **4,** 200–208.

Jacobson, M. D., Weil, M., and Raff, M. C. (1997). Programmed cell death in animal development. *Cell* **88,** 347–354.

Kamens, J., Paskind, M., Hugunin, M., Talanian, R., Allen, H., Banach, D., Bump, N., Hackett, M., Johnston, C., Li, P., Mankovich, J., Terranova, M., and Ghayur, T. (1995).

Identification and characterization of ICH-2, a novel member of the interleukin-1β-converting enzyme family of cysteine proteases. *J. Biol. Chem.* **270,** 15250–15256.

Kawane, K., Fukuyama, H., Kondoh, G., Takeda, J., Ohsawa, Y., Uchiyama, Y., and Nagata, S. (2001). Requirement of DNase II for definitive erythropoiesis in the mouse fetal liver. *Science* **292,** 1546–1549.

Kawane, K., Fukuyama, H., Yoshida, H., Nagase, H., Ohsawa, Y., Uchiyama, Y., Iida, T., Okada, K., and Nagata, S. (2003). Impaired thymic development in mouse embryos deficient in apoptotic DNA degradation. *Nat. Immunol.* **4,** 138–144.

Kawane, K., Ohtani, M., Miwa, K., Kizawa, T., Kanbara, Y., Yoshioka, Y., Yoshikawa, H., and Nagata, S. (2006). Chronic polyarthritis caused by mammalian DNA that escapes from degradation in macrophages. *Nature* **443,** 998–1002.

Khodarev, N. N., Bennett, T., Shearing, N., Sokolova, I., Koudelik, J., Walter, S., Villalobos, M., and Vaughan, A. T. (2000). LINE L1 retrotransposable element is targeted during the initial stages of apoptotic DNA fragmentation. *J. Cell. Biochem.* **79,** 486–495.

Krieser, R. J., MacLea, K. S., Longnecker, D. S., Fields, J. L., Fiering, S., and Eastman, A. (2002). Deoxyribonuclease IIa is required during the phagocytic phase of apoptosis and its loss causes lethality. *Cell Death Differ.* **9,** 956–962.

Lagarkova, M. A., Iarovaia, O. V., and Razin, S. V. (1995). Large-scale fragmentation of mammalian DNA in the course of apoptosis proceeds via excision of chromosomal DNA loops and their oligomers. *J. Biol. Chem.* **270,** 20239–20241.

Martin, S. J., Newmeyer, D. D., Mathias, S., Farschon, D. M., Wang, H.-G., Reed, J. C., Kolesnik, R. N., and Green, D. R. (1995). Cell-free reconstitution of Fas-, UV radiation- and ceramide-induced apoptosis. *EMBO J.* **14,** 5191–5200.

McIlroy, D., Sakahira, H., Talanian, R. V., and Nagata, S. (1999). Involvement of caspase 3-activated DNase in internucleosomal DNA cleavage induced by diverse apoptotic stimuli. *Oncogene* **18,** 4401–4408.

Miyanishi, M., Tada, K., Koike, M., Uchiyama, Y., Kitamura, T., and Nagata, S. (2007). Identification of Tim4 as a phosphatidylserine receptor. *Nature* **450,** 435–439.

Mukae, N., Enari, M., Sakahira, H., Fukuda, Y., Inazawa, J., Toh, H., and Nagata, S. (1998). Molecular cloning and characterization of human caspase-activated DNase. *Proc. Natl. Acad. Sci. USA* **95,** 9123–9128.

Nagase, H., Fukuyama, H., Tanaka, M., Kawane, K., and Nagata, S. (2003). Mutually regulated expression of caspase-activated DNase and its inhibitor for apoptotic DNA fragmentation. *Cell Death Differ.* **10,** 142–143.

Nakahara, M., Nagasaka, A., Koike, M., Uchida, K., Kawane, K., Uchiyama, Y., and Nagata, S. (2007). Degradation of nuclear DNA by DNase II-like acid DNase in cortical fiber cells of mouse eye lens. *FEBS J.* **274,** 3055–3064.

Newmeyer, D. D., and Wilson, K. L. (1991). Egg extracts for nuclear import and nuclear assembly reactions. *Methods Cell Biol.* **36,** 607–634.

Nishimoto, S., Kawane, K., Watanabe-Fukunaga, R., Fukuyama, H., Ohsawa, Y., Uchiyama, Y., Hashida, N., Ohguro, N., Tano, Y., Morimoto, T., Fukuda, Y., and Nagata, S. (2003). Nuclear cataract caused by a lack of DNA degradation in the mouse eye lens. *Nature* **424,** 1071–1074.

Oberhammer, F., Wilson, J. W., Dive, C., Morris, I. D., Hickman, J. A., Wakeling, A. E., Walker, P. R., and Sikorska, M. (1993). Apoptotic death in epithelial cells: Cleavage of DNA to 300 and/or 50 kb fragments prior to or in the absence of internucleosomal fragmentation. *EMBO J.* **12,** 3679–3684.

Okada, K., Iida, T., Kita-Tsukamoto, K., and Honda, T. (2005). Vibrios commonly possess two chromosomes. *J. Bacteriol.* **187,** 752–757.

Platt, N., Suzuki, H., Kurihara, Y., Kodama, T., and Gordon, S. (1996). Role for the class A macrophage scavenger receptor in the phagocytosis of apoptotic thymocytes *in vitro. Proc. Natl. Acad. Sci. USA* **93,** 12456–12460.

Sakahira, H., Enari, M., and Nagata, S. (1998). Cleavage of CAD inhibitor in CAD activation and DNA degradation during apoptosis. *Nature* **391**, 96–99.

Sakahira, H., Enari, M., and Nagata, S. (1999). Functional differences of two forms of the inhibitor of caspase-activated DNase, ICAD-L, and ICAD-S. *J. Biol. Chem.* **274**, 15740–15744.

Sakahira, H., Iwamatsu, A., and Nagata, S. (2000). Specific chaperone-like activity of inhibitor of caspase-activated DNase for caspase-activated DNase. *J. Biol. Chem.* **275**, 8091–8096.

Sakahira, H., and Nagata, S. (2002). Co-translational folding of caspase-activated DNase with Hsp70, Hsp40 and inhibitor of caspase-activated DNase. *J. Biol. Chem.* **277**, 3364–3370.

Sakahira, H., Takemura, Y., and Nagata, S. (2001). Enzymatic active site of caspase-activated DNase (CAD) and its inhibition by inhibitor of CAD. *Arch. Biochem. Biophys.* **388**, 91–99.

Savill, J., Dransfield, I., Gregory, C., and Haslett, C. (2002). A blast from the past: Clearance of apoptotic cells regulates immune responses. *Nat. Rev. Immunol.* **2**, 965–975.

Shiokawa, D., and Tanuma, S. (1999). DLAD, a novel mammalian divalent cation-independent endonuclease with homology to DNase II. *Nucleic Acids Res.* **27**, 4083–4089.

Susin, S. A., Daugas, E., Ravagnan, L., Samejima, K., Zamzami, N., Loeffler, M., Costantini, P., Ferri, K. F., Irinopoulou, T., Prevost, M. C., Brothers, G., Mak, T. W., *et al.* (2000). Two distinct pathways leading to nuclear apoptosis. *J. Exp. Med.* **192**, 571–580.

Vaux, D. L., and Korsmeyer, S. J. (1999). Cell death in development. *Cell* **96**, 245–254.

Widlak, P., Li, L. Y., Wang, X., and Garrard, W. T. (2001). Action of recombinant human apoptotic endonuclease G on naked DNA and chromatin substrates: Cooperation with exonuclease and DNase I. *J. Biol. Chem.* **276**, 48404–48409.

Woo, E.-J., Kim, Y.-G., Kim, M.-S., Han, W.-D., Shin, S., Robinson, H., Park, S.-Y., and Oh, B.-H. (2004). Structural mechanism for inactivation and activation of CAD/DFF40 in the apoptotic pathway. *Mol. Cell* **14**, 531–539.

Wyllie, A. H. (1980). Glucocorticoid-induced thymocyte apoptosis is associated with endogenous endonuclease activation. *Nature* **284**, 555–556.

Yoshida, H., Okabe, Y., Kawane, K., Fukuyama, H., and Nagata, S. (2005). Lethal anemia caused by interferon-beta produced in mouse embryos carrying undigested DNA. *Nat. Immunol.* **6**, 49–56.

Zhang, J., Liu, X., Scherer, D. C., van Kaer, L., Wang, X., and Xu, M. (1998). Resistance to DNA fragmentation and chromatin condensation in mice lacking the DNA fragmentation factor 45. *Proc. Natl. Acad. Sci. USA* **95**, 12480–12485.

CHAPTER FIFTEEN

Detection of Autophagy in Cell Death

Zahra Zakeri,* Alicia Melendez,* *and* Richard A. Lockshin[†]

Contents

* Department of Biology, Queens College and Graduate Center of the City University of New York, Flushing, New York
† Department of Biological Sciences, St. John's University, Queens, New York

Methods in Enzymology, Volume 442
ISSN 0076-6879, DOI: 10.1016/S0076-6879(08)01415-8

Abstract

Activation of autophagosomes in cell death has been described since the late 1950s as a form of cell death characterized by consumption of the bulk of the cytoplasm by lysosomal derivatives. However, it is not yet established that autophagy is a primary, causative mechanism of death rather than a response to initial problems. Methods to assess autophagic cell death are similar to those used to detect and measure autophagy, with further evidence that the affected cells are indeed dying. These methods include structural analysis using electron microscopy, examination of the activity of lysosomal enzymes, assessment of the number, size, and location of lysosomes by the uptake of fluorescent molecules; measurement of the activity of autophagy-related genes such as the cleavage and activation of LC3 by Western blotting or green fluorescent protein tagging; evaluation of the effects of inhibition of one or more lysosomal enzymes, inhibition of fusion of organelles, or inhibition of intercompartmental transfer of molecules by putatively specific inhibitors; and interference with cells in order to change the expression of components of the lysosomal system to study the effect of this change on autophagy and autophagic cell death.

1. Introduction

Classically, cell death was defined as type I (now known as apoptosis), type II lysosomal cell death (now known as autophagic cell death), and type III (now known as necrosis) (Schweichel and Merker, 1973). Today, both type I and type II are acknowledged to be active self-destruction of the cell, with at least some forms of necrosis also under biological control. However, the concept of "autophagic cell death" remains unclear. Although the concept has been prevalent since the late 1960s, what is normally identified as autophagic cell death is typically not obviously distinguishable from *autophagy,* a process by which the cell, using membranes to isolate organelles or regions of cytoplasm, eliminates damaged organelles or consumes intracellular components as resources during starvation or other limiting conditions. In *autophagic cell death* the increased number and activation of autophagic vacuoles continue until the cell dies. While there are many means of evaluating autophagy, a facet of normal homeostasis, none of these techniques specifically distinguish autophagy from autophagic cell death. Thus any experiment must also establish that cells are dying with an increase in number and activity of autophagosomes. This chapter discusses a few ways by which one can identify autophagy, keeping in mind that not all the techniques of looking at autophagy may apply to looking at autophagy in cell death.

The most interesting questions therefore become (a) what stresses or signals have led to the activation of autophagy; (b) at what point does autophagy cease to be a survival technique and lead to the destruction of the cell; and (c) why does the cell activate autophagy rather than apoptosis as a means of exit? Cells described as undergoing autophagic cell death are typically large, sedentary cells with substantial cytoplasm. They are often not mitotically active, and the triggers for autophagy are more commonly deprivation of hormones or growth factors than viral infection or mutagenic compounds. These latter insults may nevertheless also cause autophagy in dying cells. There are also some cells in which effector caspases are absent or not easily activated, in which case excessive stress on the cell leads to death through autophagy, or autophagic cell death.

For large, G2 cells, the biology is reasonable: without a high priority for the destruction of potentially dangerous DNA, much of the cell is dismantled before the cell finally undergoes apoptosis or an apoptosis-like process. Autophagy substitutes for phagocytosis. However, autophagy is seen in other situations as well, most commonly for biological or experimental reasons such as situations in which caspases are knocked out, inhibited, or otherwise restricted. When apoptosis is not an available or effective option, autophagy may be a prominent, although frequently later, pathway to death. Likewise and unsurprisingly, if cells are starved or subjected to regimens in which proteins or organelles are damaged, autophagy may also be a prominent form of cell death. In brief, for most cells, most particularly mammalian cells of hematopoietic lineage or other small, mitotically active cells, apoptosis is the preferred, highly evolved, rapid, and efficient means of exit. In other circumstances, autophagy may be a prominent part of this death; it is frequently a default when other options are not available. However, autophagy may be a survival tactic, with death ensuing only when autophagy progresses too far without rescue of the cell. We then see an increase in the number and activity of autophagosomes. Thus, autophagy may be a prominent, although frequently later, pathway to death that leads eventually to the demise of the cell, and we describe the death as autophagic—with no implication that autophagy is a cause of death rather than a response to stress, damage, or a metabolic problem. To identify this type of cell death, limited tools and markers are available, which do not differentiate autophagic death from autophagy. Death must be confirmed by other means.

In general, the measurements of function monitor the delay of death following the specific experimental treatment or conversion of the type of death to another type of death, such as necrosis. In most instances, such measurements are appropriate in defining the relative timing and importance of each form of cell death as long as it is understood that inhibition of autophagy normally does not completely prevent death. Most experimental techniques involve initiation of death sequences by unequivocally toxic materials such as cycloheximide or staurosporin, and it is extremely unlikely that cells would be able to recover from such toxicity.

Autophagy is monitored by several techniques. Most of these are classical techniques. These include structural techniques such as electron microscopy (EM); methods of evaluating the activity of lysosomes by the use of biochemical assays, histochemical assays, and uptake of fluorescent molecules; and attempts to affect the activity of lysosomes by RNAi, transfection of specific genes, and use of inhibitors (see also Klionsky *et al.*, 2007).

 ## 2. STRUCTURAL: ELECTRON MICROSCOPIC RECOGNITION OF LYSOSOMES, AUTOPHAGIC VESICLES, AND AUTOPHAGOSOMES

Electron microscopy, although laborious and limited in its capacity to survey large numbers of samples, remains the gold standard for evaluation of autophagy. EM images can reveal the steps leading to engulfment of organelles, the origin of the membranes, the sequence of steps in the transformation of an isolation membrane into a fully active, enzyme-containing autophagic vacuole, and the final destruction of the organelle or portion of cytoplasm. The methods and classifications are discussed at length in Chapter 1. Two words of caution are nevertheless appropriate. First, the relative proportion and appearance of organelles consumed by the lysosomal system may vary among tissues, cells, and circumstances, with the differences reflecting more the initial structures of the cells than differences in process. Second, the number of lysosome-related organelles observed is a measure of both the rate of formation and the rate of disappearance of each type of organelle; an increase in number may reflect failure of clearance rather than increased production. As for all methods using EM, it is rather easy to generate artifacts in fixation and staining.

Electron microscopy can detect the different types of compartments, an early autophagosome (de Duve and Wattiaux, 1966) or phagopore (Seglen, 1987), amphisomes (generated by fusion of autophagosomes with endosomes (Dunn, 1994; Gordon and Seglen, 1988), or autolysosomes (generated by fusion of amphisomes or autophagosomes with lysosomes (de Duve, 1966). The essential criterion for identifying an autophagosome by electron microscopy is the appearance of a double membrane structure containing cytoplasm (i.e., cytosol and organelles). However, there may be complications with these criteria, for example, in some tissues the transition from an autophagosome to an amphisome or to an autolysosome involves a single membrane structure (Shelburne *et al.*, 1973). EM can detect the quantity of any of the different forms and if their number changes under different conditions. Data are often scored as the percentage of autophagosome volume/cytoplasmic volume. One way to limit the bias is to count a certain region of a cell. Primary lysosomes and organelles to which they are fused may be positively identified by identifying lysosomal enzymes. The most commonly used assay is the use of Na β-glycerophosphate as a substrate for acid phosphatase and precipitating the freed phosphate using lead nitrate, but these techniques are delicate and laborious and are used more for analysis of lysosomal function than for autophagic cell death.

The methods and classifications are discussed at length in Chapter 1. There are modifications of the basic method for sample preparation depending on the sources of the samples, that is, insects, *Caenorhabditis elegans*, mammalian tissues, or cells.

2.1. Methods used to detect autophagy in tissues or embryos by electron microscopy

The following technique is satisfactory for observing autophagy in tissues. Tissues or embryos are cut into small pieces and immediately immersed in 2.5% glutaraldehyde in 1× phosphate-buffered saline (PBS), pH 7.4, for a few days (at least 24 h). The samples should be postfixed in 1% osmium tetroxide in 1× PBS, pH 7.4, for 1 to 3 h, dehydrated in graded ethanol from 50 to 100% for 10 min each, and finally embedded in Spurr resin (Zakeri *et al.*, 1994, 1996). Semithin sections can be cut by an ultramicrotome and then stained with toluidine blue to localize the orientation and where to cut the thin sections. Select areas are cut in thin sections, collected on copper grids (Ernest F. Fullam, Inc.), and stained with uranyl acetate (5% in 70% ethanol) for 15 min and then in lead citrate for 10 min. The sections are observed in an electron microscope.

2.2. Methods used to detect autophagy in cells

The following technique is satisfactory for observing autophagy in cells. Cells are fixed in 0.5% glutaraldehyde in 1× cacodylate buffer, pH 7.2, for 10 min (1× cacodylate buffer: 20.15 g sodium cacodylate trihydrate, 0.1 ml concentrated HCl, should be pH 7.2; 30 g RNase-free sucrose, 0.05 g CaCl$_2$ in 500 ml total dH$_2$O). Cells are then gently scraped off with a rubber policeman and centrifuged in microcentrifuge tubes at 12,000g for 5 min. Using a razor blade, the centrifuge tube with the pellet is sliced into 1-mm sections, which are placed in 1 ml of 2.5% glutaraldehyde in 1× cacodylate buffer, pH 7.2, overnight at room temperature. Slices containing the cells are washed three times with 1× cacodylate buffer, pH 7.2, for 5 min each. Postfixation is with 1% osmium tetroxide for 3 h at 4 °C, after which the cells are washed three times with 1× cacodylate buffer, pH 7.2, for 5 min each and dehydrated through ascending ethanol concentrations (50–100%) for 10 min each. Samples are embedded in Spurr resin, sectioned, and stained with lead citrate and aqueous uranyl acetate.

2.3. Methods used to detect autophagy in *Caenorhabditis elegans*

The following technique is satisfactory for observing autophagy in *C. elegans*. *C. elegans* is subjected to EM analysis during dauer development, a stage in which autophagic activity is high and one can detect late structures in the autophagic pathway (e.g., autolysosomes) more easily than as well as intact autophagosomes (Meléndez *et al.*, 2003). For electron microscopic analysis, a high-pressure freeze fixation method is preferred. As described in Meléndez *et al.* (2003), for electron microscopy, animals are collected in a single cohort, fixed by standard immersion fixation. Animals are loaded onto metal planchettes, packed in excess *Escherichia coli* to minimize air pockets, and loaded quickly into a Bal-Tec HPM 010 high-pressure freezing device. Animals are fast frozen within 20 to 50 μs, brought gradually up to –90 °C, and fixed by freeze substitution in 1% osmium tetroxide in acetone (McDonald, 1999). The samples are then infiltrated into epoxy resin and heat cured. These plastic blocks are thin sectioned, poststained with uranyl acetate and lead citrate, and then examined by electron microscopy. The techniques are described in detail at http://www.aecom.yu.edu/wormem/methods% 20folder/hpffix.htm.

3. IDENTIFICATION OF LYSOSOMAL ENZYME ACTIVITY

Lysosomal activity can be assessed by examining the activity of enzymes residing in the lysosomes. Lysosomes contain several acid hydrolases capable of attacking most substrates other than lipids. The most commonly measured protease is cathepsin B, which is detected using relatively specific substrates such as LLVY methyl coumarin or specific inhibitors of thiol proteases such as leupeptin or antipain, discussed later and in Stoka *et al.* (2007) and Turk and Stoka (2007). Lysosomal activity may also be measured by the conversion of LC-3 to a smaller molecular weight, marking its localization in autophagosomes, as is described later. To measure lysosomal activity, one can analyze acid phosphatase activity. Acid phosphatase is one of the acid hydrolases that normally reside in lysosomes. Because it is easy to do, using an acid phosphatase kit (Sigma, 386A or 387A), it is a classical marker for the identification of lysosomes in subcellular fractionations. One must remain aware of the fact that the acid conditions used to measure the enzymes may activate resident enzymes, not necessarily reflecting activity *in vivo*. In situations in which lysosomal enzymes may not be synthesized *de novo,* differential centrifugation may help assess localization of the activity.

3.1. *In situ* measurement

Measuring enzymatic activity is done more effectively as a biochemical test in cells, but *in situ* measurements give a more reliable indication for tissues. The *in situ* method allows examination of a single or very few cells undergoing the process among many others that are negative (Ahuja *et al.*, 1997; Zakeri and Ahuja, 1994; Zakeri *et al.*, 1994). Fixed whole transparent embryos, cells, and sections (see later for fixation and preparation of embryo and tissue sections) are passed through serial-graded solutions of 20, 10, and 5% sucrose in $1\times$ PBS for 5 min each, postfixed with a citrate–acetone–formaldehyde solution for 30 s, and washed in dH_2O for 1 min. Sections are treated with naphthol AS–BI phosphate and fast garnet stain for 1 h at 37 °C. Slides are washed in running tap H_2O for 2 min, air dried for 15 min, and counterstained with methylene blue (1:100 dilution for 1 to 2 min depending on the tissue), rinsed in dH_2O two times for 1 min each, and mounted with Crystal Mount (Fisher). The acid phosphatase activity is detected as a distinct red to purple focal precipitate. Timing is important, as sections can be under- or overstained easily. For some preparations, the dyes will penetrate living organisms or cells and, if the activity is high enough, the cells will not deteriorate during short incubations, but the necessity for and effect of fixation should be assessed for any new experimental procedure (Halaby *et al.*, 1994).

3.2. Sample preparation for mammalian embryo and tissues for measurement of lysosomal activity and other immunohistochemical purposes

Embryos and tissues are removed and immediately fixed in 4% paraformaldehyde overnight. [To make 4% paraformaldehyde (Fisher Scientific, Pittsburgh, PA), add 2 g of paraformaldehyde to 25 ml of H_2O, mix, add 5 ml of 10 M NaOH, and 5 ml of $10\times$ PBS. Stir the mixture at 55 to 65 °C (no more than 65 °C) for 2 h. Filter the solution through a 0.45-μm filter (Gelman Sciences, Ann Arbor, MI). Then adjust to pH 8.0 and H_2O to adjust the final volume to 50 ml. This solution can be stored at 4 °C no more than 1 week.]

Samples can be fixed from 2 to 24 h depending on the size of sample. To avoid tissue damage for frozen sectioning, samples can be placed in 20% sucrose for at least 12 h or when the tissue drops to the bottom. Both the fixation and the sucrose/water displacement should be at 4 °C. The samples are then immobilized in tissue freezing medium (Triangle Biomedical

Sciences, Durham, NC) embedding solution and frozen in liquid nitrogen. The frozen blocks are cut 20 μm at $-20\,^{\circ}$C with a cryostat microtome. Frozen sections (5 μm) are cut in a cryostat at $-20\,^{\circ}$C, placed on Vectabond-coated slides (Vector Laboratories, USA), and stored at $-70\,^{\circ}$C prior to use. (To make Vectabond-coated slides, place clean slides in autoclaved metal racks, dip them in 7 ml Vectabond reagent in 350 ml acetone for 5 min, wash by dipping them several times over dH$_2$O, and air dry overnight. Slides are good for at least 4 weeks. To use the slides, allow them to come to room temperature before use.) Frozen sections placed on Vectabond-coated slides can be stored for several months at $-80\,^{\circ}$C and should be brought to room temperature before use.

3.3. Biochemical measurement of lysosomal enzymes

Can be used for cells and transparent embryos. Acid phosphatase can be measured readily by the degradation of paranitrophenyl phosphate in acid conditions (normally pH 4–5) to paranitrophenol, which is yellow in alkali. A kit containing tablets of the substrate is available from Sigma-Aldrich; these tablets are very stable and can be stored at $-20\,^{\circ}$C. Cells can be collected by scraping (no trypsin), pelleted, and resuspended with 1$\times$ PBS. Transparent tissues or embryos can be homogenized gently in PBS. Equilibrate the substrate solution to 37 $^{\circ}$C. Mix 10 to 100 μl of sample with 990 to 900 μl of substrate solution, using both blanks and positive controls as provided in the kit. The samples can be used for analysis of enzymatic activity using cuvettes or multiwell plates. Incubate the tubes or plates for 5 to 30 min at 37 $^{\circ}$C. The positive control is precalibrated for a 10-min incubation. Stop the reactions with 2 ml of stop solution (0.5 N NaOH). Centrifuge if necessary, and read the absorption at 405 nm. The color is stable for several hours. The enzyme is activated readily under acid conditions and it should be remembered that, without further documentation, such as purification by differential centrifugation, the activity does not necessarily represent *in vivo* activity (Halaby *et al.*, 1994).

4. SPECIFIC UPTAKE OF MOLECULES INTO LYSOSOMES AND DETECTION OF THESE MOLECULES BY FLUORESCENCE MICROSCOPY OR CONFOCAL MICROSCOPY TO EVALUATE NUMBER, SIZE, AND LOCATION OF LYSOSOMES

There are now many fluorescent markers that are reputed to be specific for one or more forms of lysosomes, and these are commonly used for analysis by conventional fluorescence microscopy, confocal microscopy, or fluorescence-activated cell sorting. The first markers to be used are no longer popular. Acridine orange (AO) and neutral red are taken up by lysosomes. AO in lysosomes fluoresces red as opposed to the green of AO bound to DNA. However, with the development and discovery of other more effective dyes, the limitations of AO—poor penetration, relatively low and unstable fluorescence, limited specificity, and confusion with variable background from, for instance, staining of RNA—became apparent, and AO is no longer considered to be a primary or high-quality marker for lysosomes. Similarly, although chloroquine fluoresces weakly and is taken up by lysosomes, its impact on the physiology of the lysosomal compartment and its poor fluorescence restrict its utility as a lysosomal marker as opposed to its usefulness as an inhibitor (see later).

Other lysosomal markers depend variously on components typically found in one or more types of lysosome-related organelles or on the acidity within lysosomes and later lysosomal derivatives. These tend to work very reliably and effectively in the typical context in which they are used, easily grown mammalian cells from primary cultures or cell lines, often exposed to low serum for more effective or visible response to a given condition. Nevertheless, it should be remembered that the accumulation of the fluor depends on a physical feature, such as intraorganelle relative or absolute acidity or the presence or concentration of a specific class of molecules, or on biological properties, such as the ability of the dye to penetrate cell or organelle membranes or to be retained by these membranes. These latter properties depend not only on the viability of the cell, but frequently on the health of the cell and its ability to maintain ion pumps, which may be sufficiently compromised to permit drifting from ideal ion concentrations (including pH) in a still viable cell. Thus it is extremely important for an experimenter using any nonconventional system to verify, by using a range of dye concentrations, experimental times, and experimental variables, that the dye is performing as expected. A range of dyes are now available that are excited and fluoresce at wavelengths suitable for use with most microscopes. As with all experiments using fluorescence, particularly those using two or more wavelengths, the behavior of individual dyes should be verified at all wavelengths to avoid secondary or bleed-through fluorescence or even, as we have seen, fluorescence resonance energy transfer.

Acidotropic dyes: Acidotropic dyes such as Lysotracker Red DND 99 (Molecular Probes, $\lambda_{ex} = 577$ nm and $\lambda_{em} = 590$ nm) are typically cell-permeant weakly basic amines and are very effective bright markers as described by Zucker *et al.* (1998, 1999, 2000a,b) and used according to the manufacturer's instructions [5 μM in dimethyl sulfoxide (DMSO), diluted 1:100 or more into culture medium], but because the partitioning depends on the pH differential, it is very important to vary concentrations, incubation times, and number of washes for the best results. Insect cells or others that maintain lower cytoplasmic pH values than most mammalian cells typically show high backgrounds when standard conditions are used. Also, because there is some variability and instability of the dye from batch to batch, best results are obtained using fresh DMSO and new ampoules. Because the dye is acidotropic, experiments should be conducted under conditions allowing rapid uptake and visualization of the still-living cell. While the dyes may be taken up very quickly (within seconds), they will neutralize the compartments into which they partition. Therefore, results should be collected as rapidly as possible, preferably within 1 to 5 min. The staining pattern usually, but not always, survives fixation. Because there are other acidic organelles in cells into which the dye can partition, including acrosomes, some types of secretory vesicles, trans-Golgi organelles, endosomes, and some coated vesicles, results should be compared to experimental and physiological results to verify the consistency of the findings. Several fluors are available, which are taken up effectively when the external concentration is approximately 50 nM. The dyes may be assessed by flow cytometry or fluorometry, although washout of nonspecific dye remaining in the cytoplasm or autofluorescence of the cells may affect the results.

LysoTracker dyes differ from the LysoSensor probes in that protonation of the latter relieves quenching of the fluorescence caused by the weakly basic side charge. Thus LysoSensor dyes increase fluorescence when they are acidified. They are available in several colors and with sensitivities that differ in transition pH. In some circumstances they provide lower backgrounds (at least in theory) and may give an assessment of pH as well. The manufacturer (Molecular Probes) also offers a 10,000-Da dextran conjugate of the LysoSensor Yellow/Blue DND-160. This latter can be taken up by cells—it is not cell penetrant—and processed through the endocytic pathway, where it will fluoresce blue in weakly acidic organelles and yellow in very acidic cells ($\approx$pH 3.9). However, because it also neutralizes acidic organelles, results should be interpreted with caution.

The reagent N-(3-((2,4-dinitrophenyl)amino)propyl)-N-(3-aminopropyl)methylamine dihydrochloride is likewise a weakly basic amine and is taken up into acidic organelles. It is not fluorescent, but it can be detected with anti-DNP antibodies, including those labeled with colloidal gold or other reagents suitable for conventional light microscopy or electron microscopy.

These dyes are useful for studies of cultured cells and may be used for very small transparent embryos such as zebrafish embryos (especially when cells are dispersed at the end of an experiment) or organs such as *Drosophila* salivary glands, but otherwise issues of dye penetration and microscopic resolution limit extension of the technique to other situations.

4.1. Example of methods to use for cells in culture

The general technique is to dissolve the dye in saline, culture medium, or DMSO according to the manufacturer's instructions. Cells should be exposed for as short a time as possible (not more than 5–10 min) to the medium containing the fluor. Cells are then washed thoroughly, as fluorescent dyes may be adsorbed onto petri dishes or taken up diffusely in cells. If fixation is acceptable for the dye, fix with 4% paraformaldehyde and wash cells if dye is retained; otherwise examine and photograph cells as soon as possible using appropriate filters. Verify that dilutions and incubation times are optimal, running appropriate negative controls. For cells in culture, the premixed stock solution of 1.0 mM dye such as Lysotracker Red is diluted 1:1000 in media to give a final concentration of 1.0 μM. Cells are incubated at 37 °C for 30 min and washed once with 1× PBS. Cells are fixed with 4% paraformaldehyde for 10 min, washed once with 1× PBS, and covered with a coverslip for microscopic examination or prepared for FACS analysis.

4.2. Monodansylcadaverine

Monodansylcadaverine (MDC) (Sigma-Aldrich Inc.) is an autofluorescent ($\lambda_{ex} = 335$ nm; $\lambda_{em} = 518$ nm) compound that is reputed to be taken up exclusively into autophagic vacuoles and not by other acidic organelles or early or late endosomes (Biederbick et al., 1995).

Incubate cells in approximately 100 μM MDC—note that this is a relatively high concentration of dye, the impact of which should be evaluated—for periods up to 1 h. Fix in 4% paraformaldehyde if desired. Cells may be stained for other purposes, such as display of nuclei or DNA. Examine cells within 3 h of terminating initial staining.

4.3. Example: Methods using dyes in a transparent embryo such as zebrafish

Zebrafish are raised according to techniques as described in Detrich et al. (1999). Harvested eggs are maintained in embryo rearing medium (ERM) (Detrich et al., 1999) at 28.5 °C. A minimum of six embryos is used for each staining. Embryos are collected at specific intervals, and stages are confirmed using easily recognizable landmarks such as heart formation and angle of head as described by Kimmel et al. (1995). Embryos are examined using a fluorescence microscope and photographed with an appropriate digital camera. Composite images are shot in sequence and aligned using fiduciary points and calibrated using photographs of a stage micrometer. Measurements are made using the freeware "Morphometrics" available at http://life. bio.sunysb.edu/morph/.

For staining, staged embryos are placed in 100-μl rubber-bordered wells on slides with a fluorescent stain or substrate in a dark box at 30 °C. After incubation, each embryo is transferred to an individual well in an eight-well hydrophobic slide, dechorionated using No. 55 watchmaker forceps, and washed twice using fresh 0.04% tricaine in ERM (Detrich *et al.*, 1999) with as little illumination as possible. Excess tricaine is removed. Stacked and moistened coverslips supporting a covering 22 × 60-mm coverslip are used to provide a chamber for examination. Embryos 0 to 3 days of age are photographed observing their left sides, and serial optical sections are taken. Older embryos are anesthetized, mounted in molten agar at 37 to 40 °C, and photographed without coverslips. All embryos 24 h postfertilization or older should be alive with beating hearts when examined.

For dual staining, DAPI (Molecular Probes, $\lambda_{ex} = 350$; $\lambda_{em} = 470$) is added to the staining mixture at a final concentration of 2.5 μg/ml and embryos are incubated for 30 to 45 min or 5 μg/ml DAPI for 20 to 30 min. DAPI is photographed using a blue filter.

4.4. LysoTracker red

The acidotropic dye LysoTracker red DND 99 (Molecular Probes, Invitrogen), abbreviated as LR, preferentially labels ($\lambda_{ex} = 577$ nm; $\lambda_{em} = 590$ nm) spherical acidotropic compartments such as autophagic vacuoles. The protocol is modified after Zucker *et al.* (1998, 2000a,b) and, because the dye varies in quality, concentrations and washes are varied according to the signal-to-background ratio. Fresh dye is prepared in DMSO from a newly opened ampoule using 10 μM dye for 15 min. Fluorescence is identified using a red filter with embryos counterstained with acridine orange.

5. LOCALIZATION OR DIFFERENTIAL MODIFICATION OF AUTOPHAGY-RELATED GENES

LC3 appears to be the best marker at the present, as cleavage of LC3 represents formation of the autophagosome. It may be detected by a shift of molecular weight on a Western blot or by localization from a diffuse cytoplasmic distribution to a particulate distribution by immunohistochemistry. Several commercial antibodies are available, but their specificity and sensitivity vary widely, and many are reported not to work very well in specific experimental situations. All antibodies should be carefully evaluated and positive and negative controls run for novel experimental situations. The specificity of the purified antibodies may be determined in western blots of protein extracts from wild-type cells and cells lacking LC3.

Specificity of the antibodies is achieved when the signal is detected in the wild-type cell extract and not in those lacking LC3.

5.1. Analysis of activation of LC3 to detect autophagy

To examine the activity of LC3, one can use either western blotting or histochemistry. To look at activity *in situ* by histochemical methods, we can use antibodies against LC3; these show the diffuse pattern of expression of LC3 changed to punctuate distribution. Alternatively, where GFP-tagged LC3 can be used, we look at the differential localization. The GFP-tagged LC3, where it can be successfully introduced into cells or organisms, is often a more sensitive marker, as the primary and secondary antibodies often show relatively low sensitivity and specificity and high backgrounds complicate evaluations. These methods can be used in a variety of organisms and cells.

5.2. Detection of LC3 by Western blot and immunohistochemistry

Using western blotting, one can detect the cleavage of LC3; this will work well for cells or tissues in which most of the cells have activated autophagy and in which LC3 is cleaved. One would not detect the signal if only a few cells are undergoing the process. Basically, for western blotting, cells are lysed in lysing buffer and proteins are separated on SDS-polyacrylamide electrophoresis, transferred to nitrocellulose by blotting, and blocked by 5% milk. The primary antibody (anti-LC3, mouse, Medical & Biological Laboratories, LTD.) is usually used at 1:100 to 1:500 dilution overnight. The preparation is washed, exposed to the appropriately (peroxidase)-labeled secondary antibody, and visualized using enhanced chemiluminescence.

If all cells are not undergoing autophagy, then the signal may be diluted and not recognized by western blotting and the *in situ* method of immunohistochemistry is more useful. For immunohistochemistry, cells are immobilized on Vectabond-treated coverslips, fixed with 4% paraformaldehyde in 1× PBS for 10 min, washed, and blocked; diluted serum (18 μl rabbit serum in 1 ml PBST) is applied to coverslips in a humidity chamber, and incubation is at room temperature for 1 h. Approximately 50 μl primary antibody at a dilution of 1:1000 is applied; the specimens are covered with plastic coverslips, placed on a humidifying rack in 4 °C overnight or 2 h at room temperature, and then washed three times with PBST. The secondary antibody is labeled with FITC or other appropriate fluor for 1 h. Specimens are mounted with GelMount (Biomedia) and sealed *quickly* with fingernail polish, as these cells are not dehydrated and are *very* prone to desiccation until the fingernail polish dries. The slides can be then looked at with a fluorescence or confocal microscope.

5.3. LC3-GFP transfection to visualize LC3 translocation in cells

Cells are seeded on sterile glass coverslips and covered with subculturing media without antibiotics. Adding antibiotics to media will interfere with transfection efficiency and cause increased cytotoxicity of transfection reagents. Cells are washed with 1× phosphate-buffered saline prior to transfection. For transfection, Lipofectamine 2000 reagent (Invitrogen) is diluted into serum- and antibiotic-free media. Manufacturer's specifications should be followed to adjust for varying cell numbers and plate sizes. For transfection, DNA (C2-LC3-GFP plasmid) is diluted into serum- and antibiotic-free media adjusted as indicated by the manufacturer.

The dilute Lipofectamine 2000 and dilute DNA (C2-LC3-GFP plasmid) are combined, mixed gently by inverting, and incubated at room temperature for 15 to 45 min. For transfection, cells are washed thoroughly with serum-free media or 1× PBS. The transfection medium is overlaid onto washed cells. Cells are incubated for 4 to 6 h at 37 °C in a CO_2 incubator with gentle shaking every 30 min. After 4 to 6 h, antibiotic-free subculture medium is added without removing the transfection mixture. The medium is replaced 18 to 24 h following the start of transfection. The assay for GFP expression is conducted using an inverted fluorescence microscope within 18 to 24 h of the end of transfection. Increased incubation times prior to treatment will allow spontaneous LC3 activation due to overexpression of the plasmid protein. To assay for GFP expression, samples should be fixed with cold 3.5% paraformaldehyde in 1× PBS for 5 to 10 min. Immediate analysis is best, but cells may be stored briefly to allow continued collection of samples. Inactive, cytoplasmic LC3 will appear as a diffuse staining pattern throughout the cytoplasm. Activated LC3 will translocate specifically to the autophagosomes that develop during an autophagic response and will appear as bright, punctate dots within affected cells.

5.4. LC3-GFP transfection to visualize LC3 translocation in *C. elegans*

The discovery that Atg8 in yeast is a ubiquitin-like protein that undergoes lipidation and a stable association to phosphotidylethanolamine in the autophagosomal membrane has facilitated *in vivo* assays to monitor autophagy. In *C. elegans,* the Atg8 ortholog is *lgg-1;* in mammals, the Atg8 ortholog is LC3. A transgenic construct containing the promoter and coding sequences of *lgg-1* was cloned into a green fluorescent protein vector with the GFP protein located in the N terminus. Extrachromosomal (Meléndez et al., 2003) and integrated (Kang et al., 2007) versions of GFP::LGG-1 exist. The integrated version may show less variability in the expression of GFP::LGG-1 among different animals. In the absence of autophagy, GFP::LGG-1 has a diffuse cytoplasmic localization; when autophagy is induced, it associates tightly with the preautophagosomal/autophagosomal membrane and has a punctate appearance. Therefore, monitoring the intracellular localization of green fluorescent protein-tagged LGG-1 molecules is an effective and reliable method to monitor autophagy in *C. elegans* (Kang et al., 2007; Meléndez et al., 2003; Rowland et al., 2006; Samara et al., 2008; Takacs-Vellai et al., 2005).

5.5. Methods used to detect autophagy in *C. elegans* using RNAi and mutations

Caenorhabditis elegans is a genetic model system, and thus the role of autophagy can be studied using chromosomal mutations or by RNA inactivation of candidate genes. To test if autophagy genes are involved in a particular process, chromosomal mutations exist for Atg6/*bec-1*, Tor/*let-363*, Vps34/*let-512*, Atg1/*unc-51*, and Atg18/ *atgr-18*, and double mutants can be created or their phenotype may be analyzed. To test these and other orthologs that do not have a chromosomal mutation identified, RNA interference is another available method. In *C. elegans*, inactivation by RNAi can be achieved by injection, soaking the animals in the dsRNA, or feeding the animals dsRNA expressing *E. coli*. RNAi by injection has been used to uncover the role of *bec-1, unc-51, atgr-7/M7.5, lgg-1*, and *atgr-18* in the changes associated with dauer development (Meléndez et al., 2003). RNAi by injection was also performed to determine whether BEC-1 is required for the longer life span of *daf-2* mutants. In cases where the RNAi treatment is lethal, the dsRNA can be diluted and its effect weakened. DNA clones of most autophagy orthologs exist (provided by Y. Kohara, National Institute of Genetics, Japan) and can be used as templates to produce dsRNA. Template DNA can be polymerase chain reaction (PCR) amplified with Bluescript vector primers CM024 (TTGTAAAACGACGGCCAG) and CM025 (CAGATTACGCC-CAGTCTC), transcribed using an RNA transcription kit (Stratagene), and purified by an RNA quick purification column (RNeasy minikit, Qiagen) (Meléndez et al., 2003). L4 or young adult animals are microinjected with the dsRNA in the pseudocoelum or intestine and allowed to recover for 24 h. Their F1 progeny are collected in single-day cohorts for subsequent analyses. Approximately 0.5 ng of dsRNA is injected into each animal.

The dsRNA for RNAi can also be delivered by feeding, which has been accomplished successfully for *bec-1, atgr-7,* and *lgg-3* to show the role of autophagy in life span extension (Hars *et al.*, 2007; Jia and Levine, 2007), as well as for *unc-51, lgg-1, atgr-18,* and *bec-1* to show that autophagy is required for necrotic cell death (Samara *et al.*, 2008; Tóth *et al.*, 2007). The clones are available from Geneservice LTD (http://www.geneservice. co.uk/products/rnai/index.jsp) for purchasing, originally from the RNAi library of Julie Ahringer (Kamath *et al.*, 2003) or Marc Vidal (Rual *et al.*, 2004). Although orthologs to yeast Atg3, Atg5, Atg10, and Atg16 have been found, these have still to be tested for a role in the regulation of autophagy or in development. Nonspecific inactivation of genes in operons by RNAi has been observed; however, quantitative reverse transcriptase PCR can detect if the RNAi affects the mRNA levels of the gene in question.

In a similar way one can use RNAi in mammalian cells.

6. EVALUATION OF EFFECTS OF INHIBITION OF ONE OR MORE LYSOSOMAL ENZYMES, INHIBITION OF FUSION OF ORGANELLES, OR INHIBITION OF INTERCOMPARTMENTAL TRANSFER OF MATERIALS BY PUTATIVELY SPECIFIC INHIBITORS

In this experimental situation, an attempt is made to inhibit lysosomal activity and a biological consequence is monitored. 3-methyladenine (Fluka) is, like chloroquine, a cell-permeable weakly basic material that is rapidly taken up into lysosomes, where it neutralizes them, inactivating the acidic hydrolases. Thus it inhibits autophagy. Although it is used at quite high concentrations (10 mM in culture; Punnonen *et al.*, 1994), it appears not to be otherwise toxic to cells, at least in short-term experiments. For zebrafish, it can be dissolved in embryonic rearing medium and the experiment conducted as described previously. Its longer term toxicity has not been unequivocally associated with its effect on lysosomes. The techniques are essentially identical to those described earlier. Leupeptin and antipain, available from several manufacturers, can be used to inhibit cathepsin B or both cathepsin B and calpain. These materials generally penetrate cell membranes and are effective in micromolar concentrations.

REFERENCES

Ahuja, H. S., James, W., and Zakeri, Z. (1997). Rescue of the limb deformity in hammertoe mutant mice by retinoic acid-induced cell death. *Dev. Dyn.* **208,** 466–481.
Biederbick, A., Kern, H. F., and Elsässer, H. P. (1995). Monodansylcadaverine (MDC) is a specific *in vivo* marker for autophagic vacuoles. *Eur. J. Cell Biol.* **66,** 3–14.

de Duve, C., and Wattiaux, R. (1966). Functions of lysosomes. *Annu. Rev. Physiol.* **28**, 435–492.

Detrich, H. W. I., Westerfield, M., and Zon, L. I. (1999). "The Zebrafish: Biology." Academic Press, New York.

Dunn, W. A., Jr. (1994). Autophagy and related mechanisms of lysosome-mediated protein degradation. *Trends Cell Biol.* **4**, 139–143.

Gordon, P. B., and Seglen, P. O. (1988). Prelysosomal convergence of autophagic and endocytic pathways. *Biochem. Biophys. Res. Commun.* **151**, 40–47.

Halaby, R., Zakeri, Z., and Lockshin, R. A. (1994). Metabolic events during programmed cell death in insect labial glands. *Biochem. Cell Biol.* **72**, 597–601.

Hars, E. S., Qi, H., Ryazanov, A. G., Jin, S., Cai, L., Hu, C., and Liu, L. F. (2007). Autophagy regulates ageing in *C. elegans*. *Autophagy* **3**, 93–95.

Jia, K., and Levine, B. (2007). Autophagy is required for dietary restriction-mediated life span extension in *C. elegans*. *Autophagy* **3**, 597–599.

Kamath, R. S., Fraser, A. G., Dong, Y., Poulin, G., Durbin, R., Gotta, M., Kanapin, A., Le Bot, N., Moreno, S., Sohrmann, M., Welchman, D. P., Zipperlen, P., *et al.* (2003). Systematic functional analysis of the *Caenorhabditis elegans* genome using RNAi. *Nature* **421**, 231–237.

Kang, C., You, Y. J., and Avery, L. (2007). Dual roles of autophagy in the survival of *Caenorhabditis elegans* during starvation. *Genes Dev.* **21**, 2161–2171.

Kimmel, C. B., Ballard, W. W., Kimmel, S. R., Ullmann, B., and Schilling, T. F. (1995). Stages of embryonic development of the zebrafish. *Dev. Dyn.* **203**, 253–310.

Klionsky, D. J., Abeliovich, H., Agostinis, P., Agrawal, D. K., Aliev, G, Askew, D. S., Baba, M., Baehrecke, E. H., Bahr, B. A., Ballabio, A., Bamber, B. A., Bassham, D. C., *et al.* (2007). Guidelines for the use and interpretation of assays for monitoring autophagy in higher eukaryotes. *Autophagy* **4**, 151–175.

McDonald, K. (1999). High pressure freezing for preservation of high resolution fine structure and antigenicity for immunolabeling. *Methods Mol. Biol.* **117**, 77–97.

Meléndez, A., Talloczy, Z., Seaman, M., Eskelinen, E. L., Hall, D. H., and Levine, B. (2003). Autophagy genes are essential for dauer development and life-span extension in *C. elegans*. *Science* **301**, 1387–1391.

Punnonen, E. L., Marjomaki, V. S., and Reunanen, H. (1994). 3-Methyladenine inhibits transport from late endosomes to lysosomes in cultured rat and mouse fibroblasts. *Eur. J. Cell Biol.* **65**, 14–25.

Rowland, A. M., Richmond, J. E., Olsen, J. G., Hall, D. H., and Bamber, B. A. (2006). Presynaptic terminals independently regulate synaptic clustering and autophagy of GABAA receptors in *Caenorhabditis elegans*. *J. Neurosci.* **26**, 1711–1720.

Rual, J. F., Ceron, J., Koreth, J., Hao, T., Nicot, A. S., Hirozane-Kishikawa, T., Vandenhaute, J., Orkin, S. H., Hill, D. E., van den Heuvel, S., and Vidal, M. (2004). Toward improving *Caenorhabditis elegans* phenome mapping with an ORFeome-based RNAi library. *Genome Res.* **10B**, 2162–2168.

Samara, C., Syntichaki, P., and Tavernarakis, N. (2008). Autophagy is required for necrotic cell death in *Caenorhabditis elegans*. *Cell Death Differ.* **15**, 105–112.

Schweichel, J. U., and Merker, H. J. (1973). The morphology of various types of cell death in prenatal tissues. *Teratology* **7**, 253–266.

Seglen, P. O. (1987). "Regulation of Autophagic Protein Degradation in Isolated Liver Cells." Academic Press, London.

Shelburne, J. D., Arstila, A. U., and Trump, B. F. (1973). Studies on cellular autophagocytosis: The relationship of autophagocytosis to protein synthesis and to energy metabolism in rat liver and flounder kidney tubules *in vitro*. *Am. J. Pathol.* **73**, 641–670.

Stoka, V., Turk, V., and Turk, B. (2007). Lysosomal cysteine cathepsins: Signaling pathways in apoptosis. *Biol. Chem.* **388**, 555–560.

Takacs-Vellai, K., Vellai, T., Puoti, A., Passannante, M., Wicky, C., Streit, A., Kovacs, A. L., and Muller, F. (2005). Inactivation of the autophagy gene *bec-1* triggers apoptotic cell death in *C. elegans. Curr. Biol.* **15,** 1513–1517.

Tóth, M. L., Simon, P., Kovács, A. L., and Vellai, T. (2007). Influence of autophagy genes on ion-channel-dependent neuronal degeneration in *Caenorhabditis elegans. J. Cell Sci.* **120,** 1134–1141.

Turk, B., and Stoka, V. (2007). Protease signalling in cell death: Caspases versus cysteine cathepsins. *FEBS Lett.* **581,** 2761–2767.

Zakeri, Z., Quaglino, D., and Ahuja, H. S. (1994). Apoptotic cell death in the mouse limb and its suppression in the hammertoe mutant. *Dev. Biol.* **165,** 294–297.

Zakeri, Z., Quaglino, D., Latham, T., Woo, K., and Lockshin, R. A. (1996). Programmed cell death in the tobacco hornworm, *Manduca sexta:* Alteration in protein synthesis. *Microsc. Res. Tech.* **34,** 192–201.

Zakeri, Z. F., and Ahuja, H. S. (1994). Apoptotic cell death in the limb and its relationship to pattern formation. *Biochem. Cell Biol.* **72,** 603–613.

Zucker, R. M., Hunter, E. S., 3rd, and Rogers, J. M. (1999). Apoptosis and morphology in mouse embryos by confocal laser scanning microscopy. *Methods* **18,** 473–480.

Zucker, R. M., Hunter, E. S., 3rd, and Rogers, J. M. (2000a). Confocal laser scanning microscopy of morphology and apoptosis in organogenesis-stage mouse embryos. *Methods Mol. Biol.* **135,** 191–202.

Zucker, R. M., Hunter, S., and Rogers, J. M. (1998). Confocal laser scanning microscopy of apoptosis in organogenesis-stage mouse embryos. *Cytometry* **33,** 348–354.

Zucker, R. M., Keshaviah, A. P., Price, O. T., and Goldman, J. M. (2000b). Confocal laser scanning microscopy of rat follicle development. *J. Histochem. Cytochem.* **48,** 781–791.

CHAPTER SIXTEEN

METHODS FOR DISTINGUISHING APOPTOTIC FROM NECROTIC CELLS AND MEASURING THEIR CLEARANCE

Dmitri V. Krysko,[*,†] Tom Vanden Berghe,[*,†] Eef Parthoens,[†,‡] Katharina D'Herde,[§] *and* Peter Vandenabeele[*,†]

Contents

[*] Molecular Signaling and Cell Death Unit, Department for Molecular Biomedical Research, VIB, Ghent, Belgium
[†] Department of Molecular Biology, Ghent University, Ghent, Belgium
[‡] Microscopy Core, Department of Molecular Biomedical Research, VIB, Ghent, Belgium
[§] Department of Anatomy, Embryology, Histology and Medical Physics, Ghent University, Ghent, Belgium

Methods in Enzymology, Volume 442
ISSN 0076-6879, DOI: 10.1016/S0076-6879(08)01416-X

Abstract

Three major morphological types of cell death can be distinguished: type I (apoptotic cell death), type II (autophagic cell death), and type III (necrotic cell death). Details of the pathways of apoptotic and autophagic cell death have been described, and distinct biochemical markers have been identified. However, no distinct surface or biochemical markers of necrotic cell death have been identified yet, and only negative markers are available. These include absence of apoptotic parameters (caspase activation, cytochrome *c* release, and oligonucleosomal DNA fragmentation) and differential kinetics of cell death markers (phosphatidylserine exposure and cell membrane permeabilization). Moreover, a confounding factor is that apoptotic cells in the absence of phagocytosis proceed to secondary necrosis, which has many morphological features of primary necrotic cells. Secondary necrotic cells have already gone through an apoptotic stage, and so it is generally advisable in cell death research to perform time kinetics of cell death parameters. This chapter concentrates on methods that can distinguish apoptosis from necrosis on three different levels (morphological, biochemical, and analysis of cell–cell interactions) and emphasizes that only a combination of several techniques can correctly characterize cell death type. First, we describe analysis of apoptotic versus necrotic morphology by time-lapse microscopy, flow fluorocytometry, and transmission electron microscopy. We also discuss various biochemical techniques for analysis of cell surface markers (phosphatidylserine exposure versus cell permeability by flow fluorocytometry), cellular markers such as DNA fragmentation (flow fluorocytometry), caspase activation, Bid cleavage, and cytochrome *c* release (Western blotting). Next, we describe how primary and secondary necrotic cells can be distinguished by analysis of supernatant for caspases, HMGB1, and release of cytokeratin 18. Finally, we discuss cell–cell interactions during cell death and describe a quantitative method for examining dead cell clearance by flow fluorocytometry. A selection of techniques that can be used to study internalization mechanisms used by phagocytes to engulf dying cells is also presented, such as scanning and transmission electron microscopy and fluorescence microscopy.

1. INTRODUCTION

Programmed cell death (PCD) is instrumental in sculpting multicellular organisms during normal development and is important in homeostasis. The concept of PCD was introduced in 1964 by Lockshin and Williams. It was defined as a process that occurs in predictable places and at predictable times during embryogenesis, implying that cells are somehow programmed to die during the development of the organism. Later on Kerr and coauthors (1972) described the morphological characteristics of cell death during development and tissue homeostasis and suggested the term apoptosis. Three morphologically distinct types of physiological cell death were identified in tissues of mouse and rat embryos (Schweichel and Merker, 1973): type I (apoptotic cell death), type II (autophagic cell death), and type III (necrotic cell death). A major breakthrough in cell death research was identification of a biochemical marker of apoptotic cell death, namely the internucleosomal cleavage of DNA (Wyllie et al., 1984), which provided a base for unraveling the signal transduction pathways involved in the initiation and execution of apoptosis. Apoptosis (type I cell death) is characterized by a sequence of specific morphological changes in the dying cell: cellular shrinkage with condensation of the cytoplasm, sharply delineated of chromatin masses lying against the nuclear membrane, nuclear fragmentation (karyorrhexis), and the subsequent formation of membrane-confined apoptotic bodies containing a variety of cytoplasmic organelles and nuclear fragments. In apoptosis, mitochondria appear to be normal or shrunken rather than dilated or swollen (D'Herde et al., 2000; Krysko et al., 2001). Apoptotic bodies are engulfed by nonprofessional and professional phagocytes (Kerr et al., 1972; Schweichel and Merker, 1973).

Autophagy is characterized by the presence of autophagic structures with a double membrane (Clarke, 1990; Schweichel and Merker, 1973). It is important to note that autophagy is foremost a survival mechanism activated in cells undergoing different forms of cellular stress. If cellular stress continues, cell death may continue by autophagy alone, or else it may develop apoptotic or necrotic features (Maiuri et al., 2007).

Necrosis (type III cell death) is characterized by swelling of the organelles (endoplasmic reticulum, mitochondria) and the cytoplasm, followed by collapse of the plasma membrane and lysis of the cells (Schweichel and Merker, 1973). As a consequence of the prominent swelling of the cytoplasm, this type of cell death was also designed as oncosis (Majno and Joris, 1995). Necrotic cell death is often considered a passive process lacking underlying signaling events and occurring under extreme physicochemical conditions, such as abrupt anoxia, sudden shortage of nutrients, and exposure to heat or detergents. It has become evident that necrotic cell death is as well controlled and programmed as apoptotic cell death and that it results from extensive cross talk between several biochemical and molecular events at different cellular levels (for a detailed review see, Festjens et al., 2006).

Regardless of how cells die, maintenance of normal function in the multicellular organism dictates that dying cells are cleared by professional or nonprofessional phagocytes. Defective or inefficient clearance of dying cells may contribute to persistence of inflammation, excessive tissue injury, and several different human pathologies, including systemic lupus erythematosus (Munoz et al., 2005), cystic fibrosis (Vandivier et al., 2002), and chronic obstructive pulmonary disease (Hodge et al., 2003). Engulfment of apoptotic cells is regulated by a highly redundant system of receptors and bridging molecules on the dying cells and on the phagocytes. Recognition of necrotic cells by phagocytes is less clearly understood than recognition of apoptotic cells, but recent studies are highlighting its importance (Krysko et al., 2006b).

Details of the pathways of apoptotic and autophagic cell death have been described, and an array of distinct biochemical markers has been identified. However, identification of necrotic cell death has little to rely on other than cellular morphology; no distinct biochemical parameter that discriminates necrotic from apoptotic death unambiguously and positively is yet available. Even morphological evaluation of dying cells can be misleading, e.g. in the absence of phagocytosis, apoptotic cells proceed to a stage called secondary necrosis. Secondary necrotic cells resemble necrotic cells, but differ in having gone through an apoptotic stage. Therefore, it is generally advisable in cell death research to determine the time kinetics of the cell death parameters. Necrotic cell death is described mostly in passive terms as cell death that is characterized by the absence of apoptotic biochemical parameters, such as caspase activation, cytochrome *c* release and oligonucleosomal DNA fragmentation, and by differential kinetics of cell death markers such as phosphatidylserine (PS) exposure and cell membrane permeabilization. We propose that distinguishing apoptosis from necrosis should be based on three different levels: morphological, biochemical, and interactions with phagocytes.

2. ANALYSIS OF CELL DEATH PARAMETERS IN INTACT CELLS

Apoptotic cells are smaller and denser than their living counterparts (Kerr *et al.*, 1972; Kitanaka and Kuchino, 1999). First, the cell membrane starts to form blebs, which then separate from the main cell body. This is followed by shrinkage of the entire nucleus, karyorrhexis, and formation of apoptotic bodies (Earnshaw, 1995). This section describes the use of real-time imaging, flow fluorocytometry, and transmission electron microscopy (TEM) to analyze cell morphology in order to distinguish between apoptotic and necrotic cell death.

Throughout this chapter, we will use as an example the L929sA fibrosarcoma cell line, in which apoptotic as well as necrotic cell death can be induced. Stimulation of TNFR1 in L929sA cells leads to necrotic cell death, whereas in L929sA cells transfected with human Fas (L929sAhFas), the use of agonistic anti-Fas antibodies leads to clustering of Fas and induction of the apoptotic cell death pathway (Vercammen *et al.*, 1997).

2.1. Analysis of cell morphology by real-time imaging (time-lapse microscopy)

Time-lapse imaging is a technique that specifies a predefined delay between the acquisition of images. Interestingly, time-lapse microscopy had already been used to study cell death by French hematologist Marcel Bessis long before different cell death types were recognized (Bessis, 1964). He observed that leukocytes emitted pseudopodia and finally broke up with almost explosive suddenness. The scientist described this cell behavior as cell death by fragmentation. Nowadays time-lapse microscopy is the method of choice for the dynamic monitoring of individual cells for the morphological changes that occur during cell death and for appreciating the differences between apoptotic and necrotic types of cell death. Time-lapse images are recorded using a differential interference contrast (DIC) mode alone or in combination with epifluorescence optics.

2.1.1. Differential interference contrast microscopy

Differential interference contrast optics are known also as Nomarski optics, after its inventor Jerzy Nomarski, who created a virtual relief image that allows morphological analysis of transparent objects. As with phase-contrast microscopy, DIC enhances contrast by taking advantage of differences in the refractive index among parts of a specimen. However, in contrast to phase contrast, DIC optics make use of polarized light. DIC works by using a prism to direct two beams of polarized light at a specimen, with one beam slightly offset from the other. A second prism reassembles the beams to produce a shadowing effect. The result is a virtual three-dimensional image.

The combined use of DIC and time-lapse imaging makes it possible to analyze the following parameters: (1) specific morphological changes that are typical of apoptosis or necrosis and (2) the timing and kinetics of these morphological changes, as well as the subcellular events, such as the duration and onset of rounding up of cells, formation of apoptotic bodies, and chromatin condensation.

1. Two days before analysis, seed the cells at 4×10^3 cells per chamber in an eight-chambered cover glass (Lab-Tek chambered #1.0 borosilicate cover glass system, Nunc) to allow the cells to attach to the bottom and become flattened in order to obtain a better morphological profile.
2. On the day of analysis, stimulate cells with tumor necrosis factor (TNF) (1000 IU/ml) or with agonistic anti-Fas antibody (BioCheck GmbH, anti-Fas, 125 ng/ml), and put the chambered cover glass in a time-lapse microscope. Recombinant human TNF has been produced in *Escherichia coli* and purified to at least 99% homogeneity (prepared at the Department for Molecular Biomedical Research, VIB-UGent). The specific biological activity is 2.3×10^7 IU/mg as determined in a standardized cytotoxicity assay on L929sA cells.
3. Image cells using three-dimensional (3D)$(x, y,$ and $z)$ time-lapse microscopy with DIC optics.
4. Acquire time-lapse images. One can use the Leica ASMDW live cell imaging system (Leica Microsystems, Mannheim, Germany). This system includes a DM IRE2 microscope equipped with an HCX PL APO

$63\times/1.3$ glycerin-corrected 37 °C objective and a 12-bit Coolsnap HQ camera. The microscope is equipped with an incubation chamber with temperature and CO_2 controls.

The distinct morphological features of apoptotic versus necrotic cell death could be appreciated on time-lapse movies at http://www.ecdo.eu/research.htm (L929sAhFas cells, apoptosis morphology.avi; necrosis morphology.avi).

2.1.2. Combination of DIC and epifluorescence microscopy

Time-lapse microscopy can be performed simultaneously in DIC and epifluorescence mode. Coupling real-time imaging with fluorescent probes makes it possible to associate specific morphological features of cell death with particular molecular or subcellular cell death events. For example, propidium iodide (PI) or Sytox green can be used to determine plasma membrane integrity, Annexin V-Alexa Fluor 488 to visualize phosphatidylserine exposure, LysoTracker to visualize lysosomal integrity, tetramethylrosamine to measure mitochondrial depolarization, and carboxy-H_2DCFDA to detect production of reactive oxygen species. All these probes are available from Molecular Probes (Invitrogen).

1. Follow steps 1 and 2 in Section 2.1.1. To monitor morphology and membrane permeability in parallel, PI is added simultaneously with cell death stimuli to a final concentration of 3 μM. PI is a fluorescent membrane-impermeable dye that stains nuclei by intercalating between the stacked bases of nucleic acids. It is used to determine loss of cell membrane integrity, as it enters the cell only if the plasma membrane becomes permeable, and to examine the structure of chromatin. When PI is bound to nucleic acids, its absorption maximum is 535 nm and its fluorescence emission maximum is 617 nm.

2. Perform 3D (x, y, and z) time-lapse microscopy by using DIC together with epifluorescence optics. It is important to optimize the setup so that no cell death is observed in control conditions without stimulus. To reduce phototoxicity and photobleaching, lamp intensity, opening of field diaphragm, number of z sections, time frames, and camera settings (exposure time and binning) have to be optimized. Binning allows charges from adjacent pixels to be combined. The benefit of this technique is that lower exposure times are needed because sensitivity increases with the binning factor, which translates into reduced phototoxicity, albeit at the expense of reduced spatial resolution.

Secondary necrosis can be distinguished from primary necrosis by staining the nucleus with fluorescent probes such as PI. Secondary necrotic cells had passed an apoptotic cell death stage and their nuclei are fragmented or condensed; PI homogeneously stains the nucleic acids content due to its binding to DNA by intercalating between the bases. Primary necrotic cells have uncondensed nuclei with prominent nucleoli. The distinct pattern of staining with PI of apoptotic and necrotic L929sAhFas cells could be appreciated on time-lapse movies at http://www.ecdo.eu/research.htm (apoptosis_PI.avi and necrosis_PI.avi).

2.2. Analysis of cell morphology versus cell permeability by flow fluorocytometry (FACS)

Morphological changes during apoptosis and necrosis are detected accurately in most cells by their light-scattering properties in flow fluorocytometric analysis. The forward scatter reflects cell size, whereas the sideward scatter reveals the degree of granularity of the cell. The protocol described here uses flow fluorocytometry to access changes in cell size and granularity, as well as the loss of cell membrane integrity as measured by PI uptake. PI is excited with a xenon or mercury-arc lamp or with the 488-line of an argon-ion laser. PI fluorescence is detected in the FL2 or FL3 channel of the flow fluorocytometer (FACSCalibur flow fluorocytometer, Becton Dickinson). The broad emission spectrum of PI precludes combining it with other fluorochromes. Only fluorochromes with a clearly distinct emission spectrum or with a small emission spectrum overlap can be used. It is possible to compensate for small emission overlap during flow fluorocytometric measurements by changing the detector or amplifier gain of any of the fluorescence parameters (FL1, FL2, or FL3). To check or adjust the compensation, one should use control samples labeled singly with each color and unlabeled samples.

1. Seed cells at 1.5×10^5 cells/ml per well on special 24-well uncoated suspension tissue culture plates (Sarstedt) to avoid attachment of the cells to the bottom. To perform FACS analysis, it is advisable to grow adherent cells in suspension. Make sure beforehand that the adherent cells will remain viable in suspension and that they exhibit similar dose–response curves to cell death stimuli. Moreover, seed the cells only the day before analysis. In this way it is possible to work faster during kinetics and to avoid interfering manipulations, such as trypsin/EDTA treatment, which could damage or permeabilize the cells and lead to removal of antigens from their surface.

2. The next day, induce apoptotic or necrotic cell death. The features of end stage apoptosis, so-called secondary necrosis, overlap with those of necrotic cells, that is, loss of membrane integrity and leakage of proteins in the supernatant (Denecker et al., 2001). Therefore, it is advisable to perform time kinetics in order to detect the differential appearance of apoptotic or necrotic features such as caspase activation, DNA fragmentation, and release of caspase and cytokeratin 18 (discussed later).

3. Analyze the cell samples from the suspension cultures at regular time intervals on the flow fluorocytometer. Transfer 270 μl of cells in suspension from the 24-well suspension plate to a 5-ml polypropylene round-bottom tube. Prepare a $10\times$ stock of PI (from a 3 mM master stock solution) and add 30 μl directly before measurement to the samples.

4. To determine cell size and granularity, set up the flow fluorocytometer for a dot plot, with both forward and sideward scatter on linear scales (Fig. 16.1). Create a histogram in FL3 to detect PI uptake at 610 nm (far red wavelength), for example, on a FACSCalibur flow fluorocytometer (Becton Dickinson) equipped with a water-cooled argon-ion laser at 488 nm. A typical FACSCalibur flowcytometer (Becton Dickinson) can analyze five different parameters. Two detectors detect the light scatter: (I) forward scattered light (FSC, indicates cell size) and (II) side scattered light (SSC, indicates granularity). Three remaining photomultiplier tubes detect the fluorescent signals: (III) FL1 (green fluorescence; used for FITC, R123, GFP), (IV) FL2 (red fluorescence; used for PI, CMTMros, PE), and (V) FL3 (far red fluorescence, used for PI, Cy-Chrome). To highlight dead cells on the dot plot, gate PI-positive cells, for example, in red, in the FL3 histogram.

5. After measuring PI uptake, tubes with the remaining suspension cells (at least 150 μl) or freshly prepared tubes (300 μl, see step 1) can be used to determine the percentage of cells with hypoploid DNA (see Section 3.2.1).

The region of analysis is gated (R1) and it must be large enough to measure both living and dying cells. Dying cells shift to the left on the linear dot plot relative to living cells (Fig. 16.1). At time point zero, the viable cell population is detected as a PI-negative population within the region of analysis (R1). In Fas-induced apoptosis of L929 cells (Figs. 16.1A–16.1D), the membrane starts to bleb at 2 h, causing increased diffraction of the laser beam, revealed by spreading of the dots (Fig. 16.1B). The release of apoptotic particles is detected by the emergence of a population of small PI-negative dots (negative because of the absence of DNA) in the lower left corner of the dot plot (Fig. 16.1B). Condensation of the nucleus, shrinkage of the cells, and formation of DNA-containing apoptotic bodies are visualized as a population of PI-positive red dots appearing in the lower left corner of the dot plot (Fig. 16.1C). The red population in Figs. 16.1D and 16.1C represents the end stage of the apoptotic process, that is, secondary necrosis. This population clearly coincides with the population of necrotic cells (Fig. 16.1H).

2.3. Analysis of cell morphology by transmission electron microscopy

Transmission electron microscopy is considered a "golden standard" in cell death research. It provides both two- and three-dimensional (using Fourier methods) images of the cell interior and thereby enhances the understanding

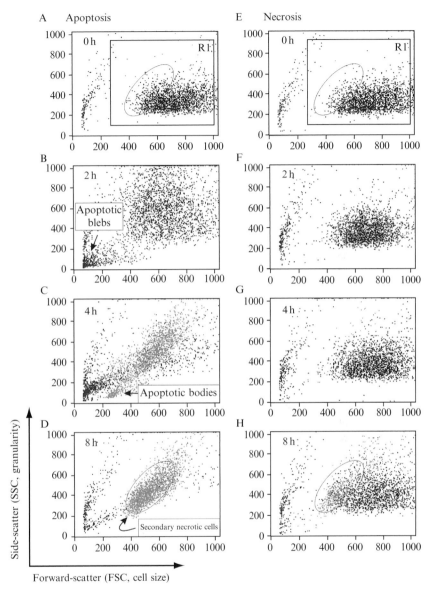

Figure 16.1 Analysis of cell morphology versus cell permeability by flow fluorocytometry. Cells were analyzed by flow fluorocytometry for changes in size (forward scatter) and granularity (side scatter), as well as for loss of membrane integrity using propidium iodide uptake (FL3). L929 fibrosarcoma cells were induced to die by apoptosis or necrosis by stimulating death receptors. The different morphological changes associated with apoptosis and necrosis can be followed over time on the dot plots (side scatter versus forward scatter). The time at which PI-positive cells appear is indicated in grey on the dot plots.

of biological structure–function relationships at (sub)cellular and molecular levels (Unwin and Ennis, 1984; Unwin and Zampighi, 1980). Compared to light microscopy, TEM is time-consuming and requires more expensive equipment, but offers much higher resolving power (0.1–0.4 nm), which provides much more detailed information about cellular morphology to distinguish apoptosis from necrosis in cell cultures. Preparation of dying cells for electron microscopy is particularly difficult because dying cells typically detach from their substrate, and spinning down these floating cells may change their original morphology. We have circumvented this problem by using macrophages attached to the bottom of tissue culture plates to capture dying apoptotic and necrotic cells.

1. Seed macrophages in adherent six-well plates at 5×10^5 cells per well and target cells in uncoated six-well suspension plates at 5×10^5 cells per well. Target cells are grown in suspension plates so that the population of dying cells can be transferred easily to the well containing attached macrophages.

2. Induce target cells (e.g., L929sAhFas) to undergo apoptosis, for example with agonistic anti-Fas antibody (125 ng/ml, for at least 1 h), or necrosis, for example with hTNF (1000 IU/ml, for at least 7 h), and collect them for coincubation with the macrophages. The times at which cells are harvested depend on the kinetics of cell death.

3. Coculture macrophages and target cells (1:1 ratio) at 37 °C, 5% CO_2 for at least 1 h.

4. Fix cocultures of macrophages and target cells in the six-well plate by immersion in TEM fixation buffer overnight at 4 °C. TEM fixation buffer: 2% glutaraldehyde containing 1 mM $CaCl_2$ and 0.1 M sucrose buffered with 0.1 M Na-cacodylate (pH 7.4).

5. Rinse 3×5 min in 100 mM Na-cacodylate containing 7.5% (w/v) sucrose.

6. Postfix in 1% (w/v) OsO_4 in the same buffer (without sucrose) overnight at 4 °C.

7. Rinse 3×5 min in 100 mM Na-cacodylate containing 7.5% (w/v) sucrose.

8. Dehydrate in a graded series of ethanol: 50% for 15 min, 70% for 20 min, 90% for 30 min, and 100% for 90 min.

9. Infiltrate with 100% ethanol + LX-112 resin (Ladd Research Industries, USA) (1:1 for 30 min, 1:2 for 30 min) and then with 100% LX-112 resin for 120 min.

10. Polymerize for at least 48 h at 60 °C.

11. Break off the plastic of the polymerized block and cut into pieces that fit into the ultramicrotome holders. Remove any remaining plastic bits from the cutting surface.

12. Cut ultrathin sections of 60 nm with a diamond knife and mount on Formvar-coated copper 100-mesh grids.

13. Evaporate the grids in a JEOL (JEC-530) autocarboncoater at 4 V during 3 s.
14. Stain with uranyl acetate (7.5% in bidistilled water, 1 drop per grid for 20 min) and Reynold's lead citrate (1 drop per grid for 10 min). To prepare Reynold's lead citrate, place 60 ml of distilled water, 3.52 g of trisodium citrate ($C_6H_5Na_3O_7 \cdot 2H_2O$), and 2.66 g of lead nitrate in a 100-ml volumetric flask. After shaking gently and letting the solution settle for 30 min, add 16 ml of 1 M NaOH slowly and mix gently. Make the volume up to 100 ml with distilled water and mix gently. The solution can be stored in plastic syringes in the refrigerator. Note that all mixing should be carried out gently to avoid introducing air into the solution, which will cause the formation of a lead carbonate precipitate that contaminates the stained sections.
15. Examine the sections by TEM.

The earliest classic ultrastructural changes detectable in apoptosis are condensation of chromatin to form uniformly dense masses lying against the nuclear envelope (Cummings et al., 1997) and persistence of the nucleolar structure until very late stages (Falcieri et al., 1994). Further features displayed by most apoptotic cells include loss of specialized surface structures such as microvilli and cell–cell contacts. The cell volume decreases due to the marked condensation of cytoplasm, and consequently cell density increases. Membrane-bound apoptotic bodies of varying size are formed; these contain well-preserved but compacted cytoplasmic organelles and/or nuclear fragments (Cummings et al., 1997; Kerr et al., 1994) (Fig. 16.2B). In contrast, cells do not fragment during necrosis (Fig. 16.2C), but their

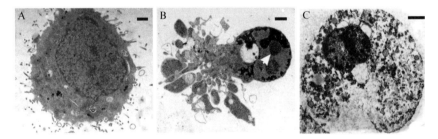

Figure 16.2 Analysis of cell morphology of apoptotic and necrotic cells by transmission electron microscopy. Unstimulated L929sAhFas fibrosarcoma cells (A) and cells exposed either to agonistic anti-Fas for 1 h (B) or to hTNF for 18 h (C). (A) The cell shows microvilli protruding from the entire surface, a smoothly outlined nucleus with heterochromatin, and well-preserved cytoplasmic organelles. (B) An apoptotic cell with sharply delineated masses of condensed chromatin, convolution of the cellular surface, and formation of apoptotic bodies. Note the nucleolus (arrow) near a cup-shaped chromatin margination. (C) A necrotic cell with clumps of chromatin with ill-defined edges, swollen mitochondria, and loss of plasma membrane integrity. Scale bars: 1 μm.

organelles swell. In contrast to cytoplasmic changes, the nuclear morphology is initially relatively unremarkable and consists of dilatation of the nuclear membrane and condensation of chromatin into small irregular circumscribed patches (Wyllie, 1981). The cytoplasm becomes increasingly translucent and, finally, the plasma membrane is disrupted, and cell contents start to leak out, provoking inflammation (Fig. 16.2C).

3. ANALYSIS OF BIOCHEMICAL CELL DEATH PARAMETERS

3.1. Analysis of surface changes during apoptosis and necrosis by flow fluorocytometry (FACS): PS exposure versus cell permeability

The exposure of phosphatidylserine (PS) on the surface of apoptotic cells is one of the best-characterized surface changes that facilitate their recognition and engulfment. Plasma membranes of viable cells exhibit substantial phospholipid asymmetry, with most of the PS residing on the inner, cytoplasmic leaflet of the membrane. Although the mechanisms involved in the translocation of PS to the external surface of the membrane are still unclear, some possibilities have been suggested, including a coordinate increase in phospholipid flip–flop (flip refers to inward and flop to outward movement) due to inactivation of the aminophospholipid translocase (Williamson and Schlegel, 2002). A novel molecular mechanism of PS exposure has been discovered in *Caenorhabditis elegans* using an RNAi reverse genetic approach and a transgenic strain expressing a GFP::annexin V reporter. Zullig *et al.* (2007) identified a *tat*-1 gene as a possible aminophospholipid transporter gene, because knocking it down abrogated PS exposure on apoptotic cells and reduced clearance of dead cell corpses. Annexin V, a Ca^{2+}-dependent phospholipid-binding protein, preferentially binds to negatively charged phospholipids such as PS and therefore is an early marker of apoptotic cell death (Denecker *et al.*, 2000). Using Alexa Fluor 488-conjugated annexin V (Invitrogen) in combination with PI uptake, determining the times at which PS exposure occurs and the membrane loses its integrity makes it possible to distinguish between apoptosis and necrosis (Denecker *et al.*, 2000). Classically, PS positivity precedes PI positivity in apoptotic cells but the two events coincide in necrotic cells (Fig. 16.3). However, it is important to mention that in some cellular systems necrotic cells, while remaining PI negative, also externalize PS (Krysko *et al.*, 2004). The reason for this is not clear, but weaker necrotic signaling probably favors the detection of PS exposure with an absence of rapid membrane permeabilization (Krysko and Vandenabeele, unpublished observations). Therefore, in some cases this PS^+/PI^- single positive staining might not discriminate between apoptosis

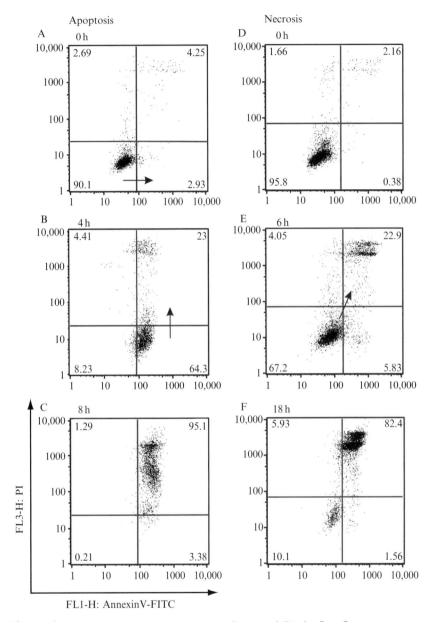

Figure 16.3 Analysis of PS exposure versus cell permeability by flow fluorocytometry. Cells were analyzed by flow fluorocytometry for loss of membrane integrity (by PI uptake) and for translocation of phosphatidylserine to the outer leaflet of the plasma membrane (by staining with annexin V-Alexa 488). L929sAhFas fibrosarcoma cells were pretreated for the indicated durations with either anti-Fas antibody or mTNF.

and necrosis. To discriminate between them, other morphological and biochemical parameters should be analyzed; these are discussed later in this overview.

   This section describes the use of flow fluorocytometry to discriminate between apoptosis and necrosis by analyzing cell membrane permeability and PS exposure. Although it does not fall within the scope of this overview, we would like to mention that PS exposure on dying cells could also be detected *in vivo*. Annexin V and its derivatives, labeled with ^{123}I, ^{125}I, ^{124}I, ^{18}F, and ^{99m}Tc, are used in a broad range of imaging applications in apoptosis research from single-photon emission computed tomography and autoradiography to positron emission tomography (Lahorte *et al.*, 2004; Van de Wiele *et al.*, 2004).

1. Follow step 1 in Section 2.2.
2. The next day, stimulate the cells, for example with TNF (10,000 IU/ml) or anti-Fas (250 ng/ml), and analyze samples of cells from the suspension cultures at regular intervals on a flow fluorocytometer.
3. Transfer 1.5 to 3×10^5 cells to an Eppendorf tube, centrifuge the cells for 5 min at 250g at 4 °C, and resuspend in 1 ml ice-cold annexin V-binding buffer (10 mM HEPES NaOH, pH 7.4, 150 mM NaCl, 5 mM KCl, 1 mM MgCl$_2$, and 1.8 mM CaCl$_2$). Keep the cells on ice to stop further biochemical activity. Detection of PS by annexin V is dramatically affected by calcium concentration, time of incubation on ice, and type of medium used (Trahtemberg *et al.*, 2007). Therefore, it is advisable to use the appropriate concentration of calcium, usually not exceeding 1.8 mM, to add calcium only 10 min before analysis of the sample, and to use a simple solution of balanced salts (containing Na, K, Cl, and Mg) for the "binding buffer." HEPES-based buffers, not phosphate buffers, should be used (Trahtemberg *et al.*, 2007).
4. Centrifuge the cells for 5 min at 250g at 4 °C and resuspend in 270 μl of annexin V-binding buffer containing 1 μg/ml of annexin V Alexa Fluor 488 (Annexin V-Alexa, Molecular Probes, Invitrogen). Incubate for 5 min on ice and protect from light.
5. Add 30 μl of PI from a 10 $\times$ substock solution (obtained from a 3 mM master stock solution) to the 270 μl of the same cell suspension directly before measurement.
6. Analyze samples by creating a dot plot showing annexin V-Alexa (FL1) versus PI (FL3), and define quadrants to divide cell populations according to annexin V and PI positivity and negativity. Also include single-stained cells in your experiment, that is, cells stained only with annexin V-Alexa (no PI added) and cells stained only with PI (no Annexin V-Alexa added) to set up compensation and quadrant markers. The absorption maximum for annexin V Alexa Fluor 488 is 495 nm and its fluorescence emission maximum is 519 nm. PI bound to nucleic acids has absorption and fluorescence emission maxima of 535 and 617 nm, respectively.

7. FL1 (520 nm) and FL3 (610 nm) channels detect annexin-V–Alexa staining and PI uptake, respectively, on a FACSCalibur flow fluorocytometer (Becton Dickinson) equipped with a water-cooled argon-ion laser at 488 nm.

A lag time between PS positivity and PI positivity is characteristic of apoptotic cells, but both events coincide in necrotic cells. Membrane changes leading to PS exposure occur rapidly in apoptotic cell death, and the cell population shifts from the lower left quadrant (PI-negative/annexin V-negative cells, Fig. 16.3A) to the lower right quadrant (PI-negative/annexin V-positive cells, Fig. 16.3B). After PS exposure to the outer leaflet of the cell membrane, cells start losing their membrane integrity, and the population shifts to the upper right quadrant (PI-positive/annexin V-positive cells, Fig. 16.3C). During necrotic cell death, PS exposure coincides with the permeabilization of the outer cell membrane. Consequently, cells immediately move from the lower left quadrant (PI-negative/annexin V-negative cells, Fig. 16.3D) to the upper right quadrant of the dot blot (PI-positive/annexin V-positive cells, Figs. 16.3E and 16.F) without passing though the intermediate PI-negative/ annexin V-positive stage.

The dynamics and interrelation between PS exposure and PI permeability during apoptosis and necrosis in L929sAhFas cells can be appreciated on time-lapse movies at http://www.ecdo.eu/research.htm (apoptosis_PS_PI. avi and necrosis_PS_PI.avi).

3.2. Analysis of cellular markers

This section describes the use of cellular markers, such as DNA fragmentation, activation of caspases, cleavage of Bid to its truncated proapoptotic form (tBid), and release of cytochrome c. These events do not occur in TNFR1-induced necrotic cell death (Denecker et al., 2001). In addition to the activation pattern of different caspases, the cleavage pattern of various substrates, such as poly(ADP-ribose) polymerase and inhibitor of caspase-activated DNase, can be detected (Lamkanfi et al., 2003). Alternatively, caspase activation can be measured using fluorogenic caspase substrates (see Section 3.2.3), although these should be used with caution, as they are not specific for particular caspases and even not for caspases in general (Darzynkiewicz and Pozarowski, 2007; Tambyrajah et al., 2007; Timmer and Salvesen, 2007). This chapter does not describe the use of dyes that are sensitive to mitochondrial membrane potential, such as tetramethylrhodamine methyl ester perchlorate, rhodamine123, JC-1 (Molecular Probes), or the use of dihydrorhodamine 123 for the measurement of reactive oxygen species (ROS), as these methods do not distinguish the mode of cell death (Kroemer et al., 1998; Metivier et al., 1998). In both cell death pathways the membrane potential of the mitochondria eventually drops and cells start to produce ROS (Denecker *et al.*, 2001; Lemasters *et al.*, 1998). It was shown that although ROS production in L929 cells occurs in both cell death pathways, it is crucial and complex I dependent only in necrotic cell death (Goossens *et al.*, 1995; Vercammen *et al.*, 1998).

3.2.1. Analysis of DNA fragmentation by flow fluorocytometry

DNA fragmentation is a hallmark of apoptosis and involves the formation of high molecular weight ($>$50 kbp) and nucleosome-sized (200 bp) DNA fragments (Nagata, 2000). An easy and quantitative way to analyze DNA fragmentation is by adding PI to the dying cell population and applying one freeze–thaw cycle to permeabilize cells and cell fragments. PI intercalates in the DNA and the size of DNA fragments appears as a hypoploid DNA histogram by flow fluorocytometry. This technique allows the discrimination between apoptotic and necrotic cell death.

1. Induce cell death as described in Section 3.1.
2. Transfer 270 μl of cells in suspension from the 24-well suspension plate to a 5-ml polypropylene round bottom tube, and add 30 μl PI from the $10 \times$ substock (prepared from a 3 mM master stock).
3. Subject the tubes to one freeze–thaw cycle by immersing them briefly in liquid nitrogen; this permeabilizes the cells, making them PI positive.
4. Analyze samples after thawing.
5. Use the same experimental setup as described in Section 3.1.

Viable nondividing cells exhibit a clear diploid DNA peak (2n, G1), whereas dividing cells give rise to a tetraploid peak (4n, G2). A nonsynchronized population of cells shows a typical biphasic peak of 2n (G1) and 4n (G2) cells (Fig. 16.4A). During apoptosis, the DNA is fragmented and is partially lost due to the formation of apoptotic bodies. As a result, apoptotic cells typically show a hypoploid DNA fluorescence pattern ("sub-G1" peak) compared to necrotic cells, which maintain their entire DNA content in the nucleus (Figs. 16.4B and 16.4C).

3.2.2. Analysis of activation of caspases by Western blotting

Caspases are a family of cysteine aspartate-specific proteases. They are synthesized as zymogens consisting of a prodomain of variable length, followed by p20 and p10 units that contain the residues essential for substrate recognition and catalytic activity. They are activated by proximity-induced autoproteolysis by interacting with a platform of adaptor proteins through a protein interaction motif in the prodomain (death effector domain or caspase recruitment domain) or by cleavage by upstream proteases in an intracellular cascade. The net result of these proteolytic activities is separation of the prodomain from the p20 and p10 subunits, which form an active heterotetramer (Lamkanfi *et al.*, 2003).

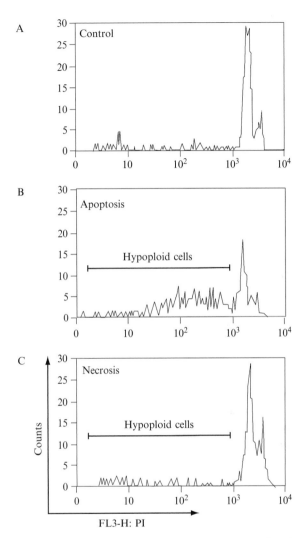

Figure 16.4 Schematic representation of DNA fragmentation analysis by flow fluoro-cytometry. (A) Control, live cells are characterized by a typical biphasic peak of 2n (G1) and 4n (G2) cells. (B) Apoptosis is characterized by the presence of hypoploid DNA, represented by the population of cells with lower fluorescence than the G1 peak. (C) Necrotic cells are characterized by the absence of hypoploid DNA.

1. Day 0: seed 2×10^5 cells per well in coated six-well tissue culture plates.
2. Day 1: stimulate the cells and collect cell samples at regular intervals in a 1.5-ml Eppendorf tube.
3. Centrifuge the cell suspension for 5 min at 250g, aspirate the medium, and wash cells with 1 ml of cold phosphate-buffered saline (PBS).

4. Centrifuge the cells for 5 min at 250g, aspirate the PBS, and lyse the cells in 80 μl of cold caspase lysis buffer. Keep the cells on ice to stop further biochemical activity of the cell. These samples can be kept frozen until they are analyzed. Cell samples should be lysed and processed in a standardized way, using equal volumes and equal numbers of the seeded cells.

5. Remove cell debris by centrifuging the lysate (10 min at 20,800g at 4 °C) and transfer the remaining cytosol to another Eppendorf tube. One should measure the protein concentration of at least the control samples and load 10 to 15 μg of protein. It is important to load the same volumes for other samples and not to correct for protein content because cells start to leak cytosolic proteins in the supernatant at later time points during the cell death process (discussed later). Add a 0.2 volume of 5 × Laemmli buffer to the cell lysate and boil for 5 to 10 min. 5 × Laemmli buffer: 312.5 mM Tris-HCl, pH 6.8, 10% SDS, 50% glycerol, and 20% β-mercaptoethanol.

6. Load equal volumes per lane on 12.5% SDS-polyacrylamide gels.

7. After SDS-PAGE electrophoresis, transfer the proteins to a nitrocellulose membrane and detect the different caspases with the appropriate antibodies: antihuman caspase-3 (BioSource) and antihuman caspase-7 (Stress-Gen Biotechnologies Corp.). Horseradish peroxidase (HRP)-coupled secondary antibodies are from Amersham Life Science (Amersham). Reveal with Chemiluminescence Reagent Plus (PerkinElmer Life Sciences, Boston, MA).

Activation of initiator procaspases results in the appearance of a prodomain and p20 and p10 subunits. Proteolytic activation of effector procaspases leads eventually to the appearance of typical p20 and p10 subunits. Depending on the antibody used, from one to all proteolytic fragments can be detected. There are also antibodies that specifically recognize the activated form of caspases (PharMingen). These antibodies can also be used in FACS analysis to detect caspase activity in cells; alternatively, cell-penetrant fluorogenic substrates can be used (Sanchez *et al.*, 2003). It is important to mention that activated caspase-3 (Cell Signaling Technology) could also be detected *in situ* by immunohistochemistry (Duan *et al.*, 2003) to discriminate between apoptosis and necrosis.

For example, in anti-Fas-induced apoptosis, the activation of caspase-3 and -7 is characterized by the appearance of the p20 subunit (Denecker *et al.*, 2001), whereas in TNFR1-induced necrosis, caspase-3 and -7 are not activated, and thus a p20 subunit does not appear (Denecker *et al.*, 2001). Procaspases and activated caspases are also released from primary and secondary necrotic cells, respectively, and can be detected in their supernatant (see Sections 3.3.2 and 3.3.3).

3.2.3. Analysis of caspase activity using fluorogenic substrates

To measure the activity of caspases instead of their proteolytic activation state, fluorogenic tetrapeptide substrates for caspases can be used. The first synthetic substrates designed and used to measure and analyze caspase activities were selected from tetrapeptide libraries ($P_4P_3P_2P_1$) in which aspartate at the P1 position (XXXD) is conjugated to a fluorogenic 7-amino-4-methylcoumarin (AMC) or 7-amino-4-trifluoromethylcoumarin (AFC) (Lamkanfi et al., 2003; Thornberry et al., 1997). In this way, acetyl(Ac)-DEVD-AMC was identified as a preferred substrate for caspase-3 and -7, Ac-LEHD-AMC for caspase-5, Ac-YVAD-AMC for caspase-1 and -4, Ac-IETD-AMC for caspase-8 and -6, and Ac-WEHD-AMC for caspase-1, -4, and -5 (Lamkanfi et al., 2003). The specificity of the substrates is not always restricted to the ones mentioned by the manufacturers (Talanian et al., 1997). The problem is loss of specificity at certain caspase concentrations. High concentrations of caspase-3 will also cleave Ac-IETD-AMC, the substrate for caspase-8 and -6. Therefore, it is appropriate to combine enzymatic measurements of caspase activities with Western blotting to identify the presence and activation status of the caspase (see Section 3.2.2). Moreover, procaspases and active caspases are released from primary and secondary necrotic cells and can be detected in their supernatant (see Sections 3.3.2 and 3.3.3).

1. Follow steps 1 to 3 in Section 3.2.2.
2. Centrifuge the cells for 5 min at 250g, aspirate the PBS, and lyse the cells in 150 μl of cold caspase lysis buffer (1% NP-40, 10 mM Tris-HCl, pH 7.4, 10 mM NaCl, and 3 mM MgCl$_2$; 1 mM phenylmethylsulfonyl fluoride, 0.3 mM aprotinin, and 1 mM leupeptin should be added directly before the cell lyses). Do not use pefablock protease inhibitors or other total protease inhibitor mixes because they interfere with caspase activity. Keep cells on ice to stop further biochemical activity. To block the catalytic cysteine of caspases and avoid the activation of caspase cascades during the lysis process, add 1 mM oxidized glutathione (GS-SG). At this stage the samples can be frozen. For the final measurement of caspase activity in the cell-free system buffer (CFS buffer), 10 mM dithiothreitol (DTT) is added to remove the GSH from the catalytic cysteine. CFS buffer: 10 mM HEPES NaOH, pH 7.4, 220 mM mannitol, 68 mM sucrose, 2 mM MgCl$_2$, 2 mM NaCl, 2.5 mM H$_2$KPO$_4$, 0.5 mM EGTA, 0.5 mM sodium pyruvate, and 0.5 mM L-glutamine. PBS: 8 g/liter NaCl, 0.2 g/liter KCl, 2.89 g/liter Na$_2$HPO$_4$·12H$_2$O, and 0.2 g/liter KH$_2$PO$_4$.
3. Remove cell debris by centrifuging the lysate (10 min at 20,800g at 4 °C) and transfer the cytosol to another Eppendorf tube (cytosolic cell lysate).
4. Measure caspase activity by incubating 10 μg cytosolic cell lysate with 50 μM of the fluorogenic substrate, for example, Ac-DEVD-AMC, in

150 μl CFS buffer. Add DTT to a final concentration of 10 mM. Caspase fluorogenic substrates: Ac-DEVD-AMC, Ac-LEHD-AMC, Ac-YVAD-AMC, Ac-IETD-AMC, and Ac-WEHD-AMC (Peptide Institute, Osaka, Japan) are prepared as 100 mM stock solutions in dimethyl sulfoxide (DMSO).

5. Monitor the release of fluorescent AMC at 37 °C every 2 min for 1 h in a fluorometer (e.g., CytoFluor, PerSeptive Biosystems) using a filter with an excitation wavelength of 360 nm and a filter with an emission wavelength of 460 nm. Express data as the increase in fluorescence per unit of time (Δfluorescence/min) (Denecker *et al.*, 2001; Vanden Berghe *et al.*, 2003).

6. Caspase activity is only detected in apoptotic conditions, both in cell lysates and in supernatants. No caspase activity has been found in necrotic cell death.

3.2.4. Analysis of Bid cleavage and cytochrome *c* release by Western blotting

To determine the role of the mitochondrial apoptotic pathway, one can investigate the engagement of mitochondria at the different stages: caspase-mediated activation of Bid, tBid translocation, and cytochrome *c* release. To detect the release of cytochrome *c* in the cytosol, the organelle and cytosolic fractions should be separated while ensuring that cytochrome *c* is not spontaneously released from mitochondria during organelle preparation. Therefore, the mild detergent digitonin is used at a concentration of 0.02%, which can permeabilize the plasma membrane, but leaves mitochondria and lysosomes intact. The appropriate concentration of digitonin and the time required for lysis depend on the cell type. As a result, a series of digitonin concentrations should be tested to select the condition that satisfies two requirements: the outer membrane is permeabilized (detection of release of cytosolic proteins, e.g., actin) and the mitochondrial or lysosomal membranes remain intact (no detection of mitochondrial proteins, e.g., COX or lysosomal proteins, e.g., hexosaminidase).

1. Follow steps 1 to 3 in Section 3.2.2.
2. Centrifuge the cells for 5 min at 250g, remove the PBS, and lyse the cells in 100 μl of 0.02% digitonin dissolved in cell-free system buffer (CFS buffer), and leave on ice for 1 min. Keep cells on ice to stop further biochemical activity.
3. Centrifuge the lysate (10 min at 20,800g at 4 °C) and transfer the cytosol to a separate Eppendorf tube. Add 0.2 volume of 5 × Laemmli buffer to the cytosolic lysate and 100 μl of 1 × Laemmli buffer to the remaining organelle fraction and boil for 5 to 10 min.
4. Load equal volumes of the organelle fraction or cytosolic cell lysate per lane on 12.5% SDS-polyacrylamide gels (see comment in step 6 of

Section 3.2.2). After SDS-PAGE electrophoresis, transfer the proteins to a nitrocellulose membrane by electroblotting and detect Bid (R&D Systems) cleavage and cytochrome c (Pharmingen) with the appropriate antibodies or antisera. Reveal with Chemiluminescence Reagent Plus (PerkinElmer Life Sciences).

Anti-Fas treatment leads to early cleavage of full-length Bid to its proapoptotic truncated form (tBid). tBid can be detected in the cytosolic fraction only briefly and in small amounts. Most of the tBid becomes rapidly associated with the organelle fraction. Therefore, the safest way to detect the engagement of Bid is to look at the disappearance of Bid from the cytosol and at its appearance in the organelle fraction. In necrotic cells, no cleavage of p22 Bid is detectable either in the cytosol or in the organelle fraction. Bid disappears from the cytosol only late in the necrotic process due to leakage, and a small amount of tBid becomes detectable in the organelle fraction, probably because of alternative Bid cleavage by lysosomal proteases (Stoka et $al.$, 2001). The appearance of tBid in the organelle fraction during anti-Fas-induced apoptosis coincides with the release of cytochrome c into the cytosol. During secondary necrosis of anti-Fas-stimulated cells, cytochrome c also accumulates in the culture supernatant. TNFR1-induced necrosis does not result in detectable release of cytochrome c into the cytosol. However, in the late necrotic phase of TNF-stimulated cells, cytochrome c starts to accumulate in the culture supernatant as soon as the plasma membrane loses its integrity. For Western blots of tBid cleavage and cytochrome c release, see Denecker et $al.$ (2001).

3.3. Analysis of supernatants from apoptotic and necrotic cells

In the absence of phagocytosis, apoptotic cells turn into secondary necrotic cells, which have many features of primary necrotic cells (discussed in Section 3.1). One such feature is a compromise of plasma membrane integrity, leading to the leakage of large amounts of intracellular contents, with consequent induction of immune responses and chemoattraction of antigen-presenting cells (Krysko and Vandenabeele, 2008). In this regard, detection of lactate dehydrogenase (LDH) release may provide an easy method for determining the extent of cell death irrespective of the type of cell death (discussed later). The detection of certain factors that are released only from primary and secondary necrotic cells in the supernatant may provide an additional tool for further characterization of these cell death types. This section describes the detection of caspase-3 and -7, a high mobility group box 1 protein (HMGB-1), and cytokeratin 18 (CK18) in the supernatant of primary and secondary necrotic cells.

3.3.1. Analysis of LDH release

Lactate dehydrogenase is a cytosolic enzyme present in all mammalian cells. Release of LDH from cells provides an accurate measure of cell membrane integrity and cell viability irrespective of the type of cell death (Rae, 1977). This method allows large-scale screening of different conditions in 96-well plates and optimization of concentrations of inhibitors or sensitizing reagents.

1. Seed cells the day before analysis at 2×10^4 cells/100 μl per well in 96-well plates.
2. Expose the cells to a serial dilution of TNF.
3. Transfer 50 μl of each cell supernatant to a 96-well plate.
4. Add 50 μl reconstituted substrate mixture (CytoTox 96 assay, Promega) and incubate for 30 min at room temperature. The CytoTox 96 assay measures LDH activity in the culture medium using a coupled two-step reaction. In the first step, LDH catalyzes the reduction of NAD^+ to NADH and H^+ by oxidation of lactate to pyruvate. In the second step of the reaction, diaphorase uses the newly formed NADH and H^+ to catalyze the reduction of a tetrazolium salt [2-p-(iodophenyl)-3-(p-nitrophenyl)-5-phenyltetrazolium chloride] to an intensely colored formazan that absorbs strongly at 490 to 520 nm. The amount of formazan formed is proportional to the amount of LDH released into the culture medium.
5. Add 50 μl stop solution to each well and measure optical density at 492 nm.
6. Results may be expressed either as optical absorbance values or as international units of enzyme (calculated from a standard curve obtained with the positive LDH control provided with the CytoTox 96 assay kit).

3.3.2. Analysis of caspases release by Western blot analysis

Secondary necrosis is characterized by release of the p20 subunits of both caspase-3 and -7 into the culture medium. In contrast, in necrosis, only procaspases-3 and -7 are detected in the supernatant (Denecker *et al.*, 2001). Therefore, detection of the p20 subunit of caspase-3 and -7 in the supernatant may be indicative of apoptotic cell death.

1. Day 0: seed 1.2×10^6 cells per 90-mm petri dish.
2. Day 1: wash cells twice with serum-free medium and induce cell death in 7 ml of serum-free medium. Collect the supernatant samples at regular time intervals in 15-ml tubes.
3. Centrifuge the supernatant for 5 min at 250g. Transfer 5 ml of the supernatant to new 15-ml tubes and centrifuge again for 5 min at 250g.
4. Distribute 4 ml of supernatant in 1.5-ml Eppendorf tubes.

5. Add 0.1 volume of deoxycholate from a stock of 5 mg/ml to each Eppendorf tube and incubate on ice for 20 min. Deoxycholic acid is added as a base (sodium salt) and precipitates after addition of the stronger trichloroacetic acid. In this protocol, it acts as a coprecipitant and helps the protein precipitate.

6. Add 0.1 volume of 100% trichloroacetic acid (Sigma) to each tube and incubate on ice for 30 min.

7. Centrifuge at 20,800g for 5 min at 4 °C . Discard the supernatant and add 500 μl of ice-cold 100% acetone.

8. Centrifuge at 20,800g for 5 min at 4 °C .

9. Dissolve each pellet in 20 to 30 μl of sample buffer (3 volumes of 2×Laemlli buffer + 1 volume of 1 M Tris-HCl) and pool the samples together. If the Laemmli buffer turns yellow, add 1 M Tris-HCl buffer 1 μl at a time until it turns blue.

10. Boil the samples for 10 min.

For analysis of caspase activation by Western blotting, follow step 4 in Section 3.2.2. During the early stages of apoptosis, caspase-3 and -7 are activated and can be detected in the cytosolic fraction (see Section 3.2.2.). From the moment when apoptotic cells lose plasma membrane integrity and proceed to the secondary necrotic stage, active caspase-3 and -7 become detectable in the supernatant. In contrast, in the late stages of necrosis, no active caspase-3 and -7 can be observed in the supernatant (Denecker *et al.*, 2001).

3.3.3. Analysis of caspases release using fluorogenic substrates

Caspase release into the supernatant can also be detected by using fluorogenic substrates (discussed in Section 3.2.3). Secondary necrosis is distinguished from primary necrosis by detecting caspase-3 and -7 activities in the supernatant. Note that there is no active caspase-3 and -7 in the supernatant in the late stages of necrosis (Denecker *et al.*, 2001).

1. Seed the cells the day before analysis at 2 × 10^4/100 μl per well in 96-well plates.

2. Expose the cells to a serial dilution of a cell death stimulus.

3. Measure caspase activity by incubating at least 20 μl supernatant with 50 μM of the fluorogenic substrate, for example, Ac-DEVD-AMC, in 150 μl CFS buffer as described in steps 4 to 6 of Section 3.2.3.

3.3.4. Analysis of HMGB-1 release by western blotting

Another protein that is released only from necrotic cells but remains bound in apoptotic cells even when they are undergoing secondary necrosis is high mobility group box 1 protein (HMGB-1) (Scaffidi *et al.*, 2002). HMGB-1 acts as an architectural chromatin-binding factor that bends DNA and promotes protein assembly on specific DNA targets (Scaffidi *et al.*, 2002).

HMGB-1 remains associated with the DNA in apoptotic cells but not during necrotic cell death. Once released from necrotic cells, HMGB-1 cells can incite inflammatory responses from macrophages (Scaffidi *et al.*, 2002).

3.3.5. Analysis of cytokeratin 18 release by ELISA

Under normal physiological conditions, cytokeratins are complexed in intermediate filaments of epithelial cells and remain insoluble (Fuchs and Weber, 1994). It has been shown that the fate of epithelial-specific intermediate filament cytokeratin (CK18) depends on the type of cell death. The M30-Apoptosense assay (Peviva AB, Bromma, Sweden) specifically measures a neoepitope formed by caspase cleavage of CK18 at Asp396 (CK18Asp396-NE M30 neoepitope) and reflects apoptotic cell death. This assay provides a specific method for discriminating between apoptotic and necrotic cell death (Cummings *et al.*, 2007). The M65 ELISA (Peviva AB, Bromma, Sweden) measures soluble CK18 released from dying cells and can be used to assess the overall death of epithelial cells due to both apoptosis and necrosis. The units of the two assays have been calibrated against identical standard material so that a ratio between caspase-cleaved and total CK18 can be calculated ("M30:M65 ratio"). Induction of apoptosis in cultured cells results in the release of caspase-cleaved CK18 and in relatively high M30:M65 ratios, whereas induction of necrosis will almost exclusively result in the release of CK18 molecules that are not caspase cleaved and in a low M30:M65 ratio. The M30:M65 ratio, therefore, reflects the type of cell death.

It is important to mention that caspase-cleaved CK 18 could also be detected *in vivo* by immunohistochemistry (Duan *et al.*, 2003). This technique has been widely applied in clinical trials as a pharmacodynamic biomarker of cell death induced by a variety of different cancer chemotherapeutic agents in a spectrum of different types of cancer (Cummings *et al.*, 2007). For the detection protocol of total CK18 and CK18Asp396-NE, refer to www.peviva.se (Peviva AB, Bromma, Sweden).

4. ANALYSIS OF CELL–CELL INTERACTIONS

The clearance of dying cells is a complex and dynamic process coordinated by interplay between ligands on dying cell, bridging molecules, and receptors on engulfing cell (Krysko *et al.*, 2006b). Efficient clearance of cells undergoing apoptotic or necrotic cell death is crucial for normal homeostasis and for the modulation of immune responses (Krysko and Vandenabeele, 2008). This section describes an *in vitro* phagocytosis assay for the analysis of interactions between macrophages and dying cells. It also describes a two-parameter flow fluorocytometry phagocytosis assay for quantifying uptake, scanning electron microscopy (SEM) to study surface changes of phagocytes during engulfment, and transmission electron and fluorescence microscopy in combination with fluid phase markers to distinguish between the internalization mechanisms used by macrophages to engulf apoptotic versus necrotic cells.

4.1. *In vitro* phagocytosis assay by flow fluorocytometry

Technical limitations and differences in experimental approaches have hindered elucidation of the process of dying cell clearance. For example, many studies use fluorescence microscopy to quantify the uptake of dying cells by counting the prestained ingested target cells inside the phagocytes in several microscopic fields. Obviously, this technique is labor-intensive, time-consuming, and has a subjective component. To overcome these limitations and to explore the recognition and engulfment of dying cells by macrophages, we developed a quantitative, objective approach. The following protocol describes a two-parameter flow fluorocytometry phago-cytosis assay to quantify the percentage of dying cell clearance, which is based on staining of target cells (L929sAhFas cells) with Cell Tracker green and phagocytes (Mf4/4) with Cell Tracker orange. These reagents pass through cell membranes, but inside the cell they are transformed into cell-impermeant reaction products. In *in vitro* phagocytosis assays we use the mouse macrophage cell line Mf4/4 as professional phagocytes, which has been characterized elsewhere (Desmedt *et al.*, 1998).

1. Detach the cells with enzyme-free cell dissociation buffer (Gibco BRL).
2. Transfer 10 to 20 × 10^6 L929sAhFas and Mf4/4 cells into separate 15-ml tubes containing complete medium, centrifuge the cells for 5 min at 250g, and resuspend them in 10 ml of serum-free medium.
3. Centrifuge the cells for 5 min at 250g and resuspend them in serum-free medium containing prediluted dyes. Stain target cells and phagocytes with 0.8 μM Cell Tracker green and 10 μM Cell Tracker orange, respectively, for 30 min at 37 °C with rotation. Cell Tracker green and Cell Tracker orange (Molecular Probes, Invitrogen): prepare as a 10 mM stock solution in sterile DMSO. Cell Tracker green and Cell Tracker orange have Abs 492 nm/Em 517 nm and Abs 541 nm/Em 565 nm, respectively.
4. Centrifuge the cells for 5 min at 250g and resuspend in 10 ml of complete medium.
5. Centrifuge the cells for 5 min at 250g and resuspend in 10 ml of complete medium. Incubate the cells for 30 min at 37 °C with rotation.
6. Count the cells and seed the day before the coculture experiment: target cells (e.g., L929sAhFas) at 2.5 × 10^5 cells per well in uncoated 24-well

suspension tissue culture plate. Target cells are grown in suspension plates so that the population of dying cells can be transferred easily to the well containing attached phagocytes (e.g., Mf4/4).

7. The following day, induce cell death in L929sAhFas target cells by stimulating them either with anti–Fas antibodies (125 ng/ml) or with mTNF (1000 IU/ml) for 2 and 5 h, respectively. To control the induction of cell death, it is advisable to seed target cells in parallel for flow fluorocytometric analysis as described in Sections 2.2 and 3.1.

8. After stimulation, collect target cells and wash twice in the medium in which macrophages will be seeded.

9. Count the target cells and add to phagocytes to obtain the desired ratio of targets to phagocytes. It is important to maintain the same ratio in all samples.

10. Incubate the coculture for 2 h. Carefully wash away target cells with PBS and detach the phagocytes with enzyme-free cell dissociation buffer (Gibco BRL). It is generally important to perform time kinetics when analyzing dead cell clearance.

11. Centrifuge for 5 min at 250g at 4 °C, resuspend in ice-cold PBS, and analyze by flow fluorocytometry.

The uptake of apoptotic cells is much faster than the uptake of necrotic cells (Brouckaert *et al.*, 2004). Figure 16.5 is a dot plot representation of a flow fluorocytometric analysis of a coculture of Mf4/4 cells stained with Cell Tracker orange with target cells stained with Cell Tracker green. In this case target cells were treated as follows: untreated control; anti–Fas 125 ng/ml, 2 h (apoptotic); and TNF 1000 IU/ml, 5 h (necrotic). Single positive Mf4/4 cells (red) and free target cells (green) accumulate in regions I and III, respectively. After engulfing dying cells, macrophages become double positive (red and green) and accumulate in region II. The percentage of double-stained macrophages out of the whole macrophage population reflects the fraction of the macrophage population involved in the clearance of target cells (percentage phagocytosis) (Fig. 16.5). Moreover, the increase in mean green fluorescence could be used as an estimate of the amount cleared by phagocytes.

4.2. Analysis of phagocyte surface changes by scanning electron microscopy

Scanning electron microscopy can be used to characterize the surface characteristics of macrophages during the internalization of dying cells. Mouse macrophages (Mf4/4) are used as professional phagocytes, and L929sAhFas as targets.

1. Stimulate target cells and establish a coculture of phagocytes with target cells at a given ratio as described in Section 4.1. Unstained cells can be

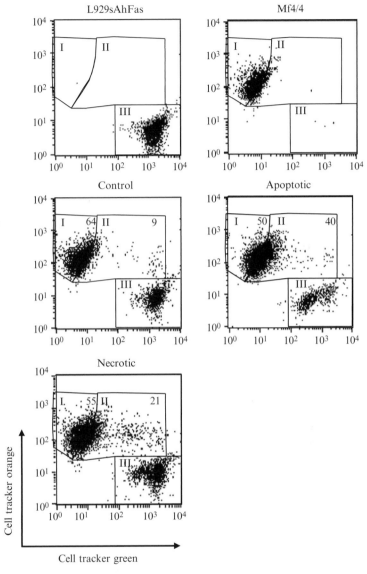

Figure 16.5 Flow fluorocytometric analysis of uptake of apoptotic and necrotic cells by the Mf4/4 macrophage cell line. Dot plot representation of flow fluorocytometric analysis of Cell Tracker orange-stained Mf4/4 cells after a 2-h coincubation with Cell Tracker green-stained target cells treated as follows: untreated (control); 2 h anti-Fas (apoptotic); and 5 h TNF (necrotic). Mf4/4 cells accumulate in regions I (no uptake) and II (uptake) and free target cells in region III. The calculated percentage of double-positive Mf4/4 cells (% uptake) for control (12%), apoptotic (44%), and necrotic (28%) indicates the extent to which macrophages are engaged in phagocytosis.

used in this setup. For control purposes, always perform the flow fluor-ocytometric assay (as described in Section 4.1) to have an idea of the percentage of engulfment.

2. Coculture macrophages and target cells in a 1:1 ratio at 37 °C, 5% CO_2 for 2 h.

3. Fix cocultures of macrophages and target cells in the 24-well plate for 1 h by immersion in prewarmed (37 °C) SEM fixation buffer (2% glutaral-dehyde containing 0.1 M sucrose buffered with 0.1 M Na-cacodylate, pH 7.2).

4. Rinse twice in 0.15 M Na-cacodylate HCl buffer.

5. Postfix for 90 min in 1% OsO_4 in 0.15 M Na-cacodylate HCl buffer at room temperature.

6. Dehydrate in a graded series of ethanol: 30% for 5 min, 50% for 5 min, 70% for 10 min, 85% for 10 min, 95% for 10 min, and 100% for 10 min.

7. Subject the specimens to a critical point drying from liquid CO_2.

8. Mount samples on SEM stubs and examine by SEM.

4.3. Analysis of internalization mechanisms

Two types of internalization processes (endocytosis) have been reported so far. Phagocytosis is the efficient uptake of large particles. The other process is fluid phase uptake of small molecules, which occurs by two distinct mechanisms: micropinocytosis is the ingestion of small vesicles via clathrin-coated pits and macropinocytosis is the ingestion of fluid via pino-somes formed by membrane ruffling (Swanson and Baer, 1995; Swanson and Watts, 1995; Torii *et al.*, 2001). Fluid-phase pinocytosis is usually studied by detecting the cellular accumulation of different soluble and impermeant probes, such as lucifer yellow (LY) and HRP (Norbury *et al.*, 1997; Norbury *et al.*, 1995; Steinman and Cohn, 1972; Steinman *et al.*, 1974; Swanson, 1989). We have developed a method involving the cocul-ture of phagocytes and target dying cells in the presence of fluid-phase markers. This technique enables us to demonstrate that apoptotic cells are internalized by a zipper-like mechanism, whereas necrotic cells are taken up by macropinocytosis (Krysko *et al.*, 2003, 2006a). It is important to mention that distinguishing between these internalization mechanisms requires the use of fluid-phase tracers (horseradish peroxidase and lucifer yellow) in combination with microscopic examination: colocalization of fluid phase markers with the ingested material in the same compartment indicates macropinocytosis. Intracellular distribution of LY can be monitored in cocultures of macrophages and target cells by fluorescence microscopy and HRP by light and transmission electron microscopy. Transmission electron microscopy is the method of choice for determining whether ingested material is localized in spacious macropinosomes or in tightly fitting phagosomes. The following sections describe the use of fluid phase

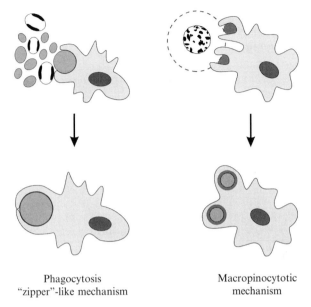

Phagocytosis   Macropinocytotic
"zipper"-like mechanism   mechanism

Figure 16.6 A schematic representation of internalization mechanisms used by macrophages to engulf apoptotic and necrotic cells. Apoptotic cells are internalized by the formation of tight-fitting phagosomes that exclude lucifer yellow, a fluid phase marker. In contrast, necrotic cell material is internalized by macrophages with the formation of spacious macropinosomes that also contain lucifer yellow (green staining).

markers to discriminate between the distinct mechanisms macrophages use to engulf apoptotic and necrotic cells (Fig. 16.6). As a quantitative control, always perform FACS assay as described in Section 4.1.

4.3.1. Transmission electron microscopy in combination with horseradish peroxidase

1. Stimulate target cells as described in Section 4.1. In this case, unstained cells may be used.
2. Add fluid phase marker HRP (1 mg/ml) and target cells simultaneously to the adherent cultures of phagocytes and coculture for 2 h at 37 °C.
3. Rinse the cocultures of macrophages and target cells in PBS buffer and fix by immersion in TEM fixation buffer for 1 h on ice. TEM fixation buffer: 2% glutaraldehyde containing 1 mM CaCl$_2$ and 0.1 M sucrose buffered with 0.1 M Na–cacodylate (pH 7.4).
4. Remove the fixative by washing several times with 0.1 M Na–cacodylate buffer (pH 7.4).
5. Reveal the presence of HRP by incubating the cocultures at 37 °C in Tris buffer (pH 7.6) containing 0.05 M 3,3-diaminobenzidine (DAB),

0.1% H_2O_2, and 1 mg aminotriazole in 9 ml of buffer to block endogenous catalase activity. Reveal the localization of HRP cytochemically by the following oxidation reaction:

$$HRP + H_2O_2 + DAB \rightarrow HRP + 2H_2O$$
$$+ \text{oxidized DAB(brownish staining)}$$

6. Wash in Tris buffer containing 7.5% sucrose and osmicate overnight in 2% OsO_4 in the same buffer without sucrose. At this step, brownish staining is converted to a black electrondense material due to the reaction of osmium tetroxide with the oxidized DAB.
7. Rinse extensively in 0.1 M Na-cacodylate buffer (pH 7.4).
8. Dehydrate in a graded series of ethanol (70, 85, 95, 100%; 10 min each).
9. Infiltrate with ethanol: LX-112 resin (Ladd Research Industries, USA) (1:1 for 30 min, 1:2 for 30 min) and then with 100% LX-112 resin for 120 min.
10. Polymerize for at least 48 h at 60 °C .
11. Break off the plastic of the polymerized block and cut into pieces that fit in the ultramicrotome holders. Remove any remaining plastic bits from the cutting surface.
12. Cut semithin sections of 2 μm and contrast them with toluidine blue (0.1%) in order to examine the quality of cocultures and to select an area for ultrathin sections.
13. Cut ultrathin sections of 60 nm with a diamond knife and mount on Formvar-coated copper 100-mesh grids.
14. Evaporate the grids in a JEOL (JEC-530) autocarboncoater at 4 V during 3 s.
15. Stain with uranyl acetate (7.5% in bidistilled water, 1 drop per grid for 20 min) and Reynold's lead citrate (1 drop per grid for 10 min). Reynold's lead citrate composition can be found in Section 2.3.
16. Examine the sections by TEM.

4.3.2. Fluorescence microscopy in combination with lucifer yellow

1. Stimulate target cells as described in Section 4.1.
2. To the nonlabeled adherent cultures of phagocytes add fluid phase markers LY CH lithium salt (Molecular Probes, Invitrogen, 1 mg/ml) simultaneously with target cells labeled with Cell Tracker orange (10 μM, prepared as described in Section 4.1) and coculture for 2 h at 37 °C . It is better to seed phagocytes on insertion glasses so that preparation for microscopy is easier.
3. Rinse adherent cocultures five times in PBS and fix in FM fixation buffer for 15 min at room temperature. FM fixation buffer: 3.8% freshly prepared paraformaldehyde in PBS (pH 7.4).

4. Rinse adherent cocultures three times in PBS.
5. Mount coverslips using Vectashield mounting medium for fluorescence with DAPI H-1200 (Vector Burlingame). Obtain DIC and fluorescence images with a confocal microscope. To detect staining use the following filter sets: for lucifer yellow BP 470/40, FT 500, 525/50; for Cell Tracker red BP 515/560, FT 580, LP 590; and for DAPI BP 340-380, FT 400, LP 425. To obtain 3D information, take images at different z levels. Carry out blind deconvolution (MLE algorithm) and 3D rotations using Leica Deblur software. Because primary and secondary necrotic cells could show background staining, quantify LY fluorescence of primary and secondary necrotic cells by determining the average fluorescence intensity and then correct all images for background staining using Metamorph 5.0 software or free software Image J (http://rsb.info.nih.gov/ij/).

ACKNOWLEDGMENTS

We thank Wim Drijvers for the artwork and Dr. Amin Bredan for editing the manuscript. Dr. Dmitri V. Krysko is paid by a postdoctoral fellowship from the BOF (Bijzonder Onderzoeksfonds 01P05807), Ghent University, and Dr. Tom Vanden Berghe is paid by a postdoctoral fellowship from FWO (Fonds Wetenschappelijk Onderzoek–Vlaanderen). This work has been supported by Flanders Institute for Biotechnology (VIB) and several grants from the European Union (EC Marie Curie Training and Mobility Program, FP6, ApopTrain, MRTN-CT-035624; EC RTD Integrated Project, FP6, Epistem, LSHB-CT-2005-019067), the Interuniversity Poles of Attraction-Belgian Science Policy (P6/18), the Fonds voor Wetenschappelijk Onderzoek–Vlaanderen (2G.0218.06 and G.0133.05), and from the Research Fund of Ghent University (Geconcerteerde Onderzoekstacties Nos. 12.0514.03 and 12.0505.02).

REFERENCES

Bessis, M. (1964). Studies on cell agony and death: An attempt at classification. *In* "Ciba Foundation Symposium of Cellular Injury" (A. V. S. de Reuck and J. Knight, Eds.), pp. 287–328. J&A Churchill, London.

Brouckaert, G., Kalai, M., Krysko, D. V., Saelens, X., Vercammen, D., Ndlovu, M., Haegeman, G., D'Herde, K., and Vandenabeele, P. (2004). Phagocytosis of necrotic cells by macrophages is phosphatidylserine dependent and does not induce inflammatory cytokine production. *Mol. Biol. Cell* **15**, 1089–1100.

Clarke, P. G. (1990). Developmental cell death: Morphological diversity and multiple mechanisms. *Anat. Embryol. (Berl).* **181**, 195–213.

Cummings, J., Ward, T. H., Greystoke, A., Ranson, M., and Dive, C. (2007). Biomarker method validation in anticancer drug development. *Br. J. Pharmacol.* **153**, 646–656.

Cummings, M. C., Winterford, C. M., and Walker, N. I. (1997). Apoptosis. *Am. J. Surg. Pathol.* **21**, 88–101.

Darzynkiewicz, Z., and Pozarowski, P. (2007). All that glitters is not gold: All that FLICA binds is not caspase. A caution in data interpretation–and new opportunities. *Cytometry A* **71,** 536–537.

Denecker, G., Dooms, H., Van Loo, G., Vercammen, D., Grooten, J., Fiers, W., Declercq, W., and Vandenabeele, P. (2000). Phosphatidyl serine exposure during apoptosis precedes release of cytochrome c and decrease in mitochondrial transmembrane potential. *FEBS Lett.* **465,** 47–52.

Denecker, G., Vercammen, D., Steemans, M., Vanden Berghe, T., Brouckaert, G., Van Loo, G., Zhivotovsky, B., Fiers, W., Grooten, J., Declercq, W., and Vandenabeele, P. (2001). Death receptor-induced apoptotic and necrotic cell death: Differential role of caspases and mitochondria. *Cell Death Differ.* **8,** 829–840.

Desmedt, M., Rottiers, P., Dooms, H., Fiers, W., and Grooten, J. (1998). Macrophages induce cellular immunity by activating Th1 cell responses and suppressing Th2 cell responses. *J. Immunol.* **160,** 5300–5308.

D'Herde, K., De Prest, B., Mussche, S., Schotte, P., Beyaert, R., Coster, R. V., and Roels, F. (2000). Ultrastructural localization of cytochrome c in apoptosis demonstrates mitochondrial heterogeneity. *Cell Death Differ.* **7,** 331–337.

Duan, W. R., Garner, D. S., Williams, S. D., Funckes-Shippy, C. L., Spath, I. S., and Blomme, E. A. (2003). Comparison of immunohistochemistry for activated caspase-3 and cleaved cytokeratin 18 with the TUNEL method for quantification of apoptosis in histological sections of PC-3 subcutaneous xenografts. *J. Pathol.* **199,** 221–228.

Earnshaw, W. C. (1995). Nuclear changes in apoptosis. *Curr. Opin Cell Biol.* **7,** 337–343.

Falcieri, E., Gobbi, P., Cataldi, A., Zamai, L., Faenza, I., and Vitale, M. (1994). Nuclear pores in the apoptotic cell. *Histochem. J.* **26,** 754–763.

Festjens, N., Vanden Berghe, T., and Vandenabeele, P. (2006). Necrosis, a well-orchestrated form of cell demise: Signalling cascades, important mediators and concomitant immune response. *Biochim. Biophys. Acta* **1757,** 1371–1387.

Fuchs, E., and Weber, K. (1994). Intermediate filaments: Structure, dynamics, function, and disease. *Annu. Rev. Biochem.* **63,** 345–382.

Goossens, V., Grooten, J., De Vos, K., and Fiers, W. (1995). Direct evidence for tumor necrosis factor-induced mitochondrial reactive oxygen intermediates and their involvement in cytotoxicity. *Proc. Natl. Acad. Sci. USA* **92,** 8115–8119.

Hodge, S., Hodge, G., Scicchitano, R., Reynolds, P. N., and Holmes, M. (2003). Alveolar macrophages from subjects with chronic obstructive pulmonary disease are deficient in their ability to phagocytose apoptotic airway epithelial cells. *Immunol Cell Biol.* **81,** 289–296.

Kerr, J. F., Winterford, C. M., and Harmon, B. V. (1994). Apoptosis: Its significance in cancer and cancer therapy. *Cancer* **73,** 2013–2026.

Kerr, J. F., Wyllie, A. H., and Currie, A. R. (1972). Apoptosis: A basic biological phenomenon with wide-ranging implications in tissue kinetics. *Br. J. Cancer* **26,** 239–257.

Kitanaka, C., and Kuchino, Y. (1999). Caspase-independent programmed cell death with necrotic morphology. *Cell Death Differ.* **6,** 508–515.

Kroemer, G., Dallaporta, B., and Resche-Rigon, M. (1998). The mitochondrial death/life regulator in apoptosis and necrosis. *Annu. Rev. Physiol.* **60,** 619–642.

Krysko, D. V., Brouckaert, G., Kalai, M., Vandenabeele, P., and D'Herde, K. (2003). Mechanisms of internalization of apoptotic and necrotic L929 cells by a macrophage cell line studied by electron microscopy. *J. Morphol.* **258,** 336–345.

Krysko, D. V., Denecker, G., Festjens, N., Gabriels, S., Parthoens, E., D'Herde, K., and Vandenabeele, P. (2006a). Macrophages use different internalization mechanisms to clear apoptotic and necrotic cells. *Cell Death Differ.* **13,** 2011–2022.

Krysko, D. V., D'Herde, K., and Vandenabeele, P. (2006b). Clearance of apoptotic and necrotic cells and its immunological consequences. *Apoptosis* **11,** 1709–1726.

Krysko, D. V., Roels, F., Leybaert, L., and D'Herde, K. (2001). Mitochondrial transmembrane potential changes support the concept of mitochondrial heterogeneity during apoptosis. *J. Histochem. Cytochem.* **49,** 1277–1284.

Krysko, D. V., and Vandenabeele, P. (2008). From regulation of dying cell engulfment to development of anti-cancer therapy. *Cell Death Differ.* **15,** 29–38.

Krysko, O., De Ridder, L., and Cornelissen, M. (2004). Phosphatidylserine exposure during early primary necrosis (oncosis) in JB6 cells as evidenced by immunogold labeling technique. *Apoptosis* **9,** 495–500.

Lahorte, C. M., Vanderheyden, J. L., Steinmetz, N., Van de Wiele, C., Dierckx, R. A., and Slegers, G. (2004). Apoptosis-detecting radioligands: Current state of the art and future perspectives. *Eur. J. Nucl. Med. Mol. Imaging* **31,** 887–919.

Lamkanfi, M., Declercq, W., Depuydt, B., Kalai, M., Saelens, X., and Vandenabeele, P. (2003). The caspase family. In "Caspases: Their Role in Cell Death and Cell Survival" (H. Walczak, Ed.), Landes Bioscience, Kluwer Academic Press, Georgetown, TX.

Lemasters, J. J., Nieminen, A. L., Qian, T., Trost, L. C., Elmore, S. P., Nishimura, Y., Crowe, R. A., Cascio, W. E., Bradham, C. A., Brenner, D. A., and Herman, B. (1998). The mitochondrial permeability transition in cell death: A common mechanism in necrosis, apoptosis and autophagy. *Biochim. Biophys. Acta* **1366,** 177–196.

Lockshin, R. A., and Williams, C. M. (1964). Programmed cell death. II. Endocrine potentiation of the breakdown of the intersegmental muscles of silkmoths. *J. Insect Physiol.* **10,** 643–649.

Maiuri, M. C., Zalckvar, E., Kimchi, A., and Kroemer, G. (2007). Self-eating and self-killing: Crosstalk between autophagy and apoptosis. *Nat. Rev. Mol. Cell Biol.* **8,** 741–752.

Majno, G., and Joris, I. (1995). Apoptosis, oncosis, and necrosis: An overview of cell death. *Am. J. Pathol.* **146,** 3–15.

Metivier, D., Dallaporta, B., Zamzami, N., Larochette, N., Susin, S. A., Marzo, I., and Kroemer, G. (1998). Cytofluorometric detection of mitochondrial alterations in early CD95/Fas/APO-1-triggered apoptosis of Jurkat T lymphoma cells: Comparison of seven mitochondrion-specific fluorochromes. *Immunol. Lett.* **61,** 157–163.

Munoz, L. E., Gaipl, U. S., Franz, S., Sheriff, A., Voll, R. E., Kalden, J. R., and Herrmann, M. (2005). SLE: A disease of clearance deficiency? *Rheumatology (Oxford)* **44,** 1101–1107.

Nagata, S. (2000). Apoptotic DNA fragmentation. *Exp. Cell Res.* **256,** 12–18.

Norbury, C. C., Chambers, B. J., Prescott, A. R., Ljunggren, H. G., and Watts, C. (1997). Constitutive macropinocytosis allows TAP-dependent major histocompatibility complex class I presentation of exogenous soluble antigen by bone marrow-derived dendritic cells. *Eur. J. Immunol.* **27,** 280–288.

Norbury, C. C., Hewlett, L. J., Prescott, A. R., Shastri, N., and Watts, C. (1995). Class I MHC presentation of exogenous soluble antigen via macropinocytosis in bone marrow macrophages. *Immunity* **3,** 783–791.

Rae, T. (1977). Tolerance of mouse macrophages in vitro to barium sulfate used in orthopedic bone cement. *J. Biomed. Mater. Res.* **11,** 839–846.

Sanchez, I., Mahlke, C., and Yuan, J. (2003). Pivotal role of oligomerization in expanded polyglutamine neurodegenerative disorders. *Nature* **421,** 373–379.

Scaffidi, P., Misteli, T., and Bianchi, M. E. (2002). Release of chromatin protein HMGB1 by necrotic cells triggers inflammation. *Nature* **418,** 191–195.

Schweichel, J. U., and Merker, H. J. (1973). The morphology of various types of cell death in prenatal tissues. *Teratology* **7,** 253–266.

Steinman, R. M., and Cohn, Z. A. (1972). The interaction of soluble horseradish peroxidase with mouse peritoneal macrophages in vitro. *J. Cell Biol.* **55,** 186–204.

Steinman, R. M., Silver, J. M., and Cohn, Z. A. (1974). Pinocytosis in fibroblasts: Quantitative studies in vitro. *J. Cell Biol.* **63,** 949–969.

Stoka, V., Turk, B., Schendel, S. L., Kim, T. H., Cirman, T., Snipas, S. J., Ellerby, L. M., Bredesen, D., Freeze, H., Abrahamson, M., Bromme, D., Krajewski, S., *et al.* (2001). Lysosomal protease pathways to apoptosis: Cleavage of bid, not pro-caspases, is the most likely route. *J. Biol.Chem.* **276**, 3149–3157.

Swanson, J. A. (1989). Phorbol esters stimulate macropinocytosis and solute flow through macrophages. *J. Cell Sci.* **94**(Pt 1), 135–142.

Swanson, J. A., and Baer, S. C. (1995). Phagocytosis by zippers and triggers. *Trends Cell Biol.* **5**, 89–93.

Swanson, J. A., and Watts, C. (1995). Macropinocytosis. *Trends Cell Biol.* **5**, 424–428.

Talanian, R. V., Quinlan, C., Trautz, S., Hackett, M. C., Mankovich, J. A., Banach, D., Ghayur, T., Brady, K. D., and Wong, W. W. (1997). Substrate specificities of caspase family proteases. *J. Biol. Chem.* **272**, 9677–9682.

Tambyrajah, W. S., Bowler, L. D., Medina-Palazon, C., and Sinclair, A. J. (2007). Cell cycle-dependent caspase-like activity that cleaves p27(KIP1) is the beta(1) subunit of the 20S proteasome. *Arch. Biochem. Biophys.* **466**, 186–193.

Thornberry, N. A., Rano, T. A., Peterson, E. P., Rasper, D. M., Timkey, T., Garcia-Calvo, M., Houtzager, V. M., Nordstrom, P. A., Roy, S., Vaillancourt, J. P., Chapman, K. T., and Nicholson, D. W. (1997). A combinatorial approach defines specificities of members of the caspase family and granzyme B: Functional relationships established for key mediators of apoptosis. *J. Biol. Chem.* **272**, 17907–17911.

Timmer, J. C., and Salvesen, G. S. (2007). Caspase substrates. *Cell Death Differ.* **14**, 66–72.

Torii, I., Morikawa, S., Nagasaki, M., Nokano, A., and Morikawa, K. (2001). Differential endocytotic characteristics of a novel human B/DC cell line HBM-Noda: Effective macropinocytic and phagocytic function rather than scavenging function. *Immunology* **103**, 70–80.

Trahtemberg, U., Atallah, M., Krispin, A., Verbovetski, I., and Mevorach, D. (2007). Calcium, leukocyte cell death and the use of annexin V: Fatal encounters. *Apoptosis* **12**, 1769–1780.

Unwin, P. N., and Ennis, P. D. (1984). Two configurations of a channel-forming membrane protein. *Nature* **307**, 609–613.

Unwin, P. N., and Zampighi, G. (1980). Structure of the junction between communicating cells. *Nature* **283**, 545–549.

Vanden Berghe, T., Kalai, M., van Loo, G., Declercq, W., and Vandenabeele, P. (2003). Disruption of HSP90 function reverts tumor necrosis factor-induced necrosis to apoptosis. *J. Biol. Chem.* **278**, 5622–5629.

Van de Wiele, C., Vermeersch, H., Loose, D., Signore, A., Mertens, N., and Dierckx, R. (2004). Radiolabeled annexin-V for monitoring treatment response in oncology. *Cancer Biother. Radiopharm.* **19**, 189–194.

Vandivier, R. W., Fadok, V. A., Hoffmann, P. R., Bratton, D. L., Penvari, C., Brown, K. K., Brain, J. D., Accurso, F. J., and Henson, P. M. (2002). Elastase-mediated phosphatidylserine receptor cleavage impairs apoptotic cell clearance in cystic fibrosis and bronchiectasis. *J. Clin. Invest.* **109**, 661–670.

Vercammen, D., Beyaert, R., Denecker, G., Goossens, V., Van Loo, G., Declercq, W., Grooten, J., Fiers, W., and Vandenabeele, P. (1998). Inhibition of caspases increases the sensitivity of L929 cells to necrosis mediated by tumor necrosis factor. *J. Exp. Med.* **187**, 1477–1485.

Vercammen, D., Vandenabeele, P., Beyaert, R., Declercq, W., and Fiers, W. (1997). Tumour necrosis factor-induced necrosis versus anti-Fas-induced apoptosis in L929 cells. *Cytokine* **9**, 801–808.

Williamson, P., and Schlegel, R. A. (2002). Transbilayer phospholipid movement and the clearance of apoptotic cells. *Biochim. Biophys. Acta* **1585**, 53–63.

Wyllie, A. H. (1981). Cell death: A new classification separating apoptosis from necrosis. *In* "Cell Death in Biology and Pathology" (I. D. Bowen and R. A. Lockshin, Eds.), pp. 9–34. Chapman & Hall, London.

Wyllie, A. H., Morris, R. G., Smith, A. L., and Dunlop, D. (1984). Chromatin cleavage in apoptosis: Association with condensed chromatin morphology and dependence on macromolecular synthesis. *J. Pathol.* **142,** 67–77.

Zullig, S., Neukomm, L. J., Jovanovic, M., Charette, S. J., Lyssenko, N. N., Halleck, M. S., Reutelingsperger, C. P., Schlegel, R. A., and Hengartner, M. O. (2007). Aminophospholipid translocase TAT-1 promotes phosphatidylserine exposure during *C. elegans* apoptosis. *Curr. Biol.* **17,** 994–999.

CHAPTER SEVENTEEN

TWO-DIMENSIONAL GEL-BASED ANALYSIS OF THE DEMOLITION PHASE OF APOPTOSIS

Alexander U. Lüthi, Sean P. Cullen, *and* Seamus J. Martin

Contents

Abstract

Apoptosis is coordinated by members of the caspase family of aspartic acid-specific proteases. To date, over 400 substrates for the apoptosis-associated caspases have been reported and there are likely to be hundreds more yet to be discovered. Global approaches toward identifying proteins cleaved by caspases during apoptosis are now possible and give a more complete perspective on the alterations to the proteome that occur during this complex process. This chapter outlines methods that have been used successfully to visualize the demolition phase of apoptosis by two-dimensional gel electrophoresis coupled with mass spectrometry. It discusses techniques used to generate cell-free

Molecular Cell Biology Laboratory, Department of Genetics, The Smurfit Institute, Trinity College, Dublin, Ireland

Methods in Enzymology, Volume 442
ISSN 0076-6879, DOI: 10.1016/S0076-6879(08)01417-1

extracts from human cells and details how these extracts can be used to activate caspases. The analysis of these extracts by two-dimensional gel electrophoresis is then described, followed by methods used to identify changes to the proteome during the demolition phase of apoptosis.

1. INTRODUCTION

Apoptosis is a complex and highly regulated process that is initiated to eliminate nonessential, aged, injured, or infected cells from the body in an ordered and controlled sequence of events (Adrain *et al.*, 2001; Kerr *et al.*, 1972). Once the decision to eliminate a cell has been made, the "execution" phase of apoptosis is carried out rapidly, with some cell types displaying characteristic morphological changes as soon as 60 min after exposure to an apoptotic stimulus. Apoptosis is coordinated by members of the caspase family of aspartic acid-specific proteases. Other members of this protease family also play essential roles in inflammation where they participate in the maturation of proinflammatory cytokines. While the mechanisms of activation of the worm, fly, and mammalian caspases are now relatively well understood (Adrain *et al.*, 2006), much less is known how caspase activation results in the characteristic apoptotic morphology. To date, over 400 substrates for the apoptosis-associated caspases have been reported (Lüthi *et al.*, 2007; for a comprehensive list, see also www.casbah.ie) and there are likely to be hundreds more yet to be discovered. Global approaches toward identifying proteins cleaved by caspases during apoptosis are now possible and give a more complete perspective on the alterations to the proteome that occur during this complex process.

Different strategies have been previously employed to identify proteins that are targeted for caspase-dependent proteolysis during apoptosis. These range from a one-dimensional approach, comparing control and apoptotic cell lysates by SDS-PAGE (Thiede *et al.*, 2005), diagonal gel analysis with recombinant caspases (Ricci *et al.*, 2004), and two-dimensional resolution of apoptotic proteomes (Thiede *et al.*, 2001). A gel-free proteomics approach has been used whereby control and apoptotic cell extracts were differentially isotope labeled, followed by fractional diagonal chromatography (COFRADIC), and differences were identified by LC/MS/MS analysis, resulting in the identification of 92 processed proteins (Van Damme *et al.*, 2005).

This chapter outlines methods used successfully to visualize the demolition phase of apoptosis by two-dimensional gel electrophoresis followed by mass spectrometry. In a first step, endogenous apoptotic caspases within cell-free extracts are activated and the resulting "degradome" is resolved by two-dimensional SDS-PAGE. Comparison with nonactivated control

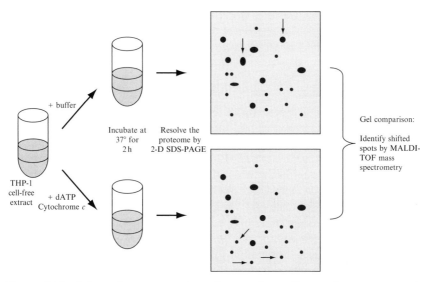

Figure 17.1 Schematic representation of comparison of control versus apoptotic cell-free extract by two-dimensional (2-D) SDS-PAGE.

extracts reveal altered protein spots that are subsequently identified by mass spectrometry (Fig. 17.1).

First we describe techniques used to generate cell-free extracts from human cells and how these extracts can be used to activate caspases. The analysis of these extracts by two-dimensional gel electrophoresis is then described, followed by methods used to identify changes to the proteome during the demolition phase of apoptosis.

2. Cell-Free Extract

Cell-free systems have proven to be a valuable and versatile tool for monitoring the degradation of the proteome during apoptosis *in vitro*. The cell-free approach for the analysis of apoptosis was first described using extracts derived from "mitotic" chicken hepatoma, cells leading to the identification of PARP as a caspase substrate (Lazebnick *et al.*, 1994). Cell extracts from *Xenopus* eggs were also used around the same time to identify regulators of caspase activation (Newmeyer *et al.*, 1994). Subsequently, the cell-free approach has been adapted for cells from different species and has proven to be a useful tool for exploring and understanding the pathways leading to caspase activation and for analyzing downstream events during apoptosis (Cullen *et al.*, 2008; Martin *et al.*, 1995).

The cell-free extract system described here is a concentrated protein preparation of the cytosolic fraction of cells devoid of nuclei, membranes, and major organelles, including mitochondria.

2.1. Preparation of THP-1 cell-free extract

For generation of cell-free extracts, 2×10^8 to 1×10^9 of healthy THP-1 cells, in exponential growth phase, are washed twice in phosphate-buffered saline (PBS) and once in buffer A [20 mM HEPES-KOH, pH 7.5, 10 mM KCl, 1.5 mM MgCl$_2$, 1 mM EDTA, 1 mM EGTA, 1 mM dithiotheritol (DTT), 100 μM phenylmethylsulfonyl fluoride, 10 μg/ml leupeptin, 2 μg/ml aprotinin] to remove cell culture media that may interfere with subsequent steps in the process. Cells are then transferred to a 2-ml Dounce-type homogenizer and compacted into a pellet at 800g for 10 min. After centrifugation, it is desirable to remove as much buffer as possible, as high salt concentrations inhibit formation of the Apaf-1 apoptosomes and interfere with subsequent isoelectric focusing of the proteins during the two-dimensional gel analysis. Cells are then resuspended in 3 volumes of hypotonic buffer A and incubated on ice for 10 to 15 min. During this step, the osmotic shock induced by buffer A weakens the cell membranes, which facilitates rupture by homogenization with 15 to 20 strokes of a B-type pestle. During this process, samples of the homogenate are visualized under a light microscope to ensure successful rupture of cell membranes. However, it is important that cellular organelles such as mitochondria are left intact as cytochrome c released from ruptured mitochondria will contaminate the extracts and automatically activate the apoptotic caspase cascade in subsequent experiments. Cell lysates are then cleared of debris and intact cells by centrifugation at 15,000g for 15 min at 4°. The cell-free extract generated using this approach (10 to 15 mg/ml of protein) is then aliquoted and stored at −75°.

2.2. Caspase activation in THP-1 cell-free extract

Caspase activation in cell-free extracts can be initiated by adding cytochrome c and dATP to the cell-free reactions, initiating the controlled activation of the apoptosome and caspase-3, -7, -6, -2, -8, and -10 downstream (Hill *et al.*, 2004; Kluck *et al.*, 1997; Li *et al.*, 1997).

THP-1 cell-free reactions consist typically of 80% THP-1 cell-free extract (described in the previous section) to which either buffer A (as the control) or cytochrome c and dATP are added. Reactions can be set up from volumes as small as 10 μl to large-scale reactions of more than 1 ml.

We typically prepare reactions in a final volume of 240 μl, which contains 192 μl of the THP-1 cell-free extract, 50 μg/ml purified heart cytochrome c, and 1 mM dATP, with buffer A added to a final volume of 240 μl. To activate caspases, reactions are incubated at 37°, and samples are

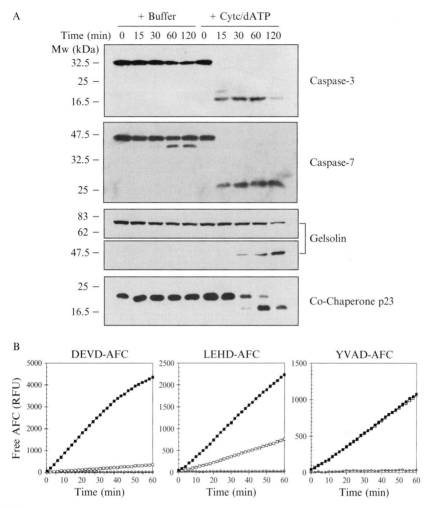

Figure 17.2 Activation of apoptotic caspases in the THP-1 cell-free extract. (A) The THP-1 cell-free extract was incubated at 37° with either buffer or 50 μg/ml cytochrome c and 1 mM dATP for the times indicated. Samples were then analyzed by Western blotting for the processing of caspase-3, -7, and the caspase substrates gelsolin and cochaperone p23. (B) The THP-1 cell-free extract was incubated at 37° for 1 h with either buffer (□) or 50 μg/ml cytochrome c and 1 mM dATP (■). Caspase activity within the extracts was then assessed using a fluorometric assay with 50 μM of the peptide substrates Ac-DEVD-AFC, Ac-LEHD-AFC, or Ac-YVAD-AFC as indicated.

taken at the various time points for SDS-PAGE/Western blot analysis. In response to the addition of cytochrome c and dATP, efficient processing of the effecter caspase-3 and -7 is readily observed in the extracts (Fig. 17.2A). Processing of the caspase substrates gelsolin and co–chaperone p23 indicated that effector caspases were active in the stimulated THP-1 cell-free extract but not in the control (Fig. 17.2A).

After 1 h, samples are taken and then diluted 1:20 or 1:40 into 50 μM of the indicated fluorogenic peptide in buffer A for fluorogenic determination of caspase activity. As expected, very efficient caspase-3 and -7 (DEVD-AFC) and -9 (LEHD-AFC) activity is detected in the activated cell-free extract. Caspase-1 (YVAD-AFC) is known to spontaneously autoprocess and become activated in THP-1 cell-free extracts under these conditions (Martinon *et al.*, 2002), although the exact mechanism leading to this activation is not known (Fig. 17.2B).

3. Two-Dimensional SDS-PAGE

For a truly global approach to the analysis of the demolition phase of apoptosis, ideally one would analyze changes within the entire proteome. Two-dimensional SDS-PAGE enables the resolution of only highly abundant proteins within the proteome, which facilitates the identification of a fraction (15–20%) of caspase-mediated processing events during apoptosis.

The THP-1 cell-free extract is prepared and activated as described previously. Loading more protein leads to a higher identification rate of the protein spots at the mass spectrometry stage, but loading less protein produces less distorted and better-focused gels. For a typical gel, 350 μg of protein is diluted in 400 μl of two-dimensional solubilization buffer (8 M urea, 4% CHAPS, 100 mM DTT, 0.05% SDS, 0.5% ampholytes pI 3–10, 0.03% bromophenol blue) and is incubated on dehydrated 17-cm immobilized pH gradient (IPG) strips for 14 to 16 h. The absorption of larger proteins can be enhanced by running a small current through the strip.

3.1. First dimension: Isoelectric focusing

Isoelectric focusing is carried out as follows.

- Linear voltage ramp to 500 V over 1 h
- 500 V for 5 h
- Linear voltage ramp to 3500 V
- 3500 V for 11 h

To achieve maximum resolution, 17-cm strips are used. Commercial IPG strips, which cover different parts of the pI range, are also available. The pI 3 to 10 range gives a good overview but lacks the detailed resolution of a pI 5 to 8 IPG strip. Proteins with a pI below 4, or above 8, are generally hard to resolve properly and appear mostly in smears or finish on the extreme ends of the strip and therefore do not resolve.

3.2. Second dimension: SDS-PAGE

The isoelectric-focused proteins strips are then incubated for 5 min in reducing buffer (6 M urea, 0.375 mM Tris-HCl, pH 8.8, 2% SDS, 20% glycerol, 2% DTT) and then for 5 min in alkylating buffer (6 M urea, 0.375 mM Tris-HCl, pH 8.8, 2% SDS, 20% glycerol, 2.5% iodoacetamide). Following this, strips are placed on top of 8 to 16% SDS-PAGE gels in agarose with a low melting point. Proteins are separated according to their molecular weights at 37.5 mA per gel in a BioRad Protein II xi electrophoresis cell. A protein ladder with markers of known molecular weight can be run alongside the gel, but is not typically necessary as precise molecular weights are subsequently determined by mass spectrometry of protein spots of interest.

3.3. Silver stain

Gels can be stained by Coomassie blue or by the more sensitive silver stain. It should be noted that a mass spectrometry-compatible silver staining protocol should be used, as the protein spots must be destained before identification by matrix-assisted laser desorption/ionization time of flight (MALDI-TOF) mass spectrometry. To achieve optimal staining of two-dimensional gels, only the highest grade chemicals and water should be used, as contaminants lead to a dramatic increase in background staining. Ultrapure chemicals from Sigma and double distilled water are used in our experiments.

Gels are fixed for at least 30 min in 50% methanol/10% acetic acid, washed for 10 min in 50% methanol, and further washed three times in H_2O for 10 min. After 1-min sensitization in 0.02% $Na_2S_2O_3$ and two brief rinses in H_2O, gels are incubated in a chilled 0.1% $AgNO_3$ stain solution for 20 min. Following two rinses in H_2O, gels are developed in 0.04% formalin/2% Na_2CO_3. Approximately three changes of the solution are required under constant rocking of the gels until a clear contrast between protein spots and background stain develops. Reactions are stopped by immediately placing the gels in a 5% acetic acid solution. Gels can now be documented (Fig. 17.3) and stored in a 1% acetic acid solution.

4. Protein Identification by Matrix-Assisted Laser Desorption/Ionization Time of Flight Mass Spectrometry

Having separated the proteome by two dimensions into single protein spots, the next step is to identify these proteins. MALDI-TOF mass spectrometry is a relatively simple method of protein identification. Other methods include Edman degradation and liquid chromatography/mass spectrometry-based analysis, but these are more complex and time-consuming. The resulting

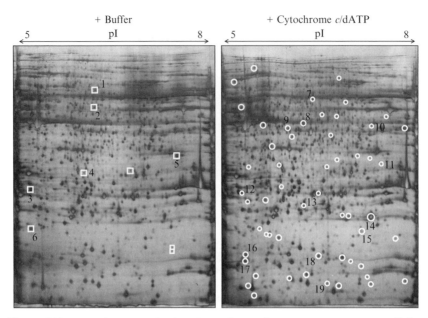

Figure 17.3 Two-dimensional gel analysis of control versus apoptotic THP-1 cell-free extract. Three hundred fifty micrograms of THP-1 cell-free extract was incubated with buffer (left) or 50 μg/ml cytochrome *c*/1 mM dATP (right) for 2 h at 37°. Protein samples were then focused on pI 5–8 IPG strips, resolved on 12% SDS-PAGE gels, and silver stained. Spots disappearing from the control gel are marked with squares, while circles denote spots that have appeared in areas where previously none were present. The numbers correspond to the proteins identified by MALDI-TOF in Table 17.1.

protein sequence can reveal the exact location of a putative caspase cleavage site but this is only achievable with the detection of numerous peptides from protein-rich spots. This section describes the MALDI-TOF MS method where we destain the excised gel spots, tryptically digest the protein in-gel, analyze the peptide mix using a MALDI-TOF mass spectrometer, and identify proteins by matching the peptide mix to an *in silico*-digested protein sequence database. We routinely use the Applied Biosystems Voyager DE-PRO, MALDI-TOF mass spectrometer, which is a relatively simple to use instrument that requires little maintenance.

4.1. Destaining and trypsin digestion of silver-stained protein spots

Protein spots of interest are excised manually from gels with a modified P1000 pipette tip and destained in 15 mM $K_3Fe(CH)_6$ and 50 mM $Na_2S_2O_3$ for 5 min in an orbital shaker at 800 rpm. This step is repeated if the gel spot is not fully destained. After five washes in 50% methanol and 10% acetic acid, the pH is adjusted by incubating the gel piece in 50 mM NH$_4$HCO$_3$ for 5 min. Gel spots are then dehydrated for 5 min in 500 μl CH$_3$CN and dried in a Speed-Vac. Dried gel pieces are rehydrated for 5 min with 3 μl digestion buffer (12.5 mM NH$_4$HCO$_3$, 0.05% n-octyl β-D-glucopyrano-side) containing 150 ng of trypsin, and then a further 10 μl of digestion buffer is added and proteins are "in-gel" digested at 37° for 14 to 16 h.

4.2. Matrix-assisted laser desorption/ionization time of flight mass spectrometry

Trypsin-digested peptides are extracted from gel pieces with 40 μl extraction buffer [66% CH$_3$CN, 0.1% trifluoracetic acid (TFA)] for 10 min in a sonicating water bath. Extraction buffer containing the peptides is combined with the supernatant of the trypsin digest and dried down in a Speed-Vac. Dried peptides are solubilized in 10 μl 5% formic acid, of which 0.5 μl is spotted onto a Teflon-coated 96-well MALDI target plate (Applied Biosystems) followed by 0.5 μl of the matrix solution (10 mg/ml α-cyano-4-hydroxy-cinnamic acid, 60% CH$_3$CN, 0.1% TFA). For more abundant protein spots, the supernatant of the trypsin digest contains sufficient washed out peptides, which are spotted directly onto MALDI target plates. Samples are allowed to crystallize before being analyzed on a Voyager DE-Pro mass spectrometer in positive reflector mode. Peptide fingerprint data of 600 to 1200 laser shots are collected per sample, deisotoped, and internally mass calibrated against known trypsin peptides. The MASCOT (http://www.matrixscience.com) and MS-Fit (http://prospector.ucsf.edu/ucsfhtml4.27.1/msfit.htm) databases are used to compare peptide mass data against theoretical trypsin-digested proteins of the NCBI or SwissProt protein database. A selection of identified spots is listed in Table 17.1 along with the matched number of peptides and the percentage coverage of the full-length proteins. For a number of the proteins identified, disappearance of their full-length forms from the control gel, as well as the appearance of cleavage fragments on the caspase-activated gel, is observed. Ideally, more than one of the resulting proteolytic fragments of the caspase substrate is identified by mass spectrometry. The peptide coverage should reveal different parts of the full-length protein, and this information can then be further used to narrow down the putative processing site(s).

5. Conclusion

Proteomic analysis of cell-free extracts using two-dimensional SDS-PAGE has been used for several years and is a powerful approach for the analysis of caspase-dependent events in apoptosis. Over the past number of

Table 17.1 List of identified appearing/disappearing protein spots of apoptotic caspase-activated THP-1 cell-free extract

Spot	Protein	Accession No.	Peptides	Coverage (%)
1	PAK2	gi_3041712	13	30
2	DOK2	gi_21618483	26	55
3	RANBP1	gi_32425497	14	46
4	PSMB3	gi_7513167	7	32
5	Calponin 2	gi_4758018	15	38
6	Bid	gi_23274161	14	66
7	NACA1	gi_5031931	10	41
8	DOK2	gi_41406050	22	52
9	Actin	gi_14250401	25	67
10	Gelsolin	gi_4504165	21	19
11	PAK2	gi_3041712	12	29
12	NACA1	gi_5031931	22	40
13	HCLS1	gi_4885405	11	21
14	Rho GDI2	gi_10835002	18	74
15	Rho GDI2	gi_10835002	8	45
16	HCLS1	gi_4885405	13	23
17	GMFG	gi_34783447	13	60
18	RGS10	gi_14424689	16	61
19	Actin	gi_15277503	15	45

[a] Appearing and disappearing protein spots from pI 3–6 and pI 5–8 gels were digested tryptically and identified by MALDI-TOF mass spectrometry. The protein name, accession number, number of matching peptides identified, and coverage of these peptides of the full-length protein are shown.

years, our laboratory has utilized this approach to study the demolition phase of apoptosis in proteomes from different species. In addition to the study of human cells (Adrain et al., 2004), we have also studied proteomes from the fly *Drosophila melanogaster* (Adrain et al., 2004; Creagh and Martin, 2008) and *Caenorhabditis elegans* (Taylor et al., 2007). Figure 17.4 illustrates a comparison of the number of proteolytic changes identified in different species using two-dimensional resolution of caspase-activated cell-free extracts. The number of proteins altered in each proteome was extrapolated according to the total number of genes present in each species.

From the analysis of two-dimensional SDS-PAGE gels it is evident that a substantial part of the proteome is altered during programmed cell death. Even though it is likely that almost half of the proteins undergoing caspase-dependent processing during apoptosis in humans have now been identified (Lüthi et al., 2007), surprisingly little is known concerning the key events that take place during the demolition phase of apoptosis.

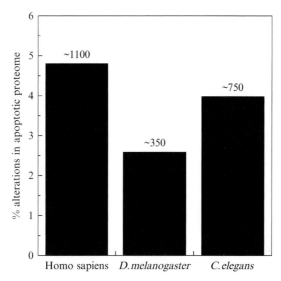

Figure 17.4 Proteome alterations in human, *D. melanogaster*, and *C. elegans* apoptotic cell-free extracts observed by two-dimensional SDS-PAGE analysis. The percentages of changes observed between control and apoptotic cell-free extracts from different species are plotted against each other. The extrapolated number of proteins cleaved by apoptotic caspases within the total proteome of each species is indicated above each bar.

ACKNOWLEDGMENTS

We thank Science Foundation Ireland for their ongoing support of work in our laboratory. SJM is a Science Foundation Ireland Principal Investigator (PI1/B038). Work in the Martin laboratory is also supported by grants from the The Wellcome Trust, the Irish Research Council for Science and Engineering Technologies (IRCSET), and the Health Research Board of Ireland.

REFERENCES

Adrain, C., Brumatti, G., and Martin, S. J. (2006). Apoptosomes: Protease activation platforms to die from. *Trends Biochem. Sci.* **31,** 243–247.

Adrain, C., Creagh, E. M., Cullen, S. P., and Martin, S. J. (2004). Caspase-dependent inactivation of proteasome function during programmed cell death in *Drosophila* and man. *J. Biol. Chem.* **279,** 36923–36930.

Adrain, C., and Martin, S. J. (2001). The mitochondrial apoptosome: A killer unleashed by the cytochrome seas. *Trends Biochem. Sci.* **26,** 390–397.

Creagh, E. M., and Martin, S. J. (2008). Bicaudal is a conserved substrate for *Drosophila* and mamalian caspases and is essential for cell survival. *J. Cell Biol.* Submitted for publication.

Cullen, S. P., Lüthi, A. U., and Martin, S. J. (2008). Analysis of apoptosis in cell-free systems. *Methods* **44,** 273–279.

Hill, M. M., Adrain, C., Duriez, P. J., Creagh, E. M., and Martin, S. J. (2004). Analysis of the composition, assembly kinetics and activity of native Apaf-1 apoptosomes. *EMBO J.* **23**, 2134–2245.

Kerr, J. F., Wyllie, A. H., and Currie, A. R. (1972). Apoptosis: A basic biological phenomenon with wide-ranging implications in tissue kinetics. *Br. J. Cancer* **26**, 239–257.

Kluck, R. M., Martin, S. J., Hoffman, B. M., Zhou, J. S., Green, D. R., and Newmeyer, D. D. (1997). Cytochrome c activation of CPP32-like proteolysis plays a critical role in a Xenopus cell-free apoptosis system. *EMBO J.* **16**, 4639–4649.

Lazebnik, Y. A., Kaufmann, S. H., Desnoyers, S., Poirier, G. G., and Earnshaw, W. C. (1994). Cleavage of poly(ADP-ribose) polymerase by a proteinase with properties like ICE. *Nature* **371**, 346–347.

Li, P., Nijhawan, D., Budihardjo, I., Srinivasula, S. M., Ahmad, M., Alnemri, E. S., and Wang, X. (1997). Cytochrome c and dATP-dependent formation of Apaf-1/caspase-9 complex initiates an apoptotic protease cascade. *Cell* **91**, 479–489.

Lüthi, A. U., and Martin, S. J. (2007). The CASBAH: A searchable database of caspase substrates. *Cell Death Differ.* **14**, 641–650.

Martin, S. J., Newmeyer, D. D., Mathias, S., Farschon, D. M., Wang, H. G., Reed, J. C., Kolesnick, R. N., and Green, D. R. (1995). Cell-free reconstitution of Fas-, UV radiation- and ceramide-induced apoptosis. *EMBO J.* **14**, 5191–5200.

Martinon, F., Burns, K., and Tschopp, J. (2002). The inflammasome: A molecular platform triggering activation of inflammatory caspases and processing of proIL-beta. *Mol. Cell* **10**, 417–426.

Newmeyer, D. D., Farschon, D. M., and Reed, J. C. (1994). Cell-free apoptosis in Xenopus egg extracts: Inhibition by Bcl-2 and requirement for an organelle fraction enriched in mitochondria. *Cell* **79**, 353–364.

Ricci, J. E., Munoz-Pinedo, C., Fitzgerald, P., Bailly-Maitre, B., Perkins, G. A., Yadava, N., Scheffler, I. E., Ellisman, M. H., and Green, D. R. (2004). Disruption of mitochondrial function during apoptosis is mediated by caspase cleavage of the p75 subunit of complex I of the electron transport chain. *Cell* **117**, 773–786.

Taylor, R. C., Brumatti, G., Ito, S., Hengartner, M. O., Derry, W. B., and Martin, S. J. (2007). Establishing a blueprint for CED-3-dependent killing through identification of multiple substrates for this protease. *J. Biol. Chem.* **282**, 15011–15021.

Thiede, B., Dimmler, C., Siejak, F., and Rudel, T. (2001). Predominant identification of RNA-binding proteins in Fas-induced apoptosis by proteome analysis. *J. Biol. Chem.* **276**, 26044–26050.

Thiede, B., Treumann, A., Kretschmer, A., Sohlke, J., and Rudel, T. (2005). Shotgun proteome analysis of protein cleavage in apoptotic cells. *Proteomics* **5**, 2123–2130.

Van Damme, P., Martens, L., Van Damme, J., Hugelier, K., Staes, A., Vandekerckhove, J., and Gevaert, K. (2005). Caspase-specific and nonspecific *in vivo* protein processing during Fas-induced apoptosis. *Nat. Methods* **2**, 771–777.

CHAPTER EIGHTEEN

METHODS TO DISSECT MITOCHONDRIAL MEMBRANE PERMEABILIZATION IN THE COURSE OF APOPTOSIS

Lorenzo Galluzzi, Ilio Vitale, Oliver Kepp, Claire Séror,
Emilie Hangen, Jean-Luc Perfettini, Nazanine Modjtahedi,
and Guido Kroemer

Contents

Abstract

In several paradigms of cell death, mitochondrial membrane permeabilization (MMP) delimits the frontier between life and death. Mitochondria control the intrinsic pathway of apoptosis and participate in the extrinsic pathway. Moreover, they have been implicated in nonapoptotic cell death modalities. Irrespective of its initiation at the inner or the outer mitochondrial membrane (IM and OM, respectively), MMP culminates in the functional (dissipation of the mitochondrial transmembrane potential, shutdown of ATP synthesis, redox imbalance) and structural (reorganization of cristae, release of toxic intermembrane space proteins into the cytosol) collapse of mitochondria. This has a profound impact on cellular metabolism, activates caspase-dependent and -independent

INSERM, U848, Institut Gustave Roussy, and Université Paris-Sud 11, Villejuif, France

Methods in Enzymology, Volume 442
ISSN 0076-6879, DOI: 10.1016/S0076-6879(08)01418-3

executioner mechanisms, and finally results in the demise of the cell. However, the partial and/or temporary permeabilization of one or both mitochondrial membranes is not always a prelude to cell death. This chapter proposes a method and several guidelines to discriminate between IM and OM permeabilization and to identify MMP that does indeed precede cell death. This approach relies on the integration of currently available techniques and may be easily introduced in the laboratory routine for a more precise detection of cell death.

1. INTRODUCTION

Mitochondria are considered the major regulators of cell death. In past years, this role was first assigned to nuclei, as during apoptosis (an evolutionarily conserved, genetically encoded mechanism through which multicellular organisms as well as unicellular populations remove unwanted cells) they display gross, easily detectable morphological changes. Although mitochondria exhibit profound cell death-related alterations, these occur at an ultrastructural/biochemical level and are more difficult to detect (Susin *et al.*, 1998). Moreover, while nuclei are involved either in the initiation (for instance, following DNA damage) or in the execution (by exhibiting chromatin condensation, pyknosis, and karyorrhexis) of apoptosis, mitochondria occupy a more central position by regulating the decision/integration phase (Green and Kroemer, 1998). At this stage, pro- and antiapoptotic signals perceived from the extracellular microenvironment or from other organelles (*e.g.*, nuclei, endoplasmic reticulum, lysosomes) reach mitochondria and are appraised to achieve a final resolution on the fate of the cell. When mitochondrial membrane permeabilization (MMP) takes place at irreparable levels, the death verdict is pronounced and an ensemble of postmortem events accounting for the phenotypic appearance of apoptosis is triggered (Kroemer *et al.*, 2007).

The lethal consequences of MMP relate to (1) the critical position occupied by mitochondria in cellular bioenergetics and (2) the release in the cytosol of proteins that are normally retained and serve vital functions within the mitochondrial intermembrane space (IMS) (Kroemer *et al.*, 2007; Ravagnan *et al.*, 2002). Proapoptotic proteins liberated as a consequence of MMP include activators of the caspase cascade (*e.g.*, cytochrome *c*, second mitochondria-derived activator of caspase/direct IAP-binding protein with a low pI, Omi stress-regulated endoprotease/high temperature requirement protein A 2), as well as caspase-independent death effectors [*e.g.*, apoptosis-inducing factor (AIF), endonuclease G] (Garrido *et al.*, 2006; Li *et al.*, 2001; Martins *et al.*, 2002; Modjtahedi *et al.*, 2006; Verhagen *et al.*, 2000). The functional impairment of cellular powerhouses associated with the activation of multiple (and partially overlapping) lethal mechanisms ultimately seals the fate of the cell. It should be kept in mind, however, that MMP sometimes affects only a fraction of mitochondria or occurs in a transient fashion, which does not lead to cell death. In these settings, which are probably connected to vital physiological mechanisms (*e.g.*, cytochrome *c* is released during megakaryocyte differentiation) and/or to adaptive stress responses (*e.g.*, oxidative insults), cells are able to reestablish the functional and structural integrity of mitochondria rather than succumbing to MMP-related apoptosis (Garrido *et al.*, 2006).

In recent years, several mechanisms have been proposed to account for MMP. So far, however, contrasting viewpoints on the causative mechanisms for MMP could not be condensed into a unified, harmonic model. At present, it appears that several distinct mechanisms are able to mediate MMP through the activity of different sets of specific molecular players that involve unequally the inner and the outer mitochondrial membrane (IM and OM, respectively) (Kroemer *et al.*, 2007).

Based on these premises, it is not surprising that several routine methods for the detection of apoptosis rely on the assessment of early mitochondrial alterations (reviewed in Galluzzi *et al.*, 2007). These techniques (including cytofluorometric as well as immunofluorescence microscopy-based protocols) present considerable advantages as compared to approaches that identify late apoptotic events (*e.g.*, internucleosomal DNA fragmentation), including the possibility to identify and purify cells that have trespassed the "point of no return" but have not yet activated executioner mechanisms. As a drawback, these methods do not discriminate between different modes of MMP induction and can yield false positive results if MMP does not lead to cell death. This chapter details an approach based on the combination of currently available protocols for a precise assessment of MMP in apoptosis.

2. MITOCHONDRIAL OUTER MEMBRANE PERMEABILIZATION

According to current beliefs, MMP may originate at the OM through at least two distinct mechanisms (for a detailed review, see Kroemer *et al.*, 2007). These include (1) the activation of proapoptotic proteins of the Bcl-2 family (*e.g.*, Bax, Bak) to build up multimeric channels, allowing for the release of IMS proteins, and (2) formation of lipidic pores due to a direct interaction between proapoptotic Bcl-2 family members (*e.g.*, Bax, truncated Bid) and the lipidic component of mitochondria (Fig. 18.1). Obviously, OM permeabilization also occurs upon its physical rupture, be it induced accidentally (for instance, in the context of necrosis) or as part of a regulated mechanism originating at the IM (the so-called mitochondrial permeability transition).

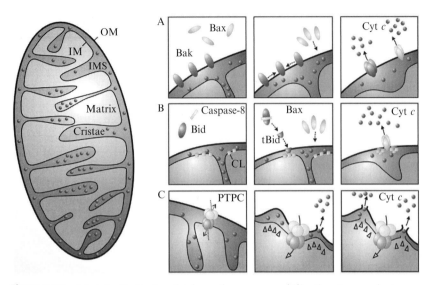

Figure 18.1 Models of mitochondrial membrane permeabilization. In some instances, mitochondrial membrane permeabilization (MMP) starts at the outer mitochondrial membrane (OM) following either the activation of proapoptotic modulators such as Bax and Bak (A) or the interaction between truncated Bid (tBid, proteolytically activated by caspase-8) and mitochondrial lipids (*e.g.*, cardiolipin, CL) (B). As an alternative, MMP is initiated at the inner mitochondrial membrane (IM) via the activation of the permeability transition pore complex (PTPC), a multiprotein structure assembled at the junctions between IM and OM that is believed to ensure the exchange of small metabolites in physiological conditions (C). In both cases, MMP results in the rupture of the OM, followed by the release of lethal intermembrane space (IMS) proteins (*e.g.*, cytochrome *c*, Cyt *c*) into the cytosol. Such released IMS proteins can activate caspases and caspase-independent death effectors to eventually induce cell death.

Basically, OM permeabilization can be ascertained via the detection of IMS proteins (*e.g.*, cytochrome *c*, AIF) in the cytosol. To this aim, several techniques may be employed, including (1) immunoblotting of subcellular fractions with antibodies that specifically recognize IMS proteins; (2) *in situ* immunofluorescence microscopy after fixation and permeabilization of cells; and (3) videomicroscopy to monitor the redistribution of IMS proteins fused to fluorescent moieties (that have been previously transfected into the cells). Immunoblotting upon fractionation is long and laborious, permits the study of a limited number of samples, and thus it is not recommended for routine determinations. Videomicroscopy provides the opportunity to observe OM permeabilization in real time from live cells, yet allows for even lower throughputs than immunoblotting. Moreover, the transfection of cells with a tagged IMS protein may be toxic and promote by itself some signs of MMP. Several other techniques have been employed for the detection of OM permeabilization (*e.g.*, cytofluorometric assessment of cytochrome *c* release, immunoelectron microscopy, ELISA-based miniaturized assays), but usually present additional drawbacks that limit their applicability [a detailed analysis of these methods would exceed the scope of this chapter, but can be found in Galluzzi *et al.* (2007)]. In conclusion, to determine OM permeabilization on a routine basis, we recommend immunofluorescence-based approaches on fixed and permeabilized samples. This ensures the possibility to (1) perform colocalization experiments, by means of organelle-specific probes and fluorochromes with distinct emission spectra; (2) store samples immediately upon fixation and stain them together at a subsequent stage (thus reducing the interstaining variability and allowing for short- and long-term kinetic studies); (3) process collectively a number of samples (50–100) that is suitable for standard laboratory routines. Immunofluorescence microscopy applied to the detection of OM permeabilization presents no major disadvantages, except for the need to follow the release of more than one IMS protein to avoid false-negative results.

2.1. Immunofluorescence detection of the release of intermembrane space proteins

1. Cells are maintained in 175-cm² flasks (BD Falcon, San Jose, CA) in appropriate growth medium (Gibco-Invitrogen, Carlsbad, CA) (see Note 1) and passaged when confluent with trypsin/EDTA (Gibco-Invitrogen)

2. For the experiments, cells are seeded in 12 (24)-well plates (Corning Inc. Life Sciences, Acton, MA) at an approximate density of 100 to 200 (50–100) $\times$ 10³ cells/well (see Note 2) in 0.5 ml of growth medium. Before seeding, 18 (12)-mm-diameter coverslips (Menzer-Gläser GmbH, Braunschweig, Germany)—previously sterilized by incubation for 15 to 30 min in 100% ethanol (Carlo Erba Reagents, Milano, Italy)—are deposited in each well (see Note 3).

3. Upon adherence (12–24 h after seeding, according to cell type, see Note 4), cells are treated with the cell death inducers of choice at an appropriate concentration (see Notes 5–7) for an additional 6 to 48 h (see Note 8).

4. Following stimulation, growth medium is removed, and cells are washed twice with phosphate-buffered saline (PBS) and incubated for 20 min at room temperature in 500 μl (300 μl for 24-well plates) of fixative solution (see Notes 9–13).

5. The fixative solution is discarded, and then cells are rinsed once in PBS and permeabilized by incubation in sodium dodecyl sulfate (SDS) (w/v in PBS, from Carlo Erba Reagents) (see Notes 14–16). After three additional washes in PBS, nonspecific binding sites are blocked by incubation of the samples in 10% fetal bovine serum (FBS) (v/v in PBS) for 20 min at room temperature (see Note 17).

6. Prior to staining with primary antibodies recognizing one (or more) IMS protein and possibly one sessile mitochondrial protein (see Notes

18 and 19), cells are rinsed once in PBS. Staining is then performed by incubating the cells with primary antibodies, diluted 1/100 to 1/500 (according to the antibody) in 200 μl (100 μl for 24-well plates) of 0.5% bovine serum albumin (BSA) (w/v in PBS, from Sigma-Aldrich, St. Louis, MO), either at room temperature for 90 min or at 4 °C overnight (see Notes 20 and 21).

7. Primary antibodies are removed (see Note 22), and cells are rinsed three times with PBS (under gentle shaking for 5 min) and incubated for 45 min at room temperature under protection from light with fluorochrome-coupled secondary antibodies (Molecular-Probes Invitrogen, Carlsbad, CA), diluted 1/300 in 200 μl (100 μl for 24-well plates) of 0.5% BSA (w/v in PBS) (see Notes 23–25).

8. Upon removal of secondary antibodies, nuclei are counterstained by incubating the coverslips with 2 μM Hoechst 33342 (Molecular-Probes Invitrogen; see Notes 26–28) for 30 min at room temperature.

9. After three additional washes in PBS (under mild shaking) (see Note 29), coverslips are mounted onto slides by means of the Fluoromount-G mounting medium (Southern Biotech, Birmingham, AL). After complete drying of the mounting medium (see Note 30), slides are ready for storage (see Note 31) or examination in a confocal fluorescence microscope (see Note 32) equipped with an oil-immersion objective allowing for 63 to 100× magnification.

10. Excitation and emission filters should be selected according to the following absorption/emission peaks: 495/519 nm for Alexa Fluor 488 (green); 578/603 nm for Alexa Fluor 568 (red); 352/461 nm for Hoechst 33342 (blue); and 358/461 nm for 4′,6 diamidino-2-phenylindole (DAPI) (blue) (see Note 33).

11. The release of IMS proteins into the cytosol is evaluated by comparing the distribution of the protein of interest with the localization of mitochondria (stained with a fluorescent mitochondrial probe or with an antibody specific for a sessile mitochondrial protein) (see Notes 34 and 35 and Fig. 18.2).

12. For each sample, the redistribution of IMS proteins from mitochondria to the cytosol should be assessed in a statistically relevant population (at least 200 cells).

3. MITOCHONDRIAL INNER MEMBRANE PERMEABILIZATION

Mitochondrial membrane permeabilization may also originate at IM. In contrast to OM, IM in healthy cells is nearly impermeable to small solutes and ions (a *conditio sine qua non* for the maintenance of the electrochemical gradient

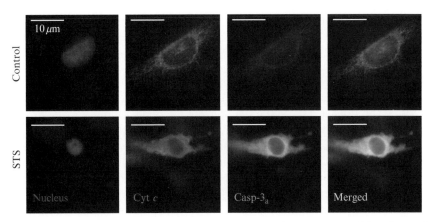

Figure 18.2 Immunofluorescence microscopy-assisted determination of the release of intermembrane space proteins. Nonsmall cell lung cancer cells (A549) were left untreated (control) or treated with 1 μM staurosporine (STS) for 12 h prior to fixation and costaining for the immunofluorescence detection of cytochrome c (Cyt c, detected with a secondary antibody emitting in red) and active caspase-3 (Casp-3$_a$, revealed by a secondary antibody fluorescing in green), as detailed in Section 2.1. Nuclei were counterstained with Hoechst 33342 (blue signal). White bars indicate picture scale (10 μm). In physiological conditions, cytochrome c exhibits a "tubular" staining (typical of healthy mitochondria), and Casp-3$_a$ cannot be detected. Following the induction of apoptosis, cytochrome c relocalizes to the cytosol (and hence exhibits a diffuse intracellular staining) where it promotes the activation of caspase-3 (which also is localized rather homogeneously throughout the cell). In merged images, an intense yellow signal clearly discriminates cells undergoing apoptosis from their healthy counterparts. Please note also the nuclear pyknosis typical of apoptotic cells. (See color insert.)

that sustains ATP synthesis). When IM impermeability is lost, for instance following the opening of the so-called permeability transition pore complex (PTPC), solutes enter the mitochondrial matrix, accompanied by a net influx of water. The resulting osmotic imbalance provokes swelling of the matrix, followed by the rupture of both mitochondrial membranes (Fig. 18.1). Altogether, this process is known as mitochondrial permeability transition (MPT). In the course of MPT, the mitochondrial transmembrane potential ($\Delta \Psi_m$) associated with the electrochemical gradient built across IM is dissipated rapidly and irreversibly, leading to a near-to-complete functional impairment of mitochondria. Temporary, rapid, and reversible $\Delta \Psi_m$ losses may occur independently from MMP. In such instances, it is likely that recovery mechanisms (not yet precisely identified) restore the normal ionic distribution across IM before the osmotic pressure raises above the physical resistance limit of IM (for a comprehensive review on MPT, see Kroemer *et al.*, 2007).

The permeability status of the IM can be assessed by means of: (1) several distinct potential-sensitive probes that accumulate in mitochondria driven by $\Delta\Psi_m$ [such as chloromethyl-X-rosamine (CMXRos), 3,3'dihexiloxalocarbocyanine iodide (DiOC$_6$(3))], or 5,5',6,6'-tetrachloro-1,1',3,3'-tetraethylbenzimidazolcarbocyanine iodide (JC-1)] (Cossarizza et al., 1993; Macho et al., 1996; Metivier et al., 1998); (2) a staining technique that is based on the differential accessibility of calcein acetoxymethyl ester (calcein AM) and its quencher, cobalt (Co^{2+}), to the mitochondrial matrix (Petronilli et al., 1999; Poncet et al., 2003). Fluorescent lipophilic cations provide a quick and rather inexpensive mean to measure $\Delta\Psi_m$ by cytofluorometry, mostly in live (but also in fixed) cells. A number of molecules with different spectral properties (thus allowing for costaining protocols, for instance in association with vital dyes for the assessment of plasma membrane integrity) are commercially available (Table 18.1). However, $\Delta\Psi_m$ is not always a reliable indicator of IM permeabilization (e.g., reduced $\Delta\Psi_m$ may result from inhibited mitochondrial respiration). On the contrary, the calcein-quenching directly allows to collect information on the IM conditions, thus avoiding false-positive results and facilitating the task to discriminate between temporary, reversible vs. definitive, apoptosis-related permeabilization events.

3.1. Cytofluorometric assessment of mitochondrial transmembrane potential

1. Cells are cultured on a routine basis, passaged, and seeded in 12 (24)-well plates as described earlier (see Section 2.1 steps 1 and 2 and Notes 1, 2, 36, and 37).
2. After 12 to 24 h (see Note 4), the putative cell death inducers are administered to the cultures (see Notes 5–7, 38–40).
3. At the end of the stimulation period, supernatants are collected in 5-ml, 12 × 75-mm fluorescence-activated cell sorter (FACS) tubes (BD Falcon) (see Note 41).
4. Cells are detached from the plate by incubation in 300 (200) µl prewarmed trypsin/EDTA (Gibco-Invitrogen) for 5 to 10 min at 37 °C (see Notes 42–45). Thereafter, cells resuspended in trypsin/EDTA solution are gathered with the corresponding supernatants (see Note 46).
5. Cells are spun at 300g for 5 min at room temperature and supernatants are discarded.
6. For staining, cell pellets are resuspended in 300 (200) µl of growth medium containing the $\Delta\Psi_m$-sensitive probe of choice at the appropriate concentration (see Table 18.1).
7. Labeling is performed at 37 °C (5% CO$_2$) under protection from light for a duration that depends on the probe of choice (see Table 18.1 and Notes 47 and 48).
8. To assess the integrity of the plasma membrane, a vital dye of choice (i.e., a probe that accumulates only in dead cells, which have lost the

Table 18.1 Common $\Delta \Psi_m$-sensitive probes for cytofluorometric analysis of MMP (Galluzzi et al., 2007)[a]

Probe	Absorption/emission[b]	Color	Channel[c]	Stock solution	Storage	Staining[d]
Rhodamine and derivatives						
Rh 123	507/529 nm	Green	FL1	100 μM in methanol	−20 °C	200 nM (37 °C, 15–30 min)
TMRE	549/575 nm	Orange	FL2	25 mM in ethanol	−20 °C	25–100 nM (37 °C, 15–30 min)
TMRM	543/573 nm	Orange	FL2	25 mM in ethanol	−20 °C	25–100 nM (37 °C, 15–30 min)
Rosamines and derivatives						
CMTMRos (MitoTracker orange)	554/576 nm	Orange	FL2	1 mM in DMSO	−20 °C	100–500 nM (37 °C, 15–45 min)
CMXRos (MitoTracker red)	579/599 nm	Red	FL1	1 mM in DMSO	−20 °C	50–500 nM (37 °C, 15–45 min)
Carbocianines						
JC-1	Monomers: 514/529 nm; J aggregates: 585/590 nm	Green; Red	FL3; FL1	1–5 mg/ml in DMSO or DMF	−20 °C	0.3–10 mg/ml (37 °C, 10–60 min)
DiOC$_6$(3)	484/501 nm	Green	FL3	10–500 μM in ethanol	−20 °C	15–50 nM (37 °C, 30–60 min)

[a] CMTMRos, chloromethyl-tetramethylrosamine; CMXRos, chloromethyl-X-rosamine; DiOC$_6$(3), 3,3′dihexiloxalocarbocyanine iodide; DMF, N,N-dimethylformamide; DMSO, dimethylsulfoxide; JC-1, 5,5′,6,6′-tetrachloro-1,1′,3,3′-tetraethylbenzimidazolcarbocyanine iodide; Rh 123, rhodamine 123; TMRE, tetramethylrhodamine ethyl ester; TMRM, tetramethylrhodamine methyl ester.

[b] Peak values.

[c] Applies to FACScan, FACSCalibur, and FACSVantage cytofluorometers (Becton-Dickinson) equipped with standard filters for fluorescence acquisition.

[d] Optimal conditions for staining may present significant variations according to the specific experimental setting.

integrity of plasma membrane) can be included in the staining protocol (see Notes 49 and 50).

9. Finally, samples are examined by means of a classic cytofluorometer, allowing for the acquisition of light-scattering data and (at least) two distinct fluorescent signals (*e.g.*, green and red) (see Note 51).

10. Upon acquisition, software-assisted analysis is performed by gating on events that exhibit normal light-scattering parameters (forward and side scatter, *i.e.*, FSC and SSC, respectively) (see Notes 52–55).

3.2. Calcein acetoxymethyl ester-based assays

1. Cells are maintained in culture, passaged, and seeded in 12 (24)-well plates as described in previous sections (see Section 2.1 steps 1 and 2 and Notes 1, 2, 36, 37, and 56).

2. At this stage, culture medium is removed and cells are incubated at 37 °C for 15 to 20 min with 1 μM calcein AM (Molecular Probes-Invitrogen) and 1 to 2 mM CoCl$_2$ (Sigma-Aldrich) in Hank's balanced salt solution (HBSS) (Gibco-Invitrogen) supplemented with 10 mM HEPES (Gibco-Invitrogen) (see Notes 57–60).

3. Thereafter, HBSS is removed and replaced by complete medium in which the molecules under study had been previously diluted to the appropriate concentration (see Notes 5–7, 39, and 61).

   a. If calcein fluorescence is analyzed by immunofluorescence microscopy, growth medium is then discarded and cells are washed, fixed, and processed as described earlier (see Section 2.1 steps 4 and 5 and Notes 9–13, 15, and 62–64).

   b. Nuclear counterstaining with Hoechst 33342 can be performed shortly before mounting and observation (or storage), as previously detailed (see Section 2.1 steps 8 and 9 and Notes 26–32).

   c. Excitation and emission filters should be chosen according to the spectral properties of the employed fluorescent dyes (see Section 2.1 step 10 and Note 33). Calcein AM is characterized by absorption and emission peaks at 494 and 517 nm, respectively.

   d. While healthy cells show bright cytosolic fluorescence with a punctuate-tubular pattern, cells that have undergone MPT during the stimulation period (from the end of labeling until fixation) exhibit a reduced signal as a result of Co^{2+}-mediated mitochondrial calcein AM quenching (see Note 65).

   e. Finally, IM permeabilization should be scored in a statistically relevant population of cells (at least 200 individuals per experimental condition) (see Note 66).

   f. When cytofluorometry is employed for the assessment of calcein fluorescence, samples are collected as described for the analysis of $\Delta \Psi_m$ (see Section 3.1 steps 3–5 and Notes 41–46).

g. Thereafter, cell pellets are resuspended in 300 (200) μl of growth medium, which may contain (or not) additional fluorescent dyes for monitoring parameters other than IM integrity (see Section 3.1 steps 6–8, Table 18.1, and Notes 47–50 and 67).

h. Finally, samples are examined as detailed earlier (see Section 3.1 steps 9 and 10 and Notes 51–55).

4. GUIDELINES FOR A PRECISE DETERMINATION OF MITOCHONDRIAL MEMBRANE PERMEABILIZATION

One of the major problems in the study of cell death is to precisely define the so-called "point of no return," the frontier between death and life, whose one-way trespassing irreversibly determines cellular demise. In several (but not all) instances, this coincides with MMP. However, as introduced earlier, MMP may occur also temporarily, partially (thus reversibly), and in nonlethal settings. By combining the aforementioned techniques, researchers should be able to characterize cell death-associated MMP with the help of the following guidelines.

- The release of (at least) two IMS proteins into the cytosol (irrespective of their reciprocal kinetics) can be documented in all samples fixed after a specific time point. It is important to follow multiple proteins as well as to make sure that the translocation is relevant and sustained (to avoid false-negative/-positive results and to exclude the partial release of some IMS proteins that has been reported to occur in specific physiological settings).

- A subpopulation of cells with reduced $\Delta\Psi_m$ (but intact plasma membranes) can be identified before the appearance of cells also exhibiting plasma membrane breakdown. This may occur before, together with, or after the cytosolic translocation of IMS, depending on the specific modality of MMP. However, it should be noted that during apoptosis $\Delta\Psi_m$ dissipation takes place at a decisional phase well before the rupture of plasma membranes, with the latter occurring as a postmortem event in the context of secondary necrosis. Thus, if cells with ruptured plasma membranes but high $\Delta\Psi_m$ are found, the researcher should be oriented toward the possibility that primary necrosis (instead of or together with apoptosis) takes place within the culture.

- IM permeabilization can be determined (directly, by means of the calcein AM quenching technique) in all samples exhibiting the loss of $\Delta\Psi_m$. For the reasons discussed previously, samples characterized by low $\Delta\Psi_m$ but intact IM (calcein^{high}) cannot be considered as apoptotic. However, because IM permeabilization is immediately followed by the dissipation of $\Delta\Psi_m$, samples in which IM are permeabilized (calcein^{low}) should not display high $\Delta\Psi_m$. Nevertheless, such a phenotype, which should *not* be considered as a prelude to apoptosis, may be encountered because of (1) transient fluctuations in the IM permeability that allows for the mixing of calcein AM and Co^{2+} ions (in this regard, it should be remembered that cells are stained with calcein AM before apoptosis induction) and (2) inappropriate manipulations during the staining protocol that lead to treatment-unrelated IM permeabilization (false-positive results) only in the sample set destined to the calcein AM assay.

As for most biological phenomena, data acquired though different techniques should be collected and combined into a systematic analysis, as the one proposed here, to achieve a thorough characterization of the apoptotic process. By following the aforementioned guidelines, researchers should be able to precisely identify the treatments that promote MMP as a prelude to apoptosis in their specific experimental system. Furthermore, nonapoptotic settings that still display an association with MMP (be it causative or not, *e.g.*, in the context of necrosis) will be promptly recognized, thus allowing for the redirection of further research to more appropriate assays and manipulations.

5. NOTES

1. Optimal growth conditions (*e.g.*, medium composition, supplements) vary according to the cell line of choice and may influence growth rates quite dramatically. As a guideline, most immortalized cell lines in culture exhibit a duplication time ranging from 18 to 72 h. The use of suboptimal growth conditions can cause cell death by starvation, which is frequently accompanied by autophagy.
2. The number of cells seeded in each well depends on their size as well as on the total duration of the assay. To avoid the appearance of overcrowding-dependent toxicity, control cultures at the end of the assay should exhibit confluence levels that never exceed 85%.
3. When scarcely adhering cells are employed, coverslips can be coated with poly-L-lysine directly upon their delivery to the plate. This is achieved by incubating the coverslips in 0.01% poly-L-lysine solution (Sigma-Aldrich) for 10 min at 37 °C, followed by rinsing in PBS. Notably, some cell types are particularly sensitive to poly-L-lysine and fail to grow on poly-L-lysine-coated coverslips.
4. Quickly adhering cell lines (*e.g.*, HeLa) can be treated after 12 h, whereas at least 24 h should be allowed for cells with longer adaptation times (*e.g.*, HCT116). Premature treatment may result in excessive toxicity due to suboptimal conditions of the culture.
5. If possible, we recommend treating the cultures with (at least) three different concentrations of each molecule under investigation (ideally

IC_{50}^{-1}, IC_{50}^{0}, IC_{50}^{1}) in order to avoid insufficient or excessive stimulation.

6. Appropriate negative controls should be always included among the treatments. Usually, negative controls are provided by the administration of the solvent used for the molecules of choice. In this regard, it should be remembered that the organic solvents commonly employed for stock solutions (*e.g.*, dimethyl sulfoxide, *N,N*-dimethylformamide) may be toxic *per se* or may promote specific cellular responses (*e.g.*, differentiation) also at very low concentrations. Such effects and the concentration of solvent at which they appear should be determined carefully for each specific experimental setting at a preliminary stage, and they should be taken into appropriate consideration when the experimental plan is defined.

7. Positive controls are also to be introduced among the treatments. For most cell lines, staurosporine (Sigma-Aldrich, 0.05–2 μM, 6–24 h) and 1-methyl-3-nitro-1-nitrosoguanidine (Sigma-Aldrich, 100–500 μM, 3–12 h) can be employed as positive controls for cytochrome c and AIF release, respectively.

8. The duration of treatment is a function of the cell type, the molecule under investigation, and its concentration. If long stimulations (>48 h) are envisioned, the number of cells initially seeded in each well should be adjusted accordingly (see also Note 2).

9. Washes should be performed by gentle aspiration and pipetting to avoid excessive detachment of cells from coverslips.

10. Fixation may be performed by different means. For a basic protocol, we recommend to employ a solution of 4% paraformaldehyde (PFA) (Sigma-Aldrich) + 0.19% picric acid (Sigma-Aldrich) in PBS.

11. PFA is toxic by inhalation and should be manipulated under an appropriate chemical hood.

12. PFA in solution has a limited stability. Therefore, it should be prepared immediately before use. Upon preparation, the solution should be kept on ice throughout the duration of the experiment.

13. Picric acid (2,3,6-trinitrophenol) is toxic and explosive. Accordingly, the stock solution should not be allowed to dry out and should always be handled with maximal care.

14. Before permeabilization, fixed samples can be stored in PBS at 4 °C for several months.

15. As an alternative, permeabilization may be achieved by incubation in 0.1% Triton-X 100 (from Sigma-Aldrich, v/v in PBS) for 15 min at room temperature or in 0.1% Tween 20 (from Sigma-Aldrich, v/v in PBS) for 5 to 10 min at room temperature.

16. To avoid staining artifacts, samples should never be allowed to totally dry out after fixation.

17. As an alternative, blocking can be performed in 3% BSA (w/v in PBS) for 20 min at room temperature.

18. To follow the redistribution of IMS proteins, mitochondria can be stained either by an antibody that recognizes a sessile mitochondrial protein (*e.g.*, heat shock 60-kDa protein, voltage-dependent anion channel) or by means of fixable, potential-sensitive probes (*e.g.*, CMXRos). In the latter case, probes should be incubated with live cells for a short period (usually 15–30 min) before fixation to allow for their accumulation in mitochondria.

19. To avoid false-negative results and to check for different kinetics in the release, we recommend following the redistribution of at least two IM proteins (*e.g.*, cytochrome *c* and AIF). To this aim, parallel samples should be envisioned, as nuclear and mitochondrial signals occupy already two of the three emission channels that are available in most fluorescence microscopes.

20. To minimize drying, it is recommended to perform staining with primary antibodies in a humid chamber. This applies to both suggested staining conditions (room temperature for 90 min and 4 °C overnight).

21. To reduce the amount of primary antibody employed for each sample, staining can be performed by removing coverslips from plates and overlying them in reverse orientation (to ensure contact between cells and antibodies) onto drops of 50 or 30 μl (for 18- and 12-mm-diameter coverslips, respectively) of primary antibody solution previously deposited over a stretch of Parafilm (Pechiney Plastic Packaging Co., Chicago, IL). In this case, coverslips should be returned to the plates at the end of the staining period.

22. Primary antibodies in 0.5% BSA are usually stable and can be stored for subsequent reutilization upon the addition of 0.02% sodium azide (Sigma-Aldrich) as a preservative agent. Sodium azide exists as an odorless white powder (toxic by ingestion) that may release a highly toxic gas upon dissolution in aqueous solutions or following contact with metals. Thus, it should always be manipulated with maximal care under an appropriate chemical hood.

23. According to the primary antibodies that have been employed, anti-rabbit, antigoat, or antimouse secondary antibodies coupled to Alexa Fluor 488 (green) or Alexa Fluor 568 (red) are used (Molecular Probes-Invitrogen).

24. To avoid excessive detachment of cells from coverslips, it is not recommended to perform staining by means of the aforementioned "drop" method (see also Note 21) more than once.

25. From this step onward, samples should be protected from light to avoid the loss of signal due to photobleaching of the fluorochromes.

26. Hoechst 33342 is mutagenic.

27. For nuclear counterstaining, Hoechst 33342 can be replaced with 300 nM DAPI (Molecular-Probes Invitrogen). In this case, coverslips should be incubated at room temperature for no longer than 5 to 10 min.

28. As an alternative, nuclei can be counterstained during the incubation of coverslips with secondary antibodies. To this aim, Hoechst 33342 should be diluted to the final concentration of 2 μM directly in the secondary antibody solution (0.5% BSA, w/v in PBS).

29. To reduce the intensity of nonspecific background signal, up to five washes can be performed.

30. Insufficient drying of the mounting medium may provoke the annoying displacement of coverslips during microscopic observation.

31. Generally, stained and mounted coverslips can be stored at 4 °C under protection from light for several weeks. However, a time-dependent decrease in the fluorescent signals will eventually result in the complete loss of fluorescence.

32. The observation can also be performed on a standard fluorescence microscope.

33. To avoid cross-channel interference from fluorochromes with wide (or partially overlapping) emission spectra, we recommend performing a separate photographic acquisition for each channel (and eventually merge the images, if necessary).

34. The colocalization between IMS proteins and mitochondria can be assessed by an experienced operator or via software-assisted analysis of fluorescent emissions from the different fluorochromes. In the latter case, the free, Java-based, open-source software ImageJ may be employed.

35. Commercially available imaging software packages allowing for single-channel and merged editing of the acquired images (*e.g.*, Adobe Photoshop, from Adobe Systems Inc., San Jose, CA) can be employed to facilitate the analysis.

36. For cytofluorometric assessments, coverslips are useless and need not be employed.

37. In physiological conditions, $\Delta\Psi_m$ is a function of both cell type and metabolic conditions. To obtain appropriate negative controls (displaying an elevated $\Delta\Psi_m$, typical of metabolically active cells), special attention should be paid for the avoidance of overcrowding, which often provokes a shutdown of mitochondrial activities.

38. For cytofluorometric determinations, treatments should be performed by administering the molecules of choice in fresh medium. This allows to exclude from the analysis the small proportion of cells that failed to adhere to the plate and underwent apoptosis before stimulation.

39. The calcium ionophore A23187 (Sigma-Aldrich, 0.5–10 μM, 3–24 h) may be employed as a positive control for IM permeabilization in most cell lines.

40. Usually, molecules affecting IM permeability and $\Delta\Psi_m$ exhibit very rapid effects. This should be kept in mind, in addition to the considerations on cell density at seeding (see also Notes 2, 8, and 37), to define

the duration of treatments. In some cases, protracted treatments may allow the cells to restore the normal conditions of IM, thus providing false-negative results. Thus, we recommend performing both short- and long-term kinetic assessments, especially for molecules not characterized previously for their effects on IM and $\Delta\Psi_m$.

41. Supernatants from treated cultures should not be discarded because they contain the fraction of cells that underwent apoptosis-dependent detachment from the substrate during the stimulation period.

42. Trypsinization time is a function of cell type, viability, and density in the well. For most cell lines, 5 min at 37 °C is sufficient to completely detach a totally confluent, healthy population. However, it is recommended to check for complete detachment of cells from the well surface by visual inspection or by rapid observation in light microscopy.

43. Slightly prolonged trypsinization may be required for some cell types or if traces of medium (containing FBS, which inhibits trypsin) remain in the wells after the collection of supernatants. This may be circumvented by washing the wells with prewarmed PBS immediately before the addition of trypsin/EDTA solution.

44. Excessive trypsinization (>10 min) should be avoided as it may result *per se* in some extent of cellular stress (thus provoking a treatment-independent loss of $\Delta\Psi_m$). As an alternative, cells can be detached by mechanical scraping. However, this may also induce $\Delta\Psi_m$ loss in more sensible cell lines.

45. Because stressed and suffering cells detach faster than healthy ones, it is important to ensure that all cells have been collected to avoid an overestimation of cells exhibiting reduced $\Delta\Psi_m$ (as well as of those that already underwent plasma membrane breakdown).

46. At this step, FBS contained in the supernatants definitely inactivates trypsin. If supernatants are scarce ($<500\ \mu l$), complete trypsin inactivation may be ensured by the addition to each 3 to 5 ml of prewarmed growth medium supplemented with 10% FBS.

47. Excessively protracted staining should be avoided, as it may result in probe-dependent toxicity.

48. When numerous samples (>12) are analyzed, it is recommended to limit the variation of the time intervening between labeling and cytofluorometric detection. To this aim, samples may be divided in groups of 10 to 12 and stained at 5- to 6-min intervals between subsequent series.

49. The choice of the vital dye should take into account its spectral properties, as well as those of the employed mitochondrial probe, so that the fluorescent emissions from the two molecules do not overlap and can be discriminated at the subsequent cytofluorometric analysis.

50. As a guideline, $DiOC_6(3)$ (which we use routinely for the assessment of $\Delta\Psi_m$ and which emits in green; absorption/emission peaks at 484/501 nm) can be associated with the vital dye propidium iodide (PI, emitting

in red; absorption/emission peaks at 535/617 nm) (Castedo *et al.*, 2002). PI staining should be performed shortly before the cytofluorometric analysis by the addition of the dye directly to the samples to a final concentration of 0.5 to 1 $\mu g/ml$. Alternatively, tetramethylrhodamine methyl ester (TMRM) (fluorescing in red; absorption/emission peaks at 543/573 nm) can be used to measure $\Delta \Psi m$ in association with 10 μM DAPI (which emits in blue; absorption/emission peaks at 358/461 nm) for the assessment of viability.

51. As an example, we use a FACScan cytofluorometer (BD, San José, CA) associated with an argon ion laser (emitting at 488 nm) for excitation. In this case, the following channels are utilized for the detection of fluorescent signals: FL1 for calcein AM and $DiOC_6(3)$; FL2 for TMRM; and FL3 for PI.

52. FSC depends on cell size, whereas SSC reflects the refractive index, which in turn is influenced by more than one parameter, including cellular shape and intracellular complexity (*i.e.*, presence of cytoplasmic organelles and granules).

53. Usually, cytofluorometers are provided with dedicated software for instrumental control and analysis. Unfortunately, the execution of (almost) all these packages is limited to the computer that controls the cytofluorometer and, very often, to a single operating system (*i.e.*, Mac). However, some analytical software packages (for both Mac- and PC-based systems) are available online free of charge. For PC users, we recommend WinMDI 2.8 (©1993–2000, Joe Trotter), originally developed for Windows 3.1 but compatible with newer versions of the operating system (freely available at http://www.cyto.purdue.edu/flowcyt/software.htm).

54. At the analytical stage, it should be kept in mind that some $\Delta \Psi_m$-sensitive probes [*e.g.*, $DiOC_6(3)$, TMRM] are influenced by mitochondrial, $\Delta \Psi_m$-unrelated (*e.g.*, mitochondrial mass) as well as by mitochondria-independent parameters (*e.g.*, cell size, plasma membrane potential or the activity of multidrug resistance pumps) (Metivier *et al.*, 1998; Zamzami *et al.*, 1995a,b). To evaluate these possibilities by an appropriate series of controls, each sample should be divided into two aliquots before centrifugation (Section 3.1 steps 4 and 5), and one of the two sets should be preincubated with 50 to 100 μM carbonyl cyanide *m*-chlorophenylhydrazone (a ionophore causing complete and irreversible $\Delta \Psi_m$ dissipation) for 5 to 10 min before staining (37 °C, 5% CO_2).

55. If $\Delta \Psi_m$-sensitive probes have been used in association with vital dyes (*e.g.*, PI), the researcher may want to limit the evaluation of $\Delta \Psi_m$ to cells with intact plasma membranes. This may be especially useful in excluding necrotic cells (primary or secondary necrosis) from the analysis. To this aim, a second gating (after the first one on events

characterized by normal FSC and SSC values; see also Note 52) on PI⁻
events can be performed.

56. Because calcein AM-based assays can be analyzed either by cytofluorometry or by fluorescence microscopy, cells should be seeded into the
most suitable plates according to the ultimate readout of choice.

57. The calcein AM staining kinetic has profound implications for the final
analysis of results. Thus, it is essential to note that cells are loaded with
calcein AM and Co^{2+} ions *before* the administration of proapoptotic
stimuli.

58. In its acetoxymethyl ester form, calcein is membrane permeant and rapidly
enters all subcellular compartments. Inside the cells, calcein AM is hydrolyzed by endogenous esterases into the highly negatively charged green
fluorescent calcein, which is retained in the cytoplasm.

59. Several metal ions are able to quench calcein fluorescence, including
Mn^{2+}, Cu^{2+}, and Co^{2+}. Co^{2+} was originally selected for the development
of this staining technique because: (1) Co^{2+} exhibits a more pronounced
quenching activity than Mn^{2+}; (2) in contrast to Cu^{2+}, Co^{2+} does not
promote peroxidative reactions, which *per se* may favor MPT; and
(3) Co^{2+} is not a substrate for the mitochondrial Ca^{2+} uniporter (so it
is excluded from the mitochondrial compartment) (Petronilli *et al.*, 1999).

60. The use of HEPES is strongly recommended to maintain the pH of the
staining medium around 7.3. At this [H^+], most calcein AM molecules are
not charged and readily cross the plasma membrane. Significant pH
variations from neutrality may shift the acid–base equilibrium to negatively (or positively) charged forms of calcein AM (that are not able to
diffuse through plasma membranes), leading to insufficient loading of cells.

61. The treatment duration should be defined as recommended earlier (see
also Notes 2, 8, 37, and 40). In addition, it should be kept in mind that
transient and rapid openings of the PTPC occur physiologically, thus
allowing some Co^{2+} ions to enter mitochondria. As a result, mitochondrial calcein fluorescence exhibits a spontaneous time-dependent
decrease, which might be misinterpreted (false-positive results) if
stimulations are excessively long.

62. Before fixation, cells can be costained with other organelle-selective
(*e.g.*, LysoTracker red) or $\Delta\Psi_m$-sensitive probes (*e.g.*, CMXRos) to
allow for the simultaneous analysis of multiple parameters (see also
Notes 63 and 64).

63. In this case, the spectral properties of the employed dyes should be
considered carefully to avoid cross-channel interference (see also Notes
49 and 50). Moreover, it is important to stress that calcein fluorescence can
be quenched by other fluorescent molecules that accumulate in mitochondria (notably TMRM),because of energy transfer processes that probably
reflect the formation of molecular complexes (Petronilli *et al.*, 1999).

64. If permeabilization is not required (*e.g.*, when only membrane-permeant dyes are used), fixed coverslips can be stored prior to observation for some weeks (in PBS at 4 °C, under protection from light).

65. The extent of signal reduction provides direct indications on the duration and extent of MPT. Thus, while transitory and/or limited IM permeabilization results in partial quenching, long-lasting openings of the PTPC almost completely abolish mitochondrial fluorescence.

66. The investigator may perform the scoring by assigning the cells to two to three groups according to the intensity of the calcein signal (*e.g.*, calcein^{high}, calcein^{intermediate}, calcein^{low}). As an alternative, the number of groups can be augmented by scoring multiple parameters simultaneously (*e.g.*, calcein^{high} $\Delta\Psi_m^{high}$, calcein^{high} $\Delta\Psi_m^{low}$, calcein^{low} $\Delta\Psi_m^{high}$, calcein^{low} $\Delta\Psi_m^{low}$).

67. Either $\Delta\Psi_m$-sensitive probes or vital dyes can be associated with calcein AM. See the previous considerations on cross-channel interference due to the overlap of emission spectra and on possible quenching phenomena of the mitochondrial calcein pool (see also Notes 49, 50, and 63).

ACKNOWLEDGMENTS

This work has been supported by a special grant from Ligue National contre le cancer (èquipe labellisée), as well as by grants from Agence Nationale de Recherche, Agence Nationale pour la Recherche sur le Sida, Cancéropôle Ile-de-France, Fondation pour la Recherche Médicale, Institut National du Cancer, European Commission (RIGHT, Active p53, Trans-Death, Death-Train, ChemoRes), and Sidaction. OK is recipient of an EMBO Ph.D. fellowship. Windows is a registered trademark of Microsoft Corporation in the United States and other countries.

REFERENCES

Castedo, M., Ferri, K., Roumier, T., Metivier, D., Zamzami, N., and Kroemer, G. (2002). Quantitation of mitochondrial alterations associated with apoptosis. *J. Immunol. Methods* **265**, 39–47.

Cossarizza, A., Baccarani-Contri, M., Kalashnikova, G., and Franceschi, C. (1993). A new method for the cytofluorimetric analysis of mitochondrial membrane potential using the J-aggregate forming lipophilic cation 5,5′,6,6′-tetrachloro-1,1′,3,3′-tetraethylbenzimidazolcarbocyanine iodide (JC-1). *Biochem. Biophys. Res. Commun.* **197**, 40–45.

Galluzzi, L., Zamzami, N., de La Motte Rouge, T., Lemaire, C., Brenner, C., and Kroemer, G. (2007). Methods for the assessment of mitochondrial membrane permeabilization in apoptosis. *Apoptosis* **12**, 803–813.

Garrido, C., Galluzzi, L., Brunet, M., Puig, P. E., Didelot, C., and Kroemer, G. (2006). Mechanisms of cytochrome c release from mitochondria. *Cell Death Differ.* **13**, 1423–1433.

Green, D., and Kroemer, G. (1998). The central executioners of apoptosis: Caspases or mitochondria. *Trends Cell Biol.* **8**, 267–271.

Kroemer, G., Galluzzi, L., and Brenner, C. (2007). Mitochondrial membrane permeabilization in cell death. *Physiol. Rev.* **87**, 99–163.

Li, L. Y., Luo, X., and Wang, X. (2001). Endonuclease G is an apoptotic DNase when released from mitochondria. *Nature* **412**, 95–99.

Macho, A., Decaudin, D., Castedo, M., Hirsch, T., Susin, S. A., Zamzami, N., and Kroemer, G. (1996). Chloromethyl-X-rosamine is an aldehyde-fixable potential-sensitive fluorochrome for the detection of early apoptosis. *Cytometry* **25**, 333–340.

Martins, L. M., Iaccarino, I., Tenev, T., Gschmeissner, S., Totty, N. F., Lemoine, N. R., Savopoulos, J., Gray, C. W., Creasy, C. L., Dingwall, C., and Downward, J. (2002). The serine protease Omi/HtrA2 regulates apoptosis by binding XIAP through a reaper-like motif. *J. Biol. Chem.* **277**, 439–444.

Metivier, D., Dallaporta, B., Zamzami, N., Larochette, N., Susin, S. A., Marzo, I., and Kroemer, G. (1998). Cytofluorometric detection of mitochondrial alterations in early CD95/Fas/APO-1-triggered apoptosis of Jurkat T lymphoma cells: Comparison of seven mitochondrion-specific fluorochromes. *Immunol. Lett.* **61**, 157–163.

Modjtahedi, N., Giordanetto, F., Madeo, F., and Kroemer, G. (2006). Apoptosis-inducing factor: Vital and lethal. *Trends Cell Biol.* **16**, 264–272.

Petronilli, V., Miotto, G., Canton, M., Brini, M., Colonna, R., Bernardi, P., and Di Lisa, F. (1999). Transient and long-lasting openings of the mitochondrial permeability transition pore can be monitored directly in intact cells by changes in mitochondrial calcein fluorescence. *Biophys. J.* **76**, 725–734.

Poncet, D., Boya, P., Metivier, D., Zamzami, N., and Kroemer, G. (2003). Cytofluorometric quantitation of apoptosis-driven inner mitochondrial membrane permeabilization. *Apoptosis* **8**, 521–530.

Ravagnan, L., Roumier, T., and Kroemer, G. (2002). Mitochondria, the killer organelles and their weapons. *J. Cell. Physiol.* **192**, 131–137.

Susin, S. A., Zamzami, N., and Kroemer, G. (1998). Mitochondria as regulators of apoptosis: Doubt no more. *Biochim. Biophys. Acta* **1366**, 151–165.

Verhagen, A. M., Ekert, P. G., Pakusch, M., Silke, J., Connolly, L. M., Reid, G. E., Moritz, R. L., Simpson, R. J., and Vaux, D. L. (2000). Identification of DIABLO, a mammalian protein that promotes apoptosis by binding to and antagonizing IAP proteins. *Cell* **102**, 43–53.

Zamzami, N., Marchetti, P., Castedo, M., Decaudin, D., Macho, A., Hirsch, T., Susin, S. A., Petit, P. X., Mignotte, B., and Kroemer, G. (1995a). Sequential reduction of mitochondrial transmembrane potential and generation of reactive oxygen species in early programmed cell death. *J. Exp. Med.* **182**, 367–377.

Zamzami, N., Marchetti, P., Castedo, M., Zanin, C., Vayssiere, J. L., Petit, P. X., and Kroemer, G. (1995b). Reduction in mitochondrial potential constitutes an early irreversible step of programmed lymphocyte death *in vivo*. *J. Exp. Med.* **181**, 1661–1672.

CHAPTER NINETEEN

Oxidative Lipidomics of Programmed Cell Death

Vladimir A. Tyurin,*,† Yulia Y. Tyurina,*,† Patrick M. Kochanek,‡
Ronald Hamilton,§ Steven T. DeKosky,¶ Joel S. Greenberger,||
Hülya Bayir,*,†,‡ and Valerian E. Kagan*,†

Contents

Abstract

Oxidized phospholipids play an important role in execution of the mitochondrial stage of apoptosis and clearance of apoptotic cells by macrophages. Therefore, the identification and quantification of oxidized phospholipids generated during apoptosis are very important. These can be achieved successfully by a newly developed approach—oxidative lipidomics, including a combination of

* Center for Free Radical and Antioxidant Health, University of Pittsburgh, Pittsburgh, Pennsylvania
† Department of Environmental and Occupational Health, University of Pittsburgh, Pittsburgh, Pennsylvania
‡ Department of Critical Care Medicine, University of Pittsburgh, Pittsburgh, Pennsylvania
§ Department of Pathology, University of Pittsburgh, Pittsburgh, Pennsylvania
¶ Department of Neurology, University of Pittsburgh, Pittsburgh, Pennsylvania
|| Department of Radiation Oncology, University of Pittsburgh, Pittsburgh, Pennsylvania

Methods in Enzymology, Volume 442
ISSN 0076-6879, DOI: 10.1016/S0076-6879(08)01419-5

electrospray ionization/mass spectrometry (ESI-MS) and fluorescence high-performance liquid chromatography techniques. Using oxidative lipidomics allows the quantification of specific phospholipids and their hydroperoxides. We characterized selective oxidation of two anionic phospholipids: cardiolipin (CL) in mitochondria and phosphatidylserine (PS) outside of mitochondria. ESI-MS analysis of cytochrome c/H_2O_2-driven tetralinoleoyl-CL (TLCL) oxidized molecular species demonstrated accumulation of products monohydroxy-TLCL; monohydroxy-monohydroperoxy-TLCL, monohydroxy-dihydroperoxy-TLCL, mono-hydroxy-trihydroperoxy-TLCL; and monohydroxy-tetrahydroperoxy-TLCL. We explored the application of oxidative lipidomics in a number of conditions in both *in vitro* and *in vivo* models where there is a known contribution of apoptosis and/or inflammation. Accumulation of CL hydroperoxides, originated from molecular species of CL containing $C_{22:6}$ after experimental traumatic brain injury, was shown. ESI-MS analysis of intestine CL in mouse after γ-irradiation detected several CL oxidized molecular species: $(C_{18:2})_3/(C_{18:2+OOH})$; $(C_{18:2})_2/(C_{18:2+OOH})_2$; $(C_{18:2})_1/(C_{18:2+OOH})_3$; and $(C_{18:2+OOH})_4$. ESI-MS analysis and tandem MS/MS experiments revealed that PS with oxidized $C_{22:6}$ [m/z 866 $(C_{18:0}/C_{22:6+OOH})$ originated from the ion at m/z 834 $(C_{18:0}/C_{22:6})$] was the major oxidized molecular species in the tested models *in vitro* and *in vivo,* including (1) cytochrome c/H_2O_2 catalyzed oxidation of rat brain PS; (2) after experimental traumatic rat brain injury in rats, (3) in postmortem brain samples from patients with Alzheimer's disease, and (4) in the small intestine in γ-irradiated mouse. We conclude that oxidative lipidomics is a powerful technique to study lipid oxidation and its role in cell death across a spectrum of tissues and insults.

1. INTRODUCTION

Oxygenated fatty acids are well-known signaling molecules: numerous oxygen–containing eicosanoids—prostaglandins, thromboxanes, and leuko-trienes, lipoxenes, resolvins and protectins—participate in regulation and coordination of cell and body metabolism (Bannenberg *et al.*, 2007; Schwab *et al.*, 2007). Their important role in cell proliferation, modulation of apoptosis, angiogenesis, inflammation, and immune surveillance has been demonstrated (Ariel and Serhan, 2007; Bannenberg *et al.*, 2007). Despite the fact that polyunsaturated phospholipids are the major target for the attack by reactive oxygen species, appreciation of the involvement and participation of oxida-tively modified phospholipids in cell signaling is just beginning to emerge. This is largely because of the lack of adequate methodologies to investigate a huge variety of oxidized phospholipids. Indeed, until recently, comprehensive analysis of different classes of lipids and their molecular species in cells, tissues, and biofluids was difficult if not impossible. However, with the advent of contemporary mass spectrometry and the development of new soft–ionization techniques, identification and structural characterization of individual phospholipid molecules in complex mixtures became possible (Watson, 2006). These technological advancements designated the emergence of a new field of research and knowledge—lipidomics. The latter can be defined as "a branch of metabolomics that includes a systems-based study of all lipids, the molecules with which they interact, and their function within the cell" (Watson, 2006). Thus, the lipidomics approach consists of analysis of entire lipidome and lipid-related metabolites, as well as characterization, and quantification of changes in the lipidome in response to internal and external stimuli such as alteration in nutritional status or under disease conditions (Gaspar *et al.*, 2007). Application of principles and approaches of lipidomics to studies of oxidatively modified lipids yielded the area of oxidative lipidomics. Similar to lipidomics, oxidative lipidomics includes identification and characterization of oxidatively modified lipids by mass spectrometry (MS) and quantitative analysis of their content in cells, tissues, and biofluids, as well as discovery of mechanisms and pathways through which they interact with other molecules and contribute to metabolism, physiological functions, and/or disease pathogenesis. Oxidative lipidomics is progressively increasing the ability to perform sensitive, quantitative, and comprehensive analysis of individual molecular species of oxygenated lipids. By utilizing oxidative lipidomics, oxidatively modified molecular species of cardiolipin (CL) in mitochondria and phosphatidylserine (PS) in plasma membrane have been characterized and shown to play an important role in signaling during programmed cell death (Kagan *et al.*, 2004, 2005). In addition, it has been confirmed that oxidized PS and oxidized phosphatidylcholine serve as preferred ligands within apoptotic cells for CD36-mediated phagocytosis by macrophages (Gao *et al.*, 2006; Greenberg *et al.*, 2006; Sun *et al.*, 2006). Moreover, oxygenated species of phosphatidylethanolamine formed after activation of human monocytes and platelets via the lipoxygenase-dependent pathway have been identified and characterized (Maskrey *et al.*, 2007).

To characterize the diversity of phospholipid oxidation products generated during apoptosis we applied oxidative lipidomics methods that include identification of polyunsaturated phospholipid molecular species as potential peroxidation substrates and their oxidation products by electrospray ionization (ESI)-MS analysis (Forrester *et al.*, 2004; Pulfer and Murphy, 2003), followed by quantitative assessments of lipid hydroperoxides using a fluorescence high-performance liquid chromatography (HPLC)-based protocol.

2. MATERIALS AND METHODS

2.1. Reagents

1,2-Diheptadecanoyl-sn-glycero-3-[phospho-l-serine] (sodium salt), 1,1′, 2,2′-tetramyristoyl-cardiolipin (sodium salt), and 1,1′,2,2′-tetralinoleoyl-cardiolipin (sodium salt) are from Avanti Polar Lipids Inc. (Alabaster, AL).

Chloroform, methanol, acetone, ammonium hydroxide, acetic acid glacial, $CaCl_2$, EDTA, SDS, microperoxidase-11, phospholipase A_2, and triphenylphosphine (TPP) are from Sigma-Aldrich (St. Louis, MO). HPTLC silica G plates are from Whatman, (Schleicher & Schuell, England). N-Acetyl-3,7-dihydroxyphenoxazine (Amplex Red) is from Molecular Probes (Eugene, OR).

2.2. *In vivo* exposures

The controlled cortical impact (CCI) model of traumatic brain injury is performed on male Sprague–Dawley rats, and the brain cortex P2 fraction is used for lipid analysis (Bayir *et al.*, 2007).

Total body irradiation (TBI) was performed in groups of C57BL/6NHsd female mice (15 Gy) using a Shephered Mark 1 Model 68 cesium irradiator, and isolated small intestine-a known target for radiation-induced injury-was used for lipid analysis (Tyurina *et al.*, 2008).

Postmortem human brain samples from control and Alzheimer's disease (AD) patients were obtained from Bank collection of UPMC Hospital of the University of Pittsburgh.

All procedures were preapproved and performed according to the protocols established by the Institutional Animal Care and Use Committee, as well as by the Institutional Review Board (human brain samples) of the University of Pittsburgh.

2.3. Lipid extraction and two-dimensional (2D) high-performance thin-layer chromatography (HPTLC) analysis

Total lipids are extracted from samples by the Folch procedure (Folch *et al.*, 1957). Lipid extracts are separated and analyzed by 2D-HPTLC (Rouser *et al.*, 1970). Lipid phosphorus is determined by a micromethod (Bottcher *et al.*, 1961).

Mass spectra of phospholipids are analyzed by direct infusion into a linear ion-trap mass spectrometer LXQ (Thermo Electron, San Jose, CA) as described previously (Tyurin *et al.*, 2008). Samples after 2D-HPTLC separation are collected, evaporated under N_2, resuspended in chloroform: methanol 1:1 v/v (20 pmol/μl), and utilized directly for acquisition of ESI mass spectra at a flow rate of 5 μl/min. The electrospray probe is operated at 5.0 kV in the negative ion mode. Source temperature is maintained at 70 or 150 °C. MS^n analysis is carried out with relative collision energy ranging from C20 to 40% and with an activation q value at 0.25 for collision-induced dissociation (CID) and a q value at 0.7 for the pulsed-Q dissociation technique. Analysis of major molecular clusters of CL was performed using multiple stage fragmentation MSn (Hsu and Turk, 2006). According to the manufacturer's instruction, caffeine, methionine–argenine–phenylalanine–alanine (MRFA), and Ultramark 1621 are used for calibration of the LXQ mass spectrometer. To create tune files, we use the following standards: $(C_{14:0})_4$-CL, $(C_{17:0})_2$-PS. Chemical structures of lipids are confirmed by using Lipid Map Data Base and ChemDraw format (www.lipidmaps.org).

Phospholipid hydroperoxides are determined quantitatively by fluorescence HPLC of products formed in the microperoxidase 11–catalyzed reaction with a fluorogenic substrate, Amplex Red, as previously described (Tyurin *et al.*, 2008). The Shimadzu LC-100AT *vp* HPLC system equipped with a fluorescence detector (RF-10Axl) and autosampler (SIL-10AD vp) are used for analysis of products separated by HPLC (Eclipse XDB-C18 column, 5 μm, 150 × 4.6 mm). The mobile phase is composed of 25 mM NaH_2PO_4 (pH 7.0)/methanol (60:40, v/v).

3. Results and Discussion

3.1. Identification of hydroxy- and hydroxy-hydroperoxy-CL molecular species formed during peroxidation of TLCL by cytochrome *c*/H_2O_2 in a model system

Oxidative stress is one of the most common triggers of apoptosis. It has been reported that CL binds cytochrome *c* avidly, unfolds the protein partially, and that the complex functions as a peroxidase, catalyzing CL oxidation essential for the release of proapoptotic factors from mitochondria (Belikova *et al.*, 2006; Kagan *et al.*, 2004, 2005; Kapralov *et al.*, 2007; Tyurin *et al.*, 2007). To characterize oxidized molecular species of CL, formed in cytochrome *c*/H_2O_2 reaction, we used TLCL. Incubation of the TLCL/cytochrome *c* complex with H_2O_2 for 30 min resulted in several oxygenated species, including those containing mono-, di-, and trihydroperoxides of CL (TLCL-OOH) and its hydroxy derivatives (Fig. 19.1). The total amount of TLCL-OOH accumulated was estimated as 170.5 ± 12.5 pmol/nmol CL and 2.8 ± 1.6 pmol/nmol CL for oxidized CL and control, respectively (Table 19.1). The structural analysis of TLCL and its multiple oxygenated species is shown in Fig. 19.1. Typical full ESI–mass spectrum in negative ion mode of nonoxidized TLCL demonstrated the presence of doubly charged [M-2H]$^{-2}$ and singly charged [M-H]$^-$ ions at *m*/*z* 723 and 1448, respectively (Fig. 19.1A, a). We performed oxidative lipidomics analysis of doubly charged species of CL because their signal intensity is higher compared to singly charged ions (Fig. 19.1A, a, b). TLCL–oxidized species revealed ions at *m*/*z* 731, 739, 747, 755, 763, 771, 779, and 787 that correspond to hydroxy-, hydroperoxy-, and hydroxy-hydroperoxy-CL molecular species (Fig. 19.1). Molecular ions of TLCL with *m*/*z* 739, 755, 771, and 787 corresponded to molecular species containing one,

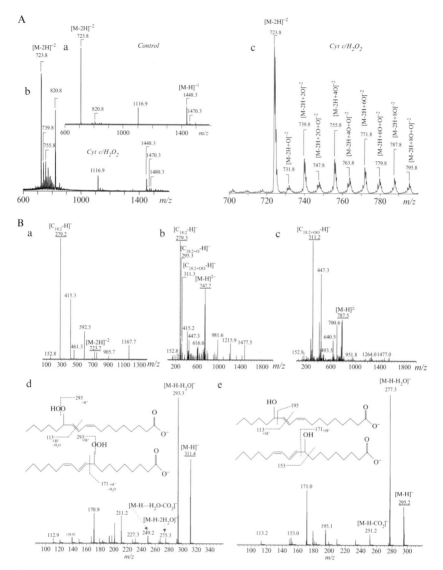

Figure 19.1 MS analysis of TLCL and its oxidation products formed during incubation with cytochrome c/H_2O_2. Typical negative ion ESI mass spectra of TLCL before (A, a) and after (A, b, c) oxidation. Tandem MS-MS experiments confirmed the structure of oxidized TLCL molecular species (B, a–d). (B, a–b) MS^2 spectra of ions: (a) with m/z 723 fragmented to an ion with m/z 279 (corresponding to nonoxidized $C_{18:2}$); (b) with m/z 747 fragmented to ions with m/z 279, 295 ($C_{18:2+OH}$), and 311 ($C_{18:2+OOH}$); (c) with m/z 787 fragmented to an ion with m/z 311 ($C_{18:2+OOH}$). (B, d, e) MS^3 spectra of ions: (d) with m/z 787 fragmented to an ion with m/z 311 ($C_{18:2+OOH}$) and with m/z 747 fragmented to an ion with m/z 295 ($C_{18:2+OH}$). The mixture of TLCL/DOPC (250 μM/250 μM) was incubated in 50 mM Na-phosphate buffer (pH 7.4) containing 100 μM DTPA in the presence of cytochrome c (5 μM) and H_2O_2 (100 μM). At the end of incubation, lipids were extracted by the Folch procedure (Folch *et al.*, 1957) and MS analysis was performed. Additionally, CL-OOH formation was confirmed after reduction of CL-OOH into CL-OH by triphenylphosphine (1 mg/ml, for 20 min at 4°). CL-OH formed was analyzed by ESI-MS (data not shown).

Table 19.1 Assessment of total content of CL and PS hydroperoxides by fluorescence HPLC in samples

Sample ID		CL-OOH pmol/nmol CL	PS-OOH pmol/nmol PS
Model system: phosphate buffer ($n = 3$)	Control	2.8 ± 1.6	1.2 ± 0.6
	PL/cytochrome c/H_2O_2	170.5 ± 12.5	35.8 ± 2.7
Mouse small Intestine ($n = 4$)	Control	6.4 ± 1.2	4.9 ± 1.2
	TBI, 15 Gy	62.6 ± 14.8	36.1 ± 15.2
Rat brain cortex ($n = 6$)	Control	6.7 ± 1.5	4.6 ± 1.2
	CCI, 24 h	110.5 ± 20.3	51.6 ± 12.7
Postmortem Human brain ($n = 8$)	Control	Not determined	8.8 ± 2.2
	AD	Not determined	65.6 ± 23.2

two, three, and four hydroperoxy groups, respectively. Doubly charged ions with m/z 731, 747, 763, and 778 were assigned to TLCL-OH; TLCL-OH-OOH, TLCL-OH-2OOH, TLCL-OH-3OOH; and TLCL-OH-4OOH molecular species, respectively.

Typical MS^2 spectra of doubly charged molecular ions of nonoxidized TLCL at m/z 723 and oxidized TLCL at m/z 747 and 787 are presented in Fig. 19.1 (B, a, b, c). Product ion at m/z 279 ($C_{18:2}$) was dominating after fragmentation of the molecular ion with m/z 723 and corresponded to nonoxidized TLCL. Molecular ions of $C_{18:2}$ (m/z 279), $C_{18:2+OH}$ (m/z 295), and $C_{18:2+OOH}$ (m/z 311) were observed in MS^2 spectrum after fragmentation of TLCL-OH-OOH at m/z 747. Similar patterns of product ions were observed when molecular ions at m/z 763 and 779 were subjected to fragmentation (data not shown). Note, product ion with m/z 311 was not found after fragmentation of molecular ion of TLCL-OH at m/z 731 (data not shown), whereas this product ion (with m/z 311, $C_{18:2+OOH}$) was dominating after fragmentation of the molecular ion at m/z 787, which corresponded to TLCL-4OOH. Molecular ions of $C_{18:2}$ (m/z 279), $C_{18:2+OOH}$ (m/z 311) and the ion with m/z 293 originated from m/z 311 after dehydration were observed in MS^2 spectrum of oxidized TLCL with m/z 739, 755, and 771 subjected to fragmentation (data not shown).

To analyze fatty acid oxidation products, we performed MS^3 assessments. CID fragmentation of $C_{18:2+OOH}$ (m/z 311) resulted in the formation of several characteristic product ions (Fig. 19.1B, d). Loss of one and two molecules of water generated two product ions with m/z 293 and 275, respectively (MacMillan and Murphy, 1995; Schneider et al., 1997). Loss of water and CO_2 resulted in the formation of m/z 249. Loss of water and cleavage of the double bond produced an ion with m/z 113—another typical for 13-s-hydroperoxy-linoleic acid product ion (MacMillan and

Murphy, 1995). A product ion with m/z 171 was formed due to loss of water and cleavage of the C_9-C_{10} bond of 9-s-hydroperoxy-linoleic acid (Schneider *et al.*, 1997). Thus MS^n analysis confirmed that 13-OOH-$C_{18:2}$-CL, as well as 9-s-OOH-$C_{18:2}$, was generated during cytochrome c-driven CL peroxidation *in vitro*. CID of the molecular ion with m/z 295 [M-H]$^-$ ($C_{18:2+OH}$) generated product ions with m/z 277 and 251 that corresponded to [M-H - H_2O] and [M-H - CO_2], respectively (Fig. 19.1B, e). Product ions with m/z 153 and 171 formed after cleavage of 9-s-OH-$C_{18:2}$ on both sides of the hydroxyl group (Kerwin and Torvik, 1996). Generation of ion with m/z 195 was associated with similar cleavage of the C_{12}-C_{13} bond of 13-s-hydroperoxy-linoleic acid (Kerwin and Torvik, 1996). In addition, cleavage of the double bond yielded a product with m/z 113—another typical for the 13-s-hydroxylinoleic acid product ion (Haeflinger and Sulzer, 2007; MacMillan and Murphy, 1995). Thus in addition to the accumulation of hydroperoxides—13-OOH-$C_{18:2}$-CL and 9-s-OOH-$C_{18:2}$-CL—we found the products of their reduction such as 13-OH-$C_{18:2}$-CL and 9-s-OH- $C_{18:2}$-CL, respectively. This demonstrates that TLCL undergoes oxidation to its hydroperoxides in the presence of cytochrome c/H_2O_2. Moreover, these results also show that cytochrome c can utilize CL-hydroperoxides as a source of oxidizing equivalents to oxidize CL and simultaneously reduce CL-OOH to CL-OH.

3.2. Identification of CL-hydroperoxides formed *in vivo*

Mitochondrial events culminating in permeability transition and release of proapoptotic factors into the cytosol are central to intrinsic apoptotic pathways triggered by different physical factors, as well as chemical agents, toxins, and drugs (Orrenius *et al.*, 2007). Previous work identified the critical role of peroxidation of CL as an essential event that follows the collapse of CL asymmetry (Kagan *et al.*, 2005) and results in the formation of CL/cytochrome c peroxidase complexes (Kagan *et al.*, 2004, 2005; Kapralov *et al.*, 2007; Tyurin *et al.*, 2007a). *In vitro*, characterization of CL peroxidation, while technically challenging, is lessened by a significant accumulation of uncleared apoptotic cells; this task becomes markedly more difficult *in vivo* mostly because of very low steady-state levels of apoptotic cells effectively phagocytized by macrophages and other professional phagocytes (Vandivier *et al.*, 2006). This is further obscured by relatively low concentrations of CL and its oxidation products in tissues (Bayir *et al.*, 2007; Kagan *et al.*, 2005; Tyurin *et al.*, 2008). Based on results *in vitro*, CL oxidation occurs very early in apoptosis, before PS externalization (Kagan *et al.*, 2005). This suggests that CL oxidation products may be good biomarkers of apoptosis *in vivo*, as their appearance and accumulation may precede clearance of apoptotic cells. Here we present examples of successful identification and quantitation of CL hydroperoxides in tissues of animal challenges with proapoptotic stimuli.

Ionizing irradiation is one of the examples of a proapoptotic stimulation triggering CL oxidation-dependent on cytochrome c levels (Kagan et al., 2005). As shown in Figs. 19.2A and 19.2B, total body irradiation of mice resulted in accumulation of oxidized molecular species of CL in intestinal mitochondria. The amount of total CL hydroperoxides was increased more that 10-fold after irradiation to 62.3 ± 10.4 vs 5.2 ± 0.8 pmol CL-OOH/nmol CL in control (Table 19.1). Molecular species of oxidized CL were represented by only two major molecular ions with m/z 1447 and 1450 for singly charged ions and m/z 723 and 724 for doubly charged ions, respectively (Fig. 19.2A, a, b). When doubly charged ions were used for structural identification of CL, the MS^2 analysis of the molecular ion with m/z 723 yielded [a]$^-$ and [b]$^-$ ions (m/z 695), which corresponded to $(C_{18:2})(C_{18:2})$-phosphatidic acid, a typical fragmentation structure of CL. Further fragmentation into ions: [a+56]$^-$ and [b+56]$^-$ with m/z 751; [a+136]$^-$ and [b+136]$^-$ with m/z 831; [a-$(C_{18:2})$]$^-$ and [b-$(C_{18:2})$]$^-$ with m/z 415, as well as ions of $(C_{18:2})$ linoleic acid ions (m/z 279) were also readily observed in MS^2 spectra[29]. In the singly charged range, the molecular ion with m/z 1167 formed after the loss of one linoleic acid residue was also detectable during fragmentation (Fig. 19.2A, b). Specifically, oxidized species of CL with m/z 1480, 1511, 1543, and 1575 for singly charged ions that corresponded to doubly charged ions wit m/z 739, 755, 771, and 787, respectively, and contained one, two, three, and four hydroperoxy groups were detectable in MS spectra (Fig. 19.2B, a). As an example of structural analysis, the MS^2 spectrum of the molecular ion with m/z 771 is shown in Fig. 19.2B, b. Pulsed-Q dissociation of the doubly charged ion (m/z 771) yielded [a]$^-$ and [b]$^-$ ions (m/z 726 and 760), which correspond to $(C_{18:2})(C_{18:2+OOH})$-phosphatidic acid and $(C_{18:2+OOH})(C_{18:2+OOH})$-phosphatidic acid, respectively. Ions with m/z 415 and 449 correspond to [a-$(C_{18:2+OOH})$]$^-$ and [b-$(C_{18:2+OOH})$]$^-$. Molecular ions of $C_{18:2}$ (m/z 279) and $(C_{18:2+OOH})$ hydroperoxy-linoleic acid (m/z 311) were also observed in MS^2 spectra (Fig. 19.2B, b).

Several hydroxy-CL species were also presented, along with signals from mono-, di-, tri-, and tetrahydroperoxy-CLs (Fig. 19.2B, a). This suggests that hydroperoxy-CLs were partially metabolized to respective hydroxy-CLs in vivo. However, the amounts of CL-OH metabolites were significantly less compared to those of CL-OOH species, based on comparisons of the intensities of MS signals with characteristic m/z (e.g., compare CL-OOH signals at m/z 1480, 1511, 1543, and 1575 with CL-OH signals at m/z 1464, 1496, 1528, and 1556).

Many acute and chronic CNS diseases are associated with apoptotic cell death and oxidative stress (Mancuso et al., 2006; Moreira et al., 2005). For example, programmed cell death contributes to the disappearance of neurons after traumatic brain injury (Graham et al., 2000). Studies of

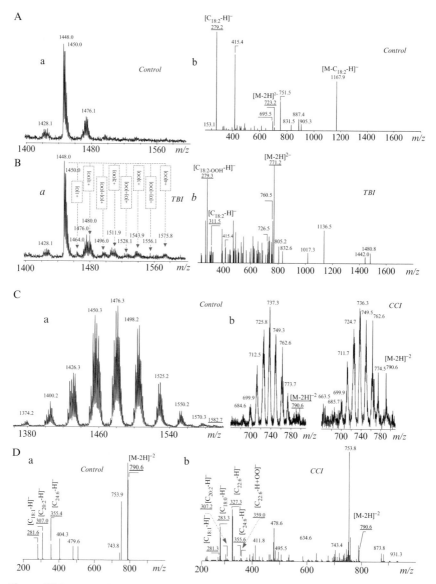

Figure 19.2 Typical negative ion ESI mass spectra of CL isolated from mouse small intestine (A, a, b—control, and B, a, b—after TBI) and from P2 fraction of rat brain cortex (C, a, b, D, a—control and C, c, D, c—after CCI). Oxidized molecular species of intestinal CL with m/z 1480.0, 1511.9, 1543.9, and 1575.8 for singly charged ions corresponded to molecular species of CL-$(C_{18:2+OOH})/(C_{18:2})/(C_{18:2})/(C_{18:2})$, $(C_{18:2+OOH})/(C_{18:2+OOH})/(C_{18:2})/(C_{18:2})$, $(C_{18:2+OOH})/(C_{18:2+OOH})/(C_{18:2+OOH})/(C_{18:2})$, and $(C_{18:2+OOH})/(C_{18:2+OOH})/(C_{18:2+OOH})/(C_{18:2+OOH})$, respectively. Typical MS^2 spectra of doubly charged $[M-2H]^{-2}$ molecular species of nonoxidized CL with m/z 723.8 (control panel A, b) and oxidized CL with m/z 771.2, which corresponds to a singly charged molecular ion $[M-H]^-$ with m/z 1543.9 (B, b). MS^2 analysis of oxidized CL (m/z 771.2) corresponding to $(C_{18:2+OOH})/(C_{18:2+OOH})/(C_{18:2+OOH})/$

CL peroxidation in the brain are particularly challenging because, in contrast to mammalian heart, liver, and intestine mitochondria where CL are composed of predominantly one or two linoleic acid-containing species (predominantly TLCL), CL from rat brain mitochondria are represented by at least nine different major molecular clusters with a variety of fatty acid residues (Fig. 19.2C, a, b, c,). These included polyunsaturated arachidonic ($C_{20:4}$) and docosahexaenoic ($C_{22:6}$) (DHA) fatty acids highly susceptible to oxidation. For example, one of the major CL molecular clusters of doubly charged CL ion at m/z 774, which corresponds to a singly charged CL ion at m/z 1550, includes at least four different CL molecular species as follows: $(C_{18:0})/(C_{22:6})/(C_{18:1})/(C_{22:6})$; $(C_{18:0})/(C_{22:6})/(C_{18:2})/(C_{22:5})$; $(C_{18:0})/(C_{20:4})/(C_{20:4})/(C_{22:5})$; and $(C_{18:2})/(C_{20:3})/(C_{20:3})/(C_{22:5})$. Traumatic brain injury resulted in a remarkably increased total amount of oxidized CL (152.4 ±19.2 pmol CL-OOH/nmol CL vs control 3.6 ± 0.6 pmol CL-OOH/nmol CL). Comparison of CL spectra before and after traumatic brain injury demonstrated an increased intensity of a doubly charged ion peak at m/z 790, which corresponds to a singly charged ion with m/z 1582 (Fig. 19.2C, b, c). MS^2 analysis of ion peak at m/z 790 showed that this ion cluster corresponds to multiple CL species with a dominant isomer of $(C_{18:1}/C_{22:6})/(C_{18:0})/(C_{22:6+OOH})$ originating from the ion at m/z 774 (see Fig. 19.2B, a, b). The structural assignment of this CL-OOH product as the one containing the hydroperoxy group in the $C_{22:6}$ residue has been confirmed by MS^n fragmentation as described earlier (data not shown).

3.3. Identification of PS-hydroperoxides in a model system containing cytochrome c/H_2O_2

The oxidized CL is required for the release of proapoptotic factors from mitochondria into the cytosol. This redox mechanism of cytochrome c is realized earlier than its other well-recognized functions in the formation of apoptosomes and caspase activation (Kagan et al., 2005). In the cytosol, released cytochrome c can interact with another anionic phospholipid, PS, and catalyze its oxidation in a similar oxygenase reaction (Kagan et al., 2005; Kapralov et al., 2007). Oxidation of PS facilitates its externalization; the effect is essential for the recognition and clearance of apoptotic cells by macrophages (Fadok et al., 1998; Kagan et al., 2003; Tyurina et al., 2007c). Thus identification and quantitation of PS oxidation products are other

---

$(C_{18:2})$-CL confirmed that it originated from a molecular cluster of CL with m/z 723.8 and contained $C_{18:2+OOH}$ (m/z 311.5). Oxidized molecular cluster of brain CL at m/z 790 [M-2H]$^{-2}$ for doubly charged ion corresponds to molecular cluster of singly charged ion [M-H]$^-$ at m/z 1582 with molecular species of CL-$(C_{18:0})/(C_{18:1})/(C_{22:6})/(C_{22:6+OOH})$. This oxidized molecular species originated from nonoxidized CL with doubly charged and singly charged ions at m/z 774 and 1550, respectively [corresponding to molecular species of CL-$(C_{18:0})/(C_{22:6})/(C_{18:1})/(C_{22:6})$].

important tasks in characterizing oxidative phospholipid signaling in apoptosis and clearance of apoptotic cells.

In a model system, we induced PS oxidation by cytochrome c/H_2O_2 and performed structural analysis of the PS oxidation products. The total amount of accumulated PS-OOH was estimated as 35.8 ± 2.7 and 1.2 ± 0.6 pmol/nmol PS for oxidation system and control, respectively (Table 19.1). Nonoxidized PS, isolated from rat brain cortex, revealed the major molecular ion at m/z 834 in the MS spectrum (in negative ionization mode) (Fig. 19.3A). Fragmentation of PS yielded a strong peak with m/z 747 caused by a loss of the serine group. Molecular fragments with m/z 283 and 327 corresponded to carboxylate anions of stearic acid ($C_{18:0}$) and DHA ($C_{22:6}$), respectively (Fig. 19.3A, a). Analysis of oxidized PS elicited several molecular species that were represented by deprotonated ions [M-H]$^-$ with m/z 850, 866, 882, and 898, which corresponded to hydroxy-, hydroperoxy-, hydroxyhydroperoxy-, and dihydroperoxy-PS (Figs. 19.3A and 19.3B). MS^2 analysis of oxidized PS molecular species showed that they originated from the molecular cluster of PS with m/z 834 ($C_{18:0}/C_{22:6}$) and contained hydroxy-DHA ($C_{22:6+OH}$), hydroperoxy-DHA ($C_{22:6+OOH}$), hydroxyhydroperoxy-DHA ($C_{22:6+OH+OOH}$), and dihydroperoxy-DHA ($C_{22:6+2OOH}$) at m/z 343, 359, 375, and 391, respectively (Fig. 19.3B, a, b, c, d). This was confirmed by the fact that similar DHA oxidation products were also formed during the incubation of free DHA in the presence of cytochrome c/H_2O_2 in phosphate buffer. The structure of DHA-oxidized species was further confirmed by the CID technique commonly used for fatty acid hydroperoxide analysis (Haeflinger and Sulzer, 2007; MacMillan and Murphy, 1995; Schneider et al., 1997). Fragmentation of the molecular ion at m/z 343, which corresponds to $C_{22:6+OH}$, resulted in product ions at m/z 325, 299, and 281, which correspond to [M-H $-$ H$_2$O]$^-$, [M-H $-$ CO$_2$]$^-$, and [M-H $-$ H$_2$O $-$ CO$_2$]$^-$, respectively (Figs. 19.3C and 19.3D, a). A product ion with m/z 121 formed after cleavage of the C_{13}-C_{14} bond was also observed in the MS^3 spectrum. Cleavage of the double bond of 13-OH-$C_{22:6}$ and loss of CO$_2$ produced m/z 188 after a two proton shift. CID of the molecular ion with m/z 359 [M-H]$^-$ ($C_{22:6+OOH}$) generated product ions at m/z 341, 315, and 297 that correspond to [M-H $-$ H$_2$O]$^-$, [M-H $-$ CO$_2$]$^-$, and [M-H $-$ H$_2$O $-$ CO$_2$]$^-$, respectively (Fig. 19.3C, b). Cleavage of the C_{13}-C_{14} bond of the 13-OOH-$C_{22:6}$ revealed ions with m/z 121 and 239 after a single proton shift (Figs. 19.3C and 19.3D, b). Additionally, product ions at m/z 149 and 165 were formed after cleavage of C_{11}-C_{12} and C_{10}-C_{11} bonds 11-OH-$C_{22:6}$, respectively (Hong et al., 2007). An ion with m/z 193 can be generated from both 13-OOH-$C_{22:6}$ and 11-OOH-$C_{22:6}$ through cleavage of the C_{12}-C_{13} bond and C_9-C_{10} double bond after a single proton shift, respectively. Thus 13-OOH-$C_{22:6}$-PS, 13-OH-$C_{22:6}$-PS, and 11-OH-$C_{22:6}$-PS were generated in the course of cytochrome c-catalyzed reaction of PS oxidation in vitro.

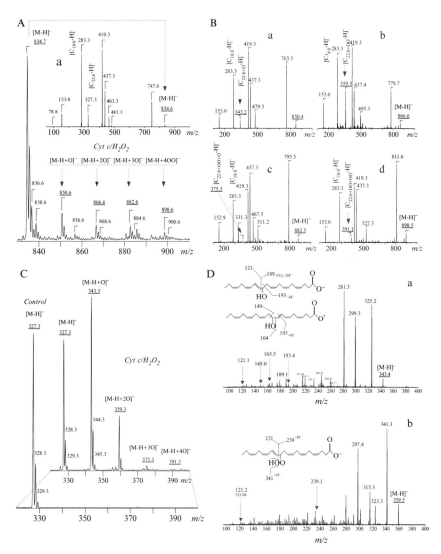

Figure 19.3 MS analysis of PS containing DHA oxidized by incubation with cytochrome c/H_2O_2. Negative ESI-MS spectrum of PS oxidized by cytochrome c/H_2O_2 (A). MS^2 spectra of nonoxidized PS (A, a) and oxidized PS molecular ions—PS-OH, PS-OOH, PS-OOH-OH, PS-OOH-OOH—with m/z 850 (B, a), 866 (B, b), 882 (B, c), and 898 (B, d), respectively. Product ions with m/z 343, 359, 373, and 391 were found after fragmentation of the molecular ion with m/z 850, 866, 882, and 898 and corresponded to $C_{22:6+OOH}$, $C_{22:6+OOH}$, $C_{22:6+OOH+OH}$, and $C_{22:6+OOH+OOH}$, respectively. Negative ESI-MS spectra of nonoxidized and oxidized DHA (after incubation with cytochrome c/H_2O_2) (C). (D, a, b) MS^3 spectra of CID-induced decomposition of molecular ions of DHA ($C_{22:6+OH}$) and ($C_{22:6+OOH}$) with m/z 343 (a) and 359 (b), respectively. Incubation conditions: 250 μM PS (isolated from P2 fraction of rat brain cortex) or 500 μM DHA were incubated during 30 min at 37° in 50 mM Na-phosphate buffer (pH 7.4) containing 100 μM DTPA in the presence of cytochrome c (5 and 10 μM for DHA) plus H_2O_2 (100 μM). At the end of incubation, lipids were extracted (Folch et al., 1957) and MS analysis was performed.

3.4. Identification of PS-hydroperoxides *in vivo*

To assess the formation of oxidized PS molecular species *in vivo* we chose to employ models that are commonly associated with triggering of cell apoptosis and enhanced production of ROS (Atamna and Boyle 2006; Epperly *et al.*, 2004; Kagan *et al.*, 2003, 2006; Kochanek *et al.*, 2000; Nunomura *et al.*, 2006; Tyurina *et al.*, 2007c). For example, ESI–MS analysis of PS in the brain cortex samples after traumatic brain injury of rats showed the enhanced formation of PS molecular species with oxidized $C_{22:6}$, PS-OOH, at m/z 866 as compared with control rats (Fig. 19.4A, a, b). MS^2 analysis of this cluster demonstrated that the ion at m/z 866 corresponded to PS–OOH with the dominating product of $(C_{18:0}/C_{22:6+OOH})$; the PS hydroperoxide was produced due to oxidation of the PS molecular species with the ion at m/z 834 $(C_{18:0}/C_{22:6})$. The total amount of hydroperoxides was increased more than 11-fold after experimental traumatic brain injury to 51.6 ± 12.7 vs 4.6 ± 1.2 pmol PS–OOH/nmol PS in control (Table 19.1).

Chronic neurodegenerative disease conditions are commonly associated with constant oxidative stress (Mancuso *et al.*, 2006; Moreira *et al.*, 2005). In Alzheimer's disease, the accumulation of isoprostanes has been established and proposed as an essential biomarker (de Jong *et al.*, 2007; Pratico *et al.*, 1998). However, the "phospholipid" origin of isoprostanes cannot be easily traced and/or linked with the apoptotic cell death (Montuschi *et al.*, 2007). By applying our oxidative lipidomics approach to postmortem brain samples from patients with Alzheimer's disease, we were able to identify PS containing $C_{22:6}$ species with hydroperoxide functions, PS–OOH (Figs. 19.2A, c, d, 19.2B, a, b). MS^2 experiments confirmed the structure of PS–OOH with oxidized $C_{22:6}$ (m/z 866). The content of PS hydroperoxides in brain samples from AD patients was fivefold higher than that in controls (Table 19.1).

Another example of the effectiveness of oxidative lipidomics in characterizing PS oxidation products is shown in Figs. 19.2C and 19.2D. We were able to demonstrate that irradiation of mice resulted in the accumulation of oxidized PS species with m/z 866 $(C_{18:0}/C_{22:6+OOH})$, 868 $(C_{18:0}/C_{22:5+OOH})$, and 870 $(C_{18:0}/C_{22:4+OOH})$ formed from molecular clusters of PS with m/z 834 $(C_{18:0}/C_{22:6})$, 836 $(C_{18:0}/C_{22:5})$, and 838 $(C_{18:0}/C_{22:4})$, respectively (Fig. 19.4, e, f). The total amounts of PS–OOH accumulated in different tissues are presented in Table 19.1.

All these examples clearly demonstrate that ESI–MS analysis can be used effectively for qualitative assessments, identification of individual molecular species, and structural characterization of phospholipids involved in oxidative modification in cells and tissues. Triggering of selective CL oxidation in mitochondria and subsequent oxidation of PS in extramitochondrial compartments are integral components of apoptotic program (Scheme 19.1).

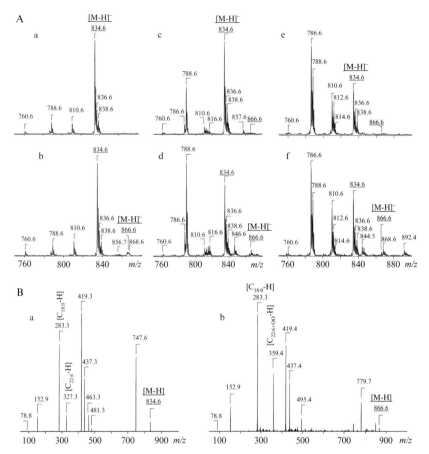

Figure 19.4 Typical negative ion ESI mass spectra of PS isolated from rat brain cortex (A, a—control, and A, b—after CCI); from midfrontal gyrus of postmortem samples from non-AD patients (A, c) and AD patients (A, d); mouse small intestine (A, e— control, and A, f—after total body irradiation, TBI). Identification of individual oxidized molecular species of PS containing $C_{22:6+OOH}$. Tandem MS-MS experiments confirmed the structure of nonoxidized and oxidized PS species (B). Typical MS^2 spectra of nonoxidized PS with m/z 834 (control panel B, a) and oxidized PS with m/z 866 (B, b) were acquired from brain human samples. MS^2 analysis of oxidized PS (m/z 866) confirmed that it originated from molecular clusters of PS with m/z 834 and contained monohydroperoxy-DHA (m/z 359).

While CL peroxidation is critical to the successful execution of the early mitochondrial segment of the apoptotic program, preferential oxidation of PS and its appearance on the surface of plasma membrane of the dying cell serve as important signals through which macrophages recognize and eliminate apoptotic cells.

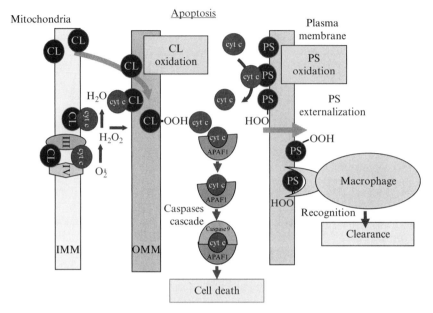

Scheme 19.1 Mitochondrial membrane lipids and cell death. Transmembrane redistribution of CL from the inner to the outer mitochondrial membrane results in increased availability of CL for binding with cytochrome *c* in the intermembrane space. Cytochrome *c*/CL complexes are formed and act as a CL-specific peroxidase catalyzing the production of CL-OOH. Selective oxidation of CL is essential for mitochondrial membrane permeabilization and release of cytochrome *c* (with or without CL), as well as other death signals into the cytosol. Cytochrome *c* is released from mitochondria in excess of its required amount to interact with apoptosis-activating factor (Apaf-1). Therefore, significant part of cytosolic cytochrome *c* can form complexes with another anionic phospholipid, PS, and catalyze its oxidation. PS oxidation plays an important role in its externalization on the surface of apoptotic cells, recognition, and phagocytosis of apoptotic cells by macrophages.

ACKNOWLEDGMENTS

Supported by grants from Pennsylvania Department of Health SAP 4100027294, DAMD 17-01-2-637, NIH U19 AIO68021, AHA0535365N, and HD057587-01A2.

REFERENCES

Ariel, A., and Serhan, C. N. (2007). Resolvins and protectins in the termination program of acute inflammation. *Trends Immunol.* **28,** 176–183.

Atamna, H., and Boyle, K. (2006). Amyloid-β peptide binds with heme to form a peroxidase: Relationship to the cytopathologies of Alzheimer's disease. *Proc. Natl. Acad. Sci. USA* **103,** 3381–3386.

Bannenberg, G., Arita, M., and Serhan, C. N. (2007). Endogenous receptor agonists: Resolving inflammation. *Sci. World J.* **7**, 1440–1462.

Bayir, H., Tyurin, V. A., Tyurina, Y. Y., Viner, R., Ritov, V., Amoscato, A. A., Zhao, Q., Zhang, X. J., Janesko-Feldman, K. L., Alexander, H., Basova, L. L., Clark, R. S. B., *et al.* (2007). Selective early cardiolipin peroxidation after traumatic brain injury: An oxidative lipidomics analysis. *Ann. Neurol.* **62**, 154–169.

Belikova, N. A., Vladimirov, Y. A., Osipov, A. N., Kapralov, A. A., Tyurin, V. A., Potapovich, M. V., Basova, L. V., Peterson, J., Kurnikov, I. V., and Kagan, V. E. (2006). Peroxidase activity and structural transitions of cytochrome c bound to cardiolipin-containing membranes. *Biochemistry* **45**, 4998–5009.

Bottcher, C. J. F., Van Gent, C. M., and Pries, C. (1961). A rapid and sensitive sub-micro phosphorus determination. *Anal. Chim. Acta* **24**, 203–204.

de Jong, D., Kremer, B. P., Olde Rikkert, M. G., and Verbeek, M. M. (2007). Current state and future directions of neurochemical biomarkers for Alzheimer's disease. *Clin. Chem. Lab. Med.* **45**, 1421–1434.

Epperly, M. W., Osipov, A. N., Martin, I., Kawai, K. K., Borisenko, G. G., Tyurina, Y. Y., Jefferson, M., Bernarding, M., Greenberger, J. S., and Kagan, V. E. (2004). Ascorbate as a "redox senser" and protector against irradiation-induced oxidative stress in 32D CL3 hematopoietic cells and subclones overexpressing human manganese superoxide dismutase. *Int. J. Radiat. Oncol. Biol. Phys.* **58**, 851–861.

Fadok, V. A., Bratton, D. L., Frasch, S. C., Warner, M. L., and Henson, P. M. (1998). The role of phosphatidylserine in recognition of apoptotic cells by phagocytes. *Cell Death Differ.* **5**, 551–562.

Folch, J., Lees, M., and Sloan-Stanley, G. H. (1957). A simple method for isolation and purification of total lipids from animal tissue. *J. Biol. Chem.* **226**, 497–509.

Forrester, J. S., Milne, S. B., Ivanova, P. T., and Brown, H. A. (2004). Computational lipidomics: A multiplexed analysis of dynamic changes in membrane lipid composition during signal transduction. *Mol. Pharmacol.* **65**, 813–821.

Gao, S., Zhang, R., Greenberg, M. E., Sun, M., Chen, X., Levison, B. S., Salomon, R. G., and Hazen, S. L. (2006). Phospholipid hydroxyalkenals, a subset of recently discovered endogenous CD36 ligands, spontaneously generate novel furan-containing phospholipids lacking CD36 binding activity *in vivo. J. Biol. Chem.* **281**, 31298–31308.

Gaspar, M. L., Aregullin, M. A., Jesch, S. A., Nunez, L. R., Villa-García, M., and Henry, S. A. (2007). The emergence of yeast lipidomics. *Biochim. Biophys. Acta* **1771**, 241–254.

Graham, S., Chen, J., and Clark, R. S. (2000). Bcl-2 family gene products in cerebral ischemia and traumatic brain injury. *J. Neurotrauma* **17**, 831–841.

Greenberg, M. E., Sun, M., Zhang, R., Febbraio, M., Silverstein, R., and Hazen, S. L. (2006). Oxidized phosphatidylserine-CD36 interactions play an essential role in macrophage-dependent phagocytosis of apoptotic cells. *J. Exp. Med.* **203**, 2613–2625.

Haeflinger, O. P., and Sulzer, J. W. (2007). Rapid LC-UV-ESI-MS method to investigate the industrial preparation of polyunsaturated fatty acid hyroperoxides in real time. *Chromatography* **65**, 435–442.

Hong, S., Lu, Y., Yang, R., Gotlinger, K. H., Petasis, N. A., and Serhan, C. N. (2007). Resolvin D1, protectin D1, and related docosahexaenoic acid–derived products: Analysis via electrospray/low energy tandem mass spectrometry based on spectra and fragmentation mechanisms. *J. Am. Soc. Mass Spectrom* **18**, 128–144.

Hsu, F. F., and Turk, J. (2006). Characterization of cardiolipin from *Escherichia coli* by electrospray ionization with multiple stage quadrupole ion-trap mass spectrometric analysis of [M - 2H + Na]⁻ ions. *J. Am. Soc. Mass Spectrom* **17**, 420–429.

Kagan, V. E., Borisenko, G. G., Serinkan, B. F., Tyurina, Y. Y., Tyurin, V. A., Jiang, J., Liu, S. X., Shvedova, A. A., Fabisiak, J. P., Uthaisang, W., and Fadeel, B. (2003).

Appetizing rancidity of apoptotic cells for macrophages: Oxidation, externalization, and recognition of phosphatidylserine. *Am. J. Physiol. Lung Cell Mol. Physiol.* **285,** L1–L17.

Kagan, V. E., Borisenko, G. G., Tyurina, Y. Y., Tyurin, V. A., Jiang, J., Potapovich, A. I., Kini, V., Amoscato, A. A., and Fujii, Y. (2004). Oxidative lipidomics of apoptosis: Redox catalytic interactions of cytochrome c with cardiolipin and phosphatidylserine. *Free Radic. Biol. Med.* **37,** 1963–1985.

Kagan, V. E., Tyurin, V. A., Jiang, J., Tyurina, Y. Y., Ritov, V. B., Amoscato, A. A., Osipov, A. N., Belikova, N. A., Kapralov, A. A., Kini, V., Vlasova, I. I., Zhao, Q., *et al.* (2005). Cytochrome *c* acts as a cardiolipin oxygenase required for release of proapoptotic factors. *Nat. Chem. Biol.* **1,** 223–232.

Kagan, V. E., Tyurina, Y. Y., Bayir, H., Chuh, C. T., Kapralov, A. A., Vlasova, I. I., Belikova, N. A., Tyurin, V. A., Amoscato, A., Epperly, M., Greenberger, J., DeKosky, S., *et al.* (2006). The "pro-apoptotic genies" get out of mitochondria: Oxidative lipidomics and redox activity of cytochrome c/cardiolipin complexes. *Chem. Biol. Interact.* **163,** 15–28.

Kapralov, A. A., Kurnikov, I. V., Vlasova, I. I., Belikova, N. A., Tyurin, V. A., Basova, L. V., Zhao, Q., Tyurina, Y. Y., Jiang, J., Bayir, H., Vladimirov, Y. A., and Kagan, V. E. (2007). Hierarchy of structural transitions induced in cytochrome c by anionic phospholipids determines its peroxidase activation and selective peroxidation during apoptosis in cells. *Biochemistry* **46,** 14232–14244.

Kerwin, J. L., and Torvik, J. J. (1996). Identification of monohydroxy fatty acids by electrospray mass spectrometry and tandem mass spectrometry. *Anal. Biochem.* **237,** 56–64.

Kochanek, P. M., Clark, R. S., Ruppel, R. A., Adelson, P. D., Bell, M. J., Whalen, M. J., Robertson, C. L., Satchell, M. A., Seidberg, N. A., Marion, D. W., and Jenkins, L. W. (2000). Biochemical, cellular, and molecular mechanisms in the evolution of secondary damage after severe traumatic brain injury in infants and children: Lessons learned from the bedside. *Pediatr. Crit. Care Med.* **1,** 4–19.

MacMillan, D. K., and Murphy, R. C. (1995). Analysis of lipid hydroperoxides and long-chain conjugated keto acids by negative ion electrospray mass spectrometry. *J. Am. Soc. Mass Spectrom.* **6,** 1190–1201.

Mancuso, M., Coppede, F., Migliore, L., Siciliano, G., and Murri, L. (2006). Mitochondrial dysfunction, oxidative stress and neurodegeneration. *J. Alzheimers Dis.* **10,** 59–73.

Maskrey, B. H., Bermudez-Fajardo, A., Morgan, A. H., Stewart-Jones, E., Dioszeghy, V., Taylor, G. W., Baker, P. R., Coles, B., Coffey, M. J., Kuhn, H., and O'Donnell, V. B. (2007). Activated platelets and monocytes generate four hydroxyphosphatidylethanolamines via lipoxygenase. *J. Biol. Chem.* **282,** 20151–20163.

Montuschi, P., Barnes, P., and Roberts, L. J. (2007). Insights into oxidative stress: The isoprostanes. *Curr. Med. Chem.* **14,** 703–717.

Moreira, P. I., Smith, M. A., Zhu, X., Nunomura, A., Castellani, R. J., and Perry, G. (2005). Oxidative stress and neurodegeneration. *Ann. N.Y. Acad. Sci.* **1043,** 545–552.

Nunomura, A., Castellani, R. J., Zhu, X., Moreira, P. I., Perry, G., and Smith, M. A. (2006). Involvement of oxidative stress in Alzheimer disease. *J. Neuropathol. Exp. Neurol.* **65,** 631–641.

Orrenius, S., Gogvadze, V., and Zhivotovsky, B. (2007). Mitochondrial oxidative stress: Implications for cell death. *Annu. Rev. Pharmacol. Toxicol.* **47,** 143–183.

Pratico, D., Lee, V. M.-Y, Trojanowski, J. Q., Rokach, J., and FitzGerald, G. A. (1998). Increased F2-isoprostanes in Alzheimer's disease: Evidence for enhanced lipid peroxidation *in vivo. FASEB J.* **12,** 1777–1783.

Pulfer, M., and Murphy, R. C. (2003). Electrospray mass spectrometry of phospholipids. *Mass Spectrom. Rev.* **22,** 332–364.

Rouser, G., Fkeischer, S., and Yamamoto, A. (1970). Two dimensional then layer chromatographic separation of polar lipids and determination of phospholipids by phosphorus analysis of spots. *Lipids* **5,** 494–496.

Schneider, C., Schreier, P., and Herderich, M. (1997). Analysis of lipoxygenase-derived fatty acid hydroperoxides by electrospray ionization tandem mass spectrometry. *Lipids* **32,** 331–336.

Schwab, J. M., Chiang, N., Arita, M., and Serhan, C. N. (2007). Resolvin 1 and protectin D1 activate inflamation-resolution programmes. *Nature* **447,** 869–874.

Sun, M., Finnemann, S. C., Febbraio, M., Shan, L., Annangudi, S. P., Podrez, E. A., Hoppe, G., Darrow, R., Organisciak, D. T., Salomon, R. G., Silverstein, R. L., and Hazen, S. L. (2006). Light-induced oxidation of photoreceptor outer segment phospholipids generates ligands for CD36-mediated phagocytosis by retinal pigment epithelium: A potential mechanism for modulating outer segment phagocytosis under oxidant stress conditions. *J. Biol. Chem.* **281,** 4222–4230.

Tyurin, V. A., Tyurina, Y. Y., Osipov, A. N., Belikova, N. A., Basova, L. V., Kapralov, A. A., Bayer, H., and Kagan, V. E. (2007a). Interactions of cardiolipin and lyso-cardiolipins with cytochrome c and tBid: Conflict or assistance in apoptosis. *Cell Death Differ.* **14,** 872–875.

Tyurin, V. A., Tyurina, Y. Y., Ritov, V. B., Lysytsya, A., Amoscato, A. A., Kochanek, P. M., Hamilton, R., DeKosky, S. T., Greenberger, J. S., Bayir, H., and Kagan, V. E. (2008). Oxidative lipidomics: Quantitative assessment of phospholipid hydroperoxides in cells and tissues. *Methods Mol. Biol.*

Tyurina, Y. Y., Basova, L. V., Konduru, N. V., Tyurin, V. A., Potapovich, A. I., Cai, P., Bayer, H., Stoyanovsky, D., Pitt, B. R., Shvedova, A. A., Fadeel, B., and Kagan, V. E. (2007c). Nitrosative stress inhibits the aminophospholipid translocase resulting in phosphatidylserine externalization and macrophage engulfment: Implications for the resolution of inflammation. *J. Biol. Chem.* **282,** 8498–8509.

Tyurina, Y. Y., Tyurin, V. A., Epperly, M. W., Greenberger, J. S., and Kagan, V. E. (2008). Oxidative lipidomics of γ-irradiation induced intestinal injury. *Free Radic. Biol. Med.* **44,** 299–314.

Vandivier, R. W., Henson, P. M., and Douglas, I. S. (2006). Burying the dead: The impact of failed apoptotic cell removal (efferocytosis) on chronic inflammatory lung disease. *Chest* **129,** 1673–1682.

Watson, A. D. (2006). The matic review series: Systems biology approaches to metabolic and cardiovascular disorders. Lipidomics: A global approach to lipid analysis in biological systems. *J. Lipid Res.* **47,** 2101–2111.

CHAPTER TWENTY

Identification and Characterization of Endoplasmic Reticulum Stress-Induced Apoptosis *In Vivo*

Kezhong Zhang* *and* Randal J. Kaufman*,†,‡

Contents

Abstract

The endoplasmic reticulum (ER) is recognized primarily as the site of synthesis and folding of secreted and membrane-bound proteins. The ER provides stringent quality control systems to ensure that only correctly folded, functional proteins are released from the ER and that misfolded proteins are degraded. The efficient functioning of the ER is essential for most cellular activities and for survival. Stimuli that interfere with ER function can disrupt ER homeostasis, impose stress to the ER, and subsequently cause accumulation of unfolded or misfolded proteins in the ER lumen. ER transmembrane proteins detect the

* Department of Biological Chemistry, The University of Michigan Medical Center, Ann Arbor, Michigan
† Department of Internal Medicine, The University of Michigan Medical Center, Ann Arbor, Michigan
‡ Howard Hughes Medical Institute, The University of Michigan Medical Center, Ann Arbor, Michigan

Methods in Enzymology, Volume 442
ISSN 0076-6879, DOI: 10.1016/S0076-6879(08)01420-1

onset of ER stress and initiate highly specific signaling pathways collectively called the "unfolded protein response" (UPR) to restore normal ER functions. However, if ER homeostasis cannot be reestablished in response to intense or prolonged ER stress, the UPR induces ER stress-associated apoptosis to protect the organism by removing the stressed cells that produce misfolded or malfunctioning proteins. This chapter summarizes current understanding of ER stress-induced apoptosis and reliable methods to examine ER stress and apoptosis in mammalian cells. Since the liver is the major organ dealing with metabolic or pathological stress and is responsible for the detoxification of chemical compounds, the experimental protocols described here focus on identification and characterization of ER stress-induced apoptosis in mouse liver.

1. THE ENDOPLASMIC RETICULUM (ER) AND THE UNFOLDED PROTEIN RESPONSE

The endoplasmic reticulum (ER) is the site of biosynthesis for sterols, lipids, membrane-bound and secreted proteins, and glycoproteins (Gaut and Hendershot, 1993; Kaufman, 1999). In higher eukaryotes, nearly all newly synthesized proteins require folding and/or assembly in the ER prior to trafficking to specific destinations to carry out their functions. As a unique protein-folding compartment and a dynamic calcium store, the ER is very sensitive to alterations in intracellular homeostasis. A number of biochemical, physiological, and pathological stimuli, such as chemicals that disrupt protein folding, nutrient depletion and hypoxia, calcium depletion, reductive or oxidative stress, expression of secretory proteins, expression of mutant difficult-to-fold proteins, unbalanced expression of subunits of protein complexes, elevated lipids or cholesterol, DNA damage, growth arrest, and viral/bacterial infection can disrupt ER homeostasis, impose stress to the ER, and subsequently cause accumulation of unfolded or misfolded proteins in the ER lumen. The cell has evolved highly specific signaling pathways called the unfolded protein response (UPR) to alter intracellular transcriptional and translational programs to deal with the accumulation of unfolded or misfolded proteins. These pathways prevent the accumulation of unfolded protein in the ER lumen by decreasing the protein-folding load, increasing the ER protein-folding capacity, and increasing the degradation of misfolded proteins through processes of ER-associated protein degradation (ERAD) or autophagy (Kaufman, 2002; Mori, 2000; Ron and Walter, 2007). In recent years, accumulating evidence suggests that the UPR signaling pathways represent an essential component of cell differentiation and function, as well as specific adaptive responses to viruses, hormones, growth factors, nutrients, and other external stimuli.

The basic UPR pathways consist of three main signaling cascades initiated by three ER-localized prototypical stress sensors: inositol-requiring 1 (IRE1), double-stranded RNA-activated protein kinase-like ER kinase (PERK), and activating transcription factor 6 (ATF6) (Ron and Walter, 2007; Schroder and Kaufman, 2005). The stress sensors IRE1, PERK, and ATF6 all have an ER lumenal domain that senses the presence of unfolded or misfolded proteins, an ER transmembrane domain that targets the protein for localization to the ER membrane, and a cytosolic functional domain. In resting cells, all three ER stress receptors are maintained in an inactive state through their association with the ER chaperone, GRP78/BiP. When cells encounter ER stress, GRP78/BiP dissociates from the three ER stress sensors, leading to activation of the UPR.

The immediate response to the accumulation of unfolded or misfolded proteins is dimerization, *trans*-autophosphorylation, and activation of PERK protein kinase to inhibit protein biosynthesis through phosphorylation of the α subunit of eukaryotic translation initiation factor (eIF2α). When eIF2α is phosphorylated, formation of the ternary translation initiation complex eIF2/GTP/Met-tRNAiMet is prevented, leading to attenuation of mRNA translation in general to reduce the protein-folding workload on the ER. Murine cells deleted in PERK or mutated at Ser51 in eIF2α to prevent phosphorylation did not attenuate protein synthesis upon ER stress. As a consequence, these cells were not able to survive ER stress (Harding et al., 2000; Scheuner, 2001). Whereas phosphorylation of eIF2α by PERK leads to attenuation of global mRNA translation, phosphorylated eIF2α selectively stimulates translation of a specific subset of mRNAs. One mRNA that requires eIF2α phosphorylation for translation encodes ATF4, a transcription factor that activates genes encoding proteins involved in amino acid biosynthesis and transport, antioxidative stress responses, and ER stress-induced apoptosis (Harding et al., 2003).

On activation of the UPR, IRE1 is also activated through its homodimerization and *trans*-autophosphorylation. In mammals, IRE1 is encoded by two homologous genes: IRE1α and IRE1β. Whereas IRE1α is apparently expressed in all cells and tissues, IRE1β expression is restricted primarily to the intestinal epithelial cells (Tirasophon et al., 1998; Wang et al., 1998). Activated IRE1α can function as an endoribonuclease to initiate removal of a 26 base small intron from X-box binding protein 1 (Xbp1) mRNA. Spliced Xbp1 mRNA encodes a potent basic leucine zipper (bZIP)-containing *trans*-activator that induces expression of genes encoding enzymes that facilitate protein folding, secretion, or degradation (Calfon et al., 2002; Lee et al., 2003; Shen et al., 2001; Yoshida et al., 2001). Studies have identified that IRE1α-mediated stress signaling is required for many physiological processes in addition to its role in the classical UPR. IRE1α can be activated by glucose in pancreatic β cells and contributes to proinsulin synthesis and β-cell differentiation and function (Lipson et al., 2006;

Scheuner and Kaufman, unpublished observation). During B-cell lympho-poesis, IRE1α is activated and required for both early and late stages of B-cell differentiation (Zhang et al., 2005). In addition, under ER stress conditions, IRE1α can serve as a scaffold to recruit protein factors, including TRAF2, BAX, and BAK, and is possibly involved in the inflammatory response and apoptosis (Hetz et al., 2006; Urano et al., 2000). ATF6 is a UPR trans-activator belonging to bZIP transcription factors of the CREB/ATF family. Under ER stress conditions, ATF6 (90 kDa) transits to the Golgi compart-ment where it is cleaved by site-1 protease (S1P) and site-2 protease (S2P) to generate a 50-kDa cytosolic bZIP-containing fragment that migrates to the nucleus to activate transcription of the target genes encoding ER-resident molecular chaperones, folding catalysts, and ERAD machinery (Nadanaka et al., 2004; Ye et al., 2000). There are two homologues of ATF6 in the mammalian genome, ATF6α and ATF6β. Whereas ATF6α mediates gene induction upon UPR activation, the role of ATF6β in the UPR is presently unknown (Wu et al., 2007; Yamamoto et al., 2007). Although ATF6α facilitates adaptation to ER stress and plays a role in assisting plasma cells to produce antibodies, it is dispensable for classical UPR signaling (Gunn et al., 2004; Wu et al., 2007). In addition, it has been noted that ATF6α is involved in lipid metabolism and the acute-phase response by dimerizing with other ER transmembrane bZIP transcription factors that are also regulated by proteolysis (Zeng et al., 2004; Zhang et al., 2006).

2. ENDOPLASMIC RETICULUM STRESS-INDUCED APOPTOSIS

In multicellular organisms, if the translational and transcriptional adaptive responses are not sufficient to relieve the unfolded or misfolded protein load, the cell enters programmed cell death—apoptosis. It has been proposed that multiple UPR pathways contribute to ER stress-induced cell apoptosis, although the mechanisms still remain largely unknown.

2.1. The PERK/eIF2α pathway

The most well studied of ER stress-induced apoptotic pathways is mediated by a transcription factor called CHOP/GADD153, which is activated primarily through the PERK/eIF2α UPR pathway. CHOP (C/EBP homologous protein) is a bZIP-containing transcription factor that was identified as a member of the CCAAT/enhancer binding protein (C/EBP) family (Ron and Habener, 1992). CHOP is also known as growth arrest and DNA damage-inducible gene 153 (GADD153), although it is induced by ER stress more than by growth arrest or DNA damage. Upon

ER stress, activated PERK phosphorylates eIF2α, which subsequently attenuates general mRNA translation through reducing the efficiency of AUG codon recognition. As a consequence, eIF2α phosphorylation can produce quantitative, as well as qualitative (through alternate initiator AUG codon recognition), changes in proteins (Kaufman, 2004). However, a few specific mRNAs that harbor regulatory sequences in their 5'-untranslated regions, for example, the internal ribosomal entry site in the cationic amino acid transporter 1 (*Cat1*) mRNA, require eIF2α phosphorylation for efficient translation (Harding *et al.*, 2000; Hinnebusch, 2000; Scheuner, 2001; Yaman *et al.*, 2003). Another mRNA that requires eIF2α phosphorylation encodes ATF4, a cAMP response element-binding transcription factor (C/EBP). Under prolonged or severe ER stress, ATF4 induces the transcription factor CHOP, a well-recognized proapoptotic factor (Harding *et al.*, 2003) (Fig. 20.1). In addition, ATF6 can also contribute to induction of the *Chop* gene (Wu *et al.*, 2007). CHOP-deficient cells are protected from ER stress-induced apoptosis (Zinszner *et al.*, 1998). Overexpression of CHOP induces cell cycle arrest or apoptosis by regulating the expression of multiple genes encoding proapoptotic factors, including death receptor 5 (DR5), Tribbles homolog 3 (TRB3), carbonic anhydrase VI (CAVI) and BCL2 family proteins (Matsumoto *et al.*, 1996; McCullough *et al.*, 2001; Ohoka *et al.*, 2005; Sok *et al.*, 1999; Yamaguchi and Wang, 2004). Dimerization of CHOP and cAMP-responsive element binding protein (CREB) suppresses the expression of *Bcl2* (McCullough *et al.*, 2001), which opposes induction of *Bcl2* by Akt/CREB (Pugazhenthi *et al.*, 2000). A decrease in the cellular level of BCL2 may increase the susceptibility of mitochondria to the proapoptotic effects of BH3-only proteins. Evidence suggests that CHOP forms heterodimers with C/EBPα under ER stress conditions to induce transcription of *Bim*, a proapoptotic BH3-only member of the BCL2 family, and is essential for ER stress-induced apoptosis in both cultured cells and within the whole animal (Puthalakath *et al.*, 2007). In addition, CHOP directly activates GADD34, which subsequently dephosphorylates eIF2α and leads to apoptosis by promoting protein synthesis in stressed cells (Marciniak *et al.*, 2004). CHOP also contributes to apoptosis through activating ERO1α, an ER oxidase that promotes hyperoxidization of the ER (Marciniak *et al.*, 2004).

2.2. The IRE1α pathway

The IRE1α/XBP1-mediated UPR pathway plays a prosurvival role through the induction of ER chaperones, ER folding catalysts, and ERAD machinery. However, in response to ER stress, IRE1α can recruit the adaptor protein tumor necrosis factor receptor-associated factor 2 (TRAF2) (Urano *et al.*, 2000). This interaction recruits c-Jun NH2-terminal inhibitory kinase (JIK), which can interact with both IRE1α and TRAF2

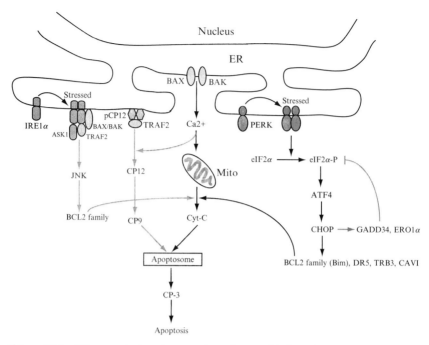

Figure 20.1 ER stress-induced apoptotic pathways. Under ER stress, PERK is activated by homodimerization and autophosphorylation. Activated PERK phosphorylates eIF2α, which subsequently induces proapoptotic factors ATF4 and CHOP. CHOP induces expression of numerous genes encoding proapoptotic factors, including DR5, TRB3, CAVI, and BCL2 family proteins that promote mitochondria-mediated apoptosis. CHOP also induces expression of GADD34 and ERO1α, leading to apoptosis by promoting protein synthesis and oxidation in the stressed ER. On activation of the UPR, IRE1α recruits TRAF2 and leads to activation of the JNK-mediated apoptotic pathway. ER stress induces BAX and BAK localization and oligomerization at the ER, which promotes calcium efflux into the cytoplasm. Calcium uptake into the mitochondrial matrix depolarizes the inner membrane and leads to transition of the outer membrane permeability pore. This causes cytochrome *c* release and Apaf-1-dependent activation of the apoptosome, leading to apoptosis. In addition, it has been proposed that ER stress induces dissociation of TRAF2 from procaspase-12 residing on the ER membrane, allowing caspase-12 activation to mediate apoptosis. Mito, mitochondria; Cyt-C, cytochrome *c*; pCP12, procaspase-12; CP12, caspase-12; CP9, caspase-9; CP-3, caspase-3.

(Yoneda *et al.*, 2001). The IRE1α – TRAF2 complex formed during ER stress can recruit the apoptosis signal-regulating kinase, which can relay various stress signals to the downstream mitogen-activated protein kinases JNK and p38 (Nishitoh *et al.*, 1998, 2002). Overexpression of JIK promotes interaction between IRE1α and TRAF2 and JNK activation in response to tunicamycin, whereas overexpression of an inactive JIK mutant inhibits JNK activation (Yoneda *et al.*, 2001). JNK activation is known to influence the cell-death machinery through phosphorylation of BCL2 family proteins, including BCL2 and BIM, which subsequently promote cell death programs (Davis, 2000). These observations led to the hypothesis that ER stress-induced, IRE1α- and TRAF2-dependent activation of JNK is one of apoptotic pathways in cells under ER stress, although there are no compelling data to confirm a requirement for the IRE1α-mediated apoptotic pathway in a physiological context (Fig. 20.1).

2.3. Caspases and BCL2 family proteins

Caspases also participate in ER stress-induced apoptosis (Fig. 20.1). In mice, procaspase-12 is localized on the cytosolic side of the ER membrane and is cleaved and activated by ER stress (Nakagawa and Yuan, 2000; Nakagawa et al., 2000). TRAF2 has been shown to interact with procaspase-12 under ER stress conditions or by overexpression of IRE1α. By this way, TRAF2 promotes the clustering of procaspase-12 at the ER membrane (Yoneda et al., 2001). It has been proposed that, during ER stress, caspase-12 activation requires the dissociation of procaspase-12 from TRAF2, which may subsequently be recruited to IRE1α (Yoneda et al., 2001). Calpains, a family of Ca^{2+}-dependent cysteine proteases, may play a role in caspase-12 activation (Nakagawa and Yuan, 2000; Rao et al., 2001). Calpain-deficient murine embryonic fibroblasts (MEFs) have reduced ER stress-induced caspase-12 activation and are resistant to ER stress–associated apoptosis (Tan et al., 2006). In addition to calpain, caspase-7 translocates from the cytosol to the cytoplasmic side of the ER membrane under ER stress and interacts with cleaved caspase-12, leading to its activation (Rao et al., 2001). Activated caspase-12 activates caspase-9, which, in turn, forms the apoptosome with released cytochrome *c* and Apaf-1 to activate cell death executor caspase-3, leading to apoptosis. Although initial studies suggest that caspase-12-deficient cells are partially resistant to ER stress-induced apoptosis (Nakagawa et al., 2000), more recent studies showed that caspase-12$^{-/-}$ mice are not protected from cell death induced by ER stress (Saleh et al., 2006). In addition, the requirement for caspase-12 in apoptosis in human cells is open to question, as the human caspase-12 gene contains several inactivating mutations (Fischer et al., 2002). It is possible that caspase-4 mediates ER stress-induced apoptosis in human cells (Hitomi et al., 2004; Kim et al., 2006). Finally, ER stress-induced apoptosis may be mediated by calpain, but not by caspases, based on the observation that calpain inhibitors, but not a pan-caspase inhibitor, blocked tunicamycin and thapsigargin-induced apoptosis (Sanges and Marigo, 2006).

Evidence suggests that ER stress–induced apoptosis is closely linked to mitochondrial mechanisms involving the BCL2 family proteins, including BCL2, BAX, and BAK, that are associated with both mitochondria and ER membranes (Krajewski et al., 1993; Zong et al., 2003). Two pathways have been proposed by which BAX and BAK may promote apoptosis in response to ER stress. During ER stress, BAX and BAK undergo conformational changes and oligomerization in the ER membrane (Zong et al., 2003). This leads to a disruption of ER Ca^{2+} stores, causing an increase in the Ca^{2+} concentration in the cytosol (Scorrano et al., 2003). Increased cytosolic Ca^{2+} flux can activate m-calpain, which cleaves and activates procaspase-12 as mentioned earlier. However, the significance of this pathway in vivo remains to be explored. The second pathway involves Ca^{2+} uptake into the mitochondrial matrix, leading to depolarization of mitochondrial inner membrane and cytochrome c release and formation of the apoptosome to activate caspase-3, DNA fragmentation, and cell death (Crompton, 1999) (Fig. 20.1).

3. METHODS USED TO IDENTIFY AND CHARACTERIZE ER STRESS-INDUCED APOPTOSIS *IN VIVO*

Most studies on ER-associated apoptosis were performed using pharmacological toxins, such as tunicamycin or thapsigargin, that cause severe ER stress. Whether those pathways indeed contribute to ER stress-induced cell death in more physiological settings, such as the mouse liver under metabolic stress, remains largely unclear. Of the pathways implicated in ER stress-induced apoptosis, only the significance of the PERK/eIF2α pathway has been confirmed in several physiological models, including pancreas β-cell death in diabetes and macrophage cell death in atherosclerotic lesions (Feng et al., 2003; Harding et al., 2001; Scheuner, 2001; Zhou et al., 2005). It has been accepted that the ER stress-inducible proapoptotic transcription factor CHOP plays a primary role in ER stress-induced apoptosis. Reliable methods have been established to characterize ER stress-induced apoptosis in vitro and in vivo. This section discusses the established experimental methods for the identification and characterization of PERK/eIF2α-mediated CHOP-dependent apoptosis in vivo. Because the liver is the major organ exposed to metabolic stress and plays a major role in drug detoxification, the approaches discussed here focus on identification and characterization of ER stress signaling and stress-induced apoptosis in the mouse liver.

3.1. Experimental protocols used to analyze UPR pathways associated with apoptosis in mouse liver

To induce ER stress in the liver, wild-type mice with C57BL/6J strain background (3 months old) are given a single intraperitoneal injection of 1 μg/g body weight of tunicamycin in 150 mM dextrose. At 24 or 36 h after injection, the mice are euthanized, and the livers are dissected for the following experiments: (1) approximately 50 mg fresh liver tissue is homogenized for the purification of liver total RNA using Trizol reagent (Invitrogen); (2) approximately 100 mg fresh liver tissue is homogenized for preparation of liver protein samples for Western blot assay or immunoprecipatation (IP)-Western blot assay; (3) a small piece of fresh liver tissue (about 50 mg) is washed briefly with phosphate-buffered saline (PBS) and then fixed in 4% PBS-buffered formalin. Fixed liver samples are paraffin embedded, and 4-μm sections are prepared and mounted on glass slides for immunohistochemical staining. The remaining liver tissue is snap frozen in liquid nitrogen and stored at 80 °C for future use.

3.1.1. Phosphorylation of IRE1α and PERK

IRE1α and PERK are primary UPR transducers, and phosphorylation of IRE1α and PERK is a hallmark of ER stress and activation of the UPR. Typically, this phosphorylation can be analyzed by an upward mobility shift upon sodium dodecyl sulfate–polyacrylamide gel electrophoresis (SDS-PAGE) and Western immunoblot analysis. Phosphorylated PERK and IRE1α can be monitored by the phospho-PERK-specific antibody (Cell Signaling) and the phospho-IRE1α specific antibody (raised in our laboratory) to probe Western blots. However, because of their low expression levels, detection of endogenous levels of IRE1α or PERK in tissues is difficult. Unfortunately, the anti-IRE1α or anti-PERK antibodies tested were not able to detect endogenous IRE1α or PERK protein in mouse liver tissue by direct immunoblot analysis. Therefore, to increase the sensitivity of detecting IRE1α or PERK by Western blot analysis, sequential immunoprecipitation and immunoblot analysis of detergent tissue lysates can be performed. We have raised murine anti-human IRE1α antibodies that can detect endogenous levels of IRE1α in murine liver tissue by IP-Western blot analysis (Kaufman et al., 2002; Tirasophon et al., 2000). In addition, a polyclonal rabbit anti-PERK PITK-289 antibody (kindly provided by Dr. Yuguang Shi, Lilly Research Laboratories, Eli Lilly and Co., Indianapolis, IN) (Shi et al., 1999) or a commercial antibody (Cell Signaling) can be used to detect endogenous PERK activation though IP-Western blot assay.

Sample preparation: (1) Approximately 100 mg of fresh liver tissue from normal mice or mice treated with tunicamycin (Tm) is transferred into an ice-cold potter homogenizer. (2) Ice cold NP-40 lysis buffer (700 μl) (1% NP-40, 50 mM Tris-HCl, pH 7.5, 150 mM NaCl, 0.05% SDS, 0.5 mM Na vanadate, 100 mM NaF, 50 mM β-glycerophosphate, and 1 mM phenylmethylsulfonyl fluoride (PMSF) supplemented with protease inhibitors (EDTA-free Complete Mini, Roche)] is added into the homogenizer containing the liver tissue. (3) The liver tissue is homogenized on ice and sonicated, and the liver tissue lysate is incubated on ice for 40 min. (4) The liver lysate is centrifuged at 4 °C at 10,000 rpm for 20 min. The supernatant is transferred into a clean, ice-cold 1.5-ml Eppendorf tube for IP-Western blot analysis for IRE1α or PERK. (6) The protein concentration of a diluted aliquot is determined by the Bradford assay.

Detection of IRE1α protein: (1) Murine anti-human IRE1α antibody (1 μl) raised in our laboratory (Tirasophon *et al.*, 2000) is incubated with 20 μl protein A beads in NP-40 lysis buffer for 1 h on an end-over-end rotator at room temperature. (2) The antibody-bound beads are washed by adding 500 μl lysis buffer in 1.5-ml Eppendorf tubes, rotating at 4 °C for 10 min, and then sedimenting in a microcentrifuge for 30 s at 8200g. The supernatant is aspirated and the tube with the bead pellet is placed on ice. The sample is washed an additional two times in a similar manner. (3) The liver tissue samples prepared as described earlier (about 700 μl) are added into the tubes with washed antibody-bound beads and are then incubated for 3 h at room temperature or overnight at 4 °C on a rotator. (4) After incubation, the beads are washed with three times with 1 ml lysis buffer and one time with 1 ml PBS. (5) The solution from the tube containing the beads is removed and then 20 μl of 2 × SDS-PAGE sample buffer (100 mM Tris, pH 6.8, 20% glycerol, 4% SDS, 0.2% bromophenol blue, 200 mM dithiothreitol) is added to the tube followed by heating at 95° for 5 min. (6) Denatured proteins are separated by SDS-PAGE on 10% Tris-glycine polyacrylamide gels and transferred to a 0.45-μm PVDF membrane (RPN1416F, GE Healthcare). (7) The blots are incubated with the same murine anti-human IRE1α antibody at a 1:1000 dilution in 0.1% Tween 20-PBS buffer (8 g NaCl, 0.2 g KCl 1.44 g Na_2HPO_4, and 0.24 g KH_2PO_4 diluted to 1000 ml distilled H_2O, pH 7.4) containing 5% skim milk or overnight at 4 °C followed by incubation with a 1:3000 dilution of horse-radish peroxidase (HRP)-conjugated anti-mouse antibody (NA9310V, GE Healthcare) for 2 h at room temperature. (8) Membrane-bound antibodies are detected by an enhanced chemiluminescence (ECL) detection reagent (RPN2106, GE Healthcare). The murine IRE1α protein is detected as one major band migrating with a molecular weight of ~120 kDa. After Tm treatment, the size of the IRE1α protein is increased slightly because of autophosphorylation.

Detection of PERK protein:(1) The rabbit anti-PERK PITK-289 anti-body (1 μl) (Shi *et al.*, 1998) or the rabbit anti-PERK antibody (1 μl) from Cell Signaling is incubated with 20 μl protein A beads in NP-40 lysis buffer for 1 h on a end-over-end rotator at room temperature. (2) Follow the steps for PERK immunoprecipitation and transfer to PVDF membranes as described earlier for IRE1α. (3) The blots are incubated with the same PERK antibody at a 1:1000 dilution for 3 h at room temperature or overnight at 4 °C, followed by incubation of HRP-conjugated anti-rabbit antibody (NA9340V, GE Healthcare) for 2 h at room temperature. (4) Membrane-bound antibodies are detected by the ECL detection reagent (RPN 2106, GE Healthcare). The murine PERK protein is detected migrating with a molecular mass of ∼ 170 kDa. After Tm treatment, the size of PERK protein is increased slightly because of autophosphorylation.

3.1.2. *Xbp1* mRNA splicing

On activation of the UPR, the primary UPR transducer IRE1α is activated and functions as an endoribonuclease to remove a 26 base small intron from the human or murine X-box binding protein 1 (*Xbp1*) mRNA to encode a potent UPR *trans*-activator (Fig. 20.2A). Because of the difficulty in detecting endogenous IRE1α protein, quantitative analysis of spliced and total *Xbp1* mRNA is the most convenient and reliable method to measure activation of the IRE1α-mediated UPR pathway.

Accurate and sensitive methods have been developed for the quantification of spliced and total *Xbp1* mRNA in mammalian cells or tissue using conventional reverse transcription (RT) polymerase chain reaction (PCR) or real-time RT-PCR: (1) Total RNA is isolated from liver tissue with the Trizol reagent (Invitrogen). (2) Synthesis of cDNA from murine total RNA is performed using a multiscribe reverse transcriptase kit (Bio-Rad). The reaction mixture

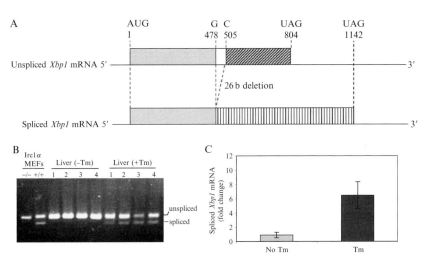

Figure 20.2 Quantitative analysis of *Xbp1* mRNA splicing. (A) Schematic representation of unspliced and spliced forms of murine *Xbp1* mRNAs and the protein coding regions. (B) Semiquantitative RT-PCR analysis of unspliced and spliced *Xbp1* mRNAs in murine liver. Wild-type C57Bl/6J mice at 3 month of age were injected intraperitoneally with 2 μg/gram body weight of tunicamycin in 150 mM dextrose. At 24 h after injection, total liver RNA samples were prepared for RT-PCR analysis. Wild-type (+/+) and *Ire1α* knockout (−/−) MEFs treated with Tm (5 μg/ml) for 8 h were included as controls. (C) Quantitative real-time RT-PCR analysis of unspliced and spliced *Xbp1* mRNAs in murine livers. The liver total RNA samples were same as described in B. Fold changes in mRNA levels were determined after normalization to internal control *β-actin* mRNA levels.

(20 μl) contains 500 ng total RNA, 4 μl 5× iScript reaction mix, and 1 μl reverse transcriptase. (3) The reverse transcription reaction is incubated at 25 °C for 10 min, followed by incubation at 48 °C for 30 min, and then reverse transcriptase is inactivated at 85 °C for 5 min. (4) The reaction mix is diluted 10-fold by the addition of 180 μl nuclease-free water. The diluted cDNA mix from the reverse transcription reaction is subjected to semiquantitative PCR or quantitative real-time PCR to determine levels of spliced and total *Xbp1* mRNAs.

For conventional semiquantitative RT-PCR, 10 μl of diluted cDNA template (25 ng) is mixed with 200 μM of each dNTP, 300 nM forward and reverse primers, 5 μl PCR reaction buffer (10 × concentration) with 15 mM MgCl$_2$, 2.6 units Taq enzyme, and nuclease-free water to a total volume of 50 μl. The PCR cycle starts with a 2-min incubation at 95 °C, then 25 cycles for 30 s at 94 °C, 30 s at 55 °C, and 45 s at 72 °C; this is followed by a final incubation at 72 °C for 7 min. The forward primer for PCR amplification of spliced and total mouse *Xbp1* mRNA is 5'-ACACGCTTGGGAATGGACAC-3', and the reverse primer is 5'-CCATGGGAAGATGTTCTGGG -3'. PCR products are separated by electrophoresis on a 3% agarose gel and visualized by ethidium bromide staining. The size of amplified unspliced *Xbp1* mRNA is 170 bp, and the size of amplified spliced *Xbp1* mRNA is 144 bp (Fig. 20.2B).

For quantitative real-time RT-PCR, a pair of real-time PCR primers is designed for the quantification of mouse *Xbp1* mRNA splicing. The forward primer sequence is 5'-GAGTCCGCAGCAGGTG-3'. This primer was designed to span the 26 base intron, thus can only anneal to the spliced *Xbp1* transcript. The reverse primer sequence is 5'-GTGTCAGAGTC-CATGGGA-3', which is 70 bases downstream of the forward primer. This pair of real-time PCR primers can specifically amplify the spliced form of murine *Xbp1* mRNA. Pairs of real-time PCR primers are also designed for quantification of the total *Xbp1* mRNA level. The forward primer sequence is 5'-AAGAACACGCTTGGGAATGG-3', and the reverse primer sequence is 5'-ACTCCCCTTGGCCTCCAC-3'. This pair of real-time PCR primers can amplify both unspliced and spliced forms of *Xbp1* mRNA. In addition, a pair of primers is used to quantitate *β-actin* mRNA by real-time RT-PCR as an internal control: the forward primer is 5'-GATCTGGCACCACACCTTCT-3' and the reverse primer is 5'-GGGGTGTTGAAGGTCTCAAA-3'.

SYBR green PCR master mix (Bio-Rad) is used to set up the quantitative real-time PCR reaction. The reaction (20 μl) contains 500 nM forward and reverse primers for *Xbp1* or *β-actin* transcripts, 12.5 ng cDNA templates made from murine liver total RNA, and 10 μl SYBR Green Supermix (50 mM KCl, 20 mM Tris-HCl, pH 8.4, 0.2 mM of each dNTP, 25U/ml iTaq DNA polymerase, 3 mM MgCl$_2$, SYBR green 1, 10 nM fluorescein and stabilizers). The thermal cycle parameters are as follow: step 1, 95 °C for 10 min; step 2, 95 °C for 15 s; and step 3, 59 °C for 1 min. Step 2 is repeated for 40 cycles. The final step is incubation at 4 °C to terminate the reaction. Data are analyzed with the iCycler iQ real-time PCR detection system (Bio-Rad) according to the manufacturer's instructions. Figure 20.2C shows quantification of the spliced *Xbp1* mRNA in murine liver with or without Tm treatment. The spliced form of *Xbp1* mRNA was increased significantly in murine liver after Tm treatment compared to that of control liver.

3.1.3. GRP78/BiP induction

It has been well established that induction of GRP78/BiP is a marker of ER stress and a central regulator of the activation of the UPR transducers (IRE1α, PERK, and ATF6). The following are methods to measure GRP78/BiP transcript and protein in murine liver.

Semiquantitative conventional RT-PCR analysis of *GRP78/BiP* transcripts: (1) Murine liver cDNA mix is prepared from total liver RNA as described in Section 3.1.2. (2) A 10-μl aliquot of diluted cDNA (25 ng) template is mixed with 200 μM of each dNTP, 300 nM forward and reverse primers, 5 μl PCR reaction buffer (10 $\times$ concentration, with 15 mM MgCl$_2$), 2.6 U Taq enzyme, and nuclease-free water to a final volume of 50 μl. (3) The PCR cycle starts with a 2-min incubation at 95 °C and then 25 cycles for 30 s at 94 °C, 30 sec at 55 °C, and 45 s at 72 °C; this is followed by a final incubation at 72 °C for 7 min. The forward primer for PCR amplification of *GRP78/BiP* is 5′-CTGGGTACATTTGATCTGACTGG-3′ and the reverse primer is 5′-GCATCCTGGTGGCTTTCCAGCCATTC-3′. The primers amplify a 397-bp mouse GRP78/BiP cDNA fragment.

Quantitative real-time RT-PCR analysis of *GRP78/BiP* transcripts: (1) A 2-μl aliquot of diluted cDNA template (12.5 ng) is mixed with 10 μl iQ SYBR Green Supermix (Bio-Rad), 150 nM forward and reverse real-time PCR primers for *GRP78/BiP* or *β-actin* (internal control), and nuclease-free water to a final volume of 20 μl. (2) The thermal cycling starts with a 10-min incubation at 95 °C, then 40 cycles for 15 s at 95 °C, and 1 min at 59 °C; this is followed by a final incubation at 4 °C to terminate the reaction. The primers used for *GRP78/BiP* quantitative real-time PCR assay are forward primer 5′-CATGGTTCTCACTAAAATGAAAGG-3′ and reverse primer 5′-GCTGGTACAGTAACAACTG-3′. Primers for the internal control *β-actin* transcript are the same as described in Section 3.1.2.

Measurement of GRP78/BiP protein level: The level of GRP78/BiP is elevated after prolonged stress treatment, although the fold of increase in protein level is usually less than that observed in mRNA levels due to the stability of the GRP78/BiP protein. An increase in the GRP78/BiP protein level is a good marker for UPR activation. However, this is not a very sensitive measure to detect low levels of ER stress and UPR activation. Therefore, it should not be concluded that the UPR is not activated if an increase in the GRP78/BiP protein is not detected. To evaluate ER stress and activation of the UPR, it is important to examine the changes in both

GRP78/BiP mRNA and protein levels. Currently, monoclonal and polyclonal antibodies against GRP78/BiP are available commercially, and the level of GRP78/BiP protein in liver tissue or cultured cells can be determined through a standard Western blot analysis. (1) The liver protein lysate is prepared as described in Section 3.1.1. (2) Denatured liver proteins (50–80 μg) are separated by SDS-PAGE on a 10% Tris-glycine gel and transferred to a 45-μm PVDF nitrocellulose membrane in 0.19 M glycine, 25 mM Tris base, and 20% methanol. (3) Nonspecific binding sites on the membrane are blocked using 5% skimmed milk in PBST (PBS containing 0.1% Tween 20) for 1 h. (4) The membrane is incubated with a rabbit polyclonal anti-GRP78 antibody (SPA-826, StressGen) at a 1:1000 dilution for 3 h at room temperature or overnight at 4°. (5) The membrane is washed with PBST and then incubated with horseradish peroxidase (HRP)-conjugated secondary antibody for 2 h at room temperature. The membrane is washed three times with PBST and chemiluminescent signals are detected with the ECL detection reagent (RPN 2106, GE Healthcare).

3.1.3. Phosphorylation of eIF2α

Upon ER stress, PERK-mediated phosphorylation of translation initiation factor eIF2α inhibits general translation to reduce the protein-folding demand of the ER, while selectively stimulating translation of the *ATF4* mRNA that encodes a transcription factor for the induction of proapoptotic factor CHOP under prolonged ER stress. Therefore, phosphorylated eIF2α is not only a reliable marker of PERK activation, but also a key transmitter of cell survival and death signals in response to ER stress (Rutkowski *et al.*, 2006). Measurement of the levels of phosphorylated and total eIF2α protein is particularly informative for evaluating activation of the PERK/eIF2α UPR pathway and ER stress-induced apoptosis. However, it must be cautioned that eIF2α phosphorylation may not reflect PERK activation because there are at least three additional eIF2α kinases—general control of nitrogen metabolism kinase 2, heme-regulated inhibitor kinase, and double-stranded RNA-activated protein kinase—that phosphorylate the same site in response to amino acid deprivation, heme deficiency, or viral infection (Kaufman, 2004).

Measurement of phosphorylated eIF2α protein: (1) Liver tissue lysate is prepared by lysis in buffer containing 1% NP-40, 50 mM Tris-HCL [pH 7.5], 150 mM NaC1, 0.05% SDS, 0.5 mM Na vanadate, 100 mM NaF, 50 mM β-glycerophosphate, and 1 mM PMSF supplemented with protease inhibitors [EDTA-free, Roche] as described in Section 3.1.1. (2) Samples (50–80 μg) of denatured liver proteins are separated on a 10% Tris-glycine polyacrylamide gel and transferred to a nitrocellulose membrane in 0.19 M glycine, 25 mM Tris base, and 20% methanol. (3) Nonspecific-binding sites on the membrane are blocked by incubation with 5% (w/v) skimmed milk in PBST (PBS containing 0.1% Tween 20) for 1 h at room temperature. (4) The membrane is incubated with a rabbit anti-Ser51-phosphorylated

eIF2α antibody (Biosource International) at a 1:1000 dilution for 3 h at room temperature or overnight at 4 °C. (5) The membrane is washed with PBST and then incubated with HRP-conjugated anti-rabbit secondary antibody for 2 h. (6) The membrane is washed with PBST three times and is then incubated with the ECL detection reagent to measure chemiluminescent signals (Fig. 20.3).

Measurement of total eIF2α protein: (1) The same amounts of denatured liver protein are separated on a 10% Tris-glycine gel and transferred to a new PVDF membrane. (2) Nonspecific binding sites on the membrane are blocked using 5% skimmed milk in PBST for 1 h. (3) The membrane is incubated with a rabbit anti-total eIF2α antibody (Cell Signaling) at a 1:1000 dilution overnight at 4 °C. (4) The membrane is washed and incubated with the secondary antibody for ECL detection as described for the detection of phosphorylated eIF2α (Fig. 20.3).

Note: The use of serine-threonine phosphatase inhibitors in the lysis buffer such as β-glycerophosphate is critical in preserving the eIF2α phosphorylation status.

3.2. Experimental protocols for identification and characterization of ER stress-inducible proapoptotic factors in the liver

As discussed earlier, ATF4 and CHOP play a proapoptotic role in ER stress-induced cell death. Indeed, CHOP is the most well-studied and well-recognized indicator for ER stress-induced apoptosis both *in vitro* and *in vivo*.

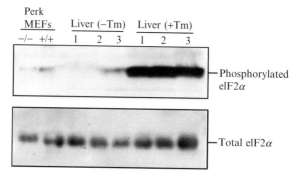

Figure 20.3 Western blot analysis of phosphorylated and total eIF2α in murine liver. Wild-type C57BL/6J mice at 3 months of age were injected intraperitoneally with 2 μg/g body weight of tunicamycin in 150 mM dextrose. At 24 h after injection, liver protein samples were prepared as described in the text for Western blot analysis. Wild-type (+/+) and *Perk*-deleted (-/-) MEFs treated with Tm (5 μg/ml) for 8 h were included as controls. The phosphorylated and total eIF2α proteins were detected using different blots for the same liver protein samples.

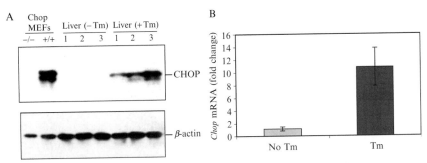

Figure 20.4 Measurement of *Chop* mRNA and protein in murine liver. (A) Western blot analysis of CHOP in murine liver. Wild-type C57BL/6J mice at 3 months of age were injected intraperitoneally with 2 μg/gram body weight of tunicamycin in 150 m*M* dextrose. At 24 h after injection, liver protein samples were prepared for Western blot analyses. Wild-type (+/+) and Chop-deleted (-/-) MEFs treated with Tm (5 μg/ml) for 30 h were included as controls. Levels of murine β-actin protein were determined as protein loading controls. (B) Quantitative real-time RT-PCR analysis of *Chop* mRNA in murine liver. The liver total RNA was prepared from the mice as described in A. The fold changes of mRNA levels were determined after normalization to internal control *β-actin* mRNA levels.

We have developed quantitative real-time RT-PCR to measure the induction of murine *Atf4* and *Chop* mRNAs in the liver upon ER stress (Fig. 20.4). In addition, polyclonal anti-ATF4 and anti-CHOP antibodies are available commercially for the detection of ATF4 and CHOP protein levels in liver tissue through Western blot analysis.

3.2.1. Quantitative real-time RT-PCR analysis of *Atf4* and *Chop* mRNAs

Quantitative real-time RT-PCR analysis of *Atf4* and *Chop* mRNAs: (1) The murine liver cDNA template is prepared from total liver RNA as described in Section 3.1.2. (2) A 2-μl aliquot of diluted cDNA template (12.5 ng) is mixed with 10 μl iQ SYBR Green Supermix, 150 n*M* forward and reverse real-time RT-PCR primers for *Atf4, Chop*, or *β-actin*, and nuclease-free water to a final volume of 20 μl. (3) The thermal cycling starts with a 10-min incubation at 95 °C, then 40 cycles for 15 s at 95 °C, and 1 min at 59 °C, followed by a final incubation at 4 °C. Primers used for *Atf4* real-time PCR assay are forward primer 5′-ATGGCCGGCTATG-GATGAT-3′ and reverse primer 5′-CGAAGTCAAACTCTTTCAGATC-CATT-3′. The primers for *Chop* are forward primer 5′-CTGCCTTT CACCTTGGAGAC-3′ and reverse primer 5′-CGTTTCCTGGGGATGA-GATA-3′. The primers for internal control *β-actin* transcript are the same as described in Section 3.1.2.

3.2.2. Measurement of ATF4 and CHOP protein levels

Western blot analysis of ATF4 or CHOP protein in liver samples: (1) Liver protein sample preparation, SDS-PAGE, and PDVF membrane transfer are the same as described in Section 3.1.1. (2) Nonspecific-binding sites on the membrane are blocked by incubation with 5% skimmed milk in PBST for 1 h. (3) To detect ATF4, the membrane is incubated with rabbit polyclonal anti–CREB2/ATF4 antibody (SC-200, Santa Cruz Biotechnologies) at a 1:500 dilution for 3 h at room temperature. To detect CHOP, the membrane is incubated with rabbit polyclonal anti–GADD153/CHOP antibody (sc-793, Santa Cruz Biotechnologies) at a 1:200 dilution for 3 h at room temperature. (4) After incubation with anti-ATF4 or anti-CHOP primary antibody, the membrane is washed with PBST and then incubated with HRP-conjugated secondary antibody for 2 h at room temperature. (5) Chemiluminescent signals on the membrane are detected by the ECL detection reagent.

3.3. Immunohistochemical staining of ER stress-induced apoptotic markers in the liver

Immunohistochemical staining is an excellent method to visualize intensity, distribution, and localization of ER stress markers and apoptosis in liver tissue. We have successfully used modified immunohistochemical staining methods to amplify and visualize the CHOP signal and apoptotic events using paraffin-embedded liver tissue sections.

3.3.1. Immunohistochemical staining of CHOP in the liver

Tyramide signal amplification to detect CHOP: To stain and amplify the CHOP signal in liver tissue, we apply a method called tyramide signal amplification that utilizes the catalytic activity of HRP to generate high-density labeling of the CHOP protein (Speel *et al.*, 1999; van Gijlswijk *et al.*, 1997). The tyramide signal amplification staining process includes (1) binding of a rabbit anti-CHOP antibody to CHOP protein in the liver tissue section followed by secondary detection of the primary antibody with an HRP-labeled anti-rabbit antibody; (2) activation of multiple copies of a dye-labeled tyramide derivative by HRP; and (3) covalent coupling of the resulting highly reactive, short-lived tyramide radicals to nucleophilic residues in the vicinity of the HRP–CHOP interaction site, resulting in minimal diffusion-related loss of signal localization (Fig. 20.5A).

To stain and amplify the signal for CHOP protein in liver tissue sections, a commercial tyramide amplification kit (Invitrogen) is used. (1) Murine liver tissue fixation, sectioning, and deparaffination are carried out according to standard immunohistochemical staining protocols. (2) The liver tissue is permeabilized by incubating with 0.2% Triton X-100 for 10 min at room

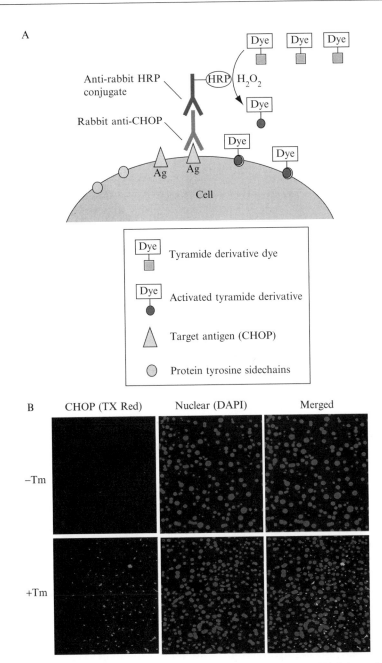

Figure 20.5 Immunohistochemical staining of CHOP in the liver. (A) Schematic representation of the tyramide signal amplification process to detect CHOP protein in the liver. (B) CHOP staining in liver sections. Wild-type C57BL/6J mice at 3 months of age were injected intraperitoneally with 2 μg/gram body weight of tunicamycin in 150 mM dextrose. At 24 h after injection, fresh liver tissues were fixed in 4% PBS-buffered formalin, paraffin embedded, and 4-μm sections prepared. (Top) Liver sections from control mice. (Bottom) Liver sections from mice treated with Tm. (See color insert.)

temperature followed by a rinse in PBS. (3) The endogenous peroxidase activity is quenched by incubating in 0.1% H_2O_2 in methanol for 1 h at room temperature. (4) The nonspecific-binding sites on the specimen are blocked by incubating the slide with 1% BSA in PBS for 1 h at room temperature. (5) The tissue section is labeled with a rabbit anti-CHOP primary antibody (sc–575, Santa Cruz Biotechnologies) at a 1:100 dilution in 1% BSA blocking reagent overnight at 4 °C. (6) After primary antibody incubation, the slide is rinsed with PBS three times and then incubated with HRP-conjugated antirabbit antibody for 60 min at room temperature. (7) The slide is rinsed with PBS three times and then incubated with the tyramide working solution (provided in the kit) for 7 min at room temperature. (8) The slide is rinsed with PBS three times and is then mounted with ProLong Gold antifade mounting medium containing 4′,6-diamidino-2-phenylindole (DAPI) to stain DNA (Molecular Probes, Invitrogen). The mounted slide is kept in the dark at room temperature for 24 h prior to analysis by fluorescence microscopy (Fig. 20.5B). Cells expressing CHOP exhibit red fluorescence when viewed at 556/573 nm (excitation/emission). DAPI-stained nuclei exhibit a blue color. The CHOP signal should be observed in nuclear, perinuclear, and cytosolic localizations.

Note: Because of the amplified detection inherent in the tyramide signal amplification process, background staining that was previously undetectable may become more prominent. Intense staining caused by nonspecific reaction of the tyramide with endogenous peroxidase in blood vessels may be observed. Because of sensitivity enhancements derived from the tyramide amplification process, dilution of the primary antibody may be increased 10-fold from the level used in conventional immunohistochemical staining of liver tissue sections.

3.3.2. Nuclear DNA fragmentation assay

During apoptosis, cells undergo many distinct morphological and biochemical changes. One striking apoptotic event involves cellular endonuclease cleavage of nuclear DNA between nucleosomes, thereby producing a mixture of DNA fragments. DNA fragmentation has been well recognized as a marker of the final stage of apoptosis (Schwartzman and Cidlowski, 1993). To detect ER stress-induced apoptosis in liver tissue sections, we have modified a DNA fragmentation assay by incorporating fluorescein-dUTP at the free 3′-hydroxyl ends of the fragmented DNA using the terminal deoxynucleotidyl transferase-mediated dUTP nick-end labeling (TUNEL) method. The fluorescein-labeled DNA can then be quantified as an indicator of apoptosis (Fig. 20.6).

Analysis of DNA fragmentation in liver tissue sections: (1) Fixation, paraffin embedding, tissue sectioning, and deparaffination are performed according to standard immunohistochemical staining protocols. (2) The TUNEL labeling process is based on a commercial DNA fragmentation

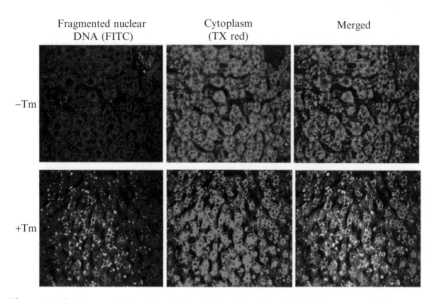

Figure 20.6 Immunohistochemical staining of nuclear DNA fragments in liver sections. Wild-type C57BL/6J mice at 3 months of age were injected intraperitoneally with 2 μg/gram body weight of tunicamycin in 150 mM dextrose. At 36 h after injection, fresh liver tissues were fixed in 4% PBS-buffered formalin, paraffin embedded, and 4-μm sections prepared. (Top) Liver sections from control mice. (Bottom) Liver sections from mice treated with Tm. (See color insert.)

assay kit (Clontech). Pretreatment of tissue sections with NaCl and fixation by formaldehyde are carried out according to the product manual. (3) The liver tissue sections are treated with 20 μg/ml proteinase K in a Coplin jar for 10 min at room temperature. (4) The slide is washed with PBS and is fixed again by 4% formaldehyde/PBS for 5 min at room temperature. (5) The slide is washed by PBS and is then incubated in an equilibration buffer (200 nM potassium cacodylate, 25 mM Tris-HCl, 0.2 mM DD, 0.25 mg/ml BSA, and 2.5 mM cobalt chloride) for 10 min at room temperature. (6) For TUNEL labeling, the tissue section is incubated with TdT reaction buffer (1 μl TdT enzyme, 5 μl nucleotide mix, 45 μl equilibration buffer) in a dark, humidified 37 °C incubator for 2 h. (7) The reaction is terminated by immersing the slides in $2 \times$ SSC (3 M NaCl, 300 mM Na_3citrate $\cdot$ H_2O) for 15 min at room temperature. (8) The slide is washed with PBS twice and then stained with propidium iodide (PI) for 10 min at room temperature. (9) The slide is washed with deionized water and mounted with ProLong Gold antifade mount media (Molecular Probes, Invitrogen). (10) The slide is kept overnight at 4 °C in the dark prior to analysis by fluorescence microscopy. Apoptotic cells exhibit strong, nuclear green fluorescence using a standard fluorescein filter set (520 ± 20 nm). All cells stained with PI exhibit strong red cytoplasmic fluorescence when viewed at more than 620 nm.

Note: If tissue section permeabilization with Triton X-100 or incubation with proteinase K is insufficient, signals may be weak or not detected. However, if the tissue sections are overdigested with proteinase K, the sample may detach from the slides. The permeabilization and proteinase K incubation times need to be optimized to avoid these problems.

ACKNOWLEDGMENTS

This work was supported by NIH Grants DK42394 (to RJK), HL52173 (to RJK), and PO1 HL057346 (to RJK) and American Heart Association Grant 0635423Z (to KZ). RJK is an Investigator of the Howard Hughes Medical Institute.

REFERENCES

Calfon, M., Zeng, H., Urano, F., Till, J. H., Hubbard, S. R., Harding, H. P., Clark, S. G., and Ron, D. (2002). IRE1 couples endoplasmic reticulum load to secretory capacity by processing the XBP-1 mRNA. *Nature* **415**, 92–96.

Crompton, M. (1999). The mitochondrial permeability transition pore and its role in cell death. *Biochem. J.* **341**, 233–249.

Davis, R. J. (2000). Signal transduction by the JNK group of MAP kinases. *Cell* **103**, 239–252.

Feng, B., Yao, P. M., Li, Y., Devlin, C. M., Zhang, D., Harding, H. P., Sweeney, M., Rong, J. X., Kuriakose, G., Fisher, E. A., Marks, A. R., Ron, D., *et al.* (2003). The endoplasmic reticulum is the site of cholesterol-induced cytotoxicity in macrophages. *Nat. Cell Biol.* **5**, 781–792.

Fischer, H., Koenig, U., Eckhart, L., and Tschachler, E. (2002). Human caspase 12 has acquired deleterious mutations. *Biochem. Biophys. Res. Commun.* **293**, 722–726.

Gaut, J. R., and Hendershot, L. M. (1993). The modification and assembly of proteins in the endoplasmic reticulum. *Curr. Opin. Cell Biol.* **5**, 589–595.

Gunn, K. E., Gifford, N. M., Mori, K., and Brewer, J. W. (2004). A role for the unfolded protein response in optimizing antibody secretion. *Mol. Immunol.* **41**, 919–927.

Harding, H. P., Zeng, H., Zhang, Y., Jungries, R., Chung, P., Plesken, H., Sabatini, D. D., and Ron, D. (2001). Diabetes mellitus and exocrine pancreatic dysfunction in perk-/-mice reveals a role for translational control in secretory cell survival. *Mol. Cell.* **7**, 1153–1163.

Harding, H. P., Zhang, Y., Bertolotti, A., Zeng, H., and Ron, D. (2000). Perk is essential for translational regulation and cell survival during the unfolded protein response. *Mol. Cell.* **5**, 897–904.

Harding, H. P., Zhang, Y., Zeng, H., Novoa, I., Lu, P. D., Calfon, M., Sadri, N., Yun, C., Popko, B., Paules, R., Stojdl, D. F., Bell, J. C., *et al.* (2003). An integrated stress response regulates amino acid metabolism and resistance to oxidative stress. *Mol. Cell* **11**, 619–633.

Hetz, C., Bernasconi, P., Fisher, J., Lee, A. H., Bassik, M. C., Antonsson, B., Brandt, G. S., Iwakoshi, N. N., Schinzel, A., Glimcher, L. H., and Korsmeyer, S. J. (2006). Proapoptotic BAX and BAK modulate the unfolded protein response by a direct interaction with IRE1alpha. *Science* **312**, 572–576.

Hinnebusch, A. (2000). "Mechanism and Regulation of Initiator Methionyl-tRNA Binding Ribosomes." Cold Spring Harbor Press, Cold Spring Harbor, NY.

Hitomi, J., Katayama, T., Eguchi, Y., Kudo, T., Taniguchi, M., Koyama, Y., Manabe, T., Yamagishi, S., Bando, Y., Imaizumi, K., Tsujimoto, Y., and Tohyama, M. (2004). Involvement of caspase-4 in endoplasmic reticulum stress-induced apoptosis and Abeta-induced cell death. *J. Cell Biol.* **165,** 347–356.

Kaufman, R. J. (1999). Stress signaling from the lumen of the endoplasmic reticulum: Coordination of gene transcriptional and translational controls. *Genes Dev.* **13,** 1211–1233.

Kaufman, R. J. (2002). Orchestrating the unfolded protein response in health and disease. *J. Clin. Invest.* **110,** 1389–1398.

Kaufman, R. J. (2004). Regulation of mRNA translation by protein folding in the endoplasmic reticulum. *Trends Biochem Sci.* **29,** 152–158.

Kaufman, R. J., Scheuner, D., Schröder, M., Shen, X., Lee, K., Liu, C. Y., and Arnold, S. M. (2002). The unfolded protein response in nutrient sensing and differentiation. *Nat. Rev. Mol. Cell Biol.* **3,** 411–421.

Kim, S. J., Zhang, Z., Hitomi, E., Lee, Y. C., and Mukherjee, A. B. (2006). Endoplasmic reticulum stress-induced caspase-4 activation mediates apoptosis and neurodegeneration in INCL. *Hum. Mol. Genet.* **15,** 1826–1834.

Krajewski, S., Tanaka, S., Takayama, S., Schibler, M. J., Fenton, W., and Reed, J. C. (1993). Investigation of the subcellular distribution of the bcl-2 oncoprotein: Residence in the nuclear envelope, endoplasmic reticulum, and outer mitochondrial membranes. *Cancer Res.* **53,** 4701–4714.

Lee, A. H., Iwakoshi, N. N., and Glimcher, L. H. (2003). XBP-1 regulates a subset of endoplasmic reticulum resident chaperone genes in the unfolded protein response. *Mol. Cell. Biol.* **23,** 7448–7459.

Lipson, K. L., Fonseca, S. G., and Urano, F. (2006). Endoplasmic reticulum stress-induced apoptosis and auto-immunity in diabetes. *Curr. Mol. Med.* **6,** 71–77.

Marciniak, S. J., Yun, C. Y., Oyadomari, S., Novoa, I., Zhang, Y., Jungreis, R., Nagata, K., Harding, H. P., and Ron, D. (2004). CHOP induces death by promoting protein synthesis and oxidation in the stressed endoplasmic reticulum. *Genes Dev.* **18,** 3066–3077.

Matsumoto, M., Minami, M., Takeda, K., Sakao, Y., and Akira, S. (1996). Ectopic expression of CHOP (GADD153) induces apoptosis in M1 myeloblastic leukemia cells. *FEBS Lett.* **395,** 143–147.

McCullough, K. D., Martindale, J. L., Klotz, L. O., Aw, T. Y., and Holbrook, N. J. (2001). Gadd153 sensitizes cells to endoplasmic reticulum stress by down-regulating Bcl2 and perturbing the cellular redox state. *Mol. Cell. Biol.* **21,** 1249–1259.

Mori, K. (2000). Tripartite management of unfolded proteins in the endoplasmic reticulum. *Cell* **101,** 451–454.

Nadanaka, S., Yoshida, H., Kano, F., Murata, M., and Mori, K. (2004). Activation of mammalian unfolded protein response is compatible with the quality control system operating in the endoplasmic reticulum. *Mol. Biol. Cell.* **15,** 2537–2548.

Nakagawa, T., and Yuan, J. (2000). Cross-talk between two cysteine protease families: Activation of caspase-12 by calpain in apoptosis. *J. Cell Biol.* **150,** 887–894.

Nakagawa, T., Zhu, H., Morishima, N., Li, E., Xu, J., Yankner, B. A., and Yuan, J. (2000). Caspase-12 mediates endoplasmic-reticulum-specific apoptosis and cytotoxicity by amyloid-beta. *Nature* **403,** 98–103.

Nishitoh, H., Matsuzawa, A., Tobiume, K., Saegusa, K., Takeda, K., Inoue, K., Hori, S., Kakizuka, A., and Ichijo, H. (2002). ASK1 is essential for endoplasmic reticulum stress-induced neuronal cell death triggered by expanded polyglutamine repeats. *Genes Dev.* **16,** 1345–1355.

Nishitoh, H., Saitoh, M., Mochida, Y., Takeda, K., Nakano, H., Rothe, M., Miyazono, K., and Ichijo, H. (1998). ASK1 is essential for JNK/SAPK activation by TRAF2. *Mol. Cell.* **2,** 389–395.

Ohoka, N., Yoshii, S., Hattori, T., Onozaki, K., and Hayashi, H. (2005). TRB3, a novel ER stress-inducible gene, is induced via ATF4-CHOP pathway and is involved in cell death. *EMBO J.* **24,** 1243–1255.

Pugazhenthi, S., Nesterova, A., Sable, C., Heidenreich, K. A., Boxer, L. M., Heasley, L. E., and Reusch, J. E. (2000). Akt/protein kinase B up-regulates Bcl-2 expression through cAMP-response element-binding protein. *J. Biol. Chem.* **275,** 10761–10766.

Puthalakath, H., O'Reilly, L. A., Gunn, P., Lee, L., Kelly, P. N., Huntington, N. D., Hughes, P. D., Michalak, E. M., McKimm-Breschkin, J., Motoyama, N., Gotoh, T., Akira, S., *et al.* (2007). ER stress triggers apoptosis by activating BH3-only protein Bim. *Cell* **129,** 1337–1349.

Rao, R. V., Hermel, E., Castro-Obregon, S., del Rio, G., Ellerby, L. M., Ellerby, H. M., and Bredesen, D. E. (2001). Coupling endoplasmic reticulum stress to the cell death program : Mechanism of caspase activation. *J. Biol. Chem.* **276,** 33869–33874.

Ron, D., and Habener, J. F. (1992). CHOP, a novel developmentally regulated nuclear protein that dimerizes with transcription factors C/EBP and LAP and functions as a dominant-negative inhibitor of gene transcription. *Genes Dev.* **6,** 439–453.

Ron, D., and Walter, P. (2007). Signal integration in the endoplasmic reticulum unfolded protein response. *Nat. Rev. Mol. Cell Biol.* **8,** 519–529.

Rutkowski, D. T., Arnold, S. M., Miller, C. N., Wu, J., Li, J., Gunnison, K. M., Mori, K., Sadighi Akha, A. A., Raden, D., and Kaufman, R. J. (2006). Adaptation to ER stress is mediated by differential stabilities of pro-survival and pro-apoptotic mRNAs and proteins. *PLoS Biol.* **4,** e374.

Saleh, M., Mathison, J. C., Wolinski, M. K., Bensinger, S. J., Fitzgerald, P., Droin, N., Ulevitch, R. J., Green, D. R., and Nicholson, D. W. (2006). Enhanced bacterial clearance and sepsis resistance in caspase-12-deficient mice. *Nature* **440,** 1064–1068.

Sanges, D., and Marigo, V. (2006). Cross-talk between two apoptotic pathways activated by endoplasmic reticulum stress: Differential contribution of caspase-12 and AIF. *Apoptosis* **11,** 1629–1641.

Scheuner, D., Song, B., McEwen, E., Lui, C., Laybutt, R., Gillespie, P., Saunders, T., Bonner-Weir, S., and Kaufman, R. J. (2001). Translational control is required for the unfolded protein response and *in vivo* glucose homeostasis. *Mol. Cell.* **7,** 1165–1176.

Schroder, M., and Kaufman, R. J. (2005). The mammalian unfolded protein response. *Annu. Rev. Biochem.* **74,** 739–789.

Schwartzman, R. A., and Cidlowski, J. A. (1993). Apoptosis: The biochemistry and molecular biology of programmed cell death. *Endocr. Rev.* **14,** 133–151.

Scorrano, L., Oakes, S. A., Opferman, J. T., Cheng, E. H., Sorcinelli, M. D., Pozzan, T., and Korsmeyer, S. J. (2003). BAX and BAK regulation of endoplasmic reticulum Ca^{2+}: A control point for apoptosis. *Science* **300,** 135–139.

Shen, X., Ellis, R. E., Lee, K., Liu, C. Y., Yang, K., Solomon, A., Yoshida, H., Morimoto, R., Kurnit, D. M., Mori, K., and Kaufman, R. J. (2001). Complementary signaling pathways regulate the unfolded protein response and are required for *C. elegans* development. *Cell* **107,** 893–903.

Shi, Y., An, J., Liang, J., Hayes, S. E., Sandusky, G. E., Stramm, L. E., and Yang, N. N. (1999). Characterization of a mutant pancreatic eIF-2alpha kinase, PEK, and co-localization with somatostatin in islet delta cells. *J. Biol. Chem.* **274,** 5723–5730.

Shi, Y., Vattem, K. M., Sood, R., An, J., Liang, J., Stramm, L., and Wek, R. C. (1998). Identification and characterization of pancreatic eukaryotic initiation factor 2 alpha-subunit kinase, PEK, involved in translational control. *Mol. Cell. Biol.* **18,** 7499–7509.

Sok, J., Wang, X. Z., Batchvarova, N., Kuroda, M., Harding, H., and Ron, D. (1999). CHOP-dependent stress-inducible expression of a novel form of carbonic anhydrase VI. *Mol. Cell. Biol.* **19,** 495–504.

Speel, E. J., Hopman, A. H., and Komminoth, P. (1999). Amplification methods to increase the sensitivity of *in situ* hybridization: Play card(s). *J. Histochem. Cytochem.* **47,** 281–288.

Tan, Y., Dourdin, N., Wu, C., De Veyra, T., Elce, J. S., and Greer, P. A. (2006). Ubiquitous calpains promote caspase-12 and JNK activation during endoplasmic reticulum stress-induced apoptosis. *J. Biol. Chem.* **281,** 16016–16024.

Tirasophon, W., Lee, K., Callaghan, B., Welihinda, A., and Kaufman, R. J. (2000). The endoribonuclease activity of mammalian IRE1 autoregulates its mRNA and is required for the unfolded protein response. *Genes Dev.* **14,** 2725–2736.

Tirasophon, W., Welihinda, A. A., and Kaufman, R. J. (1998). A stress response pathway from the endoplasmic reticulum to the nucleus requires a novel bifunctional protein kinase/endoribonuclease (Ire1p) in mammalian cells. *Genes Dev.* **12,** 1812–1824.

Urano, F., Wang, X., Bertolotti, A., Zhang, Y., Chung, P., Harding, H. P., and Ron, D. (2000). Coupling of stress in the ER to activation of JNK protein kinases by transmembrane protein kinase IRE1. *Science* **287,** 664–666.

van Gijlswijk, R. P., Zijlmans, H. J., Wiegant, J., Bobrow, M. N., Erickson, T. J., Adler, K. E., Tanke, H. J., and Raap, A. K. (1997). Fluorochrome-labeled tyramides: Use in immunocytochemistry and fluorescence in situ hybridization. *J. Histochem. Cytochem.* **45,** 375–382.

Wang, X. Z., Harding, H. P., Zhang, Y., Jolicoeur, E. M., Kuroda, M., and Ron, D. (1998). Cloning of mammalian Ire1 reveals diversity in the ER stress responses. *EMBO J.* **17,** 5708–5717.

Wu, J., Rutkowski, D. T., Dubois, M., Swathirajan, J., Saunders, T., Wang, J., Song, B., Yau, G. D., and Kaufman, R. J. (2007). ATF6alpha optimizes long-term endoplasmic reticulum function to protect cells from chronic stress. *Dev. Cell.* **13,** 351–364.

Yamaguchi, H., and Wang, H. G. (2004). CHOP is involved in endoplasmic reticulum stress-induced apoptosis by enhancing DR5 expression in human carcinoma cells. *J. Biol. Chem.* **279,** 45495–45502.

Yamamoto, K., Sato, T., Matsui, T., Sato, M., Okada, T., Yoshida, H., Harada, A., and Mori, K. (2007). Transcriptional induction of mammalian ER quality control proteins is mediated by single or combined action of ATF6alpha and XBP1. *Dev. Cell.* **13,** 365–376.

Yaman, I., Fernandez, J., Liu, H., Caprara, M., Komar, A. A., Koromilas, A. E., Zhou, L., Snider, M. D., Scheuner, D., Kaufman, R. J., and Hatzoglou, M. (2003). The zipper model of translational control: A small upstream ORF is the switch that controls structural remodeling of an mRNA leader. *Cell* **113,** 519–531.

Ye, J., Rawson, R. B., Komuro, R., Chen, X., Davé, U. P., Prywes, R., Brown, M. S., and Goldstein, J. L. (2000). ER stress induces cleavage of membrane-bound ATF6 by the same proteases that process SREBPs. *Mol. Cell.* **6,** 1355–1364.

Yoneda, T., Imaizumi, K., Oono, K., Yui, D., Gomi, F., Katayama, T., and Tohyama, M. (2001). Activation of caspase-12, an endoplastic reticulum (ER) resident caspase, through tumor necrosis factor receptor-associated factor 2- dependent mechanism in response to the ER stress. *J. Biol. Chem.* **276,** 13935–13940.

Yoshida, H., Matsui, T., Yamamoto, A., Okada, T., and Mori, K. (2001). XBP1 mRNA is induced by ATF6 and spliced by IRE1 in response to ER stress to produce a highly active transcription factor. *Cell* **107,** 881–891.

Zeng, L., Lu, M., Mori, K., Luo, S., Lee, A. S., Zhu, Y., and Shyy, J. Y. (2004). ATF6 modulates SREBP2-mediated lipogenesis. *EMBO J.* **23,** 950–958.

Zhang, K., Shen, X., Wu, J., Sakaki, K., Saunders, T., Rutkowski, D. T., Back, S. H., and Kaufman, R. J. (2006). Endoplasmic reticulum stress activates cleavage of CREBH to induce a systemic inflammatory response. *Cell* **24,** 587–599.

Zhang, K., Wong, H. N., Song, B., Miller, C. N., Scheuner, D., and Kaufman, R. J. (2005). The unfolded protein response sensor IRE1alpha is required at 2 distinct steps in B cell lymphopoiesis. *J. Clin. Invest.* **115,** 268–281.

Zhou, J., Lhoták, S., Hilditch, B. A., and Austin, R. C. (2005). Activation of the unfolded protein response occurs at all stages of atherosclerotic lesion development in apolipoprotein E-deficient mice. *Circulation* **111,** 1814–1821.

Zinszner, H., Kuroda, M., Wang, X., Batchvarova, N., Lightfoot, R. T., Remotti, H., Stevens, J. L., and Ron, D. (1998). CHOP is implicated in programmed cell death in response to impaired function of the endoplasmic reticulum. *Genes Dev.* **12,** 982–995.

Zong, W. X., Li, C., Hatzivassiliou, G., Lindsten, T., Yu, Q. C., Yuan, J., and Thompson, C. B. (2003). Bax and Bak can localize to the endoplasmic reticulum to initiate apoptosis. *J. Cell Biol.* **162,** 59–69.

CHAPTER TWENTY-ONE

ORGANELLE INTERMIXING AND MEMBRANE SCRAMBLING IN CELL DEATH

Mauro Degli Esposti

Contents

Abstract

In many cell types, intracellular organelles are involved along the progression of cell death. While many studies have focused on individual organelles such as mitochondria, evidence has accumulated that different organelles are simultaneously engaged in dynamic changes induced by death signaling before nuclear alterations are evident. This chapter examines approaches to evaluate dynamic aspects of organelle changes and intermixing during apoptosis. The methods presented here, which have been adapted from approaches used in the field of membrane traffic, enable the evaluation of mitochondrial intermixing with other organelles and the centrifugal movements of internal membranes that are associated, in particular, with death receptor-mediated apoptosis.

Faculty of Life Sciences, The University of Manchester, Manchester, United Kingdom

Methods in Enzymology, Volume 442

ISSN 0076-6879, DOI: 10.1016/S0076-6879(08)01421-3

1. INTRODUCTION

Along the path of cell destruction, diverse forms of programmed cell death (apoptosis) involve multiple intracellular organelles (Ferri and Kroemer, 2001). Mitochondria have become the most studied organelles in apoptosis since the discovery of their involvement in releasing apoptogenic factors such as cytochrome c, which activate the caspase cascade (Green and Reed, 1998). In the last decade, intense research has provided a wealth of information on how death signaling engages mitochondria (see Chapter 18), a process that also involves dynamic aspects such as the fusion and fission of mitochondrial membranes, which requires mitochondrial and cytosolic proteins, including members of the Bcl-2 family (Karbowski and Youle, 2003). However, mitochondria are also connected with the membrane traffic of other organelles that are required for either their recycling (autolysosomes) or their biogenesis and remodeling [endoplasmic reticulum (ER) and the Golgi apparatus] (Cristea and Degli Esposti, 2004). Especially in death receptor-mediated apoptosis, the membrane traffic connecting these different organelles is altered rapidly, leading to a disruption in the dynamic flow of organelle contacts. Similar to a traffic jam, local clogging then produces abnormal interactions and intermixing of membranes belonging to separated organelles such as mitochondria and Golgi (Ouasti et al., 2007). The concept of "organelle scrambling" has thus emerged to describe the intermixing of membranes that normally belong to different types of organelles but then apparently merge together following apoptosis-mediated changes in membrane traffic.

The earliest observations relating to organelle scrambling described perinuclear clustering of mitochondria in cells treated with death receptor ligands (Degli Esposti et al., 1999; De Vos et al., 1998). This clustering is now known to involve other organelles, especially the Golgi apparatus, which normally lies at one side of the nucleus (Ouasti et al., 2007) and may involve subtle changes in the contractile cortex of the cell (Huang et al., 2007). Subsequently, studies with T lymphocytes have shown that endosomal membranes can merge with mitochondria (Kawasaki et al., 2000) following the increased endocytosis associated with the stimulation of death receptors of the tumor necrosis factor superfamily (Algericas-Schimnich et al., 2002; Austin et al., 2006; Kenis et al., 2004; Lee et al., 2006; Schneider-Brachert et al., 2004). Once ligated, the prototypic death receptor Fas/CD95 induces a global alteration in the organelle membrane traffic of physiologically sensitive cells such as activated T lymphocytes. An initial wave of enhanced endocytosis is followed by a caspase-dependent movement of internal membranes toward the cell periphery that primarily involves secretory and endosomal membranes, which in part intermix with mitochondria (Ouasti et al., 2007) and may also enhance intercellular communication

(Eda *et al.*, 2004). Fas-induced changes in the membrane traffic have been documented, in particular, for T cells and their model lines such as Jurkat, which are used here as the biological system of reference to introduce methods for appreciating apoptotic changes in organelle distribution. Similar changes may also occur in other cellular contexts of cell death.

Following the temporal flow of Fas-induced changes in membrane traffic, this chapter first discusses methods to evaluate surface and endocytic changes and then methodological approaches that enable the visualization of centrifugal movements of internal membranes toward the cell surface. While several of these approaches are based on standard protocols of immunocytofluorescence with fixed cells, some of them can be integrated with quantitative measurements using florescence-based instrumentations.

2. METHODS USED TO EVALUATE INTERMIXING OF MITOCHONDRIA WITH OTHER ORGANELLES

Many surface receptors enhance endocytic pathways after triggering (Di Fiore and De Camilli, 2000). The enhanced endocytosis is often linked to rapid internalization of the activated receptors and their complexes, producing the "downmodulation" of death receptors on the cell surface (Austin *et al.*, 2006; Siegel *et al.*, 2004) and simultaneously increasing the propagation of intracellular signaling by endosome-associated receptor complexes (Di Fiore and de Camilli, 2000; Lee *et al.*, 2006; Schneider-Brachert *et al.*, 2004). Fundamentally, the internalization of activated receptors would follow the clathrin-dependent pathway of endocytosis, which is best evaluated using the established marker ferritin following its rapidly recycling receptor (e.g., Austin *et al.*, 2006). However, this is the dominant pathway of endocytosis stimulated by Fas triggering only in type I cells, which do not require mitochondria for amplifying the caspase cascade (Algericas-Schimnich *et al.*, 2002; Lee *et al.*, 2006; Peter and Krammer, 2003). Moreover, it is subsequently disrupted by caspase-mediated degradation of key components of clathrin-dependent endocytosis (Austin *et al.*, 2006). In all cells that require mitochondria for executing cell death (type II; Peter and Krammer, 2003), Fas also enhances other portals of endocytosis that contribute to receptor internalization and additionally alter membrane traffic. For instance, in T-cell lines, the internalization of death receptors may follow a route similar to that described recently for neurotropic receptors, which are internalized in long-lived vacuoles that do not overlap with clathrin-dependent endosomes (Valdez *et al.*, 2007). Because T cells lack caveolin (Orlandi and Fishman, 1998) and show negligible macropinocytosis, it is likely that they respond to Fas triggering by opening clathrin- and caveolin-independent portals of endocytosis used constitutively to traffic glycolipids and glycoproteins (Conner and Smith, 2003). These portals can be visualized at the microscopic level and also evaluated quantitatively using membrane probes such as FM1–43 (Kawasaki *et al.*, 2000; Zweifach, 2000) and glycoconjugate-recognizing proteins such as *Helix pomatia* agglutinin (HPA) (Eda *et al.*, 2004; Ouasti *et al.*, 2007). Once internalized, these probes partially also accumulate in mitochondrial membranes (Kawasaki *et al.*, 2000; Ouasti *et al.*, 2007). HPA and other lectins can also be used to decorate the surface of fixed cells and visualize the rapid alterations that death receptor induce on the cell surface as initial expression of enhanced endocytosis (Bilyy and Stoka, 2003; Ouasti *et al.*, 2007).

2.1. Cell surface staining with lectins

Early after apoptosis induction, the surface of the cell changes, forming diverse protrusions and blebbings (Weis *et al.*, 1995) that subsequently merge into apoptotic bodies (see Chapter 6). Membrane blebs are generally enriched in constituents that normally are dispersed throughout the surface of healthy cells. Several of these constituents are glycoconjugates that can be recognized by a variety of lectins having different sugar specificity (Bilyy and Stoka, 2003; Falasca *et al.*, 1996; Heyder *et al.*, 2003). The snail lectin HPA produces strong surface staining after Fas stimulation, especially in Jurkat T cells (Fig. 21.1; Ouasti *et al.*, 2007). The standard protocol used to undertake effective surface staining with fluorescent derivatives of HPA is described next; analysis is best performed with deconvolution microscopy (Ouasti *et al.*, 2007).

2.1.1. Reagents

RB buffer (modified Ringer buffer): 145 mM NaCl, 4.5 mM KCl, 2 mM MgCl2, 1 mM CaCl$_2$, 5 mM K-HEPES, pH 7.4, containing also 10 mM glucose. Human soluble recombinant FasL/CD95L (SuperFasLigand, Axxora), which is as potent as membrane-bound FasL (Holler *et al.*, 2000), is obtained as a lyophilized powder and resuspended at 0.1 mg/ml in sterile distilled water.

HPA derivatives are obtained commercially as the Alexa Fluor 488 conjugate (Molecular Probes, Invitrogen, Eugene, OR; Fig. 21.1A) and the tetramethy-rhodamine isothiocyanate (TRITC) conjugate (Sigma-Aldrich, St. Louis, Mo). Alternatively, lyophilized HPA (Sigma) can be conjugated to diverse Alexa fluorophores using appropriate labeling kits from Invitrogen. We produced blue-fluorescent HPA using the conjugation kit of Alexa Fluor 350 (Fig. 21.1B). Stock HPA solutions of 1 mg/ml in sterile phosphate-buffered saline (PBS) are stored at −20°.

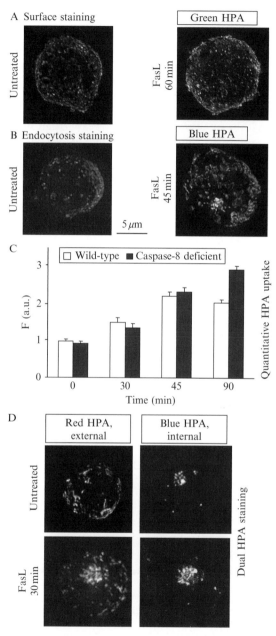

Figure 21.1 Multiple staining with HPA. (A) Surface staining of fixed Jurkat cells with Alexa Fluor 488–conjugated HPA (green HPA) before and after treatment with FasL for 1 h. (B) Staining of internalized Alexa Fluor 350–conjugated HPA (blue HPA) added externally to live Jurkat cells that were incubated in either the absence or the presence of FasL for a cumulative time of 45 min, including attachment. (C) Quantitative

2.1.2. Protocol

Jurkat cells grown to confluence in RPMI medium are washed by centrifugation and resuspended at 3 to 6 $\times$ 10^6 cells/ml in RB buffer and left either untreated or treated with 0.5 μg/ml FasL for 20 to 30 min at 37°. As a proper control, cells can be treated with a nonactivating anti-Fas antibody (for a list of suitable antibodies of this kind, see Lee *et al.*, 2006). Subsequently, cells are attached to polylysine-coated coverslips by incubation for 15 min at 37° within small plastic petri dishes and then washed twice with 1 ml PBS by dropping it gently on the side of dish. Cell attachment is checked after the first wash under an optical microscope. Subsequent cells are fixed by adding 1 ml of 4% paraformaldehyde (freshly prepared) onto each coverslip for 20 min at room temperature. After removing the fixative, coverslips are washed once with 1 ml PBS and then placed in a wet box and supplemented with 0.1 ml of 2 μg/ml of fluorescently labeled HPA dissolved in sterile PBS and incubated for 5 min at 37°. Coverslips are then washed with PBS for 5 min twice and finally placed in distilled water for 5 min before mounting them, face down, on slides containing a drop of mounting liquid. After drying for approximately 1 h, slides are fixed with nail varnish and kept at 4° until processed for microscopy.

2.1.3. Deconvolution fluorescence microscopy

Although conventional confocal microscopy can provide images with good resolution, round cells such as lymphocytes are best visualized using deconvolution microscopy. We have routinely imaged cells with DeltaVision RT (software Rx. 3.4.3, Applied Precision, Issaquah, WA) at 20° using an automated Olympus IX71 microscope with oil-immersed objectives. Once the top and bottom sections of the selected cells are found, dual-color images are obtained in stacks of >30 sections of 0.2 μm each and subsequently deconvolved with 10 cycles. The final deconvolved stacks are then projected along the z plane, which provides a high-resolution visualization of intracellular and surface details (Fig. 21.1; Ouasti *et al.*, 2007).

---

measurement of the uptake of TRITC-HPA (red HPA) in wild-type (empty histograms) and caspase-8-deficient Jurkat T cells (filled histograms) were carried out in a Fluroskan Ascent plate reader (Thermo, Basingstoke, UK) with an excitation filter centered at 544 nm and an emission filter centered at 590 nm. After incubation with 0.5 μg/ml FasL for the indicated times in the presence of 10 μg/ml of red HPA and a centrifugation wash, 50,000 cells per well were plated in quadruplicate wells and their fluorescence measured. Results showed statistically significant differences from controls (time $= 0$) after a 30-min treatment with FasL ($p < 0.05$). (D) Dual HPA staining with red HPA added first to live cells and then blue HPA added after fixation and permeabilization. Note that after a 30-min Fas stimulation with FasL there is only a limited dispersal of HPA-labeled membranes. All images were obtained with deconvolution microscopy with a 100× objective. Projected images from about 40 z sections were deconvolved for 10 cycles (Ouasti *et al.*, 2007).

2.2. Probes for apoptosis-related endocytosis: HPA

In cells such as Jurkat that require mitochondria for Fas-induced death, various probes can be used to monitor the enhanced endocytosis that follows Fas ligation. For instance, Kenis et al. (2004) used fluorescent derivatives of annexin V binding to externalized phosphatidylserine that were internalized in a clathrin-independent process in cells treated with Fas agonist antibodies. However, annexin V binding follows prior externalization of glucoconjugates that can be recognized by HPA and facilitate cell communication with macrophages (Eda et al., 2004; Yamanaka et al., 2005). HPA binds to nonterminal residues of N-acetylgalactosamine, which are enriched not only in the secretory pathway around the cis-Golgi (Perez-Villar et al., 1991), but also in heavily glycosylated surface proteins such as CD43 (Eda et al., 2004). Similar to other lectins (Vetterlein et al., 2002), the addition of external HPA to live cells leads to its rapid binding to glycosylated surface proteins, which are continuously recycled to the interior of the cell via constitutive endocytosis (Cresawan et al., 2007). HPA-binding proteins become more abundant on the cell surface after Fas stimulation and consequently surface staining with fluorescent derivatives of HPA increases by over twofold, following also endocytic changes (Fig. 21.1 and Section 2.1). TRITC-conjugated HPA and Alexa Fluor 350-conjugated HPA have been found to be most specific for labeling endosomal compartments after Fas stimulation (Ouasti et al., 2007). Images in Fig. 21.1B show the typical pattern of endocytosed HPA in control untreated and FasL-treated cells, whereas Fig. 21.1C shows the quantitative evaluation of endocytosed HPA in Jurkat T cells, which is caspase-8 independent.

2.2.1. Reagents
RB buffer (modified Ringer buffer) and other reagents are the same as those described in Section 2.1.1.

2.2.2. Protocol
Jurkat cells grown to confluence in RPMI medium are washed by centrifugation and resuspended at 3 to 6 × 10⁶ cells/ml in RB buffer containing 5 μg/ml of TRITC-HPA and left either untreated or treated with 0.5 μg/ml of FasL for 20 to 30 min at 37°. After centrifugation to remove excess external HPA, cells are resuspended in RB buffer and processed for fluorescence microscopy after attachment onto polylysine-coated coverslips as described in Section 2.1.2. Alternatively, labeled cells are transferred in quadruplicate wells of a 96-well microplate and bulk fluorescence is measured quantitatively using a fluorescence plate reader with a suitable pair of filters. For example, we have used the Ascent plate reader (Thermo, Basingstoke, UK) with excitation filter centered at 544 nm and emission filter centered at 590 nm for measuring the uptake of TRITC-HPA

(Fig. 21.1C, cf. Ouasti *et al.*, 2007). Maximal uptake of HPA is observed around 45 min after Fas stimulation. Pretreatment of cells with pan-caspase inhibitors such as benzyloxycarbonyl-Val-Ala-Asp-fluoromethylketone (z-VAD) has a negligible effect on the enhanced uptake of external HPA conjugates (Ouasti *et al.*, 2007). However, pretreatment with non-toxic concentrations (0.1 µg/ml) of *Clostridium difficile* toxin B, a general inhibitor of actin-modulating Rho-GTPases, clearly reduces the Fas-enhanced uptake of HPA. Of note, major pathways of clathrin-independent endocytosis depend on the activity of Rho-GTPases (Sabharanjak *et al.*, 2002).

2.3. The FM1–43 probe

Kawasaki *et al.* (2000) first reported that FM1–43, an amphiphilic fluorescent probe originally applied to monitor vesicle recycling in neurons (Henkel *et al.*, 1996), increased its uptake after ligation of Fas in target lymphoma cells. Subsequent work by Fomina and co-workers has established FM1–43 as a valuable probe to monitor the increased membrane traffic that follows stimulation of the T-cell receptor in T lymphocytes and Jurkat cells, which progressively involves ER, mitochondria, and Golgi membranes (Dadsetan *et al.*, 2005). A similar situation applies to T-cell lines after stimulation of Fas, in which the uptake of FM1–43 provides complementary information to the enhanced endocytosis of HPA (Ouasti *et al.*, 2007). Similar to HPA, the increase in FM1–43 uptake by Fas-stimulated cells is transient, as it is followed by a slow release via exocytic movements, which are best evaluated by the approaches described in Section 3. However, there appears to be a limited overlap between FM1–43 and HPA staining before and after early activation of Fas, also due to the broad fluorescence emission of FM1–43, encompassing green, orange, and red light. Consequently, application of FM1–43 labeling to cells costained with other probes, including most mitochondria-specific dyes, is not appropriate for high-quality fluorescence imaging. However, uptake of FM1–43 can be measured quantitatively in conventional fluorimeters (Zweifach, 2000) or plate readers, following the protocol described here.

2.3.1. Reagents

Fluorescence measurements are routinely undertaken with cells suspended in RB buffer (modified Ringer buffer, Section 2.1.1). FM1–43 from Molecular Probes/Invitrogen is resuspended at 1 mM in 50% ethanol/50% distilled water (v/v) and kept refrigerated until used. Ionomycin (Sigma) can be used as a positive control of endocytosis at the final concentration of 1 µM (Zweifach, 2000).

2.3.2. Protocol

Jurkat cells grown to confluence in RPMI medium are washed and resuspended at 5×10^6 cells/ml in RB buffer and then diluted 10-fold into 2 ml of RB, which is transferred into a plastic disposable cuvette (Kartell) placed in a fluorimeter (e.g., Perkin-Elmer LC50) with excitation set at 470 nm and emission at 565 or 580 nm (with 5-nm bandwidth). Once the background fluorescence of the cell suspension is recorded for a few minutes, FM1–43 is added to a final concentration of 3 to 4 μM and mixed rapidly. The resulting increase in fluorescence is due to partition of the probe, which is essentially nonfluorescent in aqueous media, into cell membranes and reaches a steady-state level within 20 to 30 min after reequilibration via exocytic release (Henkel et al., 1996). When cells are pretreated with FasL (or a Fas agonist antibody), the initial increase of fluorescence is generally 2-fold higher than in untreated cells, indicating an enhanced uptake of the dye due to increased endocytosis. Pretreatment of cells with pan-caspase inhibitors such as z-VAD has little effect on the initial Fas-enhanced increase of FM1–43 fluorescence, but affects subsequent reequilibration of the probe. The uptake of FM1–43 is strongly inhibited at temperatures below 15°, as for most classical markers of endocytosis.

2.4. Intermixing of mitochondria with endosomes and other organelles

An alternative method for evaluating the intermixing of different organelles during death signaling is to measure the degree of cross-contamination of membrane markers specific for one organelle in the isolated preparation of another organelle. This approach would, in principle, require extensive purification of one test organelle such as mitochondria, a hard task with lymphocytic and other cell lines used in apoptosis studies. To this end, we have developed a simple method to obtain mitochondria from lymphoma cells with purity nearly equivalent to that of conventionally prepared mitochondria from mouse and thus much higher than that achievable with other fractionation procedures (Fig. 21.2). Membrane markers specific for other organelles are usually present a low levels in these mitochondria, which are called "perinuclear," as they are obtained from the nuclear fraction of cell homogenates. However, after Fas stimulation, markers of early endosomes such as Rab5 or EEA1 and then markers of Golgi membranes such as GM130 increase their "contamination" levels in isolated mitochondria (Fig. 21.2), whereas markers of late endosomes or lysosomes decrease, if anything, their levels (Ouasti et al., 2007). In contrast, inhibition of caspases increases the distribution of markers for late endosomes and lysosomes in perinuclear (and even conventionally prepared) mitochondria after Fas stimulation, while reducing the contaminations from Golgi

A   Perinuclear mitochondria, Jurkat T-cells

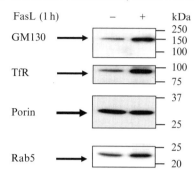

B   Conventional mitochondria, mouse liver

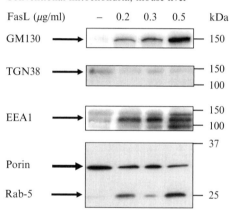

Figure 21.2 Organelle intermixing detected by Western blots of isolated mitochondria. (A) Perinuclear mitochondria from Jurkat cells (10 μg protein per lane) were separated with a 12% acrylamide SDS-PAGE and immunoblotted for GM130 (rabbit polyclonal, a kind gift of Martin Lowe, University of Manchester) and the early endosomal proteins, transferrin receptor (TfR) and Rab5 (with monoclonal antibodies from BD Pharmingen); mitochondrial loading was evaluated by coblotting with a goat polyclonal antiporin/VDAC (N-18, Santa Cruz). (B) Mitochondria (5 μg protein per lane), which had been conventionally isolated with isotonic fractionation from mouse livers treated *ex vivo* with the indicated concentrations of FasL for 1 h (Sorice *et al.*, 2004), were separated and immunoblotted as in A, except that Rab5 was detected using a monoclonal antibody from Synaptic Systems. Additionally, EEA1 was recognized with a monoclonal antibody from BD Pharmingen, whereas TGN38 (a marker overlapping late endosomal and *trans*-Golgi membranes) was recognized with a sheep polyclonal donated by Martin Lowe. Note the reduction in the mitochondrial levels of this marker in comparison with the *cis*-Golgi marker GM130.

membranes (Ouasti *et al.*, 2007). This differential intermixing of organelle-specific markers partially reflects the caspase dependence of the membrane scrambling of Golgi-related organelles with mitochondria (Ouasti *et al.*, 2007).

Moreover, it indicates that caspase inhibition could increase the content of endolysosomes within cells, in line with autophagic modes of caspase-independent death (Yu et al., 2004). The protocol for isolating both conventional and perinuclear mitochondria is presented here; although developed for T lymphoma lines such as Jurkat, they can be equally applied to a variety of other cells and tissues, including mouse liver. Typical results of Western blotting obtained as described earlier (Degli Esposti et al., 2003; Ouasti et al., 2007) are presented in Fig. 21.2, also for a comparison with mouse liver mitochondria.

2.4.1. Protocol for isolating mitochondria with conventional fractionation

Jurkat cells ($3-6 \times 10^8$/ml) are washed from growth medium and homogenized in 2 ml of isolation buffer [0.25 M mannitol, 1 mM EDTA, 10 mM K-HEPES, 0.2% bovine serum albumin (BSA), pH 7.4] containing a cocktail of protease inhibitors (Sigma) diluted 1:100 (v/v), further supplemented with 50 μM z-VAD to minimize caspase-induced degradation of proteins during fractionation. The first nuclear pellet obtained by centrifugation at 800g is rehomogenized with the supernatant and centrifuged at 800g for 10 min. The pellet is saved and frozen for the preparation of perinuclear mitochondria (see later), while the supernatant is further centrifuged at 10,000g for 10 min. The pellet is homogenized gently in 1 ml assay buffer (0.12 M mannitol, 0.08 M KCl, 1 mM EDTA, 20 mM K-HEPES, pH 7.4) and recentrifuged at 10,000g for 15 min to obtain mitochondria, which are dissolved in a minimal volume of assay buffer.

2.4.2. Protocol for isolating perinuclear mitochondria

In this procedure, extensively purified mitochondria are obtained from the nuclear pellet of the second homogenization step given earlier. After thawing, this pellet (which contains over 40% of the cellular mitochondria) is rehomogenized in assay buffer and centrifuged at 600g. The supernatant is then layered in 0.4-ml aliquots onto a cushion of 1 M mannitol (0.6 ml) covered with 0.1 ml of assay buffer containing 2% (w/v) BSA. After centrifugation at 9000g for 15 min at 4°, the pellet is resuspended in 0.5 ml of assay buffer, centrifuged at 10,000g for 10 min, and then resuspended in 25μl of assay buffer containing 1:100 (v/v) of the protease inhibitor cocktail. After evaluation of the protein concentration with a Bio-Rad assay in the presence of 0.5% Triton X-100, preparations are diluted to 0.5 to 1 mg/ml with 2× SDS sample buffer, boiled, and then stored at −80° until used for Western blotting as carried out in the experiments of Fig. 21.2.

3. Methods Used to Evaluate Caspase-Dependent Changes in Intracellular Traffic

The initial wave of Fas-enhanced endocytosis is essentially unaffected by caspase inhibition (Ouasti *et al.*, 2007) or by the absence of active caspase-8 in Jurkat and other cells requiring mitochondria for caspase amplification (unpublished data). However, the initial centripetal influx of membrane traffic is integrated with changes in the traffic of internal membranes that originates in the early secretory system, that is, specialized sections of the ER and the Golgi apparatus, which is strongly affected by caspase inhibition. Because activation of caspase-8 within cells can be detected around 30 to 40 min after Fas ligation, the concomitant Fas-induced changes in inside-out movements of internal membranes must be caused primarily by the activity of apical caspases. Austin *et al.* (2006) have reported that components of the clathrin pathway of endocytosis are cleaved by apical caspases, resulting in the inactivation of initially stimulated endocytosis. Perhaps early disruption of this pathway helps diverting membrane traffic outward, in part discharging the excess intake of membranes and fluids resulting from receptor-stimulated endocytosis, as documented for the epidermal growth factor receptor (Hamasaki *et al.*, 2004). Although the molecular reasons why internal membranes move outward and merge at the cell surface remain unclear, various approaches have been established to evaluate and visualize this phenomenon. Suitable methods are described next.

3.1. Dispersal of the Golgi apparatus and related membranes: Single and dual HPA staining

Earlier work on the involvement of the Golgi organelle during cell death has shown that several scaffolding proteins of the stacks are cleaved by caspases, leading to a caspase-dependent fragmentation of the Golgi apparatus (Hicks and Machamer, 2005; Nozawa *et al.*, 2002). Golgi-related membranes are also dispersed following apoptosis induced by ceramide treatment (Hu *et al.*, 2005) and Fas ligation (Ouasti *et al.*, 2007) as visualized by conventional immunocytochemical methods using Golgi-specific protein markers such as GM130 (Fig. 21.3A). An alternative approach is to use lectins such as HPA, which specifically label the Golgi apparatus (and related membranes) in permeabilized cells (del Valle *et al.*, 1999). Because HPA also binds to the surface of cells (see Section 2.1), additional quenching steps are required to minimize surface staining vs intracellular labeling of Golgi membranes, as described in the following protocol. We have developed the procedure further with the method of "dual HPA" staining to monitor the

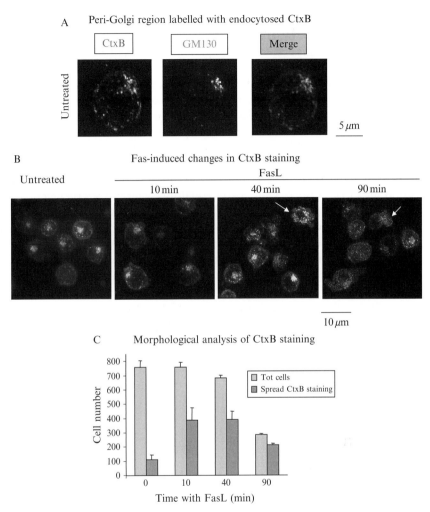

Figure 21.3 Global movements of endomembranes visualized with CtxB staining. (A) Specific immunolabeling of the Golgi complex of Jurkat cells with the membrane marker GM130 (obtained as described by Ouasti *et al.*, 2007) identifies the "peri-Golgi" region where endocytosed CTxB (Alexa Fluor 594-conjugated, red CtxB) accumulate. Projected images from about 40 z sections were obtained with a 60× objective and deconvolved for 10 cycles. (B) Grayscale images if Jurkat cells that had been labeled with Alexa Fluor 488-CtxB and then incubated for a prolonged period with FasL in concentrated solution, from which aliquots were taken 10 min before the indicated time to allow attachment onto coverslips. Images were obtained as in A and arrows indicate blebbing cells with altered morphology, an early expression of Fas-induced apoptosis (Weis *et al.*, 1995). Note the increased dispersal of CtxB staining with time. (C) Quantitative morphological analysis of multiple images obtained as in B was carried by three independent scorers in a large number of microscopy fields. The graph shows the progressive decrease in the total number of attached cells counted (total cells, light gray histograms) in equivalent sets of randomly

endocytosis of both external HPA and internal changes of Golgi membranes using differently fluorescent derivatives of HPA (Ouasti et al., 2007).

3.1.1. Protocol of single HPA staining

This protocol uses the same basic reagents as those described in Sections 2.1 and 2.2. After fixation, coverslips are washed once with 1 ml PBS and cells are permeabilized by adding 1 ml of 0.5% Triton X-100 in PBS followed by incubation for 10 to 20 min at room temperature. After taking the Triton solution off, coverslips are washed with 1 ml PBS. To attenuate HPA binding to the cell surface, 1 ml of unlabeled HPA (Sigma, 52 μg/ml in PBS) is added to the coverslips and incubated for 10 min at room temperature. After a PBS wash, 0.1 ml of 2 μg/ml fluorescent HPA is then added to stain internal membranes, placing the coverslips face down in a wet box for 5 min at room temperature. Subsequently, the coverslips are washed two times for 5 min with PBS and once with distilled water before mounting the slides. Figure 21.1D (right) shows a typical internal staining obtained with Alexa Fluor 350-conjugated HPA.

3.1.2. Protocol of dual HPA staining

This procedure is a combination of the protocol described in Section 2.2.2 with that of single HPA staining of permeabilized cells (Section 3.1.1). Live cells are first supplemented with TRITC-HPA (red HPA, Fig. 21.1D) to monitor endocytosis and then fixed and permeabilized. After quenching the cell surface with unlabeled HPA, cells are then treated with either Alexa Fluor 488-conjugated HPA (green HPA) or Alexa Fluor 350-conjugated HPA (blue HPA) to label internal Golgi membranes that become dispersed toward the cell periphery from about 30 min after Fas activation (Ouasti et al., 2007).

3.1.3. Protocol for HPA staining combined with other fluorescent markers

Prior to single staining with Alexa Fluor 488-conjugated HPA, live cells can be loaded with mitochondria-specific markers such as MitoTracker red to evaluate organelle intermixing. Our standard procedure is as follows (Degli Esposti et al., 1999). Cells are suspended in full growth medium at 1 to 2×10^6 cells/ml containing 100 nM MitoTracker red CMXRos (dissolved previously at 1 mM in dimethyl sulfoxide) and incubated at room temperature for 20 min

---

chosen fields, which reflected the increasing inability of apoptotic cells to adhere properly on coverslips, in conjunction with the loss of cells that had completely disintegrated before being seeded. Hence, data indicate the loss of viability during FasL treatment, as confirmed by Trypan blue counting. The darker histograms represent the parallel evaluation of attached cells that presented with dispersed CtxB staining (spread), i.e., loss of the normal peri-Golgi clustering (cf. B). Note that after 90 min of treatment most cells show a dispersed CtxB staining. (See color insert.)

in the dark. After centrifugation, cells are resuspended in RB buffer and all subsequent steps are carried out as described in Section 3.1.1. HPA labeling of permeabilized cells can also be combined with standard immunostaining of membrane proteins. However, caution should be exercised if whole serum is used to block permeabilized specimen before adding the primary antibody of interest because HPA conjugates bind nonspecifically to serum glycoproteins. Consequently, HPA staining of internal membranes (Section 3.1.1) should be performed before the blocking step required for specific immunostaining.

3.2. Dispersal of the Golgi apparatus and related membranes: CtxB

Two portals of clathrin-independent endocytosis have been established (Sabharanjak *et al.*, 2002). Together with the caveolin pathway, they drive the internalization of subunit B of cholera toxin (CtxB), which avidly binds to the glycolipid GM1 and is thus a valuable reporter for the traffic of membrane lipids (Cheng *et al.*, 2006; Sabharanjak *et al.*, 2002). In Jurkat cells, like most lymphocytes where caveolin is absent (Orlandi and Fishman, 1998), fluorescent derivatives of CtxB are taken up by constitutive routes of clathrin-independent endocytosis that traffic glycolipids and glycosylated proteins, which are often associated in the so-called lipid rafts (Conner and Smith, 2003; Legembre *et al.*, 2006). CtxB staining has been used predominantly to evaluate lipid rafts in studies on Fas-mediated apoptosis (Legembre *et al.*, 2006; Siegel *et al.*, 2004). We have developed a protocol of CtxB staining that enables the evaluation of Golgi dispersal and exposure of internal membranes at the cell surface, which evidences also membrane blebbing (Fig. 21.3).

3.2.1. Protocol of CtxB staining of live cells

Jurkat cells prepared as in Section 2.2 are incubated for 20 to 30 min at 37° with fluorescent cholera toxin subunit B conjugated with either Alexa Fluor 594 (red CtxB, Fig. 21.3A) or Alexa Fluor 488 (green CtxB, Fig. 21.3B). The final concentration of CtxB is 2 μg/ml, which allows complete equilibration of the internalized toxin within the peri-Golgi region of recycling endosomes (Fig. 21.3A, cf. Cheng *et al.*, 2006; Sabharanjak *et al.*, 2002). After two washes in RB buffer to remove unbound toxin, cells are treated, attached, and processed as described in Section 2.1. Images are best obtained with deconvolution microscopy (Fig. 21.3A and B), and the morphological patterns of CtxB staining (clustered vs dispersed or spread) should be evaluated by at least three independent observers using over 500 imaged cells (Fig. 21.3C). Note the transient increase in intensity and the progressive increase in dispersal of CtxB staining after the addition of FasL, allowing also the recognition of cells undergoing membrane blebbing (Fig. 21.3B, arrows), a hallmark of early Fas-induced apoptosis (Weis *et al.*, 1995).

4. OUTLOOK

This chapter introduced methodological approaches to evaluate dynamic changes in the interrelationships of organelles such as mitochondria, endosomes, and the Golgi apparatus during Fas-induced cell death. Complementary approaches include the evaluation of membrane traffic using markers for clathrin-dependent endocytosis (transferrin) or micropinocytosis (fluorescent dextran beads) (Austin *et al.*, 2006; Sabharanjak *et al.*, 2002). Changes in membrane traffic leading to organelle intermixing are only partially dependent on the activation of caspases (Ouasti *et al.*, 2007; Siegel *et al.*, 2004). In particular, when caspases are inhibited or defective, the receptor-induced increase in endocytic traffic remains while the discharge wave of outward movements of internal membranes is halted. Therefore, the immediate consequence of caspase inhibition is an accumulation of intracellular endolysosomal vacuoles, which may ultimately facilitate self-degradation of cells via the autophagy program (Yu *et al.*, 2004). In perspective, the present approaches for evaluating the global changes in the membrane traffic of intracellular organelles that occur early after Fas triggering could also be used to study the obscure interconnections between death receptor signaling and autophagic or necrotic-like cell death.

ACKNOWLEDGMENTS

For this work I am indebted to Richard Paddon, Francesca Luchetti, and Julien Tour and to the bioimaging facility of our faculty for their technical skills and support.

REFERENCES

Algeciras-Schimnich, A., Shen, L., Barnhart, B. C., Murmann, A. E., Burkhardt, J. K., and Peter, M. E. (2002). Molecular ordering of the initial signaling events of CD95. *Mol. Cell. Biol.* **22,** 207–220.

Austin, C. D., Lawrence, D. A., Peden, A. A., Varfolomeev, E. E., Totpal, K., De Maziere, A. M., Klumperman, J., Arnott, D., Pham, V., Scheller, R. H., and Ashkenazi, A. (2006). Death-receptor activation halts clathrin-dependent endocytosis. *Proc. Natl. Acad. Sci. USA* **103,** 10283–10288.

Bilyy, R. O., and Stoika, R. S. (2003). Lectinocytochemical detection of apoptotic murine leukemia L1210 cells. *Cytometry A* **56,** 89–95.

Cheng, Z. J., Singh, R. D., Sharma, D. K., Holicky, E. L., Hanada, K., Marks, D. L., and Pagano, R. E. (2006). Distinct mechanisms of clathrin-independent endocytosis have unique sphingolipid requirements. *Mol. Biol. Cell* **17,** 3197–3210.

Cresawn, K. O., Potter, B. A., Oztan, A., Guerriero, C. J., Ihrke, G., Goldenring, J. R., Apodaca, G., and Weisz, O. A. (2007). Differential involvement of endocytic compartments in the biosynthetic traffic of apical proteins. *EMBO J.* **26,** 3737–3748.

Cristea, I. M., and Degli Esposti, M. (2004). Membrane lipids and cell death: An overview. *Chem. Phys. Lipids* **129,** 133–160.

Dadsetan, S., Shishkin, V., and Fomina, A. F. (2005). Intracellular Ca^{2+} release triggers translocation of membrane marker fm1–43 from the extracellular leaflet of plasma membrane into endoplasmic reticulum in T lymphocytes. *J. Biol. Chem.* **280,** 16377–16382.

Degli Esposti, M., Cristea, I. M., Gaskell, S. J., Nakao, Y., and Dive, C. (2003). Pro-apoptotic Bid binds to monolysocardiolipin, a new molecular connection between mitochondrial membranes and cell death. *Cell Death Differ.* **10,** 1300–1309.

Degli Esposti, M., Hatzinisiriou, I., McLennan, H., and Ralph, S. (1999). Bcl-2 and mitochondrial oxygen radicals: New approaches with reactive oxygen species-sensitive probes. *J. Biol. Chem.* **274,** 29831–29837.

del Valle, M., Robledo, Y., and Sandoval, I. V. (1999). Membrane flow through the Golgi apparatus: Specific disassembly of the cis-Golgi network by ATP depletion. *J. Cell Sci.* **112,** 4017–4029.

De Vos, K., Goossens, V., Boone, E., Vercammen, D., Vancompernolle, K., Vandenabeele, P., Haegeman, G., Fiers, W., and Grooten, J. (1998). The 55-kDa tumor necrosis factor receptor induces clustering of mitochondria through its membrane-proximal region. *J. Biol. Chem.* **273,** 9673–9680.

Di Fiore, P. P., and De Camilli, P. (2001). Endocytosis and signalling: An inseparable partnership. *Cell* **106,** 1–4.

Eda, S., Yamanaka, M., and Beppu, M. (2004). Carbohydrate-mediated phagocytic recognition of early apoptotic cells undergoing transient capping of CD43 glycoprotein. *J. Biol. Chem.* **279,** 5967–5974.

Falasca, L., Bergamini, A., Serafino, A., Balabaud, C., and Dini, L. (1996). Human Kupffer cell recognition and phagocytosis of apoptotic peripheral blood lymphocytes. *Exp. Cell Res.* **224,** 152–162.

Ferri, K. F., and Kroemer, G. (2001). Organelle-specific initiation of cell death pathways. *Nat. Cell Biol.* **3,** E255–E263.

Green, D. R., and Reed, J. C. (1998). Mitochondria and apoptosis. *Science* **281,** 1309–1312.

Hamasaki, M., Araki, N., and Hatae, T. (2004). Association of early endosomal autoantigen 1 with macropinocytosis in EGF-stimulated A431 cells. *Anat. Rec. A Discov. Mol. Cell. Evol. Biol.* **277,** 298–306.

Henkel, A., Lübke, J., and Betz, W. J. (1996). FM1–43 ultrastructural localization in and release from frog motor nerve terminals. *Proc. Natl. Acad. Sci. USA* **93,** 1918–1923.

Heyder, P., Gaipl, U. S., Beyer, T. D., Voll, R. E., Kern, P. M., Stach, C., Kalden, J. R., and Herrmann, M. (2003). Early detection of apoptosis by staining of acid-treated apoptotic cells with FITC-labeled lectin from. *Narcissus pseudonarcissus. Cytometry A* **55,** 86–93.

Hicks, S. W., and Machamer, C. E. (2005). Golgi structure in stress sensing and apoptosis. *Biochim. Biophys. Acta* **1744,** 406–414.

Holler, N., Zaru, R., Micheau, O., Thome, M., Attinger, A., Valitutti, S., Bodmer, J. L., Schneider, P., Seed, B., and Tschopp, J. (2000). Fas triggers an alternative, caspase-8-independent cell death pathway using the kinase RIP as effector molecule. *Nat. Immunol.* **1,** 489–495.

Hu, W., Xu, R., Zhang, G., Jin, J., Szulc, Z. M., Bielawski, J., Hannun, Y. A., Obeid, L. M., and Mao, C. (2005). Golgi fragmentation is associated with ceramide-induced cellular effects. *Mol. Biol. Cell* **16,** 1555–1567.

Huang, P., Yu, T., and Yoon, Y. (2007). Mitochondrial clustering induced by overexpression of the mitochondrial fusion protein Mfn2 causes mitochondrial dysfunction and cell death. *Eur. J. Cell Biol.* **86,** 289–302.

Karbowski, M., and Youle, R. J. (2003). Dynamics of mitochondrial morphology in healthy cells and during apoptosis. *Cell Death Differ.* **10,** 870–880.

Kawasaki, Y., Saito, T., Shirota-Someya, Y., Ikegami, Y., Komano, H., Lee, M. H., Froelich, C. J., Shinohara, N., and Takayama, H. (2000). Cell death-associated translocation of plasma membrane components induced by CTL. *J. Immunol.* **164,** 4641–4648.

Kenis, H., van Genderen, H., Bennaghmouch, A., Rinia, H. A., Frederik, P., Narula, J., Hofstra, L., and Reutelingsperger, C. P. (2004). Cell surface-expressed phosphatidylserine and annexin A5 open a novel portal of cell entry. *J. Biol. Chem.* **279,** 52623–52629.

Lee, K. H., Feig, C., Tchikov, V., Schickel, R., Hallas, C., Schutze, S., Peter, M. E., and Chan, A. C. (2006). The role of receptor internalization in CD95 signaling. *EMBO J.* **25,** 1009–1023.

Legembre, P., Daburon, S., Moreau, P., Moreau, J. F., and Taupin, J. L. (2006). Modulation of Fas-mediated apoptosis by lipid rafts in T lymphocytes. *J. Immunol.* **176,** 716–720.

Nozawa, K., Casiano, C. A., Hamel, J. C., Molinaro, C., Fritzlerm, M. J., and Chan, E. K. (2002). Fragmentation of Golgi complex and Golgi autoantigens during apoptosis and necrosis. *Arthritis Res.* **4,** R3.

Orlandi, P. A., and Fishman, P. H. (1998). Filipin-dependent inhibition of cholera toxin: Evidence for toxin internalization and activation through caveolae-like domains. *J. Cell Biol.* **141,** 905–915.

Ouasti, S., Matarrese, P., Paddon, R., Khosravi-Far, R., Sorice, M., Tinari, A., Malorni, W., and Degli Esposti, M. (2007). Death receptor ligation triggers membrane scrambling between Golgi and mitochondria. *Cell Death Differ.* **14,** 453–461.

Perez-Vilar, J., Hidalgo, J., and Velasco, A. (1991). Presence of terminal N-acetylgalactosamine residues in subregions of the endoplasmic reticulum is influenced by cell differentiation in culture. *J. Biol. Chem.* **266,** 23967–23976.

Peter, M. E., and Krammer, P. H. (2003). The CD95(APO-1/Fas) DISC and beyond. *Cell Death Differ.* **10,** 26–35.

Sabharanjak, S., Sharma, P., Parton, R. G., and Mayor, S. (2002). GPI-anchored proteins are delivered to recycling endosomes via a distinct cdc42-regulated, clathrin-independent pinocytic pathway. *Dev. Cell* **2,** 411–423.

Schneider-Brachert, W., Tchikov, V., Neumeyer, J., Jakob, M., Winoto-Morbach, S., Held-Feindt, J., Heinrich, M., Merkel, O., Ehrenschwender, M., Adam, D., Mentlein, R., Kabelitz, D., and Schutze, S. (2004). Compartmentalization of TNF receptor 1 signaling: Internalized TNF receptosomes as death signaling vesicles. *Immunity* **21,** 415–428.

Siegel, R. M., Muppidi, J. R., Sarker, M., Lobito, A., Jen, M., Martin, D., Straus, S. E., and Lenardo, M. J. (2004). SPOTS: Signaling protein oligomeric transduction structures are early mediators of death receptor-induced apoptosis at the plasma membrane. *J. Cell Biol.* **167,** 735–744.

Sorice, M., Circella, A., Cristea, I. M., Garofalo, T., Renzo, L. D., Alessandri, C., Valesini, G., and Degli Esposti, M. (2004). Cardiolipin and its metabolites move from mitochondria to other cellular membranes during death receptor-mediated apoptosis. *Cell Death Differ.* **11,** 1133–1145.

Valdez, G., Philippidou, P., Rosenbaum, J., Akmentin, W., Shao, Y., and Halegoua, S. (2007). Trk-signaling endosomes are generated by Rac-dependent macroendocytosis. *Proc. Natl. Acad. Sci. USA* **104,** 12270–12275.

Vetterlein, M., Ellinger, A., Neumuller, J., and Pavelka, M. (2002). Golgi apparatus and TGN during endocytosis. *Histochem. Cell Biol.* **117,** 143–150.

Weis, M., Schlegel, J., Kass, G. E., Holmstrom, T. H., Peters, I., Eriksson, J., Orrenius, S., and Chow, S. C. (1995). Cellular events in Fas/APO-1-mediated apoptosis in JURKAT T lymphocytes. *Exp. Cell Res.* **219,** 699–708.

Yamanaka, M., Eda, S., and Beppu, M. (2005). Carbohydrate chains and phosphatidylserine successively work as signals for apoptotic cell removal. *Biochem. Biophys. Res. Commun.* **328,** 273–280.

Yu, L., Alva, A., Su, H., Dutt, P., Freundt, E., Welsh, S., Baehrecke, E. H., and Lenardo, M. J. (2004). Regulation of an ATG7-beclin 1 program of autophagic cell death by caspase-8. *Science* **304,** 1500–1502.

Zweifach, A. (2000). FM1–43 reports plasma membrane phospholipid scrambling in T-lymphocytes. *Biochem. J.* **349,** 255–260.

CHAPTER TWENTY-TWO

MECHANISMS AND METHODS IN GLUCOSE METABOLISM AND CELL DEATH

Yuxing Zhao, Heather L. Wieman, Sarah R. Jacobs, *and* Jeffrey C. Rathmell

Contents

Abstract

Glucose metabolism represents a critical physiological program that not only provides energy to support cell proliferation, but also directly modulates signaling pathways of cell death. With the growing recognition of regulation of cell death by glucose metabolism, many techniques that can be applied in the study have been developed. This chapter discusses several protocols that aid in the analysis of glucose metabolism and cell death and the principles in practicing them under different conditions.

1. INTRODUCTION

Cell death is a critical mechanism in maintaining tissue homeostasis, and misregulation of cell death can lead to development of a variety of diseases. For example, too little cell death contributes to cancer development

Department of Pharmacology and Cancer Biology, Department of Immunology, and Sarah W. Stedman Nutrition and Metabolism Center, Duke University, Durham, North Carolina

Methods in Enzymology, Volume 442
ISSN 0076-6879, DOI: 10.1016/S0076-6879(08)01422-5

or onset of autoimmune responses, whereas excessive cell death can cause degenerative diseases, such as neuron degeneration. Cell death is regulated by intracellular machineries that respond to changes of environment both inside and outside cells. Accumulating data have shown that metabolism plays an important role in the regulation of cell death, and studies on relationship between metabolism and cell death are expanding quickly (Gottlob *et al.*, 2001; Nutt *et al.*, 2005; Rathmell *et al.*, 2003; Zhao *et al.*, 2007) Thus, there is a growing need for the development of methods and techniques to be used in the analysis of glucose metabolism so that researchers are able to study changes and roles of metabolism in cell death. This chapter summarizes general methods developed to monitor and manipulate glucose uptake and metabolism with the background of study on cell death.

Glucose metabolism is a primary source of energy and biomaterials for the maintenance of life. In the first rate-determining step of the metabolism, glucose is transported across the plasma membrane by the facilitative glucose transporter (Glut) down its concentration gradient. Hexokinase (HK) on the mitochondria then phosphorylates glucose to glucose-6-phosphate (G6P). The product generally enters the glycolytic pathway, generating NADH, ATP, and pyruvate, or the pentose phosphate pathway (PPP). In the presence of sufficient oxygen, pyruvate from glycolysis can be fed into mitochondria and fully oxidized to produce more ATP. When oxygen is limited, however, pyruvate is disposed in the form of lactate and glycolysis becomes the main source for ATP production (Gatenby and Gillies, 2004). PPP plays an important role in the synthesis of nucleic acids for DNA and RNA, as well as generation of NADPH for the synthesis of lipids and maintenance of intracellular redox homeostasis.

There are three major forms of cell death that may be influenced by glucose metabolism: necrosis, autophagy, and apoptosis (Edinger and Thompson, 2004). Necrosis can occur when ATP levels decrease too dramatically and cells lose the ability to maintain intracellular homeostasis of ions and water, leading to rupture of the plasma membrane and leakage of the cytoplasm into the extracellular environment. Autophagy is an energy-dependent process of self-digestion that can result in cell death (Lum *et al.*, 2005). It is initiated when the uptake of extracellular nutrient sources such as amino acids, glucose, or lipids decreases and cells must use intracellular components to support mitochondrial oxidation and energy production. When autophagy occurs, double-membrane vesicles form and engulf cytoplasm and organelles (autophagosome). The autophagosome then fuses with lysosome and the compartment inside is degraded. Small molecules from this digestion can be fed into mitochondria to generate ATP. The onset of autophagy requires energy, but the outcome of autophagy can be beneficial as more ATP may be produced from mitochondria. Nutrient generation by autophagy is generally useful to cells and autophagy only becomes lethal when digestion becomes excessive and most of the cytoplasm is consumed

(Shintani and Klionsky, 2004). Nevertheless, autophagy can lead to a clean form of cell death as cells are broken down from the inside and the extracellular environment is not affected. Apoptosis is also an energy-dependent program to eliminate cells without disturbing the extracellular environment (Danial and Korsmeyer, 2004). Apoptosis is initiated through two primary pathways: the mitochondrial, or intrinsic, and the death receptor, or extrinsic (Danial and Korsmeyer, 2004). The intrinsic death pathway is regulated by Bcl-2 family proteins, and commitment to death occurs when mitochondrial cytochrome c is released. The extrinsic pathway is initiated when death ligands such as tumor necrosis factor α and Fas ligand bind to the cell membrane to initiate death receptor signaling. Both pathways eventually activate a family of cysteine–aspartic acid proteases known as caspases to execute cells. The activation and function of caspases are ATP dependent, thus energy is required for the continuance of apoptosis (Budihardjo et al., 1999). There is a specific form of cell death, eryptosis, in nonnucleate cells such as red blood cells that is apoptosis-like in morphology (Lang et al., 2006). Eryptosis requires caspases as well as Ca^{2+}-dependent channels and formation of ceramide (Lang et al., 2006), all of which can be affected by the status of glucose metabolism and levels of ATP (De Luca et al., 2005; Henquin, 2000). Therefore, eryptosis is also an energy-dependent form of cell death.

The most studied connection between metabolism and cell death is the effect of glucose metabolism on the intrinsic pathway of apoptosis. Bcl-2 family members are key regulators of this mitochondrial death pathway and include antiapoptotic, proapoptotic, and BH3-only proteins. When cells are exposed to stresses such as growth factor deprivation or DNA damage, BH3-only proteins are induced and activated and then they translocate to mitochondria to antagonize the function of antiapoptotic proteins such as Bcl-2, Bcl-xL, and Mcl-1. Proapoptotic proteins, such as Bax and Bak, then undergo conformation change and oligomerize on mitochondria to release cytochrome c. As cellular stresses signal through Bcl-2 family proteins to induce apoptosis, the influence of metabolism on the mitochondria death pathway may be reflected by its impact on Bcl-2 family members. It has been reported that loss of glucose leads to a decreased level of Mcl-1 (Alves et al., 2006; Zhao et al., 2007), induction of BH3-only proteins Noxa and Bim (Alves et al., 2006; Kuan et al., 2003), and Bax activation (Chi et al., 2000; Vander Heiden et al., 2001). These findings strongly suggest that Bcl-2 family members are regulated by glucose metabolism.

A direct interaction between glucose metabolism and apoptosis can also be found in cancer, as both metabolic and apoptotic pathways are altered in cancer cells (Gatenby and Gillies, 2004; Hanahan and Weinberg, 2000; Warburg, 1956) . Glucose metabolism is increased greatly in many cancer cells and glycolysis becomes the main source for ATP production (Gatenby and Gillies, 2004). This phenotype has been appreciated for decades as the

Warburg effect and is often associated with overexpression of glycolytic genes such as Glut1 and HKs (Smith, 2000; Warburg, 1956; Younes et al., 1997). A similar metabolic phenotype can also be found in growth factor-stimulated cells. Upon stimulation, cells undergo hypertrophy and their glucose metabolism increases dramatically to meet the energy demand for proliferation. In addition, in both cancer cells and growth factor–stimulated cells, cell death programs are suppressed to ensure cell survival. There are also reports that the status of glucose metabolism changes prior to cell death when cells are stressed (Byfield et al., 2005; Campbell et al., 2004; Haberkorn et al., 2001; Jones et al., 2005; Morissette et al., 2003; Rathmell et al., 2003a,b; Ward et al., 2007; Wolin et al., 2007; Zhao et al., 2007; Zhou et al., 2002). All these correlations have led to a hypothesis that mutual regulation may exist between these two physiological machineries (Rathmell et al., 2003a; Zhao et al., 2007; Zhou et al., 2002).

Recent findings have supported the notion that glucose metabolism regulates cell death pathways (Fig. 22.1). We have demonstrated that maintenance of glucose metabolism by overexpression of Glut1 and/or HK1 in both cell lines and primary hematopoietic cells stabilizes Mcl-1 and attenuates cell death induced by growth factor withdrawal (Zhao et al., 2007). Increased glucose metabolism protects cells by activating protein kinase C and subsequent inhibition of GSK-3, which, upon activation, promotes the degradation of Mcl-1. It has also been reported that NADPH, which is generated primarily from the PPP, mediates antiapoptotic

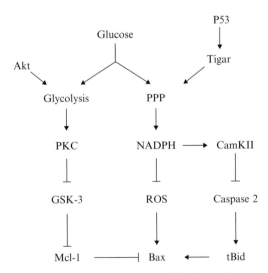

Figure 22.1 Glucose metabolism regulates cell death pathways through multiple mechanisms that culminate in the regulation of proapoptotic Bcl-2 family proteins, such as Bax, to control apoptosis.

function in a variety of conditions. For example, during DNA damage or reactive oxygen species (ROS) burst, p53 is activated to induce protein TIGAR, which lowers the flux of glucose metabolism through glycolysis and increases the relative flux through PPP (Bensaad *et al.*, 2006). As additional NADPH and glutathione are generated, ROS damages are reduced and cells are protected from cell death. Another novel antiapoptotic signal by NADPH was identified in *Xenopus* egg extracts (Nutt *et al.*, 2005). A sufficient level of NADPH is required to maintain CamKII phosphorylation of caspase-2 to inhibit its cleavage and activation. Inhibition of PPP and loss of NADPH lead to increased caspase-2 activity and cytochrome *c* release from mitochondria. In addition, loss of glucose-6-phosphate dehydrogenase (G6PD) has been associated with increased eryptosis, suggesting that PPP is important in maintaining the survival of red blood cells (Lang *et al.*, 2006).

Although the direct contribution from glucose metabolism to evasion of apoptosis in cancer cells is still not clear, it has been found that oncogenes such as Akt utilize signals from glucose metabolism to protect cells from apoptosis (Gottlob *et al.*, 2001; Rathmell *et al.*, 2003b). Akt is a master regulator of glucose metabolism. It maintains surface localization of Glut1 and mitochondrial localization of HKs and increases glucose uptake in cells (Majewski *et al.*, 2004; Wieman *et al.*, 2007). Akt also has a potent antiapoptotic function as it can inhibit Bax conformation change and cytochrome *c* release upon growth factor withdrawal (Rathmell *et al.*, 2003b). Interestingly, when glucose is limited, Akt fails to prevent cell death (Rathmell *et al.*, 2003b). This suggests that Akt needs glucose metabolism to mediate its antiapoptotic function. Other oncogenes, such as Ras and BCR-abl, also have the ability to both upregulate glucose metabolism and prevent cell death (Bentley *et al.*, 2001; Valverde *et al.*, 1998). It will be interesting to determine whether their antiapoptotic function also requires metabolic signals.

Given the importance of glucose metabolism in the regulation of cell death, it is necessary to combine the techniques from both fields and to develop novel methods to monitor the changes of glucose metabolism in stressed cells. Models and approaches to study the effects of glucose metabolism on cell death are needed to establish the role of glucose metabolism in a variety of experimental systems. This chapter describes some approaches to directly test the effect of glucose metabolism on apoptosis.

2. Methods

2.1. Metabolism and forms of cell death

Apoptosis, necrosis, and autophagy each have different metabolic requirements that can influence the cell death processes. At its extreme, necrotic cells simply break down and do not require any ATP for death. In contrast, apoptosis requires ATP-dependent caspase activation via the apoptosome to digest cells from the inside and autophagy depends on lysosomes and ATP-dependent regulation and trafficking of intracellular membranes to degrade cells. Changes in cellular nutrients or metabolic status can, therefore, change cell death pathways. Decreased glucose availability, for example, can switch apoptosis to necrosis when ATP becomes depleted (Leist *et al.*, 1997; Lieberthal *et al.*, 1998).

There are both morphological and biochemical methods to distinguish these cell death pathways. As these models of death were initially described morphologically, the most accurate approach to determine the form of cell death is to determine the morphology of dying cells by electron microscopy (EM). Rupture of cells and organelles, loss of membrane integrity, and no sign of chromatin condensation are characteristic of necrosis. If the morphology of cells appears to be membrane blebbing, nuclear condensation, fragmentation of cells into small membrane-bound vesicles, it indicates that cell death is through apoptosis. When autophagy is initiated, many double membrane vesicles will be observed in cells and organelles are often engulfed in these vesicles. The cell membrane still maintains integrity and no leakage of cytosol into the extracellular environment will be observed. Molecular approaches are also available to help determine the form of cell death. ATP content in dying cells can be measured using a luciferase-based ATP assay kit. Dramatic loss of ATP may suggest that cells do not have sufficient energy to engage ATP-dependent death pathways and may only die by necrosis. An assay of caspase activity will provide information on the executioners of cell death. Increased caspase activity in dying cells suggests that cells die of apoptosis, whereas necrotic cells do not show activity of caspases. Modification of LC3 will indicate autophagy, which can be monitored by altered mobility on SDS-PAGE or a punctuated distribution of LC3–GFP by confocal microscopy. Results from individual approaches, however, may not be conclusive and are most useful when combined or with EM analysis of cell death.

2.2. How to monitor glucose metabolism

When the effect of glucose metabolism on cell death is studied, it is important to know what changes happen to glucose metabolism after cell stresses. After uptake, glucose is phosphorylated by HK1 to become G6P, which can enter either the glycolysis pathway or the PPP. The following protocols can be used to measure the uptake of glucose and flux of glucose to glycolysis. Since only live cells should be used to measure glucose metabolism, these experiments should be done before cells commit to die or live cells need to be sorted before the performance of these experiments.

2.2.1. Glucose uptake

Glucose is essential for the survival of most cells, and glucose homeostasis can be monitored by measuring glucose uptake levels. Levels of glucose uptake are dynamic and regulated, which can be seen as cells display increased glucose uptake during growth and proliferation and decreased glucose uptake when cells atrophy and die. Glucose uptake can be measured by exposing cells to a radiolabeled glucose analog 2-deoxyglucose (2-DOG), which can be taken into cells via facilitative glucose transporters and phosphorylated by hexokinases to glucose-6-phosphate but is not metabolized further.

2.2.1.1. Reagents

Kreb's Ringer HEPES (KRH) buffer: 136 mM NaCl, 4.7 mM KCl, 1.25 mM CaCl$_2$, 1.25 mM MgSO$_4$, 10 mM HEPES, pH 7.4

2-Deoxy-D-[H^3]glucose (2 μCi/rxn)

Phloretin 200 μM

Dow Corning 550 silicon fluid

Dinonyl phthalate

NaOH 1 M

Phosphate-buffered saline (PBS)

2.2.1.2. Protocol

To measure glucose uptake in nonadherent cells, cells are washed once in PBS and resuspended in KRH. 2-Deoxy-D-[H^3]glucose (2 μCi/rxn) is then added for a period of either 1 or 5 min at 37°. The reactions are quenched by the addition of ice-cold 200 μM phloretin and centrifugation through an oil layer, which consists of a 1:1 ratio of Dow Corning 550 silicon fluid and dinonyl phthalate. After centrifugation, the cell pellets are washed with KRH and solubilized in 1 M NaOH, and radioactivity is measured with a scintillation counter. Figure 22.2 demonstrates that levels of glucose uptake can be measured with endogenous levels of glucose transporters in both the presence and the absence of growth factor. The assay also indicates a threefold increase in glucose uptake when FLAG-tagged Glut1 is stably overexpressed. For adherent cells, simply wash cells in PBS, culture in KRH with radiolabeled glucose for 1 or 5 min, rinse three times in PBS, and lyse cells in the dish to determine the amount of radiolabeled glucose internalized. The measurement of glucose uptake, however, does not differentiate between surface and total glucose transporter levels and glucose transporter activity, which are all key factors in the amount of glucose uptake. In order to decipher between these factors, total and surface levels of glucose transporters can be measured.

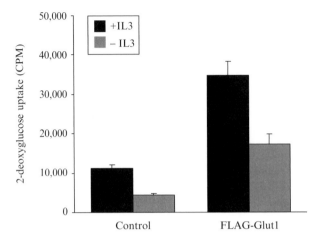

Figure 22.2 Glucose uptake can be raised artificially by the expression of glucose transporters. Glucose uptake was measured in control FL5.12 cells and FL5.12 cells stably expressing FLAG-Glut1 that were deprived growth factor (IL3) for 6 h.

2.2.2. Glut1 protein levels

In addition to measuring glucose uptake, it is important to quantify the total amount of Glut1 in cells. Glut1 is highly hydrophobic, with 12 transmembrane segments, and can form large aggregates upon cell lysis. In addition, Glut1 is glycosylated on an asparagine residue on the first exofacial loop (Asano *et al.*, 1991) that causes even nonaggregated Glut1 to run as a smear on immunoblots. Using a standard cell lysis, 10 min on ice in a RIPA lysis buffer (150 mM NaCl, 10 mM Tris, pH 7.2, 0.1% SDS, 1.0% Triton X-100, 1% deoxycholate, 5 mM EDTA, protease inhibitors) followed by centrifugation for 10 min at 4° to preclear the lysates, and boiling for 10 min in 5 × sample buffer (50% glycerol, 250 mM Tris, pH 6.8, 10% SDS, 0.05% bromophenol blue) led to a Glut1 smear at a high molecular weight by immunoblot (Fig. 22.3, lane 1). Glut1 aggregation can be minimized by using an alternative cell lysis protocol.

2.2.2.1. Reagents

Glut1 lysis buffer: 1% Triton X-100, 0.1% SDS, protease inhibitors
Glut1 5 × sample buffer: 1.56 ml 2 *M* Tris-HCl, pH 6.8, 1 g SDS, 5 ml
    glycerol, 2.5 ml 2-mercaptoethanol, 5 mg bromophenol blue
NP-40
PNGaseF kit (New England Biolabs)

2.2.2.2. Protocol
Cells are lysed in Glut1 lysis buffer for 1 h on ice and then precleared by centrifugation for 10 min at 4°. Glut1 sample buffer is then added at 1 × and samples are incubated at room temperature for

1   2   3

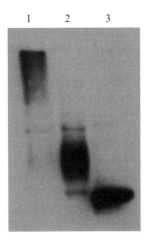

Figure 22.3   The Glut1 protein is sensitive to cell lysis conditions and runs as a smear on SDS-PAGE unless treated properly. Immunoblot of FLAG–Glut1 using cell lysates is treated as follows: lane 1, RIPA lysis buffer (smear around 150 kDa); lane 2, Glut1 lysis buffer (smear around 55 kDa); and lane 3, Glut1 lysis buffer followed by PNGaseF treatment (sharp band at 40 kDa).

30 min (Fig. 22.3, lane 2). To fully collapse the Glut1 band, it is also necessary to remove Glut1 glycosylation by treating the cells with PNGaseF before loading the samples on a gel as seen in Fig. 22.3 (lane 3). Cell lysates that have been prepared via the Glut1 lysis method can be used for PNGaseF treatment. For the treatment as indicated by the manufacturer, dilute 10 μg protein in 9 μl of H_2O, add 1 μl of denaturing buffer, and incubate for 30 min at room temperature. After incubation, add 4 μl of H_2O, 2 μl NP-40, 2 μl G7 reaction buffer, and 2 μl PNGaseF for 1 h at 37° followed by the addition of 1 × Glut1 sample buffer and an additional 30-min incubation at room temperature before loading the gel. The ability to collapse the Glut1 band allows precise quantitation of total Glut1 protein.

2.2.3. Glut1 trafficking

Glut1 must be on the cell surface to transport glucose into the cell, and approaches to measure cell surface Glut1 levels are critical in understanding how cell stresses and death pathways affect glucose uptake. Surface staining of endogenous Glut1 may be accomplished using labeled HTLV-env proteins, which bind Glut1 to serve as a coreceptor for the virus, or anti-Glut1 extracellular domain-specific antibodies. Binding of HTLV-env does not depend solely on Glut1, however, and we have found the alternative approach of using antibodies directed against the extracellular domain of Glut1 to provide poor sensitivity. In addition, while these techniques can be successful for human cells, it is less clear how well such approaches work with murine cells. To bypass these concerns and allow highly specific tracking of Glut1 surface levels and trafficking, we have generated an epitope-tagged version of Glut1. A tandem 2X-FLAG tag is inserted in the first exofacial loop of Glut1, allowing surface levels of Glut1 to be detected via an anti-Flag antibody and flow cytometry analysis. While this approach does not measure changes to endogenous Glut1 trafficking, we have found that FLAG–Glut1 trafficking effectively represents the trafficking pathway of endogenous Glut1 and, therefore, provides an effective tool in determining how Glut1 trafficking is affected by various cell stresses. To measure surface FLAG–Glut1 levels, stably transfected or retrovirally infected cells are washed once in PBS/2% FBS and blocked with anti-Fc γ III/II (to block background staining from Fc receptors in hematopoietic cells) and 5% rat serum for 10 min on ice. Primary rabbit anti-FLAG is added at 1:100 for an additional 20 min. Cells are then washed twice with PBS/2% FBS to wash off the unbound FLAG antibody. Cells are then incubated in PBS/2% FBS, 5% rat serum, and R-PE donkey anti-rabbit at 1:100 for 20 min on ice. The remaining secondary antibody is washed off with PBS/2% FBS and cells are fixed in 1% paraformaldehyde in PBS to allow subsequent analysis of surface levels of FLAG–Glut1 via flow cytometry. Figure 22.4 shows a histogram of the mean surface FLAG–Glut1

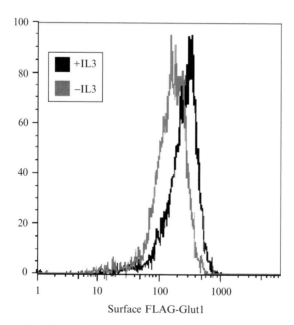

Figure 22.4 Surface Glut1 trafficking can be modeled by expression and analysis of tagged Glut1. Flow cytomeric analysis of surface FLAG–Glut1 levels in FL5.12 cells in the presence of growth factor (IL3) or after 6 h withdrawal from growth factor.

levels of FL5.12 cells in the presence or absence of the cytokine interleukin-3 (IL3). The histogram plot indicates a loss (shift to the left) of surface FLAG–Glut1 when the cytokine is withdrawn from the cells. It is important to note that since the total levels of Glut1 can also be a factor in glucose metabolism, it is necessary to normalize the surface levels of FLAG–Glut1 to total levels of FLAG–Glut1. This can be done by intracellular flow cytometry of parallel samples or by measurement of the pixel density on an immunoblot. Measuring glucose uptake, total Glut1 protein levels, and surface levels of Glut1 together provides a detailed picture of glucose metabolism and glucose transporters in cell stress and death.

2.2.4. Measurement of glycolysis

After glucose uptake, glucose flux through glycolysis can be highly regulated and impact cell fate by the control of cellular ATP levels. When cells undergo death upon stresses such as DNA damage, an increase in glycolysis and ATP levels may favor the activation of caspases and lead to apoptotic cell death (Zamaraeva et al., 2005). Conversely, decreased glycolysis is likely to induce autophagy to supplement cells with ATP from self-digestion, and as a consequence, cells may die of apoptosis and/or autophagy, depending on the extent of autophagy and ATP availability (Lum et al., 2005). Glycolytic flux can be determined by analyzing the conversion of [5-^{3}H]glucose to ^{3}H$_2$O (Vander Heiden et al., 2001). ^{3}H on C5 of glucose is released in the form of H$_2$O at the second to last step of glycolysis, when 2-phosphogycerate is converted to phosphoenolpyruvate by enolase.

2.2.4.1. Reagents

Krebs buffer: 115 mM NaCl, 2 mM KCl, 25 mM NaHCO$_3$, 1 mM MgCl$_2$,
    2 mM CaCl$_2$, 0.25% FBS, pH 7.4
[5-^{3}H]Glucose
HCl 0.2 M

2.2.4.2. Protocol
To measure the rate of glycolytic flux and to determine if cell stresses affect this pathway, cells are washed in PBS once and resuspended in Krebs buffer. After a 30-min incubation, Krebs buffer containing glucose and [5-^{3}H]glucose is added to cells to make final concentration of 10 mM glucose containing [5-^{3}H]glucose (10 μCi/ml) in 0.5 ml and incubated for 1 h at 37°. After 1 h, 0.1 ml of 0.2 M HCl is then added to the mixture to stop the reaction. Then 0.2 ml of the reaction mixture is transferred to a small open tube that is placed in a scintillation vial that contains 0.5 ml water. The scintillation vial is then sealed to allow ^{3}H-H$_2$O to evaporate from the tube and condense in the 0.5 ml water in the bottom of the scintillation vial. In time, ^{3}H$_2$O will establish an equilibrium between the cell lysate and the water in the scintillation vial. In contrast, ^{3}H-glucose

will not evaporate and will remain in the cell lysate. After at least 24 h, ^{3}H in water and the cell lysate are counted to determine the rate of glycolytic flux. The fraction of conversion of glucose to H_2O is calculated as the following:

$$F = \left(\text{activity}_{\text{vial}} / V_{\text{H2O in vial}} * V_{\text{tube}} \right.$$
$$\left. + \text{activity}_{\text{vial}}\right) / \left(\text{activity}_{\text{tube}} + \text{activity}_{\text{vial}}\right)$$

2.2.5. Mitochondrial potential

Mitochondria are key organelles of ATP production. Products from glycolysis such as pyruvate are fed into mitochondria to generate ATP through oxidative phosphorylation, which is driven by transmembrane electrical potential, known as mitochondrial potential. When glucose metabolism is altered, the availability of substrates changes and mitochondrial potential will change accordingly. Cytochrome c release from mitochondria, such as occurs in apoptosis, also decreases mitochondrial potential because cytochrome c is a key component of the electron transportation chain. Thus, mitochondrial potential can be used as an indicator of the status of glucose metabolism, and a decrease in mitochondrial potential suggests a loss of glucose metabolism and a possible onset of apoptosis.

Mitochondrial potential can be measured with fluorescent dyes such as tetramethylrhodamine ethylester (TMRE). There are a number of other dyes that may fulfill this purpose, but TMRE is known to be highly sensitive to small changes in potential that may occur prior to apoptosis. TMRE is a cell-permeable lipophilic cation that can accumulate in mitochondria, and its fluorescence increases in proportion to mitochondrial potential. To stain cells with TMRE, cells are treated by death stimuli over a certain time course and 150 nM of TMRE is added to the culture for 30 min at 37°. For negative staining control, 50 μM of carbonyl cyanide 3-chlorophenylhydrazone (CCCP) is added together with TMRE. CCCP is a protonophore and can diminish mitochondrial potential to show the background staining of TMRE that is not due to mitochondrial potential. After incubation, cells are washed with PBS once and resuspended in PBS containing 2% FBS with 15 nM of TMRE to maintain TMRE equilibrium. Cells are then subject to flow cytometry to analyze the relative intensity of red fluorescence.

2.3. Overexpression of glycolytic genes to increase glucose metabolism

Loss of glucose metabolism may contribute to cell death or simply occur concurrently with cell death. Methods to modify glucose metabolism are needed, therefore, to study the effect of glucose metabolism on cell death. One approach is to promote elevated glucose uptake by expression of a glucose transporter and to determine the effects of this alteration on cell metabolism on cell death. In the absence of this or a similar gain-of-function experiment, it is impossible to fully ascertain the role of loss of metabolism on cell death. In order to increase glucose metabolism without disturbing other signal pathways, we overexpressed glycolytic genes in nontransformed, growth factor-dependent cell lines to boost metabolic pathways (Rathmell et al., 2003b). In particular, Glut1 and HK1 were chosen because they regulate the first rate-limiting steps of the whole glucose metabolism. Glut1 transports glucose down its concentration gradient into cells and HK1 phosphorylates glucose to glucose-6-phosphate, which is the starting material for both glycolysis and the PPP. Glut1 is highly regulated at localization, trafficking, and protein half-life, and the activity of HK1 is sensitive to localization and glucose concentration (Edinger et al., 2003; Rathmell et al., 2000; Wieman et al., 2007; Wilson, 2003). Upon growth factor deprivation, Glut1 is internalized and degraded in lysosomes, leading to decreased glucose uptake (Edinger et al., 2003; Rathmell et al., 2000; Wieman et al., 2007). Overexpression of Glut1 can overwhelm this regulation and, thus, maintain the inflow of glucose under conditions when Glut1 is normally internalized and glucose uptake is reduced (Rathmell et al., 2003b; Zhao et al., 2007). Overexpression of HK1 can "trap" glucose inside cells because phosphorylated glucose cannot be transported out of cells, but can only enter downstream metabolic pathways.

There are several advantages of overexpressing glycolytic genes to increase glucose metabolism compared to simply increasing the concentration of glucose in media, which also can promote glucose uptake. First, glucose uptake rates are often more limited by surface Glut1 levels than the availability of glucose based on the low K_m of Glut1 and the relatively high levels of glucose normally available in vivo. Second, because Glut1/HK1 overexpression does not require a change of glucose concentration to increase glucose uptake, cells will not be affected by increased osmotic pressure. High extracellular glucose can also increase ROS levels in cells, which can affect cell death by inducing proapoptotic BH3-only proteins (Callaghan et al., 2005; Sade and Sarin, 2004). We have observed Glut1/HK1 cells, however, to have lower levels of ROS than control cells (unpublished results). Third, overexpression of Glut1/HK1 reflects a physiological model of high cell intrinsic glucose metabolism, such as occurs in cancer or activated lymphocytes (Frauwirth and Thompson, 2004; Semenza et al., 2001). In contrast, increased extracellular glucose for sustained periods models another pathological condition, diabetes, in which loss of systemic metabolic control may have an important impact on individual cell physiology and fate. Selecting which approach, therefore, may be dictated by the disease process in question, with hyperglycemia leading itself to increased glucose levels and altered signaling of cancer leading itself to overexpression of glucose transporters.

When overexpressing glycolytic genes to promote glucose metabolism, several criteria should be kept in mind. Glucose transporters and HKs are preferred candidates because they determine the first rate-limiting steps in glucose metabolism. Increased G6P from the phosphorylation of glucose by HKs is often sufficient to drive the progress of the pathways of glucose metabolism for a certain period of time. Increased levels of downstream genes, however, may not have a similar effect if the need for increased glucose uptake cannot be met. It is also important to determine which Gluts and HKs to overexpress, as they have different properties in different cells. For example, Glut1 is expressed ubiquitously, has a low K_m and high transport rate, and often localizes to the cell surface. Glut3, however, may localize differently, such as to intracellular compartments, thus making it not suitable for maintaining glucose uptake. HK1 and HK2 have a similar K_m for glucose at 0.05 mM, whereas glucokinase (GK), which is also known as hexokinase 4 and is expressed mainly in the liver, has a much higher K_m for glucose at 5 mM (Wilson, 2003). Because intracellular glucose rarely reaches a concentration of 5 mM, the physiologic concentration of glucose in blood, GK will have very low activity in normal culture conditions and is not ideal for catalyzing glucose phosphorylation in cell culture. The role that specific glucose transporters may play in specific pathological conditions is unclear, but the different cell surface trafficking patterns of different transporters may profoundly influence the overall cellular glucose uptake and cell death pattern.

2.4. Inhibition of glucose metabolism

Decreasing glucose metabolism can be an alternative approach to study effects of glucose metabolism on cell death. There are several ways to limit glucose metabolism, the most direct method being decreasing the concentration of glucose in culture or even depleting glucose from media. The concentration of glucose ranges from 3 to 5 mM physiologically in normal blood, but may go much lower in tissues or poorly vascularized areas. It has been suggested that cells can behave normally when the concentration of glucose goes down to 0.1 mM (Vander Heiden *et al.*, 2001). When the glucose concentration is lower than 0.1 mM, substrates for glucose metabolism become limited and cell growth is slowed. Using a glucose analog to replace glucose in media is another method used to limit glucose metabolism. 2-DOG is often used as an analog of glucose and can be taken up by cells through glucose transporters. However, after phosphorylation by HK, 2-DOG cannot be metabolized further by glycolysis. Phospho-2-DOG then accumulates, resulting in feedback inhibition of HKs. Thus, 2-DOG limits overall glucose metabolism. Despite its inability to progress through glycolysis, it should be noted that 2-DOG may be metabolized through a few steps in the PPP (Mourrieras *et al.*, 1997), providing a limited amount of NADPH for cells. The few NADPH generated by metabolizing 2-DOG may provide a survival advantage for cells, as NADPH may inhibit caspase-2 activation and ROS accumulation. It should also be noted that low glucose or 2-DOG can also induce an unfolded protein response via hypoglycosylation of proteins in the ER. Apoptosis can be induced by ER stress through upregulation and activation of BH3-only proteins (Puthalakath et al., 2007; Reimertz et al., 2003). Thus, when studying effects of low glucose on cell death, contributions from ER stress should be considered.

Chemical inhibitors can also be used in cell culture to inhibit glucose metabolism. Trifluoromethoxy carbonyl cyanide phenylhydrazone (FCCP) is a protonophore and can uncouple mitochondrial oxidative phosphorylation. Use of FCCP will decrease mitochondrial potential and ATP production, and increase ROS generation in cells. Glycolysis and PPP may still go on as they do not proceed in mitochondria. Dehydroepiandrosterone (DHEA) and 6-aminonicotinamide (6-AN) are inhibitors of G6PD, which catalyzes the first step of the PPP. Inhibition of G6PD with DHEA or 6-AN stops the PPP, thus limiting the generation of NADPH and ribosugar for the synthesis of DNA and RNA. Iodoacetate can be added to the culture to stop glycolytic flux, as it inhibits the function of glyceraldehyde-3-phosphate dehydrogenase, a critical enzyme in glycolysis. The advantages of using chemical inhibitors are easy access, low cost, and relatively fast screening. The general concern, however, is their specificity, and the results from chemical inhibitors can be misleading. Thus, data from inhibitors are not conclusive and should be used only as supportive evidence. Ultimately, it will be necessary to confirm the result of chemical inhibitors by downregulation of target genes by RNAi.

2.5. Bcl-2 markers for glucose metabolism

Ultimately, cell death occurs via the Bcl-2 family of proteins when glucose metabolism is prevented. Several Bcl-2 family members have been reported to regulate cell death in response to changes in metabolic status and their levels can be analyzed when cells are exposed to stresses that may affect metabolism. Bax changes it conformation and becomes activated when cells are deprived of glucose (Chi et al., 2000; Vander Heiden et al., 2001). To monitor the conformation change of Bax, a conformation-specific antibody (6A7) that only recognizes active Bax can be used to immunoprecipitate Bax and the precipitates can be subjected to immunoblot to determine the levels of active Bax. Alternatively, stressed cells can be fixed and immunostained with the antibody, and active Bax can be analyzed by flow cytometry. Levels of Mcl-1 decrease when glucose becomes limited, whereas increased glucose metabolism protects Mcl-1 from degradation (Alves et al., 2006; Zhao et al., 2007). In BH3-only proteins, Noxa has been reported to be induced and initiate cell death in T cells upon loss of glucose (Alves *et al.*, 2006). Thus, Mcl-1 and Noxa can be analyzed by immunoblot to determine the effect of glucose metabolism on cell death.

3. Conclusion

This chapter described general approaches that can be applied to the study of glucose metabolism and cell death. With these methods, glucose metabolism in cells can be increased or decreased to analyze its effect on cell death programs. Glucose metabolism can also be monitored to provide information on metabolic changes in response to cell stresses. These protocols should be subject to modifications to fit different experimental settings and data from them should be interpreted in combination with information on forms of cell death. In particular, there is a need for single cell assays and biochemical approaches that utilize minimal material to allow more accurate analysis of small sample size. As more and more researchers begin to focus on metabolic regulation of cell death, this field is expanding quickly and new techniques and methods are expected to emerge to meet the growing need for accurate and easy analysis on both glucose metabolism and cell death.

ACKNOWLEDGMENTS

We thank members of the Rathmell laboratory for their technical and scientific contribution to the development of this manuscript. This work was funded by National Cancer Institute R01CA123350.

REFERENCES

Alves, N. L., Derks, I. A., Berk, E., Spijker, R., van Lier, R. A., and Eldering, E. (2006). The Noxa/Mcl-1 axis regulates susceptibility to apoptosis under glucose limitation in dividing T cells. *Immunity* **24,** 703–716.

Asano, T., Katagiri, H., Takata, K., Lin, J. L., Ishihara, H., Inukai, K., Tsukuda, K., Kikuchi, M., Hirano, H., Yazaki, Y., and Oka, Y. (1991). The role of N-glycosylation of GLUT1 for glucose transport activity. *J. Biol. Chem.* **266,** 24632–24636.

Bensaad, K., Tsuruta, A., Selak, M. A., Vidal, M. N., Nakano, K., Bartrons, R., Gottlieb, E., and Vousden, K. H. (2006). TIGAR, a p53-inducible regulator of glycolysis and apoptosis. *Cell* **126,** 107–120.

Bentley, J., Walker, I., McIntosh, E., Whetton, A. D., Owen-Lynch, P. J., and Baldwin, S. A. (2001). Glucose transport regulation by p210 Bcr-Abl in a chronic myeloid leukaemia model. *Br. J. Haematol.* **112,** 212–215.

Budihardjo, I., Oliver, H., Lutter, M., Luo, X., and Wang, X. (1999). Biochemical pathways of caspase activation during apoptosis. *Annu. Rev. Cell Dev. Biol.* **15,** 269–290.

Byfield, M. P., Murray, J. T., and Backer, J. M. (2005). hVps34 is a nutrient-regulated lipid kinase required for activation of p70 S6 kinase. *J. Biol. Chem.* **280,** 33076–33082.

Callaghan, M. J., Ceradini, D. J., and Gurtner, G. C. (2005). Hyperglycemia-induced reactive oxygen species and impaired endothelial progenitor cell function. *Antioxid. Redox Signal* **7,** 1476–1482.

Campbell, M., Allen, W. E., Sawyer, C., Vanhaesebroeck, B., and Trimble, E. R. (2004). Glucose-potentiated chemotaxis in human vascular smooth muscle is dependent on cross-talk between the PI3K and MAPK signaling pathways. *Circ. Res.* **95,** 380–388.

Chi, M. M., Pingsterhaus, J., Carayannopoulos, M., and Moley, K. H. (2000). Decreased glucose transporter expression triggers BAX-dependent apoptosis in the murine blastocyst. *J. Biol. Chem.* **275,** 40252–40257.

Danial, N. N., and Korsmeyer, S. J. (2004). Cell death: Critical control points. *Cell* **116,** 205–219.

De Luca, T., Morre, D. M., Zhao, H., and Morre, D. J. (2005). NAD+/NADH and/or CoQ/CoQH2 ratios from plasma membrane electron transport may determine ceramide and sphingosine-1-phosphate levels accompanying G1 arrest and apoptosis. *Biofactors* **25,** 43–60.

Edinger, A. L., and Thompson, C. B. (2004). Death by design: Apoptosis, necrosis and autophagy. *Curr. Opin. Cell Biol.* **16,** 663–669.

Edinger, A. L., Cinalli, R. M., and Thompson, C. B. (2003). Rab7 prevents growth factor-independent survival by inhibiting cell-autonomous nutrient transporter expression. *Dev. Cell* **5,** 571–582.

Frauwirth, K. A., and Thompson, C. B. (2004). Regulation of T lymphocyte metabolism. *J. Immunol.* **172,** 4661–4665.

Gatenby, R. A., and Gillies, R. J. (2004). Why do cancers have high aerobic glycolysis? Nat. *Rev. Cancer* **4,** 891–899.

Gottlob, K., Majewski, N., Kennedy, S., Kandel, E., Robey, R. B., and Hay, N. (2001). Inhibition of early apoptotic events by Akt/PKB is dependent on the first committed step of glycolysis and mitochondrial hexokinase. *Genes Dev.* **15,** 1406–1418.

Haberkorn, U., Altmann, A., Kamencic, H., Morr, I., Traut, U., Henze, M., Jiang, S., Metz, J., and Kinscherf, R. (2001). Glucose transport and apoptosis after gene therapy with HSV thymidine kinase. *Eur. J. Nucl. Med.* **28,** 1690–1696.

Hanahan, D., and Weinberg, R. A. (2000). The hallmarks of cancer. *Cell* **100,** 57–70.

Henquin, J. C. (2000). Triggering and amplifying pathways of regulation of insulin secretion by glucose. *Diabetes* **49,** 1751–1760.

Jones, R. G., Plas, D. R., Kubek, S., Buzzai, M., Mu, J., Xu, Y., Birnbaum, M. J., and Thompson, C. B. (2005). AMP-activated protein kinase induces a p53-dependent metabolic checkpoint. *Mol. Cell* **18,** 283–293.

Kuan, C. Y., Whitmarsh, A. J., Yang, D. D., Liao, G., Schloemer, A. J., Dong, C., Bao, J., Banasiak, K. J., Haddad, G. G., Flavell, R. A., Davis, R. J., and Rakic, P. (2003). Absence of excitotoxicity-induced apoptosis in the hippocampus of mice lacking the Jnk3 gene. *Proc. Natl. Acad. Sci. USA* **100,** 15184–15189.

Lang, F., Lang, K. S., Lang, P. A., Huber, S. M., and Wieder, T. (2006). Mechanisms and significance of eryptosis. *Antioxid. Redox Signal* **8,** 1183–1192.

Leist, M., Single, B., Castoldi, A. F., Kuhnle, S., and Nicotera, P. (1997). Intracellular adenosine triphosphate (ATP) concentration: A switch in the decision between apoptosis and necrosis. *J. Exp. Med.* **185,** 1481–1486.

Lieberthal, W., Menza, S. A., and Levine, J. S. (1998). Graded ATP depletion can cause necrosis or apoptosis of cultured mouse proximal tubular cells. *Am. J. Physiol.* **274,** F315–F327.

Lum, J. J., Bauer, D. E., Kong, M., Harris, M. H., Li, C., Lindsten, T., and Thompson, C. B. (2005). Growth factor regulation of autophagy and cell survival in the absence of apoptosis. *Cell* **120,** 237–248.

Lum, J. J., DeBerardinis, R. J., and Thompson, C. B. (2005). Autophagy in metazoans: Cell survival in the land of plenty. *Nat. Rev. Mol. Cell Biol.* **6**, 439–448.

Majewski, N., Nogueira, V., Bhaskar, P., Coy, P. E., Skeen, J. E., Gottlob, K., Chandel, N. S., Thompson, C. B., Robey, R. B., and Hay, N. (2004). Hexokinase-mitochondria interaction mediated by Akt is required to inhibit apoptosis in the presence or absence of Bax and Bak. *Mol. Cell* **16**, 819–830.

Morissette, M. R., Howes, A. L., Zhang, T., and Heller Brown, J. (2003). Upregulation of GLUT1 expression is necessary for hypertrophy and survival of neonatal rat cardiomyo-cytes. *J. Mol. Cell. Cardiol.* **35**, 1217–1227.

Mourrieras, F., Foufelle, F., Foretz, M., Morin, J., Bouche, S., and Ferre, P. (1997). Induction of fatty acid synthase and S14 gene expression by glucose, xylitol and dihy-droxyacetone in cultured rat hepatocytes is closely correlated with glucose 6-phosphate concentrations. *Biochem. J.* **326**(Pt 2), 345–349.

Nutt, L. K., Margolis, S. S., Jensen, M., Herman, C. E., Dunphy, W. G., Rathmell, J. C., and Kornbluth, S. (2005). Metabolic regulation of oocyte cell death through the CaM-KII-mediated phosphorylation of caspase-2. *Cell* **123**, 89–103.

Puthalakath, H., O'Reilly, L. A., Gunn, P., Lee, L., Kelly, P. N., Huntington, N. D., Hughes, P. D., Michalak, E. M., McKimm-Breschkin, J., Motoyama, N., Gotoh, T., Akira, S., *et al.* (2007). ER stress triggers apoptosis by activating BH3-only protein Bim. *Cell* **129**, 1337–1349.

Rathmell, J. C., Elstrom, R. L., Cinalli, R. M., and Thompson, C. B. (2003a). Activated Akt promotes increased resting T cell size, CD28-independent T cell growth, and development of autoimmunity and lymphoma. *Eur. J. Immunol.* **33**, 2223–2232.

Rathmell, J. C., Fox, C. J., Plas, D. R., Hammerman, P., Cinalli, R. M., and Thompson, C. B. (2003b). Akt-directed glucose metabolism can prevent Bax conforma-tion change and promote growth factor-independent survival. *Mol. Cell. Biol.* **23**, 7315–7328.

Rathmell, J. C., Vander Heiden, M. G., Harris, M. H., Frauwirth, K. A., and Thompson, C. B. (2000). In the absence of extrinsic signals, nutrient utilization by lymphocytes is insufficient to maintain either cell size or viability. *Mol. Cell* **6**, 683–692.

Reimertz, C., Kogel, D., Rami, A., Chittenden, T., and Prehn, J. H. (2003). Gene expression during ER stress-induced apoptosis in neurons: Induction of the BH3-only protein Bbc3/PUMA and activation of the mitochondrial apoptosis pathway. *J. Cell Biol.* **162**, 587–597.

Sade, H., and Sarin, A. (2004). Reactive oxygen species regulate quiescent T-cell apoptosis via the BH3-only proapoptotic protein BIM. *Cell Death Differ.* **11**, 416–423.

Semenza, G. L., Artemov, D., Bedi, A., Bhujwalla, Z., Chiles, K., Feldser, D., Laughner, E., Ravi, R., Simons, J., Taghavi, P., and Zhong, H. (2001). 'The metabolism of tumours': 70 years later. *Novartis Found Symp* **240**, 251–260; discussion 60–64.

Shintani, T., and Klionsky, D. J. (2004). Autophagy in health and disease: A double-edged sword. *Science* **306**, 990–995.

Smith, T. A. (2000). Mammalian hexokinases and their abnormal expression in cancer. *Br. J. Biomed. Sci.* **57**, 170–178.

Valverde, A. M., Navarro, P., Benito, M., and Lorenzo, M. (1998). H-ras induces glucose uptake in brown adipocytes in an insulin- and phosphatidylinositol 3-kinase-independent manner. *Exp. Cell Res.* **243**, 274–281.

Vander Heiden, M. G., Plas, D. R., Rathmell, J. C., Fox, C. J., Harris, M. H., and Thompson, C. B. (2001). Growth factors can influence cell growth and survival through effects on glucose metabolism. *Mol. Cell. Biol.* **21**, 5899–5912.

Warburg, O. (1956). On the origin of cancer cells. *Science* **123**, 309–314.

Ward, M. W., Huber, H. J., Weisova, P., Dussmann, H., Nicholls, D. G., and Prehn, J. H. (2007). Mitochondrial and plasma membrane potential of cultured cerebellar neurons during glutamate-induced necrosis, apoptosis, and tolerance. *J. Neurosci.* **27**, 8238–8249.

Wieman, H. L., Wofford, J. A., and Rathmell, J. C. (2007). Cytokine stimulation promotes glucose uptake via phosphatidylinositol-3 kinase/Akt regulation of Glut1 activity and trafficking. *Mol. Biol. Cell* **18,** 1437–1446.

Wilson, J. E. (2003). Isozymes of mammalian hexokinase: Structure, subcellular localization and metabolic function. *J. Exp. Biol.* **206,** 2049–2057.

Wolin, M. S., Ahmad, M., Gao, Q., and Gupte, S. A. (2007). Cytosolic NAD(P)H regulation of redox signaling and vascular oxygen sensing. *Antioxid. Redox Signal* **9,** 671–678.

Younes, M., Brown, R. W., Stephenson, M., Gondo, M., and Cagle, P. T. (1997). Over-expression of Glut1 and Glut3 in stage I nonsmall cell lung carcinoma is associated with poor survival. *Cancer* **80,** 1046–1051.

Zamaraeva, M. V., Sabirov, R. Z., Maeno, E., Ando-Akatsuka, Y., Bessonova, S. V., and Okada, Y. (2005). Cells die with increased cytosolic ATP during apoptosis: A bioluminescence study with intracellular luciferase. *Cell Death Differ* **12,** 1390–1397.

Zhao, Y., Altman, B. J., Coloff, J. L., Herman, C. E., Jacobs, S. R., Wieman, H. L., Wofford, J. A., Dimascio, L. N., Ilkayeva, O., Kelekar, A., Reya, T., and Rathmell, J. C. (2007). Glycogen synthase kinase 3alpha and 3beta mediate a glucose-sensitive antiapoptotic signaling pathway to stabilize Mcl-1. *Mol. Cell. Biol.* **27,** 4328–4339.

Zhou, R., Vander Heiden, M. G., and Rudin, C. M. (2002). Genotoxic exposure is associated with alterations in glucose uptake and metabolism. *Cancer Res.* **62,** 3515–3520.

Author Index

Subject Index

A

AAD, *see* 7-Amino actinomycin D

Acridine orange, lysosomal membrane
permeabilization detection, 186–188

Activation-induced cell death, flow cytometry
assays in human immunodeficiency
virus, 69–70

AICD, *see* Activation-induced cell death

7-Amino actinomycin D, flow cytometry assays
cell-mediated cytotoxicity assay, 78
plasma membrane integrity, 71–74

Annexin-V
Bid function studies, 247–248
flow cytometry assays of apoptosis, 68–70,
318, 320–321
granzyme-mediated apoptosis assay, 228–229

Apaf-1, *see* Apoptosome

APO-1, *see* CD95 death-inducing
signaling complex

Apoptosis
definition, 3
glucose metabolism, 440–441, 443
morphologic characteristics, 4
nucleases, *see* Caspase-activated DNase;
DNase II
pathways, 53
supernatant analysis, 327–330

Apoptosome
caspase-9 activation apoptosome
Apaf-1 structure and function, 144–146
assembly, 146–147
detection and assembly assays, 148–149
functional models, 147–148
history of study, 144
Ced-4 apoptosome
Ced-3 activation pathway, 150
detection and assembly assays, 151–152
structure, 150–151
Dronc activation apoptosome assembly
analysis, 149–150
types, 143

ASC pyroptosome
activity assay, 267–268
ASC–green fluorescence protein construct
confocal time-lapse bioimaging, 257–258
retroviral transduction, 255–257
vector construction, 254–255
assembly and purification

ASC–green fluorescence protein-
containing pyroptosomes, 261–263
endogenous pyroptosomes, 263–264
recombinant pyroptosomes
assembly from occlusion bodies, 266–267
expression in bacteria, 265
purification of occlusion bodies, 265–266
biochemical assay of formation, 258–260
caspase-1 activation, 252–253
cross-linking studies, 260–261

ATF4
polymerase chain reaction of messenger
RNA, 409–410
Western blot, 411

ATF6, unfolded protein response role, 397–398

Autophagy
Caenorhabditis elegans studies using RNA
interference and mutation, 303–304
definition, 3, 290, 309
electron microscopy studies
Caenorhabditis elegans, 294
cells, 293–294
overview, 292–293
tissues and embryos, 293
glucose metabolism, 440–441, 443
LC3 marker of autophagosome formation
green fluorescent protein fusion for
translocation visualization
Caenorhabditis elegans, 302–303
cells, 302
immunohistochemistry of
cleavage, 301–302
overview, 300–301
Western blot of cleavage, 301
lysosomal enzyme assays
biochemical assays, 296
inhibition effects on autophagy, 304
overview, 294–295
sample preparation for
immunohistochemistry, 295–296
in situ measurement, 295
lysosome analysis with fluorescence
microscopy
cell culture, 299
dyes, 297–298
Lysotracker red, 300
monodansylcadaverine, 299
zebrafish studies, 299–300
morphologic characteristics, 4

485

O

Osmium tetroxide, fixation of electron
    microscopy specimens, 15–16

P

Palmitoylation, *see* CD95 death-inducing
    signaling complex
PARP, *see* Poly(ADP)ribose polymerase
PCR, *see* Polymerase chain reaction
Perforin
    activity detection
        fluorescence microscopy, 219, 221
        hemoglobin release assay, 219
    granzyme release mediation, 214
    purification from rat leukemia cell line
        cell growth in rats, 215–216
        cytotoxic granule isolation, 216–217
        immobilized metal affinity
            chromatography, 217–219
    titration for granzyme-mediated apoptosis
        assay, 227
PERK
    endoplasmic reticulum stress–induced
        apoptosis pathway, 398–399
    phosphorylation analysis, 403–405
    unfolded protein response role, 397
Phagocytosis
    flow cytometry assay, 331–332
    scanning electron microscopy of surface
        changes, 332–334
Phosphatidylserine
    annexin-V assays, *see* Annexin-V
    hydroperoxides, *see* Phospholipid oxidation,
        apoptosis
Phospholipid oxidation, apoptosis
    lipidomics
        cardiolipin hydroperoxide identification
            model system, 379–382
            in vivo studies, 382–385
        high-performance liquid
            chromatography, 379
        lipid extraction, 378
        mass spectrometry, 378–389
        materials, 377–378
        mouse total body irradiation, 378
        phosphatidylserine hydroperoxide
            identification
            model system, 385–387
            in vivo studies, 388–389
        two-dimensional high-performance
            thin-layer chromatography, 378
    overview, 376–377
Poly (ADP) ribose polymerase, caspase cleavage
    assay, 176–177
Polymerase chain reaction

Atf4 messenger RNA analysis in endoplasmic
    reticulum stress–induced
    apoptosis, 409–410
Chop messenger RNA analysis in endoplasmic
    reticulum stress–induced
    apoptosis, 409–410
GRP78/BiP induction analysis in unfolded
    protein response, 407–408
Xbp1 messenger RNA splicing analysis in
    unfolded protein response, 405–407
Propidium iodide, granzyme-mediated apoptosis
    assay, 228–229
Proteomics, *see* Mass spectrometry;
    Two-dimensional gel electrophoresis
Pyroptosis
    definition, 252
    pyroptosome, *see* ASC pyroptosome

R

RNA interference, autophagy studies in
    Caenorhabditis elegans, 303–304

S

Sucrose density gradient centrifugation, CD95
    death-inducing signaling complex
    advantages, 91
    calibration of gradients, 90
    equipment, 85–86
    fraction harvesting, 90
    gradient setup, 88–89
    materials, 87–88
    postnuclear lysate preparation, 89
    principles, 85–86
    ultracentrifugation, 89
    Western blot, 90

T

Terminal deoxynucleotidyl transferase-mediated
    duTP nick end labeling
    cytospin preparation, 10–11
    endoplasmic reticulum stress–induced
        apoptosis analysis, 413–415
    flow cytometry assays, 66–67
    principles, 5, 10
Tetramethylrhodamine ethyl ester, mitochondrial
    membrane potential single-cell
    measurement, 29–31
TG2, *see* Transglutaminase type 2
Thin-layer chromatography
    lipid raft coimmunoprecipitation
        analysis, 132
    phospholipid oxidation lipidomics in
        apoptosis with two-dimensional
        high-performance thin-layer
        chromatography, 378

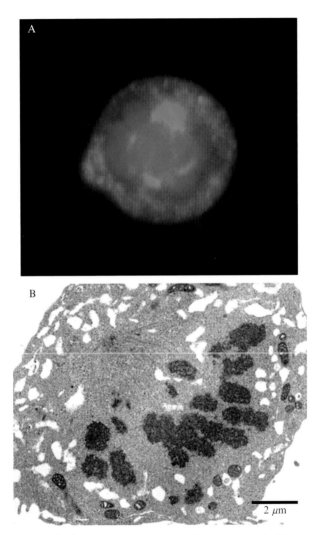

Antonella Tinari *et al.*, Figure 1.5   Micrographs illustrating typical features of mitotic catastrophe as detected by fluorescence microscopy (A) and electron microscopy (B). (A) Hoechst staining (in blue) indicates the chromatin derangement, whereas microtubules are labeled in green. (B) Note signs of defective arrangement of chromosomes and cell damage.

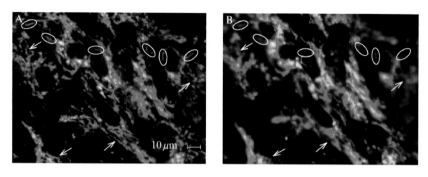

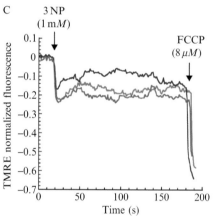

Soraya S. Smaili _et al._, Figure 2.1 3NP induces a decrease in $\Delta\Psi_m$. Astrocyte cultures from B6xCBA/F1 mice loaded with TMRE (50 nM) for 15 min, before (A) and after (B) the addition of 1 mM 3NP. Images were acquired using a CCD camera and 40× oil immersion objective with a delay of 6 s. Circles indicate mitochondria that lost $\Delta\Psi_m$, and arrows are regions that previously presented high density of mitochondria that went out of focus due to the increase in TMRE in the cytosol. (C) $\Delta\Psi_m$ fluorescence traces extracted from digital images and during stimulation of cells with 3NP followed by FCCP (8 μM). Traces presented were obtained from three different mitochondria. Similar measurements were made in five different experiments, and results were compiled with the percentage of mitochondria showing decrease, increase, or no change in $\Delta\Psi_m$ (adapted from Rosenstock _et al._, 2004).

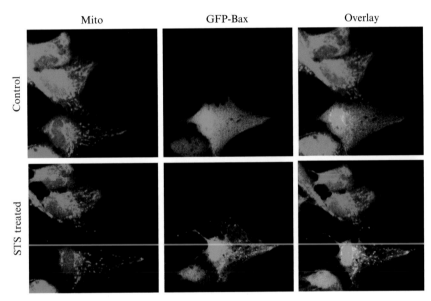

Soraya S. Smaili *et al.*, **Figure 2.2**   Bax translocation during apoptosis induction. Primary cultured astrocytes transfected previously with GFP–Bax were treated with 500 n*M* staurosporine (STS), loaded with Mitotracker red CMXRos (MTR, 20 n*M*), and tracked over time by confocal microscopy for 4 h. The GFP–Bax overexpressing cell displayed a diffused pattern, indicative of its cytosolic location. Upon apoptosis induction with STS, GFP–Bax shifted from a diffuse to a punctate pattern that mainly colocalized with mitochondria (MTR staining) as shown in the overlay panels.

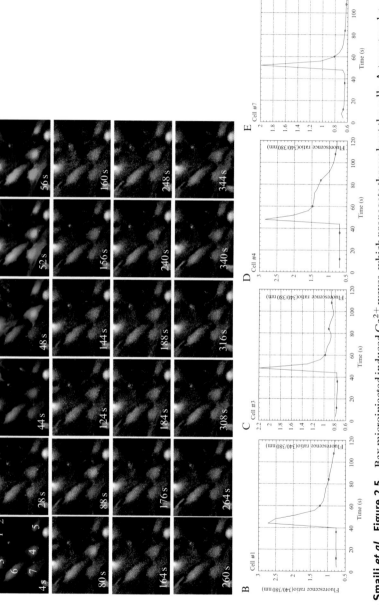

Soraya S. Smaili et al., Figure 2.5 Bax microinjected induced Ca²⁺ waves, which propagate throughout the cells. Astrocytes plated on coverslips were loaded with Fura-2 AM (10 μM) in the microscopy buffer. Cytosolic Ca²⁺ levels in isolated cells were analyzed using high-resolution digital microscopy with an inverted microscope coupled to a cooled CCD camera and controlled by a computer software. For each experiment, one cell in the field was used for microinjection of recombinant Bax (rBax, 10 ng/ml) that was injected in bolus. (A) Images were collected at 3-s intervals and rBax was injected in cell #1 at 40 s when changes in fluorescence intensity were observed (red pseudocolor represents an increase in the 340/380 ratio, indicating an elevation in cytosolic Ca²⁺). (B) Graphs show that the increase in cytosolic Ca²⁺ after the rBax injection occurred not only in injected cell #1, but also in adjacent cells #2 (C), #3 (D), and #4 (E) at different intervals. Cells showed a transient peak, which reached a maximum response after 2 to 4 s from the beginning of the effect (adapted from Carvalho et al., 2004).

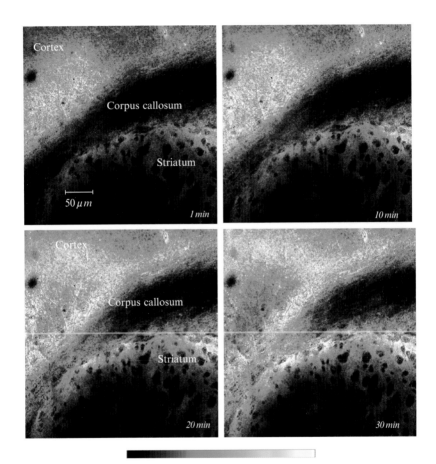

Soraya S. Smaili *et al.*, Figure 2.6 Glutamate induced an increase in Ca^{2+} in brain slices. A brain slice (200 μm thickness) from a 2-month-old B6xCBA/F1 male mouse was loaded with calcium dye, Fluo-3 AM. Images acquired by confocal microscope showed few details of the brain structures of interest (cortex and striatum). An increase in cytosolic Ca^{2+} occurred a few minutes after glutamate addition. In addition, there was a Ca^{2+} wave from the cortex and striatum to the corpus callosum direction. The bar represents the scale of fluorescence intensities (0–255). The colors were attributed.

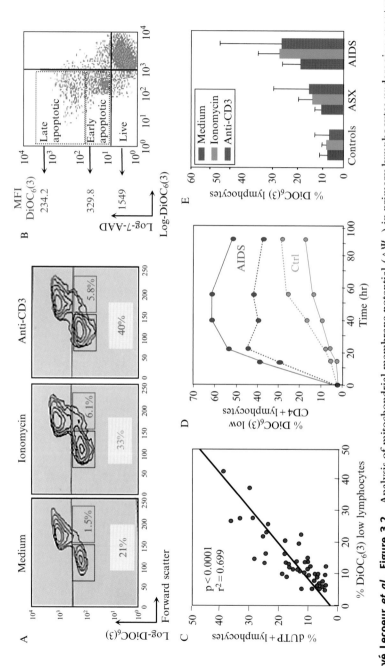

Hervé Lecoeur et al., Figure 3.2 Analysis of mitochondrial membrane potential ($\triangle\Psi_m$) in primary lymphocytes undergoing apoptosis. (A) Measure of $\triangle\Psi_m$ collapse in PBMC from an AIDS patient, cultured overnight in medium or stimulated with ionomycin or anti-CD3 mAbs. Biparametric FSC/DiOC$_6$(3) density plot are shown. The percentage of DiOC$_6$(3) low cells is indicated. Red boxes correspond to the fraction of lymphocytes that dropped $\Delta\Psi_m$ but did not show any cell shrinkage (first apoptosis step). Black boxes correspond to cells more engaged toward the apoptotic process, with both a collapsed $\Delta\Psi_m$ and cell shrinkage. (B) Combined analysis of $\Delta\Psi_m$ collapse and loss of PM integrity in PBMC from an AIDS patient stimulated overnight with ionomycin. A biparametric $DiOC_6(3)$/7-AAD dot plot is presented. Early and late apoptotic cells are defined according to the extent of 7-AAD staining (7-AAD^{Lo} and 7-AAD^{Hi}, respectively). The mean fluorescence intensity (MFI) of the $DiOC_6(3)$ staining is indicated in blue. (C) Correlation between the percentage of $DiOC_6(3)^{Lo}$ and dUTP$^+$ (i.e., with fragmented DNA, TUNEL assay) in CD4$^+$ T cells from an AIDS patient cultured overnight in medium. A combined CD4 staining was performed for each assay, and apoptosis was determined on gated CD4$^+$ Tcells. (D) Kinetics of apoptosis induction in PBMC from a control donor (Ctrl, green lines) and an AIDS patient (brown lines). PBMC were cultured in medium (plain lines) or in the presence of anti-CD3 mAbs (red line). A dual CD4/$DiOC_6(3)$ staining was performed. CD4$^+$ cells were gated, and $\Delta\Psi_m$ collapse was analyzed in this subset. (E) Quantification of $DiOC_6(3)^{Lo}$ cells in CD4$^+$ lymphocytes from controls ($n = 10$), ASX ($n = 37$), and AIDS ($n = 13$) subjects following overnight culture in medium, ionomycin, or anti-CD3 mAbs. Mean and standard deviations are shown.

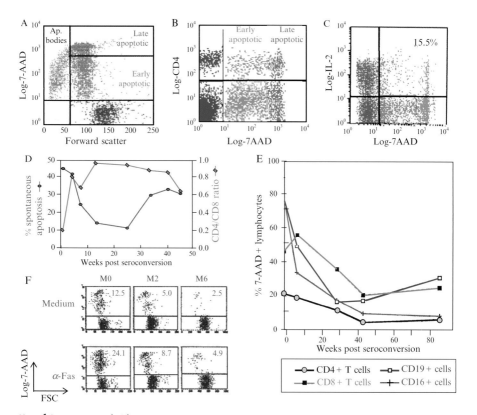

Hervé Lecoeur *et al.*, Figure 3.5 Detection of the loss of plasma membrane integrity in apoptotic lymphocytes by the 7-AAD assay. (A) Biparametric 7-AAD/FSC dot plot of PBMC from an AIDS patient undergoing spontaneous apoptosis (48-h culture). Early apoptotic (green dots), late apoptotic (red dots) lymphocytes, and apoptotic bodies/debris (orange dots) can be discriminated easily from live lymphocytes (blue dots) according to combined cell size and 7-AAD staining. (B) Biparametric 7-AAD/CD4 dot plot PBMC from an AIDS patient undergoing spontaneous apoptosis (24-h culture). Apoptotic bodies/cell debris were discarded from the analysis. Early and late apoptotic lymphocytes were discriminated easily through the intensity of 7-AAD staining. (C) Triple staining CD3/IL-2 /7-AAD of PBMC from an HIV[+] subject stimulated overnight in the presence of PMA (50 ng/ml), ionomycin (300 ng/ml), and PHA-A (100 ng/ml). Brefeldin A (10 μg/ml) was added during the last 12 h of culture. Analysis was performed on gated CD3 Tcells. The percentage of apoptosis in IL-2-secreting CD3[+] Tcells is indicated. (D) Kinetic of spontaneous apoptosis and CD4/CD8 ratio in a subject following HIV-specific seroconversion. Spontaneous apoptosis (24-h culture) was quantified in total PBMC with the 7-AAD assay (red line), the CD4/CD8 ratio was determined *ex vivo* by flow cytometry (green line). (E) Kinetic of spontaneous apoptosis in different lymphocyte subsets from the same donor. (F) Spontaneous and Fas-induced apoptosis in CD4[+] Tcells from an HIV-infected patient at baseline and 2 and 6 months following initiation of antiretroviral therapy. FSC/7-AAD dot plots on gated CD4[+] Tcells are shown, and the percentage of apoptotic CD4 Tcells is indicated.

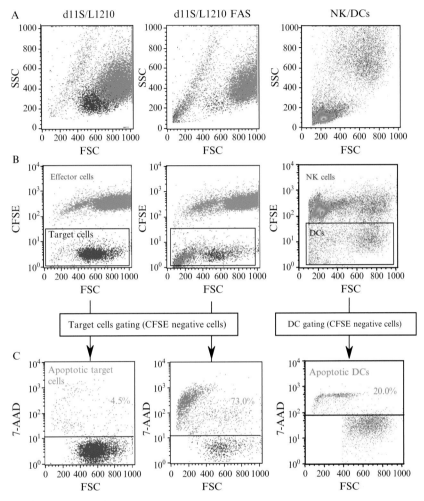

Hervé Lecoeur *et al.*, Figure 3.6 Analysis of cell-mediated cytotoxicity by the cyto-fluorimetric CFSE/7-AAD assay. (A– C) Cytotoxicity assay between d11s effector and L1210 or L1210Fas target cells. (A) Biparametric FSC/SSC dot plots of the cell coculture (4-h contact) corresponding to an E:T ratio of 7:1. Effectors (CFSE-stained d11s cells) appear in green, live target cells in blue, and apoptotic target cells in red (7-AAD positive cells). (B) Biparametric FSC/CFSE dot plots of the coculture. Target cells (CFSE nega-tive) can be specifically gated. (C) Biparametric FSC/7-AAD dot plots of selected target cells. Target cell lysis is quantified by the percentage of 7-AAD positive cells (red dots). (D–F) Cytotoxicity assay between human NK cells and DC. After CFSE staining activated NK cells have been cocultured with DC. Gated DC (CFSE negative) showed 20% lysis, quantified by 7-AAD staining.

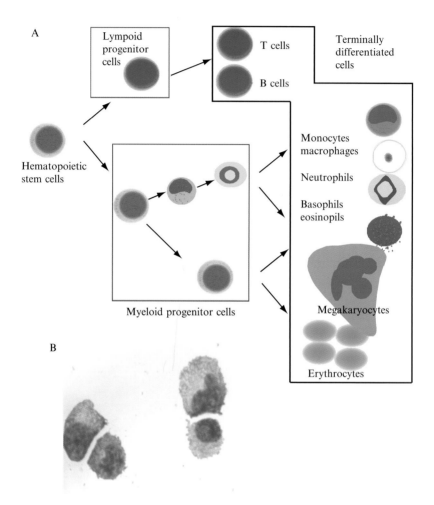

Sandra S. Zinkel, Figure 12.2 (A) Hematopoiesis. All lineages of hematopoietic cells arise from a puripotent stem cell. Progenitor cells then commit to the myeloid or lymphoid lineage (red boxes) prior to terminal differentiation (black box). (B) Cytospin of Hox11-immortalized myeloprogenitor cells. One thousand cells in 50 μl of PBS were centrifuged at 700 rpm for 7 min in a cytospin. Cells were allowed to dry and were then stained with May–Grunwald–Giemsa stain (Sigma).

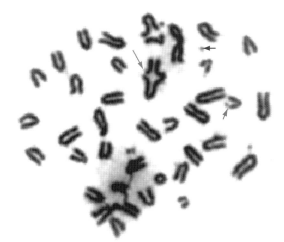

Sandra S. Zinkel, Figure 12.3 Metaphase spread of mitomycin c-treated $Bid^{-/-}$ MPCs. Cells were treated with 100 nM mitomycin c for 24 h. The black arrow indicates a quadriradial, the blue arrow indicates a chromosome fragment, and the red arrow indicates a chromosome break.

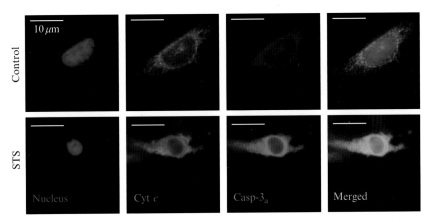

Lorenzo Galluzzi *et al.*, Figure 18.2 Immunofluorescence microscopy-assisted determination of the release of intermembrane space proteins. Nonsmall cell lung cancer cells (A549) were left untreated (control) or treated with 1 μM staurosporine (STS) for 12 h prior to fixation and costaining for the immunofluorescence detection of cytochrome c (Cyt c, detected with a secondary antibody emitting in red) and active caspase-3 ($Casp-3_a$, revealed by a secondary antibody fluorescing in green), as detailed in Section 2.1. Nuclei were counterstained with Hoechst 33342 (blue signal). White bars indicate picture scale (10 μm). In physiological conditions, cytochrome c exhibits a "tubular" staining (typical of healthy mitochondria), and $Casp-3_a$ cannot be detected. Following the induction of apoptosis, cytochrome c relocalizes to the cytosol (and hence exhibits a diffuse intracellular staining) where it promotes the activation of caspase-3 (which also is localized rather homogeneously throughout the cell). In merged images, an intense yellow signal clearly discriminates cells undergoing apoptosis from their healthy counterparts. Please note also the nuclear pyknosis typical of apoptotic cells.

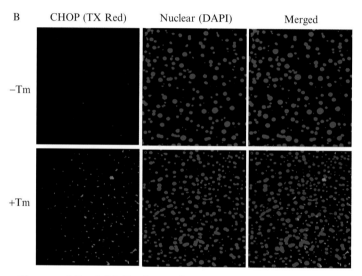

Kezhong Zhang and Randal J. Kaufman, Figure 20.5B Immunohistochemical staining of CHOP in the liver. CHOP staining in liver sections. Wild-type C57BL/6J mice at 3 months of age were injected intraperitoneally with 2 μg/gram body weight of tunicamycin in 150 mM dextrose. At 24 h after injection, fresh liver tissues were fixed in 4% PBS-buffered formalin, paraffin embedded, and 4-μm sections prepared. (Top) Liver sections from control mice. (Bottom) Liver sections from mice treated with Tm.

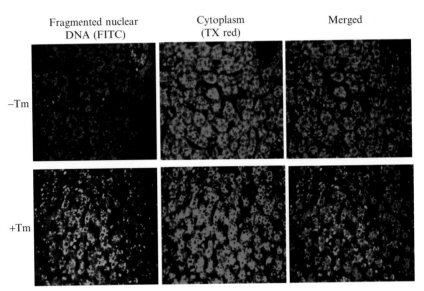

Kezhong Zhang and Randal J. Kaufman, Figure 20.6 Immunohistochemical staining of nuclear DNA fragments in liver sections. Wild-type C57BL/6J mice at 3 months of age were injected intraperitoneally with 2 μg/gram body weight of tunicamycin in 150 mM dextrose. At 36 h after injection, fresh liver tissues were fixed in 4% PBS-buffered formalin, paraffin embedded, and 4-μm sections prepared. (Top) Liver sections from control mice. (Bottom) Liver sections from mice treated with Tm.

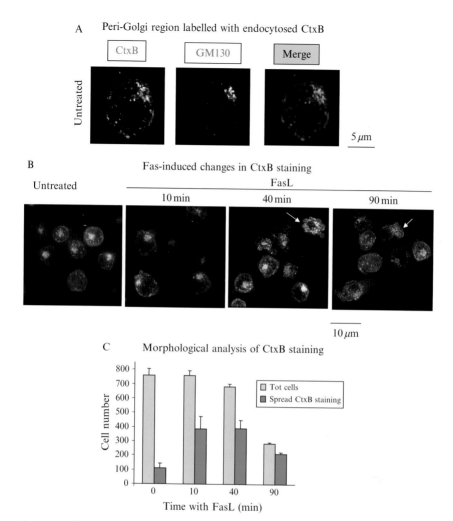

Mauro Degli Esposti, Figure 21.3 Global movements of endomembranes visualized with CtxB staining. (A) Specific immunolabeling of the Golgi complex of Jurkat cells with the membrane marker GM130 (obtained as described by Ouasti *et al.*, 2007) identifies the "peri-Golgi" region where endocytosed CTxB (Alexa Fluor 594-conjugated, red CtxB) accumulate. Projected images from about 40 z sections were obtained with a $60\times$ objective and deconvolved for 10 cycles. (B) Grayscale images if Jurkat cells that had been labeled with Alexa Fluor 488-CtxB and then incubated for a prolonged period with FasL in concentrated solution, from which aliquots were taken 10 min before the indicated time to allow attachment onto coverslips. Images were obtained as in A and arrows indicate blebbing cells with altered morphology, an early expression of Fas-induced apoptosis (Weis *et al.*, 1995). Note the increased dispersal of CtxB staining with time. (C) Quantitative morphological analysis of multiple images obtained as in B was carried by three independent scorers in a large number of microscopy fields. The graph shows the progressive decrease in the total number of attached cells counted (total cells, light gray histograms) in equivalent sets of randomly chosen fields, which reflected the increasing inability of apoptotic cells to adhere properly on coverslips, in conjunction with the loss of cells that had completely disintegrated before being seeded. Hence, data indicate the loss of viability during FasL treatment, as confirmed by Trypan blue counting. The darker histograms represent the parallel evaluation of attached cells that presented with dispersed CtxB staining (spread), i.e., loss of the normal peri-Golgi clustering (cf. B). Note that after 90 min of treatment most cells show a dispersed CtxB staining.

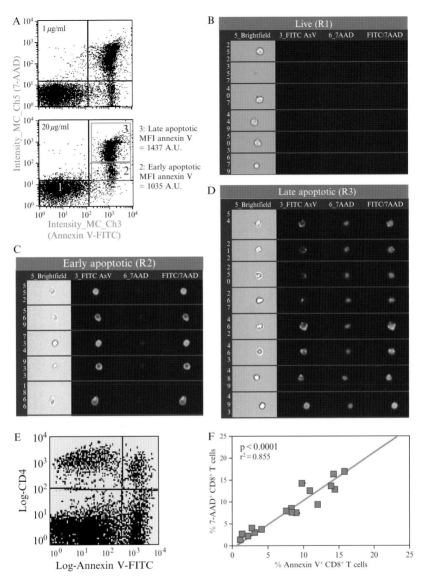

Hervé Lecoeur et al., Figure 3.4 Annexin-V staining of apoptotic cells and correlation with 7-AAD staining. (A) Biparametric annexin-V/7-AAD dot plot of murine thymocytes incubated overnight in culture medium at 37 °C. The co-staining annexin-V-FITC/7-AAD has been performed under two experimental conditions: 1 μg/ml and 20 μg/ml of 7-AAD. Early apoptotic cells are annexin-V^+ and 7-AAD^{low}. (B, C, D) Analysis of such thymocytes by multi-spectral imaging in flow. Cells were stained by annexin-V-FITC and 20 μg/ml 7-AAD, unfixed and were analysed with an Image Stream 100 (Amnis). Live, early and late apoptotic cells were gated (gates R1, R2 and R3 respectively) and representative thymocytes from these gates are presented in parts B, C and D. Bright field, annexin-V staining (channel 3), 7-AAD staining (channel 6) and the overlapping of these two parameters are presented for each cell. (E) PBMC from an ASX HIV-infected patient were incubated overnight in culture medium à 37 °C. Apoptosis of CD4 T cells was quantified following a co-staining with annexin-V and anti-CD4 mAb. The percentage of apoptosis in CD4 T cells in indicated on the figure. (F) PBMC from ASX HIV-infected patients ($n = 17$) were incubated overnight in culture medium à 37 °C. Apoptosis of CD8 T+ cells was quantified following a co-staining with annexin-V, or 7-AAD, and anti-CD8 mAb. The linear regression curve between both variables is shown. The coefficient of correlation and the p-value are indicated.